Advanced Electrical Engineering

Advanced Electrical Engineering

A H Morton
BSc, CEng, FIEE, MIERE
Formerly Head of Department of Electrical Engineering,
Paisley College of Technology

Longman Scientific & Technical
Longman Group UK Limited
Longman House, Burnt Mill, Harlow,
Essex CM20 2JE, England
and Associated Companies throughout the world.

First published in Great Britain by Pitman 1966
Reprinted 1967, 1969, 1973, 1976, 1977, 1979,
1980, 1981, 1982, 1984 and 1985
Reprinted by Longman Scientific & Technical 1986

ISBN 0-582-99462-4

Produced by Longman Singapore Publishers Pte Ltd
Printed in Singapore

BOUGHT ON 2.2.87 BY DAVID J.B. [illegible]
FOR £10.95.

Preface

Lewis Carroll, the genius of gentle humour, writes in *The Three Voices,*

> Or, stretched beside some babbling brook,
> To con, with inexpressive look,
> An unintelligible book.

While it is not anticipated that this present volume will be read in such pastoral surroundings, it is to be hoped that it will not prove unintelligible to those for whom it is intended. The aim has been to present in a clear manner some of the basic theoretical concepts which should be known to all professional electrical and electronic engineers.

The text is based on lectures given to students preparing for the Advanced Electrical Engineering paper in the Part III Examination of the Institution of Electrical Engineers, and to final year undergraduates preparing for the external B.Sc. degree of the University of London.

The theoretical ideas are illustrated by numerous worked examples, and a wide selection of problems with answers is given at the end of each chapter, so that the book can be used as a working textbook.

In the first section, on circuit theory, an introduction is given to matrix methods of analysis, and to the use of the Laplace transformation for dealing with transient problems. The chapter on transmission lines includes a treatment of the Smith circle diagram, for both loss-free and low-loss lines. The second section of the book is concerned with field theory, and the final section deals with principles of electrical measurements.

The units used throughout are those of the Système Internationale (the S.I. units), which incorporates the M.K.S.A. units.

Many of the examples and problems are taken from the examination papers of London University, the Institution of Electrical Engineers, and the Royal Radar Establishment College of Electronics, Malvern. The author acknowledges, with thanks, the ready permission obtained from these bodies to make use of such material. The answers given are, however, the sole responsibility of the author, helped by the (often unwitting) efforts of many students, to whom thanks are also due. The following abbreviations are appended to examples and problems—

A.E.E. Advanced Electrical Engineering paper, Part III, Institution of Electrical Engineers

I.E.E. A specialist paper, Part III, Institution of Electrical Engineers

L.U. An external B.Sc. paper, University of London

R.R.E. Royal Radar Establishment College of Electronics

My personal thanks are due to Principal J. Gray, R.R.E. College of Electronics, for his encouragement in the early stages of the work; to Ken Donaldson and Bill Craig of Paisley College of Technology, for reading the manuscript and drawing many of the diagrams; and to those who assisted in the massive typing effort required. A special word of thanks must also go to my two friends Iain Shepherd (who read the galley proofs) and Lyndon Spence, in co-operation with whom the original idea of this book was conceived, and whose unfailing help has been a constant source of strength.

A.H.M.

Paisley, 1965

Contents

CHAPTER 1

Linear Circuit Analysis

IN a linear circuit each element has a value which is independent of the magnitudes or directions of currents or potential differences. The passive elements are denoted by their impedances, Z, or their corresponding admittances, Y. In this chapter only linear networks will be considered, that is networks made up of linear resistive, inductive or capacitive elements. Sources of electrical energy will be shown as either constant-voltage or constant-current sources.

The manipulation of the equations involved in the study of networks is greatly facilitated by the use of matrix methods. Accordingly the first section of the chapter is devoted to a simple consideration of matrix algebra and its application to networks. This is followed by a section which deals with circuit theorems and general analysis.

1.1. Introduction to Matrix Algebra

1.1.1. MATRIX NOTATION

A *matrix* is an ordered array of numbers (or symbols) arranged in rows and columns. Two matrices are equal only if each element of one is identically equal to the corresponding element of the other. A matrix must not be confused with a determinant, which is effectively only a single number.

In general, a matrix has m rows and n columns and is referred to as an m by n matrix, or a matrix of order m by n. A matrix which has equal numbers of rows and columns is called a *square* matrix. A matrix which consists of a column only is called a *column* matrix (or sometimes a column vector), while a *row* matrix (or row vector) has one row only. Frequently a matrix is represented by a single symbol enclosed in square brackets. Thus

$$\begin{bmatrix} a_{11} & a_{12} & a_{13} \\ a_{21} & a_{22} & a_{23} \\ a_{31} & a_{32} & a_{33} \end{bmatrix} = [a_{rs}] \text{ is a square } 3 \times 3 \text{ matrix}$$

$$\begin{bmatrix} i_{11} \\ i_{21} \\ i_{31} \end{bmatrix} = [i] \text{ is a column matrix}$$

and $$[b_{11} \quad b_{12} \quad b_{13}] = [b] \text{ is a row matrix}$$

The subscripts indicate the row and column of each element.

The *transpose* of a matrix is the array which results when the rows and columns of a matrix are interchanged. Thus, if

$$[A] = \begin{bmatrix} a_{11} & a_{12} \\ a_{21} & a_{22} \\ a_{31} & a_{32} \end{bmatrix}$$

then the transpose is

$$[A]_t = \begin{bmatrix} a_{11} & a_{21} & a_{31} \\ a_{12} & a_{22} & a_{32} \end{bmatrix}$$

A *symmetric* matrix is one which is unchanged by transposition. Thus

$$\begin{bmatrix} 3 & 1 \\ 1 & 3 \end{bmatrix} \text{ is a symmetric matrix}$$

It follows from the definition of equal matrices that, for example, if

$$\begin{bmatrix} (i_1 z_{11} + i_2 z_{12}) \\ (i_1 z_{21} + i_2 z_{22}) \end{bmatrix} = \begin{bmatrix} v_1 \\ v_2 \end{bmatrix} \quad . \quad . \quad (1.1)$$

then this gives the two equations

$$i_1 z_{11} + i_2 z_{12} = v_1 \quad \text{and} \quad i_1 z_{21} + i_2 z_{22} = v_2 \quad . \quad (1.2)$$

1.1.2. Matrix Multiplication

The fundamental operation of matrix algebra is the process called matrix multiplication. This process is only possible between two matrices in which the number of columns of the first matrix is equal to the number of rows of the second matrix. The product of two matrices is itself a matrix, whose elements are obtained from the following definition. The element of the ith row and jth column of the product matrix is obtained by summing the products of each element of the ith row of the first matrix with the corresponding elements of the jth column of the second matrix. Thus the element in the ith row and jth column of the product $[a]$ $[b]$ is

$$a_{i1}b_{1j} + a_{i2}b_{2j} + a_{i3}b_{3j} + \ . \ . \ .$$

In the simple case of a 2×2 square matrix multiplied by a 2×1 column matrix we obtain

$$\begin{bmatrix} z_{11} & z_{12} \\ z_{21} & z_{22} \end{bmatrix} \begin{bmatrix} i_1 \\ i_2 \end{bmatrix} = \begin{bmatrix} (z_{11}i_1 + z_{12}i_2) \\ (z_{21}i_1 + z_{22}i_2) \end{bmatrix} \quad . \quad . \quad (1.3)$$

If the first matrix is of order $m \times n$ (m rows and n columns) and the second is of order $n \times p$ (n rows and p columns), the product matrix will have m rows and p columns. In the example given above, the first matrix has 2 rows and 2 columns, the second has 2 rows and 1 column, and the product has 2 rows and 1 column.

It follows from the definition of matrix multiplication that this is a non-commutative process, i.e.

$$[A][B] \neq [B][A]$$

This is obvious in eqn. (1.3), since if the order of the multiplication is reversed, the process of multiplication is not even possible. The reader is invited to verify that for 2×2 square matrices (where reversal of the order of multiplication is possible) the non-commutative property is demonstrable. Thus,

$$\begin{bmatrix} 1 & 2 \\ 2 & 3 \end{bmatrix} \times \begin{bmatrix} 2 & 1 \\ 3 & 1 \end{bmatrix} = \begin{bmatrix} (2+6) & (1+2) \\ (4+9) & (2+3) \end{bmatrix} = \begin{bmatrix} 8 & 3 \\ 13 & 5 \end{bmatrix}$$

while

$$\begin{bmatrix} 2 & 1 \\ 3 & 1 \end{bmatrix} \times \begin{bmatrix} 1 & 2 \\ 2 & 3 \end{bmatrix} = \begin{bmatrix} 4 & 7 \\ 5 & 9 \end{bmatrix}$$

The first factor in a matrix product is called the pre-multiplier and the second is called the post-multiplier.

Combining eqns. (1.1) and (1.3), we may write

$$\begin{bmatrix} z_{11} & z_{12} \\ z_{21} & z_{22} \end{bmatrix} \begin{bmatrix} i_1 \\ i_2 \end{bmatrix} = \begin{bmatrix} v_1 \\ v_2 \end{bmatrix} \quad . \quad . \quad (1.4)$$

to represent the two simultaneous equations (1.2). In the same way, the matrix equation representing three simultaneous equations is

$$\begin{bmatrix} z_{11} & z_{12} & z_{13} \\ z_{21} & z_{22} & z_{23} \\ z_{31} & z_{32} & z_{33} \end{bmatrix} \begin{bmatrix} i_1 \\ i_2 \\ i_3 \end{bmatrix} = \begin{bmatrix} v_1 \\ v_2 \\ v_3 \end{bmatrix} \quad . \quad . \quad . \quad (1.5)$$

Eqns. (1.4) and (1.5) may both be represented in the shortened form as

$$[z][i] = [v]$$

Note that the order of the multiplication must be preserved, due to its non-commutative nature.

Following the definition of matrix multiplication we can define the unit matrix as one which, when pre-multiplying a matrix $[a]$, gives a product equal to the original matrix $[a]$. The unit matrix consists of unit terms on the major diagonal and zeros elsewhere, the number of columns of the unit matrix corresponding to the number of rows of the post-multiplier. Thus

$$[1][a] = [a] \quad . \quad . \quad . \quad . \quad (1.6)$$

For example,

$$\begin{bmatrix} 1 & 0 & 0 \\ 0 & 1 & 0 \\ 0 & 0 & 1 \end{bmatrix} \begin{bmatrix} 2 \\ 5 \\ 9 \end{bmatrix} = \begin{bmatrix} 2 \\ 5 \\ 9 \end{bmatrix}$$

The reader is invited to verify that for square matrices either pre- or post-multiplication by a unit matrix leaves the matrix unchanged. For example,

$$\begin{bmatrix} 1 & 0 \\ 0 & 1 \end{bmatrix} \begin{bmatrix} 2 & -1 \\ 5 & 7 \end{bmatrix} = \begin{bmatrix} 2 & -1 \\ 5 & 7 \end{bmatrix} = \begin{bmatrix} 2 & -1 \\ 5 & 7 \end{bmatrix} \begin{bmatrix} 1 & 0 \\ 0 & 1 \end{bmatrix}$$

1.1.3. Multiplication by a Scalar

If a matrix is multiplied by a scalar quantity, then *each term* in the matrix is multiplied by the scalar—

$$k[a_{rs}] = [ka_{rs}] \quad . \quad . \quad . \quad . \quad (1.7)$$

Thus

$$10\begin{bmatrix} 3 & 2 \\ -1 & 2 \end{bmatrix} = \begin{bmatrix} 30 & 20 \\ -10 & 20 \end{bmatrix}$$

Note that this is a different rule from that which applies to the multiplication of a determinant by a constant, in which only *one* row or column is multiplied by the constant.

1.1.4. Co-factors

The *co-factor* of an element a_{rs} in a square matrix is defined as the determinant of the elements remaining in the matrix when the row and column containing a_{rs} are removed, multiplied by $(-1)^{r+s}$. The sign is readily obtained from the pattern

$$\begin{matrix} + & - & + & . & . & . \\ - & + & - & . & . & . \\ + & - & + & . & . & . \end{matrix}$$

For the 2 × 2 matrix

$$\begin{bmatrix} A & B \\ C & D \end{bmatrix}$$

the co-factor of A is D, of B is $-C$, of C is $-B$ and of D is A, and the array of co-factors is

$$\begin{bmatrix} D & -C \\ -B & A \end{bmatrix}$$

For the 3 × 3 matrix

$$\begin{bmatrix} 1 & 3 & 4 \\ 2 & 2 & 3 \\ 1 & 1 & -2 \end{bmatrix}$$

the co-factor of the 2 in the first column is

$$-\begin{vmatrix} 3 & 4 \\ 1 & -2 \end{vmatrix} = -(-10) = 10$$

and that of the 2 in the second column is

$$\begin{vmatrix} 1 & 4 \\ 1 & -2 \end{vmatrix} = -6$$

The array of co-factors may thus be built up by finding the co-factors of each element in the original array.

1.1.5. The Inverse Matrix

The *inverse* of a square matrix $[a]$ is defined as that matrix which, when pre- or post-multiplied by $[a]$, yields a unit matrix. It is given the symbol $[a]^{-1}$.

$$[a][a]^{-1} = [a]^{-1}[a] = [1] \qquad (1.8)$$

That such a matrix exists may be shown by first considering that different inverse matrices exist such that

$$[a][a']^{-1} = [1] \quad \text{and} \quad [a'']^{-1}[a] = [1]$$

Then

$$[a'']^{-1}[a][a']^{-1} = [a'']^{-1}$$

But since

$$[a'']^{-1}[a] = [1]$$

then

$$[a']^{-1} = [a'']^{-1}$$

Now let us see if we can derive an expression for the inverse of a 2×2 matrix. Consider

$$\left.\begin{aligned} z_{11}i_1 + z_{12}i_2 = v_1 \\ z_{21}i_1 + z_{22}i_2 = v_2 \end{aligned}\right\} \quad \text{or} \quad [z][i] = [v]$$

Solving for i_1 and i_2 in the usual way gives

$$i_1 = \frac{(z_{22}v_1 - z_{12}v_2)}{(z_{11}z_{22} - z_{12}z_{21})}$$

and

$$i_2 = \frac{(-z_{21}v_1 + z_{11}v_2)}{(z_{11}z_{22} - z_{12}z_{21})}$$

where $z_{11}z_{22} - z_{12}z_{21}$ is the *determinant*, D, of the matrix array. Hence

$$i_1 = \frac{z_{22}}{D} v_1 - \frac{z_{12}}{D} v_2$$

and

$$i_2 = \frac{-z_{21}}{D} v_1 + \frac{z_{11}}{D} v_2$$

Hence, in matrix form,

$$\begin{bmatrix} i_1 \\ i_2 \end{bmatrix} = \frac{1}{D}\begin{bmatrix} z_{22} & -z_{12} \\ -z_{21} & z_{11} \end{bmatrix}\begin{bmatrix} v_1 \\ v_2 \end{bmatrix} \quad . \quad . \quad . \quad (1.9)$$

or

$$[i] = [y][v] \quad \text{where} \quad [y] = \frac{1}{D}\begin{bmatrix} z_{22} & -z_{12} \\ -z_{21} & z_{11} \end{bmatrix}$$

But if

$$[z][i] = [v]$$

then

$$[z]^{-1}[z][i] = [z]^{-1}[v] \quad . \quad . \quad . \quad (1.10)$$

where $[z]^{-1}$ is the inverse of $[z]$, so that $[z]^{-1}[z] = [1]$, i.e.

$$[i] = [z]^{-1}[v] \quad . \quad . \quad . \quad (1.11)$$

Comparing eqns. (1.11) and (1.9), it will be seen that

$$[y] = [z]^{-1} = \frac{1}{D}\begin{bmatrix} z_{22} & -z_{12} \\ -z_{21} & z_{11} \end{bmatrix} \quad . \quad . \quad . \quad (1.12)$$

This means that inversion is possible only if the determinant D is not zero. The matrix is then said to be *non-singular*.

In general terms, the inverse of a square matrix is found by—

1. *Transposing the matrix;* e.g. for the 2×2 matrix $[z]$ the transpose, $[z]_t$, is

$$[z]_t = \begin{bmatrix} z_{11} & z_{21} \\ z_{12} & z_{22} \end{bmatrix}$$

2. *Replacing each element of the transposed matrix by its co-factor;* e.g. for the 2×2 matrix we obtain

$$\begin{bmatrix} z_{22} & -z_{12} \\ -z_{21} & z_{11} \end{bmatrix}$$

3. *Dividing the resulting matrix by the determinant, D, of the original matrix.*

For example, consider the matrix

$$[A] = \begin{bmatrix} 3 & -1 \\ 5 & 4 \end{bmatrix}$$

then

$$[A]_t = \begin{bmatrix} 3 & 5 \\ -1 & 4 \end{bmatrix}$$

and the array of co-factors of $[A]$ is

$$\begin{bmatrix} 4 & 1 \\ -5 & 3 \end{bmatrix}$$

The determinant, D, of the original matrix is $D = 3 \times 4 - (-1) \times 5 = 17$, and hence

$$[A]^{-1} = \begin{bmatrix} \frac{4}{17} & \frac{1}{17} \\ -\frac{5}{17} & \frac{3}{17} \end{bmatrix}$$

The reader can readily verify that

$$[A][A]^{-1} = [1] = [A]^{-1}[A]$$

Example 1.1. Find the inverse of the 3×3 matrix

$$[A] = \begin{bmatrix} 1 & 2 & 0 \\ -1 & 2 & 1 \\ 3 & 0 & 3 \end{bmatrix}$$

The transpose is

$$[A]_t = \begin{bmatrix} 1 & -1 & 3 \\ 2 & 2 & 0 \\ 0 & 1 & 3 \end{bmatrix}$$

The array of co-factors of $[A]_t$ is

$$\begin{bmatrix} 6 & -6 & 2 \\ 6 & 3 & -1 \\ -6 & 6 & 4 \end{bmatrix}$$

The determinant of $[A]$ is $6 + 6 + 6 = 18$, and hence the inverse of $[A]$ is

$$[A]^{-1} = \frac{1}{18}\begin{bmatrix} 6 & -6 & 2 \\ 6 & 3 & -1 \\ -6 & 6 & 4 \end{bmatrix}$$

As a check,

$$[A][A]^{-1} = \frac{1}{18}\begin{bmatrix} 1 & 2 & 0 \\ -1 & 2 & 1 \\ 3 & 0 & 3 \end{bmatrix}\begin{bmatrix} 6 & -6 & 2 \\ 6 & 3 & -1 \\ -6 & 6 & 4 \end{bmatrix} = \frac{1}{18}\begin{bmatrix} 18 & 0 & 0 \\ 0 & 18 & 0 \\ 0 & 0 & 18 \end{bmatrix} = [1]$$

Example 1.2. Find the inverse of the diagonal matrix

$$[z] = \begin{bmatrix} z_{11} & 0 & 0 \\ 0 & z_{22} & 0 \\ 0 & 0 & z_{33} \end{bmatrix}$$

$$[z]_t = \begin{bmatrix} z_{11} & 0 & 0 \\ 0 & z_{22} & 0 \\ 0 & 0 & z_{33} \end{bmatrix}$$

i.e. the transpose of a diagonal matrix is the same as the matrix itself. The array of co-factors of $[z]_t$ is

$$\begin{bmatrix} z_{22}z_{33} & 0 & 0 \\ 0 & z_{11}z_{33} & 0 \\ 0 & 0 & z_{11}z_{22} \end{bmatrix}$$

The determinant of $[z]$ is $z_{11}z_{22}z_{33}$, and hence

$$[z]^{-1} = \frac{1}{z_{11}z_{22}z_{33}}\begin{bmatrix} z_{22}z_{33} & 0 & 0 \\ 0 & z_{11}z_{33} & 0 \\ 0 & 0 & z_{11}z_{22} \end{bmatrix}$$

$$= \begin{bmatrix} \frac{1}{z_{11}} & 0 & 0 \\ 0 & \frac{1}{z_{22}} & 0 \\ 0 & 0 & \frac{1}{z_{33}} \end{bmatrix} \quad . \quad . \quad . \quad . \quad (1.13)$$

1.1.6. Addition and Subtraction of Matrices

Matrices add and subtract in the same way as vectors, i.e. corresponding components are added together. This is possible only if the matrices are of the same order. Thus

$$\begin{bmatrix} 7 & 2 \\ 3 & 1 \end{bmatrix} + \begin{bmatrix} 3 & 1 \\ 1 & 2 \end{bmatrix} = \begin{bmatrix} 10 & 3 \\ 4 & 3 \end{bmatrix}$$

and

$$\begin{bmatrix} -7 \\ 3 \end{bmatrix} - \begin{bmatrix} 4 \\ -1 \end{bmatrix} = \begin{bmatrix} -11 \\ 4 \end{bmatrix}$$

It is sometimes useful to be able to write a column matrix involving summation terms in the form of a product. Consider

$$\begin{bmatrix} (V_{10} + V_a) \\ (V_{10} + V_b) \\ (V_{10} + V_c) \end{bmatrix}$$

There are four separate terms in this matrix, and we start by writing these as a column matrix. The column matrix is then pre-multiplied in this case by a 3×4 matrix consisting of units and zeros which are located by inspection to give the desired result. Thus

$$\begin{bmatrix} 1 & 1 & 0 & 0 \\ 1 & 0 & 1 & 0 \\ 1 & 0 & 0 & 1 \end{bmatrix} \begin{bmatrix} V_{10} \\ V_a \\ V_b \\ V_c \end{bmatrix} = \begin{bmatrix} (V_{10} + V_a) \\ (V_{10} + V_b) \\ (V_{10} + V_c) \end{bmatrix} \quad . \quad . \quad (1.14)$$

The 3×4 matrix with unit elements is called a *connexion* matrix.

1.2. Network Theorems and Analysis

In many cases the use of appropriate theorems simplifies the analysis of networks. Some important theorems will be proved in the following sections.

Before proceeding, it is necessary to define the terms which will be used in connexion with complex notation. Sinusoidal voltages and currents, and other quantities which can be regarded as complex, will be called *complexors*, and the Argand diagram in which they can be represented will be called a *complexor diagram*. (Some authors use the terms *phasor* and *phasor diagram*.) The components of a complexor are, in general, $a + jb$, where a will be called the *reference* (or real) component, along the reference (or real) axis, and b will be called the *quadrate* (or imaginary) component, along the quadrate (or imaginary) axis. It has previously been the custom to call such complex quantities *vectors*, or at any rate to say that they can be represented by vectors on a vector diagram, but they are not really vector quantities, which have direction as well as magnitude. The term vector quantity will therefore be reserved for mechanical force, electric and magnetic flux density, voltage gradient, magnetic field strength, current density, etc., all of which have both direction and magnitude.

1.2.1. Node Analysis—The Floating Admittance Matrix

Consider a network with four nodes as shown in Fig. 1.1, each node being joined to every other node by an admittance y_{mn}, where

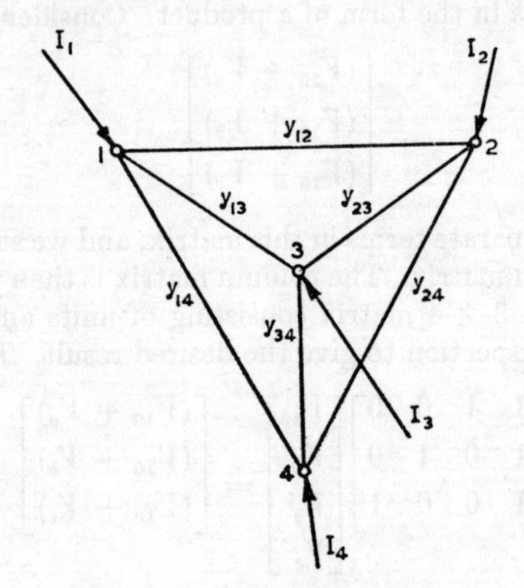

Fig. 1.1. A Four-node Circuit

the subscripts refer to the two nodes joined by y. (Note that $y_{mn} = y_{nm}$.) Assume no mutual-inductance couplings, and let the potentials of the nodes be V_1, V_2, V_3, V_4, with currents I_1, I_2, I_3, I_4 entering the nodes from constant-current sources.

The node equation for node 1 is

$$I_1 = y_{12}(V_1 - V_2) + y_{13}(V_1 - V_3) + y_{14}(V_1 - V_4)$$

or

$$I_1 = (y_{12} + y_{13} + y_{14})V_1 - y_{12}V_2 - y_{13}V_3 - y_{14}V_4$$

$$= y_{11}V_1 - y_{12}V_2 - y_{13}V_3 - y_{14}V_4$$

where y_{11} is the sum of the admittances at node 1. In the same way, for node 2,

$$I_2 = -y_{12}V_1 + y_{22}V_2 - y_{13}V_3 - y_{24}V_4$$

for node 3,

$$I_3 = -y_{13}V_1 - y_{23}V_2 + y_{33}V_3 - y_{34}V_4$$

and for node 4,

$$I_4 = -y_{14}V_1 - y_{24}V_2 - y_{34}V_3 + y_{44}V_4$$

These equations may be represented in matrix form as

$$\begin{bmatrix} I_1 \\ I_2 \\ I_3 \\ I_4 \end{bmatrix} = \begin{bmatrix} y_{11} & -y_{12} & -y_{13} & -y_{14} \\ -y_{12} & y_{22} & -y_{23} & -y_{24} \\ -y_{13} & -y_{23} & y_{33} & -y_{34} \\ -y_{14} & -y_{24} & -y_{34} & y_{44} \end{bmatrix} \begin{bmatrix} V_1 \\ V_2 \\ V_3 \\ V_4 \end{bmatrix} \quad . \quad . \quad (1.15)$$

or simply

$$[I] = [y][V] \quad . \quad . \quad . \quad . (1.15a)$$

The y-matrix is called the *floating admittance* matrix, since all the nodes have been assumed "floating." Note that (*a*) the sum of the currents $I_1 + I_2 + I_3 + I_4$ must be zero, (*b*) the sum of each row and column in the floating admittance matrix is zero, and (*c*) adding a constant voltage to *each* node does not alter the currents.

If any node is earthed the potential of that node is zero. The admittance matrix of the network then reduces to the floating admittance matrix with the row and column corresponding to the earth node omitted.

Obviously the results can be extended to any number of nodes. Nodal analysis is appropriate when there are more meshes than nodes in a circuit.

Example 1.3. The equivalent circuit of a transistor is shown in Fig. 1.2. Determine the current I_i if the input at E is a constant current of 2 mA, B is earthed and $r_e = 25\ \Omega$, $r_b = 1{,}000\ \Omega$, $r_c = 1\ \text{M}\Omega$, $a = 0{\cdot}98$, $R_g = R_L = 10\ \text{k}\Omega$.

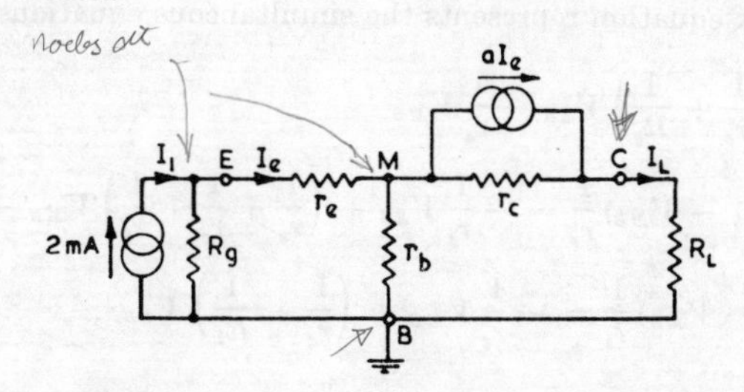

Fig. 1.2. Equivalent A.C. Circuit of a Transistor

Although this problem may readily be solved by mesh current analysis, nodal analysis will be used here to illustrate the method. The floating admittance matrix is

$$\begin{bmatrix} I_1 \\ I_M \\ I_C \\ I_B \end{bmatrix} = \begin{bmatrix} \left(\frac{1}{r_e} + \frac{1}{R_g}\right) & -\frac{1}{r_e} & 0 & -\frac{1}{R_g} \\ -\frac{1}{r_e} & \left(\frac{1}{r_e} + \frac{1}{r_c} + \frac{1}{r_b}\right) & -\frac{1}{r_c} & -\frac{1}{r_b} \\ 0 & -\frac{1}{r_c} & \left(\frac{1}{r_c} + \frac{1}{R_L}\right) & -\frac{1}{R_L} \\ -\frac{1}{R_g} & \frac{1}{r_b} & -\frac{1}{R_L} & \left(\frac{1}{R_g} + \frac{1}{r_b} + \frac{1}{R_L}\right) \end{bmatrix} \begin{bmatrix} V_E \\ V_M \\ V_C \\ V_B \end{bmatrix}$$

where the currents are those entering the nodes, and the voltages are the node potentials. Since B is earthed, and $I_M = -aI_e$, $I_C = aI_e$, this simplifies to

$$\begin{bmatrix} I_1 \\ -aI_e \\ aI_e \end{bmatrix} = \begin{bmatrix} \left(\frac{1}{r_e} + \frac{1}{R_g}\right) & -\frac{1}{r_e} & 0 \\ -\frac{1}{r_e} & \left(\frac{1}{r_e} + \frac{1}{r_c} + \frac{1}{r_b}\right) & -\frac{1}{r_c} \\ 0 & -\frac{1}{r_c} & \left(\frac{1}{r_c} + \frac{1}{R_L}\right) \end{bmatrix} \begin{bmatrix} V_{EB} \\ V_{MB} \\ V_{CB} \end{bmatrix}$$

This matrix equation represents the simultaneous equations

$$I_1 = \left(\frac{1}{r_e} + \frac{1}{R_g}\right) V_{EB} - \frac{1}{r_e} V_{MB}$$

$$-a(V_{EB} - V_{MB}) \frac{1}{r_e} = -\frac{1}{r_e} V_{EB} + \left(\frac{1}{r_e} + \frac{1}{r_c} + \frac{1}{r_b}\right) V_{MB} - \frac{1}{r_c} V_{CB}$$

$$a(V_{EB} - V_{MB}) \frac{1}{r_e} = -\frac{1}{r_c} V_{MB} + \left(\frac{1}{r_c} + \frac{1}{R_L}\right) V_{CB}$$

since $I_e = (V_{EB} - V_{MB})/r_e$.

Substituting numerical values, with admittances in millimhos and currents in milliamperes,

$$2 = 40{\cdot}1\ V_{EB} - 40\ V_{MB}$$
$$0{\cdot}8V_{EB} = 1{\cdot}8\ V_{MB} - 0{\cdot}001\ V_{CB}$$
$$39{\cdot}2\ V_{EB} - 39{\cdot}2\ V_{MB} = 0{\cdot}101\ V_{CB}$$

The solution for the output voltage is $V_{CB} = 19{\cdot}2$ V, so that

$$I_L = V_{CB}/R_L = \underline{\underline{1{\cdot}92\text{ mA}}}$$

The reader may care to verify this result by mesh current analysis (section 1.2.4.).

Note that if the input is taken from a constant-voltage source this should first be converted to the equivalent constant-current source (Section 1.2.9). Similar manipulation can be used when either of the other nodes is made common to the input and output circuits to give the common-emitter or common-collector configuration, this common node being taken as the reference potential point.

1.2.2. Millman's Theorem

Millman's theorem (parallel-generator theorem) states that, if any number of linear bilateral admittances Y_1, Y_2, Y_3, . . . Y_n are connected to a common node 0′, and if the voltages V_{10}, V_{20}, . . . V_{n0} of the free ends of the admittances with respect to a common point 0 are known, then the voltage of 0′ with respect to 0 is

$$V_{0'0} = \frac{\sum_{k=1}^{n} Y_k V_{k0}}{\sum_{k=1}^{n} Y_k} \quad . \quad . \quad . \quad (1.16)$$

Consider the star of admittances shown in Fig. 1.3. The floating admittance matrix gives

$$\begin{bmatrix} I_1 \\ I_2 \\ \cdot \\ \cdot \\ \cdot \\ I_n \\ I_{0'} \end{bmatrix} = \begin{bmatrix} Y_1 & 0 & 0 & \cdots & -Y_1 \\ 0 & Y_2 & 0 & \cdots & -Y_2 \\ \cdot & & & & \\ \cdot & & & & \\ \cdot & & & & \\ 0 & 0 & 0 & \cdots & -Y_n \\ -Y_1 & -Y_2 & -Y_3 & \cdots & (Y_1 + Y_2 + \ldots Y_n) \end{bmatrix} \begin{bmatrix} V_{10} \\ V_{20} \\ \cdot \\ \cdot \\ \cdot \\ V_{n0} \\ V_{0'0} \end{bmatrix}$$

where I_1, I_2, etc., are the external currents entering the network at nodes 1, 2, etc.

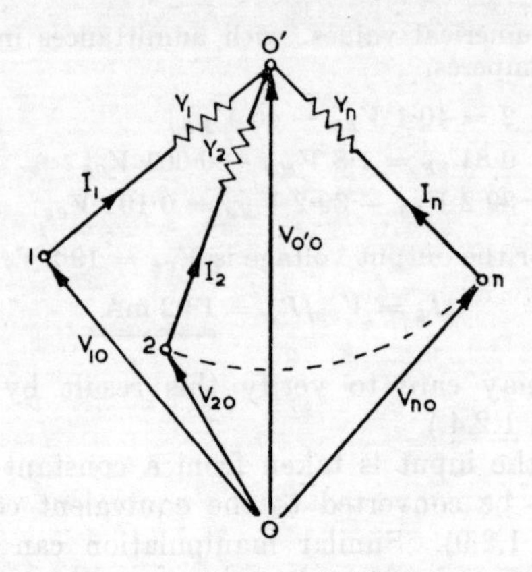

Fig. 1.3. Millman's Theorem

Hence

$$\begin{bmatrix} I_1 \\ I_2 \\ \cdot \\ \cdot \\ \cdot \\ I_{0'} \end{bmatrix} = \begin{bmatrix} (V_{10}Y_1 - V_{0'0}Y_1) \\ (V_{20}Y_2 - V_{0'0}Y_2) \\ \cdot \\ \cdot \\ \cdot \\ (-V_{10}Y_1 - V_{20}Y_2 - \ldots + \Sigma Y V_{0'0}) \end{bmatrix}$$

But the external current at node 0′ is zero, since there is no external connexion, and hence, equating the bottom elements in each column,

$$I_{0'} = 0 = -V_{10}Y_1 - V_{20}Y_2 - \ldots + \Sigma Y V_{0'0}$$

and so Millman's theorem is verified since

$$V_{0'0} = \frac{V_{10}Y_1 + V_{20}Y_2 + \ldots + V_{n0}Y_n}{\Sigma Y}$$

Equating corresponding elements in each matrix gives the currents in the corresponding admittance elements, thus

$$\left.\begin{aligned} I_1 &= V_{10}Y_1 - V_{0'0}Y_1 \\ I_2 &= V_{20}Y_2 - V_{0'0}Y_2 \end{aligned}\right\} \quad . \quad . \quad . \quad (1.17)$$

etc.

Example 1.4. Three impedances $Z_1 = 10\ \Omega$, $Z_2 = j20\ \Omega$ and $Z_3 = -j25\ \Omega$ are connected in star to the lines R, Y and B respectively of a symmetrical 3-wire 400-V 3-phase system of phase sequence RYB. Calculate the currents in the three lines.

If 0 is the star point of the supply, and 0′ is the star point of the unbalanced load, then taking the R phase voltage of the supply as the reference complexor,

$$V_{RO} = (400/\sqrt{3})\underline{/0^\circ} = 231\underline{/0^\circ}$$

and $V_{YO} = 231\underline{/240^\circ}$ (i.e. lagging behind V_{RO} by 120°)

$$= -115 - j200$$

$$V_{BO} = 231\underline{/120^\circ} = -115 + j200$$

so that, from Millman's theorem,

$$V_{0'0} = \frac{(231\underline{/0^\circ} \times 0{\cdot}1) + (231\underline{/240^\circ} \times 0{\cdot}05\underline{/-90^\circ}) + (231\underline{/120^\circ} \times 0{\cdot}04\underline{/90^\circ})}{0{\cdot}1 - j0{\cdot}05 + j0{\cdot}04}$$

$$= \frac{23{\cdot}1 + 11{\cdot}55 \cos 150^\circ + j11{\cdot}55 \sin 150^\circ + 9{\cdot}24 \cos 210^\circ + j9{\cdot}24 \sin 210^\circ}{0{\cdot}1 - j0{\cdot}01}$$

$$= \frac{5{\cdot}1 + j1{\cdot}165}{0{\cdot}1 - j0{\cdot}01} = 52{\cdot}0\underline{/18{\cdot}3^\circ} = 49{\cdot}3 + j16{\cdot}3$$

Hence

$$I_R = (V_{RO} - V_{0'0})Y_R = (182 - j16{\cdot}3)0{\cdot}1$$
$$= \underline{\underline{18{\cdot}2\underline{/-5^\circ} \text{ A}}}$$

$$I_Y = (V_{YO} - V_{00})Y_Y$$
$$= (-115 - j200 - 49{\cdot}3 - j16{\cdot}3)(-j0{\cdot}05)$$
$$= \underline{\underline{13{\cdot}6\underline{/143^\circ} \text{ A}}}$$

$$I_B = (V_{BO} - V_{0'0})Y_B$$
$$= (-115 + j200 - 49{\cdot}3 - j16{\cdot}3)(j0{\cdot}04)$$
$$= \underline{\underline{9{\cdot}9\underline{/222^\circ} \text{ A}}}$$

1.2.3. Rosen's Theorem

Rosen's theorem is a generalization of the star-delta transformation, whereby a star network of n elements may be replaced by a mesh of $\frac{1}{2}n(n-1)$ branches. It states that, if n terminals are connected by admittances $Y_1, Y_2, \ldots Y_n$ to a common star point, the network is equivalent to a network of admittances $Y_{12}, Y_{13}, \ldots Y_{pq} \ldots$ joining each pair of terminals, where

$$Y_{pq} = \frac{Y_p Y_q}{\sum_{k=1}^{n} Y_k} \quad . \quad . \quad . \quad (1.18)$$

The proof follows from Millman's theorem. Thus, if the voltages of the terminals 1, 2, 3, . . . are $V_{10}, V_{20}, V_{30}, \ldots$, with respect to an external point, 0, and the currents entering these terminals are I_1, I_2, . . ., then for the star network of Fig. 1.4 (*a*),

$$V_{0'0} = \frac{Y_1 V_{10} + Y_2 V_{20} + \ldots + Y_n V_{n0}}{Y_1 + Y_2 + \ldots + Y_n}$$

(from eqn. (1.16))

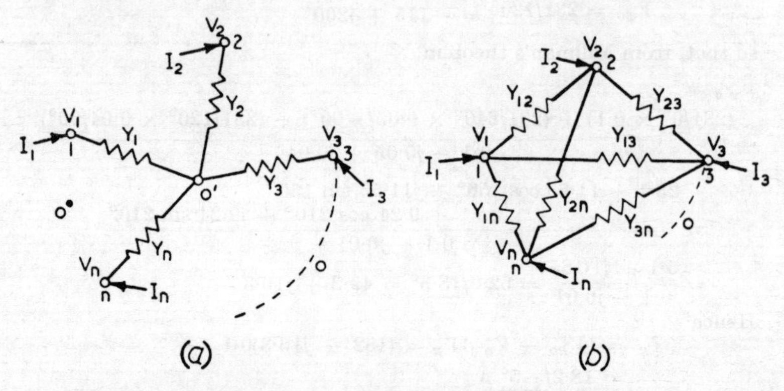

FIG. 1.4. ROSEN'S STAR-MESH EQUIVALENCE THEOREM

Also, from eqn. (1.17),

$$I_1 = V_{10}Y_1 - V_{0'0}Y_1$$

$$= V_{10}Y_1 - \frac{(V_{10}Y_1 + V_{20}Y_2 + \ldots + V_{n0}Y_n)Y_1}{\sum_{k=1}^{n} Y_k}$$

$$= \left\{\frac{(V_{10} - V_{20})Y_2 + (V_{10} - V_{30})Y_3 + \ldots + (V_{10} - V_{n0})Y_n}{\Sigma Y_k}\right\}Y$$

$$= (V_{10} - V_{20})\frac{Y_1Y_2}{\Sigma Y_k} + (V_{10} - V_{30})\frac{Y_1Y_3}{\Sigma Y_k} + \ldots + (V_{10} - V_{n0})\frac{Y_1Y_n}{\Sigma Y_k}$$

This is the current which would flow into terminal 1 if it were connected by admittances $Y_1Y_2/\Sigma Y_k$, $Y_1Y_3/\Sigma Y_k$, . . . $Y_1Y_n/\Sigma Y_k$, etc., to terminals 2, 3, . . ., *n*.

Similarly, the current entering terminal 2 is equivalent to that into a network of admittances $Y_1Y_2/\Sigma Y_k$, $Y_2Y_3/\Sigma Y_k$, . . . connecting terminal 2 to the other terminals, and so on.

Hence the star network may be replaced by the mesh of Fig. 1.4 (*b*), where

$$Y_{12} = \frac{Y_1 Y_2}{\Sigma Y_k}, \quad Y_{13} = \frac{Y_1 Y_3}{\Sigma Y_k}, \quad \ldots, \quad Y_{pq} = \frac{Y_p Y_q}{\Sigma Y_k}, \quad \text{etc.}$$

There is no corresponding general theorem for converting mesh networks to star networks, since for $n > 3$ the mesh has more elements than the equivalent star.

(*a*) *Star–Delta Transformation.* From the above analysis the 3-element star of admittances Y_1, Y_2, Y_3 has an equivalent delta made up of admittances

$$Y_{12} = \frac{Y_1 Y_2}{Y_1 + Y_2 + Y_3} \quad . \quad . \quad . \quad (1.19a)$$

$$Y_{23} = \frac{Y_2 Y_3}{Y_1 + Y_2 + Y_3} \quad . \quad . \quad (1.19b)$$

$$Y_{31} = \frac{Y_3 Y_1}{Y_1 + Y_2 + Y_3} \quad . \quad . \quad . \quad (1.19c)$$

This is the *star–delta transformation.* It may be represented as an impedance transformation by writing,

$$Z_{12} = \frac{Y_1 + Y_2 + Y_3}{Y_1 Y_2} = \frac{1}{Y_1} + \frac{1}{Y_2} + \frac{Y_3}{Y_1 Y_2}$$

i.e. $$Z_{12} = Z_1 + Z_2 + \frac{Z_1 Z_2}{Z_3}$$

Similarly $$Z_{23} = Z_2 + Z_3 + \frac{Z_2 Z_3}{Z_1} \quad . \quad . \quad . \quad (1.20)$$

and $$Z_{31} = Z_3 + Z_1 + \frac{Z_3 Z_1}{Z_2}$$

(*b*) *Delta–Star Transformation.* In the delta network, the mesh has the same number of elements as the equivalent star, and an inverse transformation is possible. Thus, dividing eqn. (1.19*c*) by eqn. (1.19*b*),

$$\frac{Y_1}{Y_2} = \frac{Y_{31}}{Y_{23}} \quad . \quad . \quad . \quad (1.19d)$$

The admittance between terminals 1 and 2 (with terminal 3 open) is $Y_1Y_2/(Y_1 + Y_2)$ for the star network, and $Y_{12} + Y_{31}Y_{23}/(Y_{31} + Y_{23})$ for the delta. For equivalence,

$$\frac{Y_1Y_2}{Y_1 + Y_2} = Y_{12} + \frac{Y_{31}Y_{23}}{Y_{31} + Y_{23}}$$

Substituting for Y_2 from eqn. (1.19*d*),

$$\frac{Y_1^{\,2}Y_{23}/Y_{31}}{Y_1(1 + Y_{23}/Y_{31})} = Y_{12} + \frac{Y_{31}Y_{23}}{Y_{31} + Y_{23}}$$

Hence

$$Y_1 = Y_{12} + Y_{13} + \frac{Y_{12}Y_{31}}{Y_{23}} \qquad (1.21a)$$

In the same way,

$$Y_2 = Y_{12} + Y_{23} + \frac{Y_{12}Y_{23}}{Y_{31}} \qquad (1.21b)$$

and

$$Y_3 = Y_{23} + Y_{31} + \frac{Y_{23}Y_{31}}{Y_{12}} \qquad (1.21c)$$

This is the *delta–star transformation.*

Alternatively,

$$Z_1 = \frac{1}{Y_1} = \frac{Y_{23}}{Y_{12}Y_{31} + Y_{12}Y_{23} + Y_{31}Y_{23}}$$

$$= \frac{1}{Z_{23}}\left\{\frac{1}{\dfrac{1}{Z_{12}Z_{31}} + \dfrac{1}{Z_{12}Z_{23}} + \dfrac{1}{Z_{31}Z_{23}}}\right\}$$

i.e.

$$\left.\begin{aligned} Z_1 &= \frac{Z_{12}Z_{31}}{Z_{12} + Z_{31} + Z_{23}} \\ Z_2 &= \frac{Z_{23}Z_{12}}{Z_{12} + Z_{31} + Z_{23}} \\ Z_3 &= \frac{Z_{23}Z_{31}}{Z_{12} + Z_{31} + Z_{23}} \end{aligned}\right\} \qquad (1.22)$$

(Similarly, for Z_2; and for Z_3.)

Example 1.5. Transform the stray capacitances shown in the a.c. bridge network of Fig. 1.5(*a*) to an equivalent mesh.

At (*a*) the stray capacitances form a 4-element star; the equivalent mesh is shown at (*b*). From eqn. (1.18),

$$Y_{12} = j\omega C_{12} = \frac{j\omega C_1\, j\omega C_2}{j\omega C_1 + j\omega C_2 + j\omega C_3 + j\omega C_4}$$

so that
$$C_{12} = \frac{3 \times 3{\cdot}5}{3 + 3{\cdot}5 + 2{\cdot}5 + 2} = \underline{\underline{0{\cdot}955 \text{ pF}}}$$

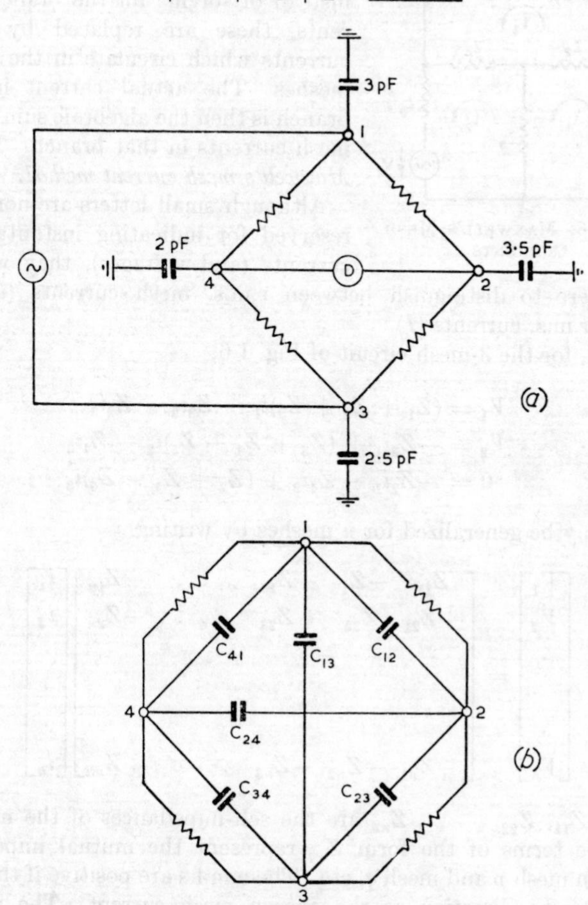

Fig. 1.5. Stray Capacitances in an A.C. Bridge

The other mesh capacitances are found in the same way.

Note that the capacitances C_{12}, C_{23}, C_{34} and C_{41} are across the arms of the bridge, and hence affect the balance conditions. C_{13} is in parallel with the a.c. supply, and C_{24} is across the detector, so that neither affects the balance of the bridge.

1.2.4. Mesh Current Analysis

The current in any branch of a network may be found by writing down the mesh equations and solving the resulting simultaneous equations. The number of equations required may be reduced by one, if instead of solving for the branch currents, these are replaced by mesh currents which circulate in the closed meshes. The actual current in any branch is then the algebraic sum of the mesh currents in that branch. This is *Maxwell's mesh current method.*

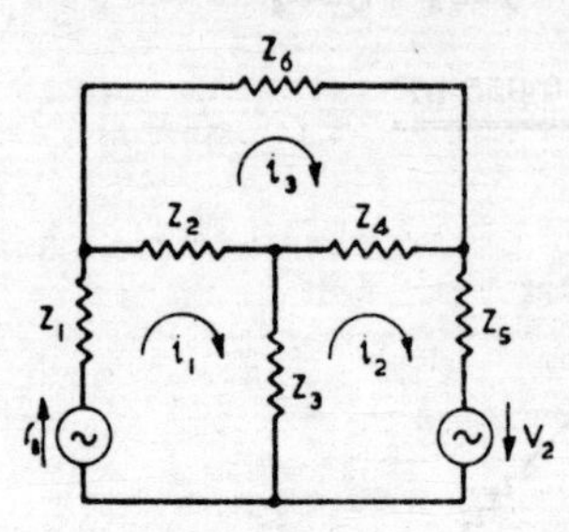

Fig. 1.6. Maxwell's Mesh Currents

Although small letters are normally reserved for indicating instantaneous currents (and voltages), they will be used here to distinguish between r.m.s. mesh currents (i) and actual r.m.s. currents (I).

Thus, for the 3-mesh circuit of Fig. 1.6,

$$\begin{aligned} V_1 &= (Z_1 + Z_2 + Z_3)i_1 - Z_3 i_2 - Z_2 i_3 \\ V_2 &= -Z_3 i_1 + (Z_3 + Z_4 + Z_5)i_2 - Z_4 i_3 \\ 0 &= -Z_2 i_1 - Z_4 i_2 + (Z_2 + Z_4 + Z_6)i_3 \end{aligned}$$

This may be generalized for n meshes by writing

$$\begin{bmatrix} V_1 \\ V_2 \\ \cdot \\ \cdot \\ \cdot \\ V_n \end{bmatrix} = \begin{bmatrix} Z_{11} & -Z_{12} & -Z_{13} & \cdots & -Z_{1n} \\ -Z_{21} & Z_{22} & -Z_{23} & \cdots & -Z_{2n} \\ \cdot & & & & \\ & & & & \\ & & & & \\ -Z_{n1} & -Z_{n2} & -Z_{n3} & \cdots & Z_{nn} \end{bmatrix} \begin{bmatrix} i_1 \\ i_2 \\ \cdot \\ \cdot \\ \cdot \\ i_n \end{bmatrix} \tag{1.23}$$

where Z_{11}, Z_{22}, . . . , Z_{nn} are the self-impedances of the meshes, and the terms of the form Z_{pq} represent the mutual impedance between mesh p and mesh q, etc. The e.m.f.s are positive if they act in the same direction as the chosen mesh current. The matrix equation may be written briefly as

$$[V] = [Z][i]$$

If mutual-inductive coupling exists between branches, this must be included in the mesh equations, care being taken with the signs of the mutual inductances. It is convenient to use the dot notation in such cases: dots are placed at the ends of coupled coils to indicate that when a current enters the dotted end of one coil increasing positively, it gives rise to an e.m.f. directed out of the dotted ends of all coupled coils.

Example 1.6. Write down the mesh matrix for the circuit shown in Fig. 1.7(*a*) if all the coils are wound on the same former and the winding

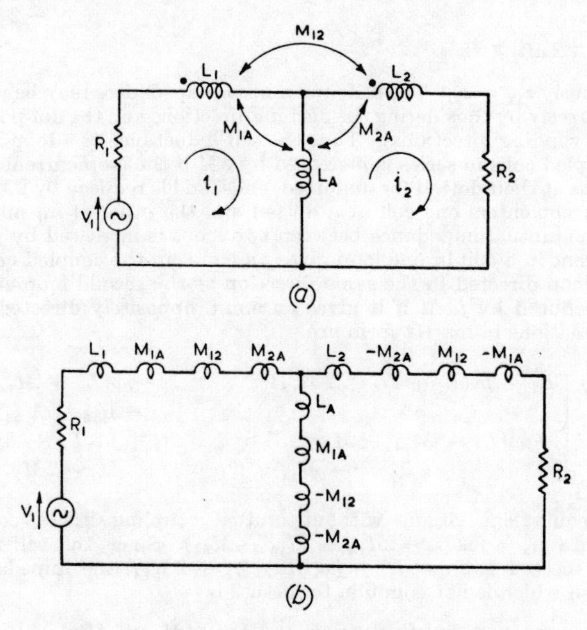

FIG. 1.7. MESH INCLUDING MUTUAL COUPLING

directions are as indicated by the dots. Hence draw an equivalent circuit with no mutual couplings.

Because the coils are wound together, each coil is mutually coupled to every other one.

The current in L_1 gives rise to an e.m.f. in L_A equal to $j\omega M_{1A}i_1$ and directed out of the dotted end of L_A; the current in L_A gives an e.m.f. $j\omega M_{1A}(i_1 - i_2)$ directed out of the dotted end of L_1; and the current in L_2 gives an e.m.f. $j\omega M_{12}i_2$ directed out of the dotted end of L_1 and an e.m.f. $j\omega M_{2A}i_2$ directed out of the dotted end of L_A. Hence the loop equation for mesh 1 is

$$V_1 - j\omega M_{1A}i_1 - j\omega M_{1A}(i_1 - i_2) - j\omega M_{12}i_2 - j\omega M_{2A}i_2 = (R_1 + j\omega(L_1 + L_A))i_1 - j\omega L_A i_2$$

$$\text{or } V_1 = \{R_1 + j\omega(L_1 + L_A + 2M_{1A})\}\, i_1 - j\omega(L_A + M_{1A} - M_{12} - M_{2A})i_2$$

$$= z_{11}i_1 - z_{12}i_2$$

For mesh 2, the loop equation is

$$0 + j\omega M_{1A}i_1 - j\omega M_{2A}i_1 - j\omega M_{12}i_1 + j\omega M_{2A}i_2 + j\omega M_{2A}i_2 = - j\omega L_A i_1 + \{R_2 + j\omega(L_A + L_2)\}i_2$$

$$\text{or } 0 = j\omega(L_A + M_{1A} - M_{12} - M_{2A})i_A + \{R_2 + j\omega(L_A - L_2 - 2M_{2A})\}i_2$$

$$= - z_{21}i_1 + z_{22}i_2$$

Obviously $z_{12} = z_{21}$. The self- and mutual impedances may be written down directly by considering the current directions and the dot positions (i.e. the winding directions). Thus the self-inductance of a loop having two coupled coils in series is increased by $2M$ if the loop current enters both coils at their dotted (or undotted) ends, and is reduced by $2M$ if the loop current enters one coil at a dotted and the other at an undotted end. The mutual impedance between two loops is increased by $j\omega M$ if the current in a coil in one loop gives an e.m.f. in the coupled coil in a second loop directed in the same direction as the second loop current, and is reduced by $j\omega M$ if it gives an e.m.f. oppositely directed. The mesh equations in matrix form are

$$\begin{bmatrix} V_1 \\ 0 \end{bmatrix} = \begin{bmatrix} R_1 + j\omega(L_1 + L_A + 2M_{1A}) & -j\omega(L_A + M_{1A} - M_{12} - M_{2A}) \\ -j\omega(L_A + M_{1A} - M_{12} - M_{2A}) & R_2 + j\omega(L_A + L_2 - 2M_{2A}) \end{bmatrix} \begin{bmatrix} i_1 \\ i_2 \end{bmatrix}$$

The equivalent circuit without mutual coupling has a common impedance $z_{12} = j\omega(L_A + M_{1A} - M_{12} - M_{2A})$. Since the self-impedance of mesh 1 is $z_{11} = R_1 + j\omega(L_1 + L_A + 2M_{1A})$ the impedance of this mesh which is not common to mesh 2 is

$$z_{11} - z_{12} = R_1 + j\omega(L_1 + M_{1A} + M_{12} + M_{2A})$$

Similarly for mesh 2, the impedance which is not common to mesh 1 is

$$z_{22} - z_{12} = R_2 + j\omega(L_2 - M_{2A} + M_{12} - M_{1A})$$

The equivalent circuit with no mutual couplings is shown in Fig. 1.7(*b*).

It is left to the reader to verify the equivalent circuits of Fig. 1.8.

1.2.5. The Superposition Theorem

For a network with n loops, eqn. (1.23) gives the matrix equation as

$$[V] = [Z][i] \qquad (1.23a)$$

Hence

$$[i] = [Z]^{-1}[V]$$

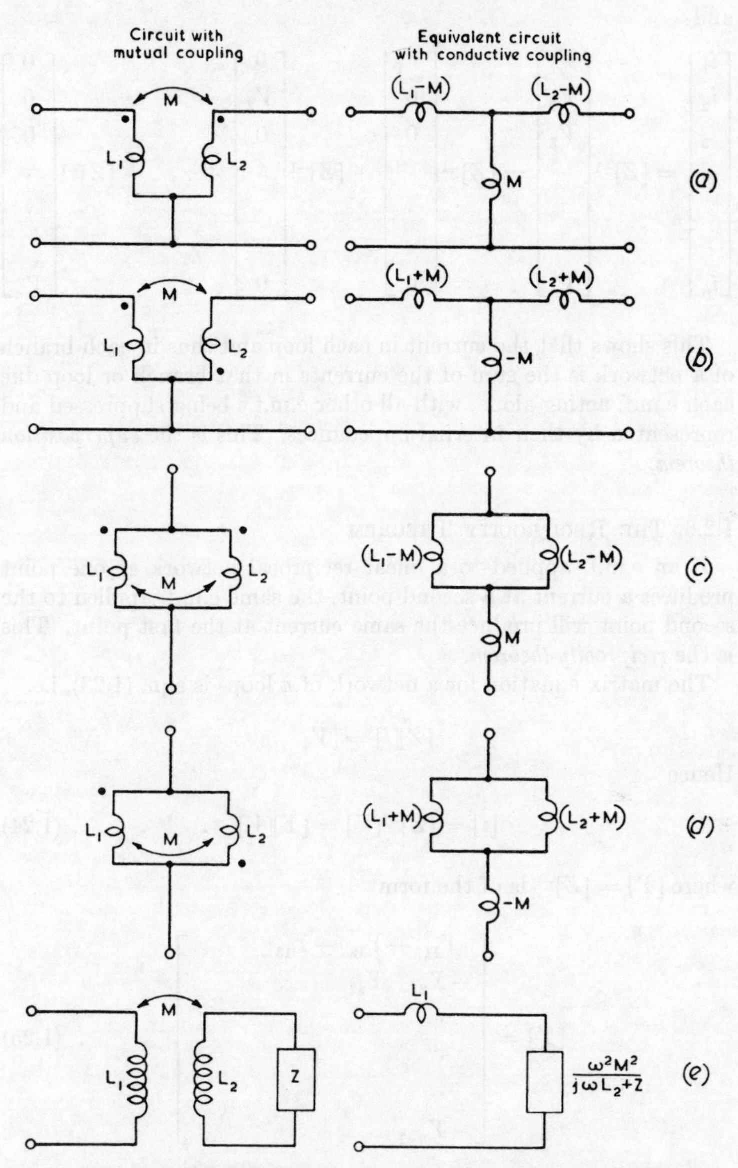

FIG. 1.8. EQUIVALENT CIRCUITS FOR MUTUAL INDUCTANCE

and

$$\begin{bmatrix} i_1 \\ i_2 \\ i_3 \\ \cdot \\ \cdot \\ \cdot \\ i_n \end{bmatrix} = [Z]^{-1} \begin{bmatrix} V_1 \\ V_2 \\ V_3 \\ \cdot \\ \cdot \\ \cdot \\ V_n \end{bmatrix} = [Z]^{-1} \begin{bmatrix} V_1 \\ 0 \\ 0 \\ \cdot \\ \cdot \\ \cdot \\ 0 \end{bmatrix} + [Z]^{-1} \begin{bmatrix} 0 \\ V_2 \\ 0 \\ \cdot \\ \cdot \\ \cdot \\ 0 \end{bmatrix} + \ldots + [Z]^{-1} \begin{bmatrix} 0 \\ 0 \\ 0 \\ \cdot \\ \cdot \\ \cdot \\ V_n \end{bmatrix}$$

This shows that the current in each loop and thus in each branch of a network is the sum of the currents in that branch or loop due each e.m.f. acting alone, with all other e.m.f.s being suppressed and represented by their internal impedances. This is the *superposition theorem.*

1.2.6. The Reciprocity Theorem

If an e.m.f. applied to a linear reciprocal network at one point produces a current at a second point, the same e.m.f. applied to the second point will produce the same current at the first point. This is the *reciprocity theorem.*

The matrix equation for a network of n loops is eqn. (1.23), i.e.

$$[Z][i] = [V]$$

Hence

$$[i] = [Z]^{-1}[V] = [Y][V] \qquad (1.24)$$

where $[Y] = [Z]^{-1}$ is of the form

$$[Y] = \begin{bmatrix} Y_{11} & -Y_{12} & -Y_{13} & \cdots \\ -Y_{21} & Y_{22} & \cdots & \\ \cdot & & & \\ \cdot & & & \\ \cdot & & & \\ -Y_{n-1} & & & \end{bmatrix} \qquad (1.25)$$

Owing to the reciprocal nature of the mutual impedances Y_{12} is equal to Y_{21}, and so on. Suppose an e.m.f. V is inserted into the first loop, with all other e.m.f.s suppressed; then

$$\begin{bmatrix} i_1 \\ i_2 \\ \cdot \\ \cdot \\ \cdot \\ i_n \end{bmatrix} = \begin{bmatrix} Y_{11} & -Y_{12} & -Y_{13} & \cdot & \cdot & \cdot \\ -Y_{21} & Y_{22} & & & & \\ \cdot & & & & & \\ \cdot & & & & & \\ \cdot & & & & & \\ -Y_{n1} & \cdot & \cdot & \cdot & \cdot & Y_{nn} \end{bmatrix} \begin{bmatrix} V \\ 0 \\ \cdot \\ \cdot \\ \cdot \\ 0 \end{bmatrix}$$

The current in loop 2 is therefore

$$i_2 = -Y_{21}V = -Y_{12}V$$

If now the e.m.f. is inserted into loop 2 we have

$$\begin{bmatrix} i_1' \\ i_2' \\ \cdot \\ \cdot \\ \cdot \\ i_n' \end{bmatrix} = \begin{bmatrix} Y_{11} & -Y_{12} & \cdot & \cdot & \cdot & \cdot \\ -Y_{21} & Y_{22} & & & & \\ \cdot & & & & & \\ \cdot & & & & & \\ \cdot & & & & & \\ -Y_{n1} & & & & & Y_{nn} \end{bmatrix} \begin{bmatrix} 0 \\ V \\ \cdot \\ \cdot \\ \cdot \\ 0 \end{bmatrix}$$

so that $i_1' = -Y_{12}V = i_2$. This result obviously applies to any two loops, and so the theorem is proved.

1.2.7. Input Admittance

Suppose that we wish to determine the input admittance of a network at a pair of terminals such as those in Fig. 1.9 (*a*). The network is complete when the terminals are short-circuited, and the matrix equation (1.24) applies. To determine the input admittance at the terminals, the e.m.f.s of all sources within the network are suppressed, but not their internal impedances, and an e.m.f. V_p is applied across the two terminals.

Combining eqns. (1.23), (1.24) and (1.25),

$$\begin{bmatrix} i_1 \\ \cdot \\ \cdot \\ \cdot \\ i_p \\ \cdot \\ \cdot \\ \cdot \\ i_n \end{bmatrix} = \begin{bmatrix} Y_{11} & -Y_{12} & \cdot & \cdot & \cdot & \cdot & \cdot \\ \cdot & & & & & & \\ \cdot & & & & & & \\ \cdot & & & & & & \\ -Y_{p1} & -Y_{p2} & \cdot & \cdot & \cdot & Y_{pp} & \cdot \; \cdot \; \cdot \\ \cdot & & & & & & \\ \cdot & & & & & & \\ \cdot & & & & & & \\ -Y_{n1} & \cdot & \cdot & \cdot & \cdot & \cdot & Y_{nn} \end{bmatrix} \begin{bmatrix} 0 \\ 0 \\ \cdot \\ \cdot \\ V_p \\ \cdot \\ \cdot \\ \cdot \\ 0 \end{bmatrix}$$

Hence

$$i_p = Y_{pp} V_p \quad \text{so that} \quad Y_{pp} = \frac{i_p}{V_p}$$

Thus the input admittance at the terminals is Y_{pp}.

Example 1.7. Find the input impedance at the terminals AB of the network shown in Fig. 1.9(*b*).

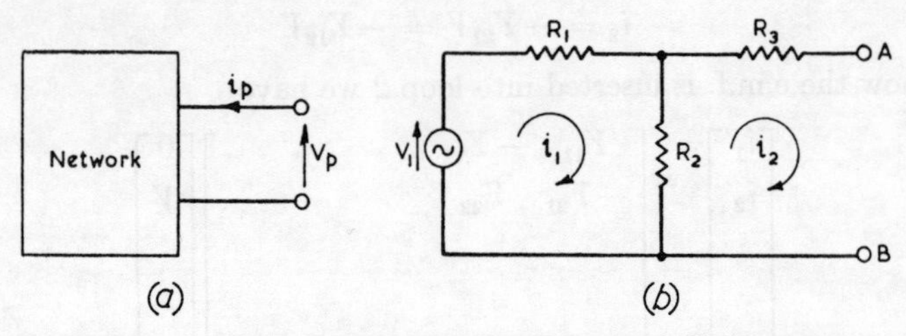

Fig. 1.9. Input Admittance at a Pair of Terminals

With AB short-circuited the impedance matrix is

$$[Z] = \begin{bmatrix} R_1 + R_2 & -R_2 \\ -R_2 & R_2 + R_3 \end{bmatrix}$$

Hence

$$[Z]^{-1} = \frac{1}{(R_1R_2 + R_2R_3 + R_1R_3)} \begin{bmatrix} R_2 + R_3 & R_2 \\ R_2 & R_1 + R_2 \end{bmatrix}$$

(from eqn. (1.12))

so that

$$Y_{22} = \frac{R_1 + R_2}{R_1R_2 + R_2R_3 + R_1R_3}$$

and the input impedance is $Z_{in} = R_3 + R_1R_2/(R_1 + R_2)$, as can be verified from simple circuit theory.

1.2.8. Thévenin's Theorem

There are several forms of this important circuit theorem. One form is that "the current in any branch of a network is equal to the current which would be produced by the voltage appearing across a break in the branch acting on the internal impedance at the break."

For a network of n meshes,

$$[V] = [Z][i] \quad \text{or} \quad [i] = [Z]^{-1}[V] = [Y][V]$$

or

$$\begin{bmatrix} i_1 \\ \cdot \\ \cdot \\ \cdot \\ i_p \\ \cdot \\ \cdot \\ i_n \end{bmatrix} = \begin{bmatrix} Y_{11} & -Y_{12} & \cdot & \cdot & \cdot \\ \cdot & & & & \\ \cdot & & & & \\ \cdot & & & & \\ -Y_{p1} & -Y_{p2} & \cdot & \cdot & \cdot \\ \cdot & & & & \\ \cdot & & & & \\ -Y_{n1} & \cdot & \cdot & \cdot & \cdot \cdot \cdot \end{bmatrix} \begin{bmatrix} V_1 \\ V_2 \\ \cdot \\ \cdot \\ V_p \\ \cdot \\ \cdot \\ V_n \end{bmatrix} \quad \cdot \quad \cdot \quad (1.26)$$

The current in the pth mesh is

$$i_p = -Y_{p1}V_1 - Y_{p2}V_2 - \cdot \cdot \cdot + Y_{pp}V_p - \cdot \cdot \cdot - Y_{pn}V_n$$

Suppose that the mesh currents are so organized that only i_p flows through one branch of the pth loop, and that a break is made in this branch. If a constant-voltage generator with an e.m.f. V_{oc} equal to the open-circuit voltage across the break is introduced in the break, then the current i_p would be reduced to zero, and the matrix equation of the network would be

$$\begin{bmatrix} i_1' \\ \cdot \\ \cdot \\ i_p' \\ \cdot \\ \cdot \\ \cdot \\ i_n' \end{bmatrix} = [Y] \begin{bmatrix} V_1 \\ \cdot \\ \cdot \\ (V_p - V_{oc}) \\ \cdot \\ \cdot \\ \cdot \\ V_n \end{bmatrix}$$

In this case,

$$i_p' = 0 = -Y_{p1}V_1 - Y_{p2}V_2 - \ldots + Y_{pp}(V_p - V_{oc}) - \ldots - Y_{pn}V_n$$

Hence

$$\begin{aligned} Y_{pp}V_{oc} &= -Y_{p1}V_1 - Y_{p2}V_2 - \ldots + Y_{pp}V_p - \ldots - Y_{pn}V_n \\ &= i_p \quad \text{(when } V_{oc} \text{ is not present)} \end{aligned}$$

This gives

$$i_p = Y_{pp}V_{oc}$$

and thus proves *Thévenin's theorem,* since from Section 1.2.7, Y_{pp} is the input admittance at the break.

A useful alternative form of Thévenin's theorem states that any system with two terminals can be replaced by a constant-voltage source, V_T, in series with an impedance Z_T, where V_T is the terminal open-circuit voltage and Z_T is the impedance looking into the terminals with the sources of e.m.f.—but not their internal impedances—short-circuited. It follows that the internal impedance of a network at its output terminals is obtained by dividing the open-circuit voltage across these terminals by the short-circuit current through them.

Care must be exercised in the use of Thévenin's theorem in equivalent circuit networks which contain dependent voltage sources, i.e. sources whose e.m.f.s depend on voltages or currents within the network, such as equivalent transistor circuits or equivalent valve circuits which include feedback. In these cases it is not admissible to short-circuit such sources in calculating internal impedance.

Thévenin's theorem is also known as the Helmholtz theorem, or the Helmholtz–Thévenin theorem, on historical grounds.

1.2.9. Norton's Theorem

Norton's theorem is the constant-current form of Thévenin's theorem, and states that any network with two terminals may be replaced by a source of constant current, I_s, in parallel with an

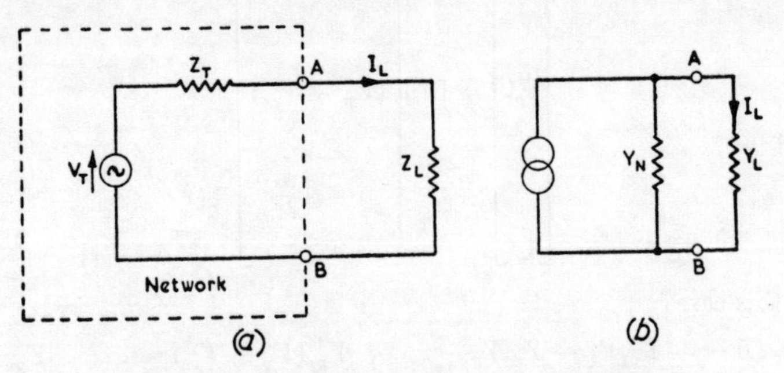

Fig. 1.10. Illustrating Norton's Theorem

admittance Y_T, where I_s is the terminal short-circuit current, and Y_T is the admittance looking into the terminals with all e.m.f.s suppressed and "represented" by their internal impedances.

The proof follows directly from Thévenin's theorem. Thus the network behind any two terminals may be represented by an e.m.f. V_T in series with an impedance Z_T (Fig. 1.10).

If an impedance Z_L is connected across AB, then I_L is given by

$$I_L = \frac{V_T}{Z_T + Z_L} = \frac{V_T}{1/Y_T + 1/Y_L} = V_T \frac{Y_T Y_L}{Y_T + Y_L}$$

$$= I_s \frac{Y_L}{Y_T + Y_L} \quad . \quad . \quad . \quad . \quad (1.27)$$

where $I_s = V_T Y_T$ is the terminal short-circuit current.

But this is the current through an admittance Y_L $(= 1/Z_L)$ from a source of constant current, I_s, in parallel with an admittance $Y_T = 1/Z_T$, and so proves Norton's theorem.

It follows directly that any constant-voltage source of e.m.f. V and internal impedance Z can be represented as far as any external circuit is concerned by a constant-current source of current $I_s = V/Z$ in parallel with an admittance $Y = 1/Z$.

Bibliography

Tropper, *Matrix Theory for Electrical Engineering Students* (Harrap).
Bode, *Network Analysis and Feedback Amplifier Design* (Van Nostrand).
Braae, *Matrix Algebra for Electrical Engineers* (Pitman).

Problems

1.1. Show that

(i)
$$\begin{bmatrix} 3 & 2 \\ 5 & -1 \end{bmatrix} \begin{bmatrix} 2 & 1 \\ 1 & 2 \end{bmatrix} = \begin{bmatrix} 8 & 7 \\ 9 & 3 \end{bmatrix}$$

and (ii)
$$\begin{bmatrix} 2 & 1 \\ 1 & 2 \end{bmatrix} \begin{bmatrix} 3 & 2 \\ 5 & -1 \end{bmatrix} = \begin{bmatrix} 11 & 3 \\ 13 & 0 \end{bmatrix}$$

1.2. Show that

$$\begin{bmatrix} 3 & 2 & 0 \\ 1 & 5 & 4 \\ 0 & 2 & 2 \end{bmatrix} \begin{bmatrix} 1 & -3 \\ 0 & 1 \\ 6 & 4 \end{bmatrix} = \begin{bmatrix} 3 & -7 \\ 25 & 18 \\ 12 & 10 \end{bmatrix}$$

1.3. Show that the inverse of

$$\begin{bmatrix} 7 & -3 \\ -4 & 2 \end{bmatrix} \quad \text{is} \quad \begin{bmatrix} 1 & 1{\cdot}5 \\ 2 & 3{\cdot}5 \end{bmatrix}$$

1.4. (i) Show that the inverse of

$$\begin{bmatrix} 3 & -1 & 4 \\ -1 & 2 & -3 \\ 4 & -3 & 2 \end{bmatrix} \quad \text{is} \quad \begin{bmatrix} 0{\cdot}2 & 0{\cdot}4 & 0{\cdot}2 \\ 0{\cdot}4 & 0{\cdot}4 & -0{\cdot}2 \\ 0{\cdot}2 & -0{\cdot}2 & -0{\cdot}2 \end{bmatrix}$$

(ii) Check the results by matrix multiplication.

1.5. Show that

$$\begin{bmatrix} 1 & 1 & 0 & 0 & 0 \\ 0 & 1 & 0 & 1 & 0 \\ 1 & 0 & 0 & 0 & 1 \end{bmatrix} \begin{bmatrix} V_1 \\ V_2 \\ V_3 \\ V_4 \\ V_5 \end{bmatrix} = \begin{bmatrix} (V_1 + V_2) \\ (V_2 + V_4) \\ (V_1 + V_5) \end{bmatrix}$$

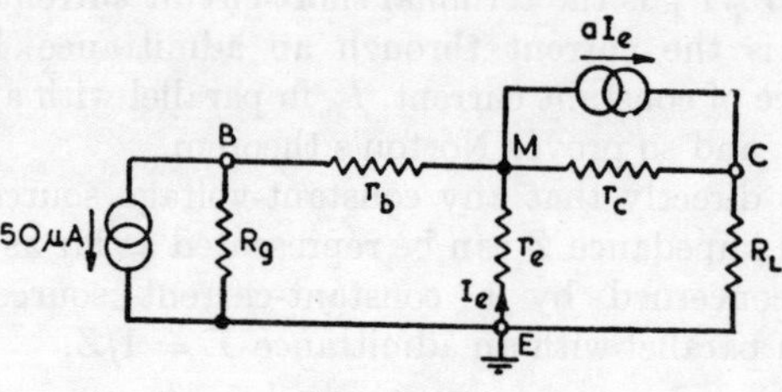

FIG. 1.11

1.6. For the transistor circuit shown in Fig. 1.11 $r_e = 10\ \Omega$, $r_b = 200\ \Omega$, $r_c = 200\ \text{k}\Omega$, $a = 0{\cdot}98$, $R_g = 10\ \text{k}\Omega$, $R_L = 1{,}000\ \Omega$. If node E is common to the input and output circuits and the constant-current generator supplies 50 μA, determine the current I_L.

(*Ans.* 2·12 mA.)

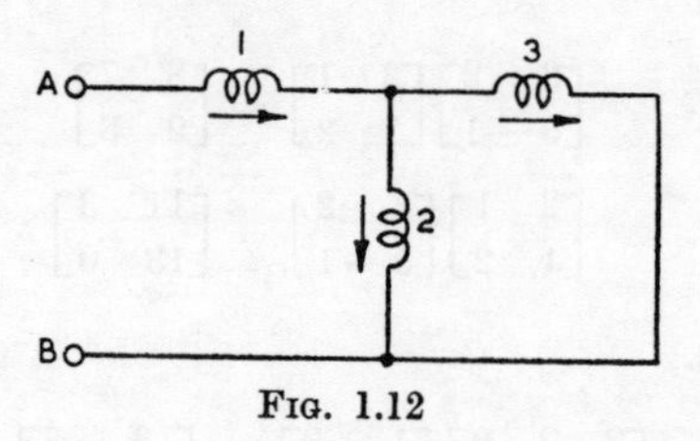

FIG. 1.12

1.7. Three mutually coupled coils, 1, 2 and 3, of negligible resistance, are connected as sh wn in Fig. 1.12. The self-inductances are $L_1 = 2$ H, $L_2 = 3$ H, $L_3 = 5$ H and for the directions of current shown in the diagram the mutual inductances are $M_{12} = 1$ H, $M_{13} = 2$ H and, $M_{23} = 2$ H.

Find the value of the equivalent inductance of the network between the terminals A and B. (*L.U. Part III Theory*, 1959)

(*Ans.* 7 H.)

1.8. Prove the general star-mesh transformation.

One end of each of five resistors, having resistances of 4, 4, 2, 8 and 8 Ω, respectively, is joined to a common point Q. The potentials of the other ends of the resistors, relative to a common point P, are 240 V, 220 V, 230 V, 250 V and 240 V, respectively. Using the star-mesh transformation, or otherwise, calculate the current in each resistor and state its direction.

Calculate the potential of Q relative to that of P.

(*Ans.* 1·75 A, −3·25 A, −0·75 A, 2·125 A, 0·875 A, 233 V.)

(*L.U. Part III Theory*, 1962)

1.9. Three voltmeters having resistances of 10, 10 and 5 kΩ, respectively, are connected in star to a balanced 3-phase 3-wire supply. The line voltage is 440 V. Determine the readings of the three voltmeters.

(*A.E.E.*, 1959.)

(*Ans.* 190 V, 290 V, 290 V.)

1.10. Derive an expression for the voltage V_{sn} existing between the neutral point N of a balanced 3-phase star-connected generator of V volts per phase and the star point S of its unbalanced load consisting of star-connected *admittances* Y_a, Y_b, Y_c; the phase sequence is A, B, C. Such an ideal generator of 240 V per phase supplies star-connected *impedances* Z_a, Z_b, Z_c, of $(2 + j20)$, $(11 + j16)$ and $(13 + j0)$, ohms, respectively. Calculate (i) V_{sn} and (ii) the power in phase A.

(*I.E.E. Supply*, 1961.)

(*Ans.* 143 V, 685 W.)

1.11. Give a proof of the delta-star transformation.

Three resistances of 5, 10 and 15 Ω are connected in delta across a 3-phase supply. Find the values of the three resistors, which if connected in star across the same supply, would take the same line currents.

If this star-connected load is supplied from a 4-wire 3-phase system with 260 V between lines, calculate the current in the neutral.

(*L.U. Part II Theory*, 1958.)

(*Ans.* 2·5 Ω, 1·67 Ω, 5 Ω; 52 A.)

1.12. State the Helmholtz–Thévenin theorem.

A linear resistance network contains steady d.c. sources; P and Q are two points in the network. The p.d. between P and Q is measured (i) by an electrostatic voltmeter, and (ii) by a permanent-magnet moving-coil voltmeter having a resistance of 12,500 Ω; readings of 120 V and 114 V, respectively, are obtained. Determine the reading which should be obtained if a dynamometer [electrodynamic] voltmeter were used, having a resistance of 4,000 Ω. (*L.U. Part II Theory*, 1961.)

(*Ans.* 103 V.)

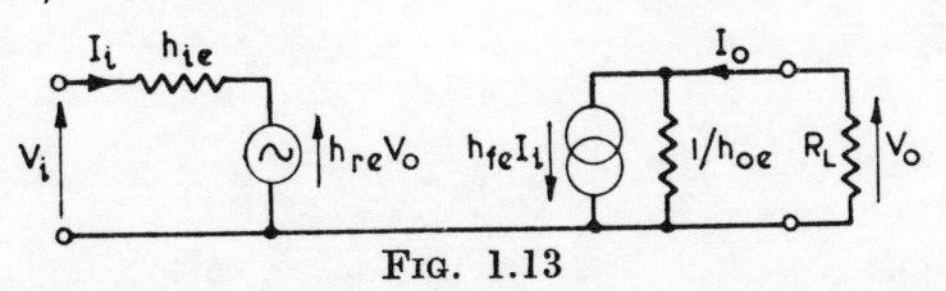

FIG. 1.13

1.13. For the h-parameter small-signal equivalent circuit of a transistor shown in Fig. 1.13, determine the current gain, I_o/I_i, and the input resistance, if $h_{ie} = 1{,}000\ \Omega$, $h_{re} = 4 \times 10^{-4}$, $h_{fe} = 70$, $h_{oe} = 50\ \mu$mhos, and $R_L = 5$ kΩ.

(*Ans.* 56; 888 Ω.)

1.14. State the Helmholtz-Thévenin theorem for a.c. networks involving a number of sources and interconnexions.

A network of impedances and sources of alternating e.m.f.s has two output terminals. The open-circuit voltage at the terminals is 260 V. The current flowing when the terminals are short-circuited is 20 A, and

13 A when connected through a coil of 11 Ω reactance and negligible resistance.

Determine the components of the equivalent circuit feeding the terminals. What value of load impedance would give maximum power output? (*L.U. Part II Theory*, 1958.)

(*Ans.* $(12 + j5)\ \Omega$, $(12 - j5)\ \Omega$.)

CHAPTER 2

Two-port Networks

A TWO-PORT NETWORK is a network which has two input and two output terminals. It may also be called a *four-terminal network* or *quadripole network*. Two-port networks are of importance as transmission elements, and it is therefore necessary to analyse the relationships between input and output voltages and currents. Matrix algebra gives a convenient method whereby the operation of complicated networks can be built up from the analysis of more simple elements, which will all be assumed to be linear, reciprocal and passive.

2.1. ABCD Parameters

For any linear two-port network, there will be a linear relationship between input voltage and current (V_1, I_1), and output voltage and current (V_2, I_2), which may be expressed in the form

$$\left.\begin{aligned} V_1 &= \mathrm{A}V_2 + \mathrm{B}I_2 \\ I_1 &= \mathrm{C}V_2 + \mathrm{D}I_2 \end{aligned}\right\} \quad . \quad . \quad . \quad (2.1)$$

or

$$\begin{bmatrix} V_1 \\ I_1 \end{bmatrix} = \begin{bmatrix} \mathrm{A} & \mathrm{B} \\ \mathrm{C} & \mathrm{D} \end{bmatrix} \begin{bmatrix} V_2 \\ I_2 \end{bmatrix} \quad . \quad . \quad . \quad (2.1a)$$

where A and D are dimensionless constants, B is a constant with the dimensions of an impedance, and C is a constant with the dimensions of an admittance. The constants A, B, C, D are sometimes called the *linear parameters* or *auxiliary constants* of the network, and the matrix

$$\begin{bmatrix} \mathrm{A\,B} \\ \mathrm{C\,D} \end{bmatrix}$$

the *transfer matrix* of the network. The directions of the currents and voltages in eqn. (2.1) are taken to be those shown in Fig. 2.1 (*a*).

For two networks in cascade (Fig. 2.1 (*b*)) we can write the input of the first in terms of its output as

$$\begin{bmatrix} V_1 \\ I_1 \end{bmatrix} = \begin{bmatrix} A_1 & B_1 \\ C_1 & D_1 \end{bmatrix} \begin{bmatrix} V_x \\ I_x \end{bmatrix}$$

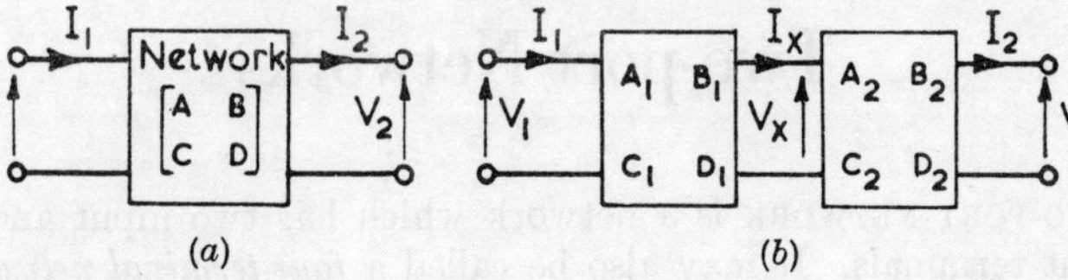

FIG. 2.1. GENERAL TWO-PORT NETWORK AND CASCADED NETWORKS

But V_x and I_x are the input voltage and current of the second network, so that

$$\begin{bmatrix} V_x \\ I_x \end{bmatrix} = \begin{bmatrix} A_2 & B_2 \\ C_2 & D_2 \end{bmatrix} \begin{bmatrix} V_2 \\ I_2 \end{bmatrix}$$

whence

$$\begin{bmatrix} V_1 \\ I_1 \end{bmatrix} = \begin{bmatrix} A_1 & B_1 \\ C_1 & D_1 \end{bmatrix} \begin{bmatrix} A_2 & B_2 \\ C_2 & D_2 \end{bmatrix} \begin{bmatrix} V_2 \\ I_2 \end{bmatrix} \quad . \quad . \quad . \quad (2.2)$$

and, from Section 1.2,

$$\begin{bmatrix} V_1 \\ I_1 \end{bmatrix} = \begin{bmatrix} (A_1A_2 + B_1C_2) & (A_1B_2 + B_1D_2) \\ (C_1A_2 + D_1C_2) & (C_1B_2 + D_1D_2) \end{bmatrix} \begin{matrix} V_2 \\ I_2 \end{matrix} = \begin{bmatrix} A & B \\ C & D \end{bmatrix} \begin{bmatrix} V_2 \\ I_2 \end{bmatrix}$$

where A B C D are the linear parameters of the combined network.

2.2. Simple Transmission Networks

In many cases a transmission network is made up of cascaded series and shunt elements, and the ABCD parameters can be readily evaluated in terms of these elements.

(*a*) *Series Impedance.* For the single series impedance transmission network shown in Fig. 2.2 (*a*) we may write, by inspection,

$$V_1 = V_2 + ZI_2$$

and

$$I_1 = I_2$$

or

$$\begin{bmatrix} V_1 \\ I_1 \end{bmatrix} = \begin{bmatrix} 1 & Z \\ 0 & 1 \end{bmatrix} \begin{bmatrix} V_2 \\ I_2 \end{bmatrix} \quad . \quad . \quad . \quad (2.3)$$

Hence the linear parameters are A = 1, B = Z, C = 0 and D = 1.

(*b*) *Shunt Admittance.* Fig. 2.2 (*b*) shows a transmission network with a single shunt admittance. Again, by inspection,

$$\begin{aligned} V_1 &= V_2 \\ I_1 &= YV_2 + I_2 \end{aligned} \quad \text{or} \quad \begin{bmatrix} V_1 \\ I_1 \end{bmatrix} = \begin{bmatrix} 1 & 0 \\ Y & 1 \end{bmatrix} \begin{bmatrix} V_2 \\ I_2 \end{bmatrix} \quad . \quad (2.4)$$

so that the linear parameters are A = D = 1, B = 0, C = Y.

(*c*) *Half-T Network.* The half-T network (Fig. 2.2 (*c*)) can be considered as the cascade connexion of a single series impedance

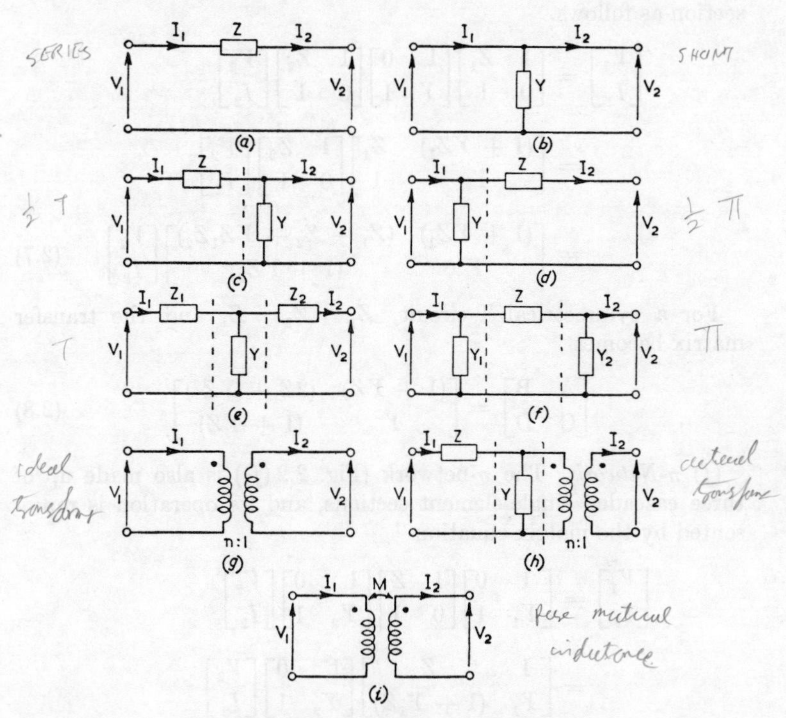

FIG. 2.2. TYPICAL TWO-PORT NETWORKS

and a single shunt admittance. Hence, from eqns. (2.2), (2.3) and (2.4), we may write

$$\begin{bmatrix} V_1 \\ I_1 \end{bmatrix} = \begin{bmatrix} 1 & Z \\ 0 & 1 \end{bmatrix} \begin{bmatrix} 1 & 0 \\ Y & 1 \end{bmatrix} \begin{bmatrix} V_2 \\ I_2 \end{bmatrix} = \begin{bmatrix} (1 + YZ) & Z \\ Y & 1 \end{bmatrix} \begin{bmatrix} V_2 \\ I_2 \end{bmatrix} \quad (2.5)$$

following the rules for matrix multiplication given in Section 1.2.

(*d*) *Half-π Network.* The half-π network (Fig. 2.2 (*d*)) is simply a shunt-admittance network in cascade with a series-impedance network. Hence,

$$\begin{bmatrix} V_1 \\ I_1 \end{bmatrix} = \begin{bmatrix} 1 & 0 \\ Y & 1 \end{bmatrix} \begin{bmatrix} 1 & Z \\ 0 & 1 \end{bmatrix} \begin{bmatrix} V_2 \\ I_2 \end{bmatrix} = \begin{bmatrix} 1 & Z \\ Y & (1+YZ) \end{bmatrix} \begin{bmatrix} V_2 \\ I_2 \end{bmatrix} \quad (2.6)$$

(*e*) *T-Network.* The T-network shown in Fig. 2.2 (*e*) is made up of three cascaded sections, and hence the overall parameters are obtained from the matrix product of the transfer matrices of each section as follows.

$$\begin{bmatrix} V_1 \\ I_1 \end{bmatrix} = \begin{bmatrix} 1 & Z_1 \\ 0 & 1 \end{bmatrix} \begin{bmatrix} 1 & 0 \\ Y & 1 \end{bmatrix} \begin{bmatrix} 1 & Z_2 \\ 0 & 1 \end{bmatrix} \begin{bmatrix} V_2 \\ I_2 \end{bmatrix}$$

$$= \begin{bmatrix} (1+YZ_1) & Z_1 \\ Y & 1 \end{bmatrix} \begin{bmatrix} 1 & Z_2 \\ 0 & 1 \end{bmatrix} \begin{bmatrix} V_2 \\ I_2 \end{bmatrix}$$

$$= \begin{bmatrix} (1+YZ_1) & (Z_1+Z_2+YZ_1Z_2) \\ Y & (1+YZ_2) \end{bmatrix} \begin{bmatrix} V_2 \\ I_2 \end{bmatrix} \quad (2.7)$$

For a symmetrical T-circuit, $Z_1 = Z_2 = Z$, and the transfer matrix becomes

$$\begin{bmatrix} \mathrm{A} & \mathrm{B} \\ \mathrm{C} & \mathrm{D} \end{bmatrix} = \begin{bmatrix} (1+YZ) & (2Z+YZ^2) \\ Y & (1+YZ) \end{bmatrix} . \quad (2.8)$$

(*f*) *π-Network.* The π-network (Fig. 2.2 (*f*)) is also made up of three cascaded single-element sections, and its operation is represented by the matrix equation

$$\begin{bmatrix} V_1 \\ I_1 \end{bmatrix} = \begin{bmatrix} 1 & 0 \\ Y_1 & 1 \end{bmatrix} \begin{bmatrix} 1 & Z \\ 0 & 1 \end{bmatrix} \begin{bmatrix} 1 & 0 \\ Y_2 & 1 \end{bmatrix} \begin{bmatrix} V_2 \\ I_2 \end{bmatrix}$$

$$= \begin{bmatrix} 1 & Z \\ Y_1 & (1+Y_1Z) \end{bmatrix} \begin{bmatrix} 1 & 0 \\ Y_2 & 1 \end{bmatrix} \begin{bmatrix} V_2 \\ I_2 \end{bmatrix}$$

$$= \begin{bmatrix} (1+Y_2Z) & Z \\ (Y_1+Y_2+Y_1Y_2Z) & (1+Y_1Z) \end{bmatrix} \begin{bmatrix} V_2 \\ I_2 \end{bmatrix} . \quad (2.9)$$

For the symmetrical π-network, $Y_1 = Y_2 = Y$, and hence the transfer matrix is

$$\begin{bmatrix} \mathrm{A} & \mathrm{B} \\ \mathrm{C} & \mathrm{D} \end{bmatrix} = \begin{bmatrix} (1+YZ) & Z \\ (2Y+Y^2Z) & (1+YZ) \end{bmatrix} . \quad (2.10)$$

Note that the π-network which is equivalent to a given T-network may be found by the star-delta transformation (eqn. (1.19)), and the T which is equivalent to a given π, by the delta-star transformation (eqn. (1.21)). Alternatively, in both cases the equivalent networks may be found by equating the elements in the transfer matrices of eqns. (2.7) and (2.9).

(*g*) *Ideal Transformer.* An ideal transformer is assumed to be loss-free, and gives a transformation ratio of $n:1$ (Fig. 2.2 (*g*)). Thus

$$V_1 = nV_2$$

and

$$I_1 = I_2/n$$

so that

$$\begin{bmatrix} V_1 \\ I_1 \end{bmatrix} = \begin{bmatrix} n & 0 \\ 0 & 1/n \end{bmatrix} \begin{bmatrix} V_2 \\ I_2 \end{bmatrix} \quad . \quad . \quad . \quad (2.11)$$

(*h*) *Actual Transformer.* The imperfections in an actual transformer may be represented by a series impedance, Z, and a shunt admittance, Y (referred to the primary), followed by an ideal transformer (Fig. 2.2 (*h*)). The overall transfer matrix is then that for the three separate elements in cascade.

$$\begin{bmatrix} V_1 \\ I_1 \end{bmatrix} = \begin{bmatrix} 1 & Z \\ 0 & 1 \end{bmatrix} \begin{bmatrix} 1 & 0 \\ Y & 1 \end{bmatrix} \begin{bmatrix} n & 0 \\ 0 & 1/n \end{bmatrix} \begin{bmatrix} V_2 \\ I_2 \end{bmatrix}$$

$$= \begin{bmatrix} n(1 + YZ) & Z/n \\ nY & 1/n \end{bmatrix} \begin{bmatrix} V_2 \\ I_2 \end{bmatrix} \quad . \quad . \quad (2.12)$$

(*i*) *Pure Mutual Inductance.* It is assumed that the self-inductance and resistance of each coil shown in Fig. 2.2 (*i*) can be represented as external series impedances. The relations for the mutual coupling alone are

$$V_1 = -j\omega M I_2$$

and

$$I_1 = V_2/j\omega M$$

Hence

$$\begin{bmatrix} V_1 \\ I_1 \end{bmatrix} = \begin{bmatrix} 0 & -j\omega M \\ 1/j\omega M & 0 \end{bmatrix} \begin{bmatrix} V_2 \\ I_2 \end{bmatrix} \quad . \quad . \quad (2.13)$$

Example 2.1. A 3-phase transmission line has the following linear parameters: A = D = $0{\cdot}96\underline{/2^\circ}$; B = $55\underline{/65^\circ}$ ohms per phase; C = $0{\cdot}0005\underline{/80^\circ}$ mho per phase. Determine the sending-end voltage and power factor when the line supplies a load of 45 MW at 132 kV and 0·8 power factor lagging.

It is convenient to work in phase values. Hence, with V_2 as the reference complexor,

$$V_2 = (132/\sqrt{3})\underline{/0^\circ} = 76{\cdot}3\underline{/0^\circ} \text{ kV}$$

Also $I_2 = \frac{45{,}000}{\sqrt{3} \times 132 \times 0{\cdot}8} \angle{-\cos^{-1} 0{\cdot}8} = 246\angle{-37°}$ A

Hence $V_1 = AV_2 + BI_2$

$$= (0{\cdot}96\angle{2°} \times 76{,}300\angle{0°}) + (55\angle{65°} \times 246\angle{-37°})$$

$$= 85{,}600\angle{6{\cdot}6°}$$

and the sending-end line voltage is $\sqrt{3} \times 85{\cdot}6 = \underline{\underline{148 \text{ kV.}}}$

To find the sending-end power factor we must determine the sending-end current from the equation

$$I_1 = CV_2 + DI_2$$

$$= (0{\cdot}0005\angle{80°} \times 76{,}300\angle{0°}) + (0{\cdot}96\angle{2°} \times 246\angle{-37°})$$

$$= 224\angle{-26{\cdot}2°} \text{ A}$$

The phase angle between the sending-end current and voltage is therefore $6{\cdot}6° + 26{\cdot}2° = 32{\cdot}8°$, and the power factor is

$$\cos 32{\cdot}8° = \underline{\underline{0{\cdot}84}} \text{ lagging}$$

2.3. ABCD Relations for a Passive Network

The reciprocity theorem may be used to deduce an extremely important relationship which exists between the linear parameters of a passive two-port network. First consider a voltage V, applied to the input terminals of a two-port network whose output terminals

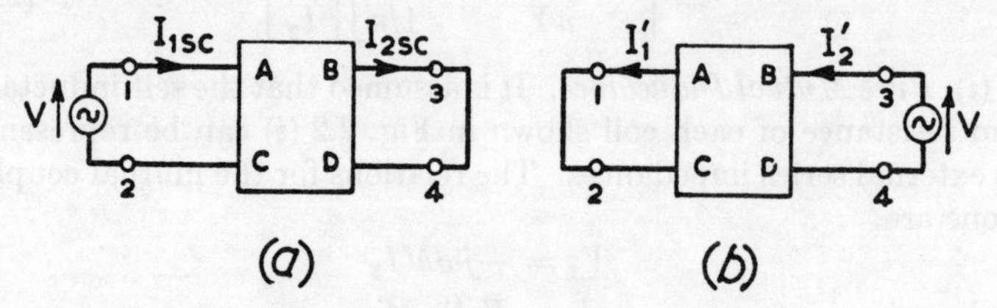

FIG. 2.3. CIRCUITS TO DEDUCE THAT AD − BC = 1

are short-circuited (Fig. 2.3 (*a*)). Since $V_2 = 0$, eqn. (2.1) gives

$$V = BI_{2sc} \quad . \quad . \quad . \quad . \quad (2.14)$$

and

$$I_{1sc} = DI_{2sc} \quad . \quad . \quad . \quad . \quad (2.15)$$

If the voltage, V, is applied to the output terminals, and the input is short-circuited, then taking account of the reversed current directions, and since $V_1 = 0$,

$$0 = AV - BI_2' \quad \text{or} \quad I_2' = AV/B \quad . \quad . \quad (2.16)$$

and

$$-I_1' = CV - DI_2' \quad . \quad . \quad . \quad (2.17)$$

But, by the reciprocity theorem, if the network contains no voltage or current sources, $I_1' = I_{2sc}$. Hence, from eqns. (2.16) and (2.17),

$$-I_{2sc} = \mathrm{C}V - \mathrm{DA}V/\mathrm{B}$$

But, from eqn. (2.14), $I_{2sc} = V/\mathrm{B}$, and hence

$$-V/\mathrm{B} = \mathrm{C}V - \mathrm{DA}V/\mathrm{B}$$

or

$$\mathrm{AD} - \mathrm{BC} = 1 \quad . \quad . \quad . \quad (2.18)$$

Note that (AD − BC) is the determinant of the transfer matrix.

A further relation can be deduced for symmetrical networks (which are identical when looked at from either the input or the output terminals). In this case the input impedance for the short-circuited network of Fig. 2.3 (*a*) is obtained from eqns. (2.14) and (2.15) as

$$Z_{1\,in} = V/I_{1sc} = \mathrm{B}/\mathrm{D}$$

For the network supplied at the "output" terminals, as in Fig. 2.3 (*b*), eqn. (2.16) gives

$$Z_{2\,in} = V/I_2' = \mathrm{B}/\mathrm{A}$$

But, for a symmetrical network, $Z_{1\,in} = Z_{2\,in}$, and hence

$$\mathrm{A} = \mathrm{D} \quad \text{for symmetrical networks} \quad . \quad . \quad (2.19)$$

Also eqn. (2.18) becomes

$$\mathrm{A}^2 - \mathrm{BC} = 1 \quad . \quad . \quad . \quad . \quad (2.20)$$

These results may be checked on the transfer matrices which were derived in Section 2.2.

2.4. Output in Terms of Input

For a passive two-port network, as in Fig. 2.1 (*a*), eqn. (2.1) gives

$$\begin{bmatrix} V_1 \\ I_1 \end{bmatrix} = \begin{bmatrix} \mathrm{A} & \mathrm{B} \\ \mathrm{C} & \mathrm{D} \end{bmatrix} \begin{bmatrix} V_2 \\ I_2 \end{bmatrix}$$

Hence

$$\begin{bmatrix} V_2 \\ I_2 \end{bmatrix} = \begin{bmatrix} \mathrm{A} & \mathrm{B} \\ \mathrm{C} & \mathrm{D} \end{bmatrix}^{-1} \begin{bmatrix} V_1 \\ I_1 \end{bmatrix}$$

$$= \frac{1}{(\mathrm{AD} - \mathrm{BC})} \begin{bmatrix} \mathrm{D} & -\mathrm{B} \\ -\mathrm{C} & \mathrm{A} \end{bmatrix} \begin{bmatrix} V_1 \\ I_1 \end{bmatrix} \quad \text{(from eqn. (1.10))}$$

$$= \begin{bmatrix} \mathrm{D} & -\mathrm{B} \\ -\mathrm{C} & \mathrm{A} \end{bmatrix} \begin{bmatrix} V_1 \\ I_1 \end{bmatrix} \quad (\text{since AD} - \text{BC} = 1) \quad . \quad (2.21)$$

This equation gives the output voltage and current in terms of the input quantities and the ABCD parameters.

2.5. Evaluation of ABCD Parameters from O.C. and S.C. Tests

If a two-port network has its output terminals open-circuited, then since $I_2 = 0$, eqn. (2.1) yields

$$V_1 = \mathrm{A}V_2 \quad \text{or} \quad \mathrm{A} = V_1/V_2\big|_{output\ o.c.} \qquad (2.22)$$

and

$$I_1 = \mathrm{C}V_2 \quad \text{or} \quad \mathrm{C} = I_1/V_2\big|_{output\ o.c.} \qquad (2.23)$$

Suppose that the output is short-circuited so that $V_2 = 0$; we now obtain from eqn. (2.1)

$$V_1 = \mathrm{B}I_2 \quad \text{or} \quad \mathrm{B} = V_1/I_2\big|_{output\ s.c.} \qquad (2.24)$$

and

$$I_1 = \mathrm{D}I_2 \quad \text{or} \quad \mathrm{D} = I_1/I_2\big|_{output\ s.c.} \qquad (2.25)$$

These relationships enable the ABCD constants of any network to be determined, even when the networks do not consist of simple cascaded sections.

2.6. ABCD Parameters of a Symmetrical Lattice

A symmetrical lattice network is shown in Fig. 2.4 (*a*). The same network is redrawn at (*b*) as a bridge network. The ABCD para-

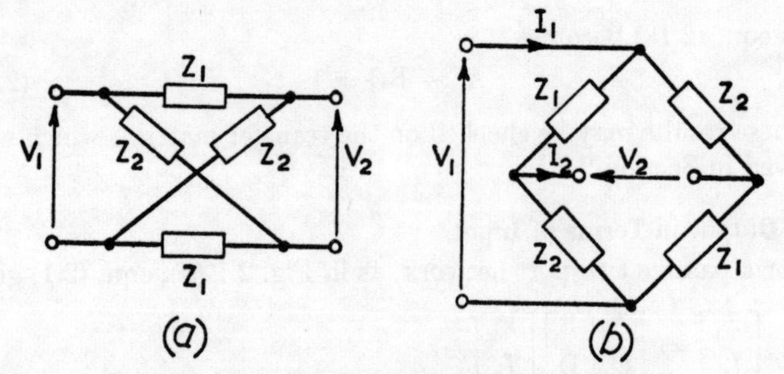

Fig. 2.4. The Symmetrical Lattice Network

meters may be found by applying the results of the previous section. Thus for an open-circuited output, eqn. (2.22) gives

$$\mathrm{A} = \frac{V_1}{V_2} = \frac{V_1}{\dfrac{V_1}{Z_1 + Z_2} Z_2 - \dfrac{V_1}{Z_1 + Z_2} Z_1} = \frac{Z_2 + Z_1}{Z_2 - Z_1}$$

Since the network is symmetrical this is also the constant D. It should be noted that, if $Z_1 > Z_2$ there will be a phase reversal at the output, and the lattice will have no *realizable* equivalent π- or T-network.

Eqn. (2.23) gives, for this same open-circuited condition,

$$C = \frac{I_1}{V_2} = \frac{\dfrac{2V_1}{Z_1 + Z_2}}{\dfrac{V_1}{Z_2 + Z_1} Z_2 - \dfrac{V_1}{Z_2 + Z_1} Z_1} = \frac{2}{Z_2 - Z_1}$$

Now, for a symmetrical network, $A^2 - BC = 1$, and so

$$B = \frac{A^2 - 1}{C} = \frac{(A + 1)(A - 1)}{C}$$

$$= \frac{\left(\dfrac{Z_2 + Z_1}{Z_2 - Z_1} + 1\right)\left(\dfrac{Z_2 + Z_1}{Z_2 - Z_1} - 1\right)}{2/(Z_2 - Z_1)}$$

$$= \frac{2Z_1Z_2}{Z_2 - Z_1}$$

The transfer matrix for the symmetrical lattice is therefore

$$\begin{bmatrix} A & B \\ C & D \end{bmatrix} = \begin{bmatrix} \dfrac{Z_2 + Z_1}{Z_2 - Z_1} & \dfrac{2Z_1Z_2}{Z_2 - Z_1} \\ \dfrac{2}{Z_2 - Z_1} & \dfrac{Z_2 + Z_1}{Z_2 - Z_1} \end{bmatrix} \quad . \quad . \quad (2.26)$$

Example 2.2. If, in Fig. 2.5(a), ${}_1R_1 = 30\ \Omega$ and ${}_2R_2 = 40\ \Omega$, find the T-circuit equivalent to the symmetrical lattice.

Since the lattice is symmetrical, the equivalent T-circuit (Fig. 2.2(e)) must also be symmetrical, and hence for this circuit eqn. (2.8) applies and we may write, for the equivalence of eqns. (2.8) and (2.26),

$$\begin{bmatrix} (1 + YZ) & (2Z + YZ^2) \\ Y & (1 + YZ) \end{bmatrix} = \begin{bmatrix} \dfrac{R_2 + R_1}{R_2 - R_1} & \dfrac{2R_1R_2}{R_2 - R_1} \\ \dfrac{2}{R_2 - R_1} & \dfrac{R_2 + R_1}{R_2 - R_1} \end{bmatrix}$$

Equating this term by term,

$$Y = 2/(R_2 - R_1) = 2/10 = \underline{\underline{0{\cdot}2 \text{ mho}}}$$

and $\quad 1 + YZ = (R_2 + R_1)/(R_2 - R_1) = 70/10 = 7$

so that $\quad Z = 6/0{\cdot}2 = \underline{\underline{30\ \Omega}}$

The equivalent circuit is shown in Fig. 2.5(b). Obviously, if $R_1 > R_2$ the shunt arm would be a negative resistance, and a T-equivalent circuit would not be practical.

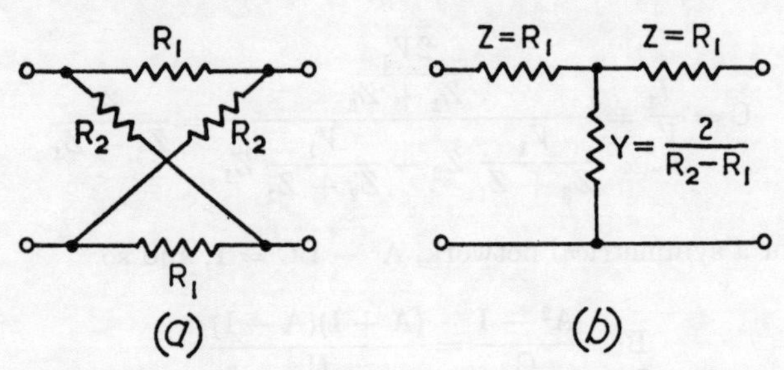

FIG. 2.5. T-EQUIVALENT OF A SYMMETRICAL LATTICE

2.7. Networks in Parallel

In order to evaluate the ABCD parameters of networks in parallel it is assumed that the presence of the second network does not alter the transfer characteristics of the first, and vice versa. Thus a single series impedance cannot be connected in parallel with a single shunt admittance, since the direct connexion between the input and output for the shunt element would simply short-circuit the series element.

Consider two compatible networks whose inputs and outputs can be connected in parallel as shown in Fig 2.6 (a). The overall ABCD

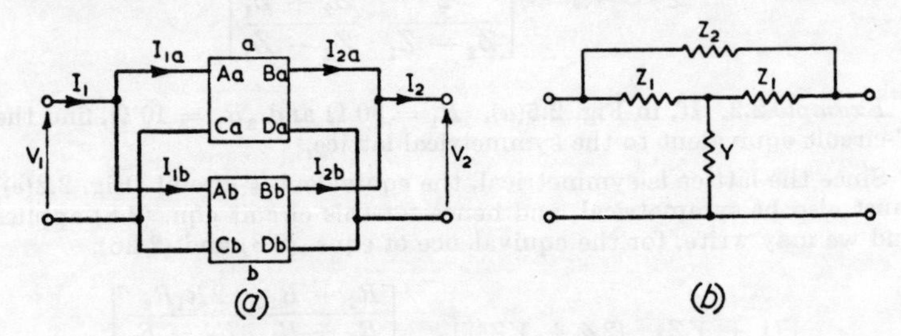

FIG. 2.6. NETWORKS IN PARALLEL

parameters of the parallel combination can be obtained by first developing a new set of parameters, called the *y-parameters*, which express the input and output currents in terms of the input and output voltages. Thus for network (a),

$$V_1 = \mathrm{A}_a V_2 + \mathrm{B}_a I_{2a}$$

and

$$I_{1a} = \mathrm{C}_a V_2 + \mathrm{D}_a I_{2a}$$

Hence $I_{2a} = \frac{1}{B_a} V_1 - \frac{A_a}{B_a} V_2 = y_{21}' V_1 + y_{22}' V_2$

and $I_{1a} = \frac{D_a}{B_a} V_1 + \left(C_a - \frac{A_a D_a}{B_a}\right) V_2 = \frac{D_a}{B_a} V_1 - \frac{1}{B_a} V_2$

since $B_a C_a - A_a D_a = -1$.

This latter equation may in turn be written

$$I_{1a} = y_{11}' V_1 + y_{12}' V_2$$

In matrix form,

$$\begin{bmatrix} I_{1a} \\ I_{2a} \end{bmatrix} = \begin{bmatrix} D_a/B_a & -1/B_a \\ 1/B_a & -A_a/B_a \end{bmatrix} \begin{bmatrix} V_1 \\ V_2 \end{bmatrix} = \begin{bmatrix} y_{11}' & y_{12}' \\ y_{21}' & y_{22}' \end{bmatrix} \begin{bmatrix} V_1 \\ V_2 \end{bmatrix}$$

In the same way the input and output currents for network (*b*) can be written

$$\begin{bmatrix} I_{1b} \\ I_{2b} \end{bmatrix} = \begin{bmatrix} D_b/B_b & -1/B_b \\ 1/B_b & -A_b/B_b \end{bmatrix} \begin{bmatrix} V_1 \\ V_2 \end{bmatrix} = \begin{bmatrix} y_{11}'' & y_{12}'' \\ y_{21}'' & y_{22}'' \end{bmatrix} \begin{bmatrix} V_1 \\ V_2 \end{bmatrix}$$

For the two networks in parallel the total current entering is $I_1 = I_{1a} + I_{1b}$, and the total current leaving is $I_2 = I_{2a} + I_{2b}$, so that addition of the corresponding matrices gives

$$\begin{bmatrix} I_1 \\ I_2 \end{bmatrix} = \begin{bmatrix} (I_{1a} + I_{1b}) \\ (I_{2a} + I_{2b}) \end{bmatrix} = \begin{bmatrix} (y_{11}' + y_{11}'') & (y_{12}' + y_{12}'') \\ (y_{21}' + y_{21}'') & (y_{22}' + y_{22}'') \end{bmatrix} \begin{bmatrix} V_1 \\ V_2 \end{bmatrix}$$

$$= \begin{bmatrix} \left(\frac{D_a}{B_a} + \frac{D_b}{B_b}\right) & \left(-\frac{1}{B_a} - \frac{1}{B_b}\right) \\ \left(\frac{1}{B_a} + \frac{1}{B_b}\right) & \left(-\frac{A_a}{B_a} - \frac{A_b}{B_b}\right) \end{bmatrix} \begin{bmatrix} V_1 \\ V_2 \end{bmatrix}$$

This matrix equation can be written as the two simultaneous equations

$$I_1 = \left(\frac{D_a}{B_a} + \frac{D_b}{B_b}\right) V_1 - \left(\frac{1}{B_a} + \frac{1}{B_b}\right) V_2 \quad \cdot \quad \cdot \quad (2.27)$$

and $$I_2 = \left(\frac{1}{B_a} + \frac{1}{B_b}\right) V_1 - \left(\frac{A_a}{B_a} + \frac{A_b}{B_b}\right) V_2 \quad \cdot \quad \cdot \quad (2.28)$$

From these two equations, the ABCD parameters of the two networks in parallel can be found as follows, using the results of Section 2.5.

$$\mathrm{A} = \left.\frac{V_1}{V_2}\right|_{I_2=0} = \frac{\mathrm{A}_a\mathrm{B}_b + \mathrm{B}_a\mathrm{A}_b}{\mathrm{B}_a + \mathrm{B}_b} \quad . \quad . \quad . \quad . \quad (2.29a)$$

$$\mathrm{B} = \left.\frac{V_1}{I_2}\right|_{V_2=0} = \frac{\mathrm{B}_a\mathrm{B}_b}{\mathrm{B}_a + \mathrm{B}_b} \quad . \quad . \quad . \quad . \quad (2.29b)$$

$$\mathrm{D} = \left.\frac{I_1}{I_2}\right|_{V_2=0} = \frac{\mathrm{D}_a\mathrm{B}_b + \mathrm{D}_b\mathrm{B}_a}{\mathrm{B}_a + \mathrm{B}_b} \quad . \quad . \quad . \quad . \quad (2.29c)$$

and
$$\mathrm{C} = \left.\frac{I_1}{V_2}\right|_{I_2=0} = \left(\frac{\mathrm{D}_a}{\mathrm{B}_b} + \frac{\mathrm{D}_b}{\mathrm{B}_a}\right)\frac{(\mathrm{A}_a\mathrm{B}_b + \mathrm{A}_b\mathrm{B}_a)}{(\mathrm{B}_a + \mathrm{B}_b)} - \frac{\mathrm{B}_a + \mathrm{B}_b}{\mathrm{B}_a\mathrm{B}_b}$$

$$= \mathrm{C}_a + \mathrm{C}_b + \frac{(\mathrm{A}_a - \mathrm{A}_b)(\mathrm{D}_b - \mathrm{D}_a)}{\mathrm{B}_a + \mathrm{B}_b} \quad . \quad . \quad (2.29d)$$

For example, the bridged-T network of Fig. 2.6 (*b*) may be considered to be a symmetrical-T in parallel with a single series element. From eqns. (2.3), (2.4) and (2.29), the linear parameters of the bridged-T are

$$\begin{bmatrix} \mathrm{A} & \mathrm{B} \\ \mathrm{C} & \mathrm{D} \end{bmatrix} = \frac{1}{(2Z_1 + Z_2 + Z_1^2Y)}\begin{bmatrix} (2Z_1 + Z_1^2Y + Z_2 + Z_1Z_2Y) & (2Z_1Z_2 + Z_1^2Z_2Y) \\ (Y(2Z_1 + Z_2 + Z_1^2Y) - Y^2Z_1^2) & (2Z_1 + Z_1^2Y + Z_2 + Z_1Z_2Y) \end{bmatrix}$$

Example 2.3. A 3-phase transmission line has the following circuit parameters: $\mathrm{A}_a = 0{\cdot}97\underline{/0{\cdot}6^\circ}$, $\mathrm{B}_a = 60\underline{/70^\circ}$ ohms per phase. If a second line having parameters $\mathrm{A}_b = 0{\cdot}97\underline{/0{\cdot}4^\circ}$ and $\mathrm{B}_b = 50\underline{/76^\circ}$ ohms per phase is connected in parallel with the first, determine the sending-end voltage when delivering 50 MW at 132 kV and 0·8 lagging power factor at the receiving end. (*I.E.E. Supply*, 1961.)

The phase voltage, V_2, at the receiving end is $132/\sqrt{3} = 76{\cdot}3$ kV.
The line current at the receiving end is

$$I_2 = \frac{50{,}000}{\sqrt{3} \times 132 \times 0{\cdot}8}\underline{/-\cos^{-1} 0{\cdot}8} = 273\underline{/-37^\circ}\ \mathrm{A}$$

(The receiving-end voltage, V_2, is taken as the reference complexor.)
From eqn. (2.29) the combined A and B parameters are

$$\mathrm{A} = \frac{\mathrm{A}_a\mathrm{B}_b + \mathrm{A}_b\mathrm{B}_a}{\mathrm{B}_b + \mathrm{B}_a} = \frac{48{\cdot}5\underline{/76{\cdot}6^\circ} + 58{\cdot}2\underline{/70{\cdot}4^\circ}}{60\underline{/70^\circ} + 50\underline{/76^\circ}} = 0{\cdot}97\underline{/0{\cdot}5^\circ}$$

and $\quad \mathrm{B} = \dfrac{\mathrm{B}_a\mathrm{B}_b}{\mathrm{B}_a + \mathrm{B}_b} = \dfrac{3000\underline{/146^\circ}}{109{\cdot}8\underline{/72{\cdot}7^\circ}} = 27{\cdot}3\underline{/73{\cdot}3^\circ}$

Hence $V_1 = (0{\cdot}97\underline{/0{\cdot}5^\circ} \times 76{,}300\underline{/0^\circ}) + (27{\cdot}3\underline{/73{\cdot}3^\circ} \times 273\underline{/-37^\circ})$

$= 82{\cdot}3\underline{/3{\cdot}5^\circ}$ kV

The sending-end line voltage is therefore **142·4 kV.**

2.8. The Loaded Two-port Network

When an impedance Z_L is connected across the output terminals of a two-port network (Fig. 2.7), the relationship between V_2 and

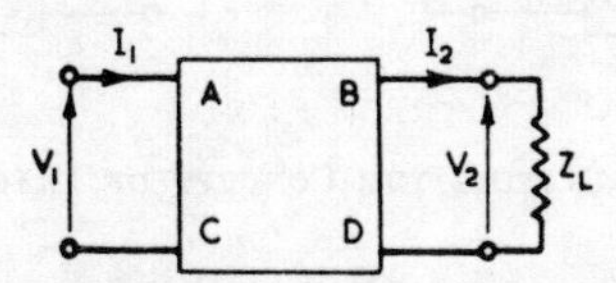

FIG. 2.7. THE LOADED TWO-PORT NETWORK

I_2 is $I_2 = V_2/Z_L$. Then eqn. (2.1) gives

$$\left.\begin{aligned} V_1 &= \mathrm{A}V_2 + \mathrm{B}V_2/Z_L = (\mathrm{A} + \mathrm{B}/Z_L)V_2 \\ I_1 &= \mathrm{C}V_2 + \mathrm{D}V_2/Z_L = (\mathrm{C} + \mathrm{D}/Z_L)V_2 \end{aligned}\right\} \qquad (2.30)$$

and

The input impedance of the network is

$$Z_{in} = \frac{V_1}{I_1} = \frac{\mathrm{A} + \mathrm{B}/Z_L}{\mathrm{C} + \mathrm{D}/Z_L} = \frac{\mathrm{A}Z_L + \mathrm{B}}{\mathrm{C}Z_L + \mathrm{D}} \qquad (2.31)$$

2.9. Image Impedances

When a load is connected to a generator through a two-port network (e.g. a filter, an equalizer, or an impedance-matching network in a communication circuit) it is usually desirable that maximum power shall be delivered by the generator to the network input terminals. The *maximum power transfer theorem* states that, for a generator with resistive internal impedance, maximum power is transferred when a resistive load at the generator terminals has a resistance equal to the generator resistance. Also maximum power is received by a load which is supplied through a transmission network from some source when the resistance seen at the load terminals is equal to the load resistance. Two-port transmission networks designed to comply with these requirements are said to be designed on an image basis. A network working between its *image impedances* is said to be matched for reflections (*see* Chapter 7).

We may define two image impedances for a two-port network, the input image impedance Z_{I1} and the output image impedance Z_{I2}. These impedances are functions of the network such that, when the network is terminated at the output by Z_{I2}, the impedance seen at the input is Z_{I1}; and when the input terminals are closed by an impedance Z_{I1}, the impedance measured at the output terminals is Z_{I2}. This is illustrated in Fig. 2.8.

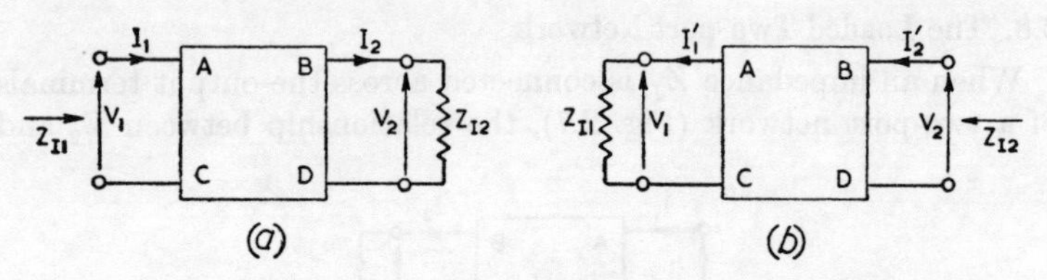

Fig. 2.8. Illustrating the Concept of Image Impedance

The image impedances may be found in terms of the linear parameters as follows. For the circuit of Fig. 2.8 (*a*), the input impedance is, by definition, Z_{I1}. Hence, from eqn. (2.31),

$$Z_{I1} = Z_{in\,1} = \frac{V_1}{I_1} = \frac{AZ_{I2} + B}{CZ_{I2} + D} \quad . \quad . \quad (2.32)$$

For the network of Fig. 2.8 (*b*), from eqn. (2.21),

$$\begin{bmatrix} V_2 \\ -I_2' \end{bmatrix} = \begin{bmatrix} A & B \\ C & D \end{bmatrix}^{-1} \begin{bmatrix} V_1 \\ -I_1' \end{bmatrix} = \begin{bmatrix} D & -B \\ -C & A \end{bmatrix} \begin{bmatrix} V_1 \\ -I_1' \end{bmatrix}$$

Hence
$$\begin{bmatrix} V_2 \\ I_2' \end{bmatrix} = \begin{bmatrix} D & B \\ C & A \end{bmatrix} \begin{bmatrix} V_1 \\ I_1' \end{bmatrix} \quad . \quad . \quad . \quad (2.33)$$

where $I_1' = V_1/Z_{I1}$.

Therefore
$$Z_{I2} = \frac{V_2}{I_2'} = \frac{DZ_{I1} + B}{CZ_{I1} + A} \quad . \quad . \quad . \quad (2.34)$$

Substituting in eqn. (2.32),

$$Z_{I1} = \frac{ADZ_{I1} + AB + BCZ_{I1} + AB}{CDZ_{I1} + CB + CDZ_{I1} + AD}$$

whence
$$Z_{I1}^2(2CD) = 2AB$$

so that
$$Z_{I1} = \sqrt{\frac{AB}{CD}} \quad . \quad . \quad . \quad . \quad (2.35)$$

Similarly, substituting from eqn. (2.32) in eqn. (2.34), and then solving for Z_{I2},

$$Z_{I2} = \frac{ADZ_{I2} + BD + BCZ_{I2} + BD}{CAZ_{I2} + CB + ACZ_{I2} + AD}$$

$$= \sqrt{\frac{DB}{CA}} \quad . \quad . \quad . \quad . \quad . \quad . \quad (2.36)$$

Note that for a *symmetrical network* A = D, so that

$$Z_{I1} = Z_{I2} = \sqrt{\frac{B}{C}} \quad . \quad . \quad . \quad . \quad (2.37)$$

Example 2.4. The components of the T-network shown in Fig. 2.9 at a particular frequency are $X_1 = 50\ \Omega$, $X_2 = 80\ \Omega$, and $B_1 = 0{\cdot}01$ mho. Find the generator internal resistance and load resistance for which the

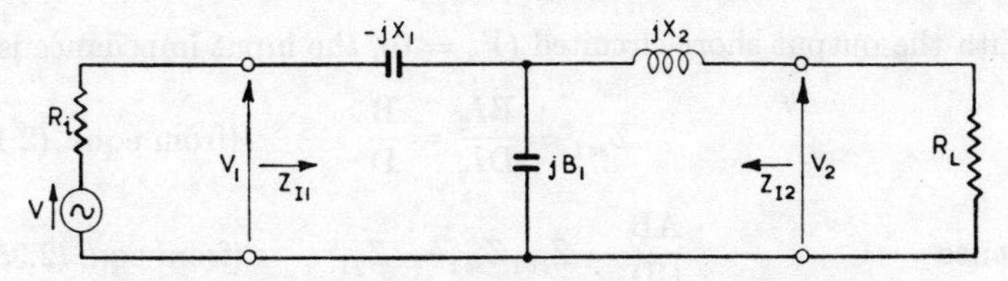

FIG. 2.9. MATCHING A RESISTIVE LOAD TO A GENERATOR

circuit will give correct matching. Calculate the phase shift between V_1 and V_2 when the network works between these resistances.

From eqn. (2.7), the ABCD parameters of the T-network are

$$A = 1 + Z_1Y = 1 + X_1B_1 = 1{\cdot}5$$
$$B = Z_1 + Z_2 + Z_1Z_2Y = -jX_1 + jX_2 + jX_1X_2B_1 = j70$$
$$C = Y = jB_1 = j0{\cdot}01$$
$$D = 1 + Z_2Y = 1 - X_2B_1 = 0{\cdot}2$$

For matching $R_i = Z_{I1}$ and $R_L = Z_{I2}$. From eqn. (2.35),

$$R_i = Z_{I1} = \sqrt{\frac{AB}{CD}} = \sqrt{\frac{1{\cdot}5 \times j70}{j0{\cdot}01 \times 0{\cdot}2}} = \underline{\underline{229\ \Omega}}$$

Similarly $$Z_{I2} = \sqrt{\frac{DB}{CA}} = \sqrt{\frac{0{\cdot}2 \times j70}{j0{\cdot}01 \times 1{\cdot}5}} = \underline{\underline{30{\cdot}6\ \Omega}}$$

When the network works between these resistances, the power delivered by the generator to the input terminals is $V^2/4R_i$, and since there is no dissipation in the T-network, this is also the power in R_L. This may be checked by finding V_2 in terms of V_1. The load power is then V_2^2/R_L.

With $R_L = 30{\cdot}6\ \Omega$, the load current is $V_2/30{\cdot}6$, and eqn. (2.1) gives

$$V_1 = \mathrm{A}V_2 + \mathrm{B}I_2 = 1{\cdot}5V_2 + j70V_2/30{\cdot}6 = V_2 \times 2{\cdot}74\underline{/57^\circ}$$

Hence V_2 lags behind V_1 by $\underline{\underline{57^\circ}}$.

Note that above the frequency for which X_2B_1 becomes equal to unity, D is negative, both Z_{I1} and Z_{I2} become purely reactive, and the T-network cannot match two resistive elements. Obviously when $X_2B_1 = 1$, $Z_1 = \infty$ and there can be no transmission. This will be discussed further in Section 2.13.

2.10. Image Impedance in Terms of Z_{oc} and Z_{sc}

With the output port of a two-port network open-circuited ($I_2 = 0$), the input impedance is

$$Z_{oc1} = \frac{V_1}{I_1} = \frac{\mathrm{A}V_2}{\mathrm{C}V_2} = \frac{\mathrm{A}}{\mathrm{C}} \qquad \text{(from eqns. (2.1))}$$

With the output short-circuited ($V_2 = 0$), the input impedance is

$$Z_{sc1} = \frac{\mathrm{B}I_2}{\mathrm{D}I_2} = \frac{\mathrm{B}}{\mathrm{D}} \qquad \text{(from eqns. (2.1))}$$

Hence

$$\frac{\mathrm{AB}}{\mathrm{CD}} = Z_{oc1}Z_{sc1} = Z_{I1}^2 \qquad \text{(from eqn. (2.35))}$$

or

$$Z_{I1} = \sqrt{(Z_{oc1}Z_{sc1})} \qquad (2.38)$$

Also the impedance seen looking into the output terminals with the input terminals open-circuited is (from eqn. (2.33) with $I_1 = 0$)

$$Z_{oc2} = \frac{V_2}{I_2'} = \frac{\mathrm{D}}{\mathrm{C}}$$

and with the input terminals short-circuited ($V_1 = 0$ in eqn. (2.33)),

$$Z_{sc2} = \frac{\mathrm{B}}{\mathrm{A}}$$

Hence

$$\frac{\mathrm{DB}}{\mathrm{AC}} = Z_{oc2}Z_{sc2} = Z_{I2}^2 \qquad \text{(from eqn. (2.36)}$$

or

$$Z_{I2} = \sqrt{(Z_{oc2}Z_{sc2})} \qquad (2.39)$$

2.11. Iterative Impedance

It is always possible to determine an impedance which, when terminating a network, gives rise to an input impedance equal to itself. This is called the *iterative impedance*, Z_{it}, of the network, and is illustrated in Fig. 2.10. If the network is unsymmetrical there will be a second iterative impedance Z_{it}' obtained by looking into the output terminals with the input terminals terminated in Z_{it}'.

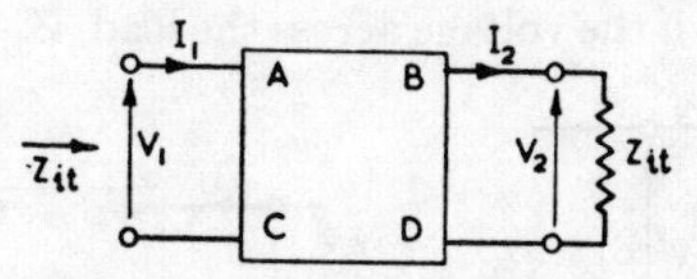

FIG. 2.10. ILLUSTRATING ITERATIVE IMPEDANCE

The iterative impedance at the input may be expressed in terms of the ABCD constants as follows. Consider the circuit of Fig. 2.10, in which $I_2 = V_2/Z_{it}$. Eqns. (2.1) give

$$V_1 = AV_2 + BV_2/Z_{it} \quad \text{and} \quad I_1 = CV_2 + DV_2/Z_{it}$$

and hence

$$Z_{in} = \frac{V_1}{I_1} = \frac{AZ_{it} + B}{CZ_{it} + D}$$

$$= Z_{it}, \text{ by definition}$$

From this,

$$CZ_{it}^2 + (D - A)Z_{it} - B = 0$$

and

$$Z_{it} = \frac{A - D \pm \sqrt{[(D - A)^2 + 4BC]}}{2C} \quad . \quad . \quad (2.40)$$

For a symmetrical network, A = D and so

$$Z_{it} = \sqrt{\frac{B}{C}} \quad . \quad . \quad . \quad . \quad (2.41)$$

Thus for a symmetrical network the iterative impedance is equal to the image impedance (from eqn. (2.37)). In this case the impedances are usually referred to as the *characteristic impedance*, Z_0, of the symmetrical network, where

$$Z_0 = Z_{it} = Z_I = \sqrt{\frac{B}{C}} \quad . \quad . \quad . \quad (2.42)$$

and, from eqn. (2.38),

$$Z_0 = \sqrt{Z_{oc}Z_{sc}} \quad . \quad . \quad . \quad . \quad (2.43)$$

The iterative impedance of a network is most useful in the design of attenuators, as will be seen in the following sections.

2.12. Insertion-loss Ratio

The *insertion-loss* ratio, A_L, of a two-port network is defined as the ratio of voltage across the load with the network removed to the voltage across the load with the network inserted. This is illustrated in Fig. 2.11. Thus if the voltage across the load, Z_L, with the network

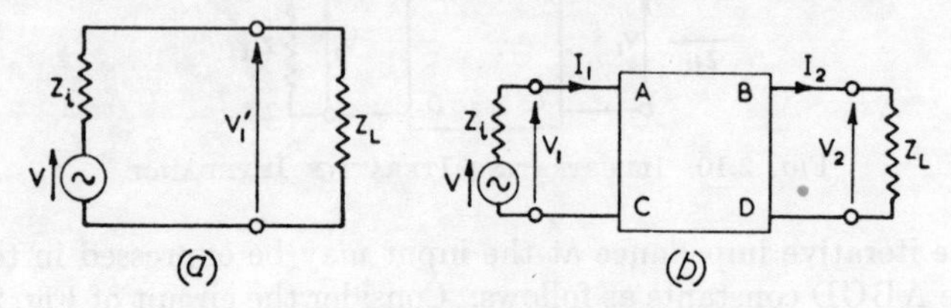

FIG. 2.11. ILLUSTRATING INSERTION LOSS

omitted is V_1' (Fig. 2.11 (*a*)), and with the network inserted is V_2 (Fig. 2.11 (*b*)), then

$$A_L = \frac{V_1'}{V_2} \qquad (2.44)$$

This will be a complex number in the general case, with an insertion phase shift of (arg V_1' − arg V_2) and an insertion loss of

$$A_L = 20 \log_{10} |V_1'/V_2| \text{ decibels} \qquad (2.45)$$

If the transmission network is designed for iterative operation, its input impedance is equal to Z_L, and hence V_1' will be equal to V_1. In this case

$$A_L = \frac{V_1}{V_2} \qquad (2.46)$$

and the insertion loss will therefore be a function of the network alone, and will not depend on the source impedance. This is why attenuators are designed for iterative operation, since in this case a given network will have a known insertion loss when feeding its iterative impedance, independently of the source impedance.

Example 2.5. Design a half-π attenuator section for a 10-dB attenuation and an iterative impedance of 600 Ω. If this network is inserted between a 75-Ω generator and a 75-Ω load, find the insertion loss.

The circuit is shown in Fig. 2.12. For iterative operation, $R_L = 600$ Ω, and the resistance seen looking into terminals 11′ is then also 600 Ω. Hence

$$600 = \frac{R_2(R_1 + 600)}{R_1 + R_2 + 600} \qquad \text{(i)}$$

Also for 10-dB insertion loss,

$$10 = 20 \log |V_1/V_2|$$

and therefore

$$\left|\frac{V_1}{V_2}\right| = 3{\cdot}16 = \frac{V_1}{\dfrac{V_1}{R_1 + 600} \times 600} = \frac{R_1 + 600}{600}$$

From this $\quad R_1 = 1{,}896 - 600 = \underline{\underline{1{,}296\ \Omega}}$

Substituting in eqn. (i),

$$600 = \frac{R_2 - 1{,}896}{R_2 + 1{,}896} \quad \text{or} \quad R_2 = \underline{\underline{878\ \Omega}}$$

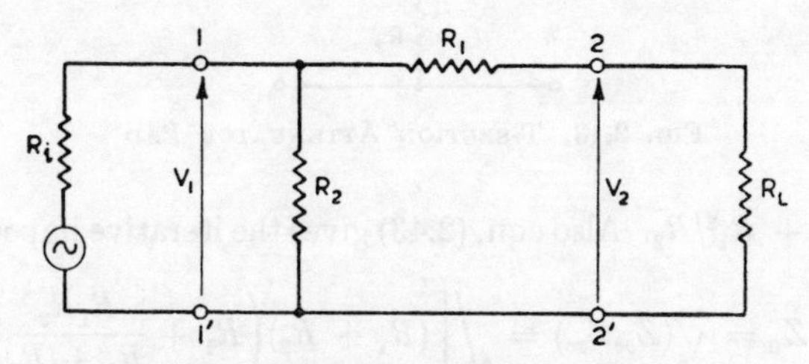

FIG. 2.12. HALF-π ATTENUATOR SECTION

This network will always give an attenuation of 10 dB when $R_L = 600\ \Omega$ irrespective of the source impedance. Although this analysis shows no dependence on frequency, and no insertion phase shift, stray capacitance and residual inductance will in fact limit the upper frequency for which the attenuation remains at 10 dB.

When the network is inserted between a 75-Ω generator and a 75-Ω load the operation will not be iterative. In this case, if the source e.m.f. is V volts, then when the 75-Ω load is connected direct to the generator the terminal voltage is $V_1' = V/2$.

With the network inserted,

$$V_2 = \frac{V}{R_i + \dfrac{R_2(R_1 + R_L)}{R_1 + R_2 + R_L}} \frac{R_2}{R_1 + R_2 + R_L} R_L$$

Substituting numerical values, $V_2 = 0{\cdot}048$ V.

The insertion-loss ratio is $A_L = 0{\cdot}5V/0{\cdot}048V = 10{\cdot}45$, and the insertion loss is

$$20 \log_{10} 10{\cdot}45 = \underline{\underline{20{\cdot}4\ \text{dB}}}$$

For iterative operation, the insertion-loss ratio can be easily expressed in terms of the ABCD parameters. Since the load is the iterative impedance, Z_{it},

$$I_2 = V_2/Z_{it} \qquad \text{(Fig. 2.11)}$$

and eqn. (2.1) gives

$$V_1 = \mathrm{A}V_2 + \mathrm{B}V_2/Z_{it}$$

or $\quad A_L = V_1/V_2 = \mathrm{A} + \mathrm{B}/Z_{it}$ (iterative operation only) $\quad$ (2.47)

2.13. Symmetrical-T Attenuator Pad

The constants A and B of the symmetrical-T attenuator section shown in Fig. 2.13 are given by eqn. (2.8) as $\mathrm{A} = 1 + R_1/R_2$, and

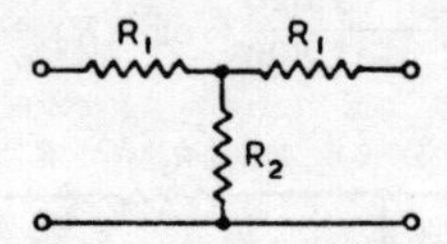

FIG. 2.13. T-SECTION ATTENUATOR PAD

$\mathrm{B} = 2R_1 + R_1^2/R_2$. Also eqn. (2.43) gives the iterative impedance as

$$Z_{it} = Z_0 = \sqrt{(Z_{oc}Z_{sc})} = \sqrt{\left[(R_1 + R_2)\left(R_1 + \frac{R_1R_2}{R_1 + R_2}\right)\right]}$$

$$= \sqrt{(R_1^2 + 2R_1R_2)} \qquad (2.48)$$

Hence $\quad R_2 = (Z_0^2 - R_1^2)/2R_1 \qquad$ (2.49)

The insertion-loss ratio for iterative operation is

$$A_L = \mathrm{A} + \mathrm{B}/Z_{it} = 1 + \frac{R_1}{R_2} + \frac{R_1^2 + 2R_1R_2}{R_2Z_0}$$

$$= 1 + \frac{R_1}{R_2} + \frac{Z_0}{R_2} \quad \text{(from eqn. (2.48))}$$

Substituting from eqn. (2.49),

$$A_L = 1 + \frac{2R_1^2}{Z_0^2 - R_1^2} + Z_0\frac{2R_1}{Z_0^2 - R_1^2}$$

$$= \frac{Z_0^2 + 2Z_0R_1 + R_1^2}{Z_0^2 - R_1^2} = \frac{Z_0 + R_1}{Z_0 - R_1}$$

Hence

$$Z_0(A_L - 1) = R_1(A_L + 1)$$

and $\quad R_1 = Z_0\dfrac{A_L - 1}{A_L + 1} \qquad$ (2.50)

From eqn. (2.49),

$$R_2 = \frac{Z_0^2}{2R_1} - \frac{R_1}{2} = \frac{Z_0^2}{2Z_0}\frac{A_L+1}{A_L-1} - \frac{Z_0}{2}\frac{A_L-1}{A_L+1}$$

$$= Z_0\frac{2A_L}{A_L^2-1} \quad . \quad . \quad . \quad . \quad . \quad . \quad (2.51)$$

These equations enable the network to be designed for a given characteristic impedance and insertion-loss ratio.

2.14. Symmetrical-π Attenuator Pad

It is left as an exercise for the reader to verify that for a symmetrical π-section attenuator, with an insertion-loss ratio of A_L and a characteristic impedance of Z_0,

$$R_1 = Z_0\frac{A_L^2-1}{2A_L} \quad \text{and} \quad R_2 = Z_0\frac{A_L+1}{A_L-1} \quad . \quad (2.52)$$

where R_1 is the series arm resistance, and R_2 is the resistance of each shunt arm.

It should be noted that when a network is terminated in its iterative impedance, the output current, I_2, is V_2/Z_{it}, and the input current, I_1, is V_1/Z_{it}, so that

$$A_L = \frac{V_1}{V_2} = \frac{I_1}{I_2} \quad . \quad . \quad . \quad . \quad (2.53)$$

2.15. Propagation Coefficient

The *propagation coefficient*, γ, of a *symmetrical* two-port network is defined as the natural logarithm of the ratio of input to output voltage or current when the network is *terminated in its characteristic impedance*—

$$\gamma = \log_e\frac{I_1}{I_2} = \log_e\frac{V_1}{V_2} \quad . \quad . \quad . \quad (2.54)$$

Generally this ratio will be complex, say $k\underline{/\beta}$ or $ke^{j\beta}$. Hence

$$\gamma = \log_e ke^{j\beta} = \log_e k + j\beta = \alpha + j\beta$$

α is called the *attenuation coefficient* and is equal to $\log_e |V_1/V_2|$ or $\log_e |I_1/I_2|$. It is measured in *nepers*. Two voltage or current levels differ by 1 neper when $\log_e |V_1/V_2| = 1$.

Since the network is terminated in its characteristic impedance, the input impedance will be equal to the output impedance, and hence the attenuation coefficient can be expressed in decibels. Thus

$$\text{Voltage ratio in decibels} = 20 \log_{10} \left| \frac{V_1}{V_2} \right|$$

Hence

$$\text{Voltage ratio in nepers} = \frac{\log_e 10 \times \text{ratio in decibels}}{20}$$

$$= \frac{\text{Ratio in decibels}}{8{\cdot}68}$$

β is called the *phase-change coefficient*, and gives the phase difference between input and output.

The relationship between the propagation coefficient of a symmetrical network and the ABCD parameters may be found as follows. For a two-port network terminated in its characteristic impedance, Z_0, the output current is $I_2 = V_2/Z_0$, and from eqn. (2.1) the input voltage is

$$V_1 = AV_2 + BV_2/Z_0$$

Hence $\gamma = \log_e V_1/V_2 = \log_e (A + B/Z_0)$

But for a symmetrical network, $Z_0 = \sqrt{(B/C)}$, from eqn. (2.41), and so

$$\frac{B}{Z_0} = \sqrt{(BC)} = \sqrt{(AD - 1)} \quad \text{(from eqn. (2.18))}$$

$$= \sqrt{(A^2 - 1)} \quad \text{since } A = D \qquad (2.55)$$

This gives

$$\gamma = \log_e [A + \sqrt{(A^2 - 1)}] = \cosh^{-1} A \qquad (2.56)^*$$

and hence

$$A = \cosh \gamma$$

$$= \cosh (\alpha + j\beta)$$

$$= \cosh \alpha \cos \beta + j \sinh \alpha \sin \beta \qquad (2.57)$$

Alternatively,

$$\cosh (\alpha + j\beta) = \tfrac{1}{2}(e^{(\alpha + j\beta)} + e^{-(\alpha + j\beta)})$$

$$= \tfrac{1}{2}e^{\alpha}\underline{/\beta} + \tfrac{1}{2}e^{-\alpha}\underline{/-\beta} \qquad (2.58)$$

Since the network is assumed to be symmetrical, A = D and hence $D = \cosh \gamma$.

* If $\cosh \gamma = x$, then $\sinh \gamma = \sqrt{(x^2 - 1)}$ and $\cosh \gamma + \sinh \gamma = e^{\gamma}$ (Euler's formula). Hence $x + \sqrt{(x^2 - 1} = e^{\gamma}$, or $\nu = \log_e [x + \sqrt{(x^2 - 1)}] = \cosh^{-1} x$.

From eqn. (2.55),

$$B = Z_0\sqrt{(A^2 - 1)} = Z_0\sqrt{(\cosh^2\gamma - 1)} = Z_0 \sinh\gamma \quad (2.59)$$

and since AD − BC = 1,

$$C = \frac{A^2 - 1}{B} = \frac{\sinh^2\gamma}{Z_0 \sinh\gamma} = \frac{\sinh\gamma}{Z_0} \quad . \quad . \quad (2.60)$$

The transmission equations (2.1) can therefore be written in the form

$$\begin{bmatrix} V_1 \\ I_1 \end{bmatrix} = \begin{bmatrix} \cosh\gamma & Z_0 \sinh\gamma \\ \dfrac{1}{Z_0}\sinh\gamma & \cosh\gamma \end{bmatrix} \begin{bmatrix} V_2 \\ I_2 \end{bmatrix}$$

$$. \quad (2.61)$$

When the network is not terminated in its characteristic impedance, this equation gives the input-output relations. Also from those relations we can write the input impedance of the network under open-circuited or short-circuited output conditions as

$$Z_{sc} = \frac{V_1}{I_1}\bigg|_{V_2=0} = Z_0 \tanh\gamma$$

and

$$Z_{oc} = \frac{V_1}{I_1}\bigg|_{I_2=0} = Z_0/\tanh\gamma$$

This gives

$$Z_{oc}Z_{sc} = Z_0^2 \quad . \quad . \quad . \quad . \quad (2.62)$$

and

$$\frac{Z_{sc}}{Z_{oc}} = \tanh^2\gamma \quad . \quad . \quad . \quad . \quad (2.63)$$

or

$$\gamma = \tanh^{-1}\sqrt{(Z_{sc}/Z_{oc})} = \tfrac{1}{2}\log_e \frac{1 + \sqrt{(Z_{sc}/Z_{oc})}}{1 - \sqrt{(Z_{sc}/Z_{oc})}} \quad . \quad (2.63a)^*$$

The propagation coefficient enables us to determine the transfer characteristics of a symmetrical network. For networks which are designed on an image-impedance basis, and which are not symmetrical, it is useful to define an *image-transfer coefficient*, θ, as one-half of the natural logarithm of the complex ratio of input to output apparent power (volt-amperes) when the network is operated

* Let $\tanh\ \gamma = x = (e^{\gamma} - e^{-\gamma})/(e^{\gamma} + e^{\gamma-\gamma}) = (e^2 - 1)/(e^{2\gamma} + 1)$. Then $e^{2\gamma} = (1 + x)/(1 - x)$, or $\gamma = \frac{1}{2}\log_e (1 + x)/(1 - x)$.

between its image impedances. This formulation emphasizes the fact that image operation is used when we want maximum power transfer. Thus

$$\theta = \tfrac{1}{2}\log_e \frac{V_1 I_1}{V_2 I_2} \qquad (2.64)$$

The image-transfer coefficient will generally be a complex quantity. If Z_{I1} and Z_{I2} are the input and output image impedances, then $I_1 = V_1/Z_{I1}$ and $I_2 = V_2/Z_{I2}$, so that

$$\theta = \tfrac{1}{2}\log_e \frac{V_1{}^2 Z_{I2}}{V_2{}^2 Z_{I1}} = \log_e \frac{V_1}{V_2}\sqrt{\frac{Z_{I2}}{Z_{I1}}} \qquad (2.64a)$$

When $Z_{I1} = Z_{I2}$ (i.e. when the network is symmetrical), eqn. (2.64*a*) reduces to eqn. (2.54), and the image-transfer coefficient is identical with the propagation coefficient.

Example 2.6. Determine the characteristic impedance of the lattice network shown in Fig. 2.14. Show that, when $Z_1 = X_1$ and $Z_2 = X_2$ and the network is terminated by its characteristic impedance, the propagation coefficient, γ, is given by $\tanh \gamma/2 = \sqrt{(X_1/X_2)}$. Hence deduce that, if X_1 and X_2 are of opposite sign, the network has zero attenuation at all frequencies.

(*A.E.E.*, 1959.)

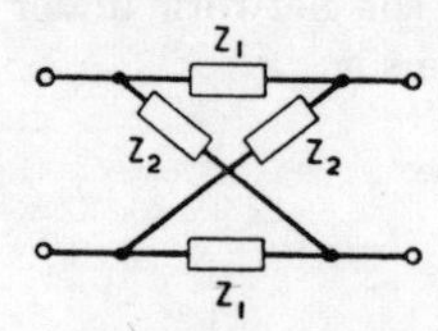

FIG. 2.14. LATTICE NETWORK

From eqn. (2.43), the characteristic impedance is

$$Z_0 = \sqrt{(Z_{sc}Z_{oc})} = \sqrt{\left(2\frac{Z_1 Z_2}{Z_1 + Z_2}\,\frac{Z_1 + Z_2}{2}\right)} = \sqrt{(Z_1 Z_2)}$$

If Z_1 and Z_2 are pure reactances,

$$Z_0 = \sqrt{(\pm X_1 X_2)}$$

From eqn. (2.63*a*),

$$\gamma = \frac{1}{2}\log_e \frac{1 + \sqrt{(Z_{sc}/Z_{oc})}}{1 - \sqrt{(Z_{sc}/Z_{oc})}}$$

$$= \frac{1}{2}\log_e \frac{1 + \sqrt{(4Z_1Z_2)}/(Z_1 + Z_2)}{1 - \sqrt{(4Z_1Z_2)}/(Z_1 + Z_2)}$$

$$= \frac{1}{2}\log_e \frac{(\sqrt{Z_2} + \sqrt{Z_1})^2}{(\sqrt{Z_2} - \sqrt{Z_1})^2} = \log_e \frac{1 + \sqrt{(Z_1/Z_2)}}{1 - \sqrt{(Z_1/Z_2)}}$$

$$= 2\tanh^{-1}\sqrt{(Z_1/Z_2)}$$

Hence

$$\tanh\frac{\gamma}{2} = \sqrt{\frac{Z_1}{Z_2}} = \sqrt{\pm\frac{X_1}{X_2}}$$

If X_1 and X_2 are of opposite sign,

$$\gamma = \log_e \frac{1 + j\sqrt{(X_1/X_2)}}{1 - j\sqrt{(X_1/X_2)}}$$

$$= \log_e 1/\underline{\beta} \quad \text{where } \beta = 2 \tan^{-1}\sqrt{(X_1/X_2)}$$

$$= \log_e e^{j\beta} = j\beta$$

Since the propagation coefficient has no reference term there can be no attenuation.

2.16. Prototype π-section Constant-k Low-pass Filter

A *low-pass filter* is a network which is designed to have zero attenuation up to a given frequency (called the *cut-off frequency*) and a large attenuation above this. It is composed of pure reactance

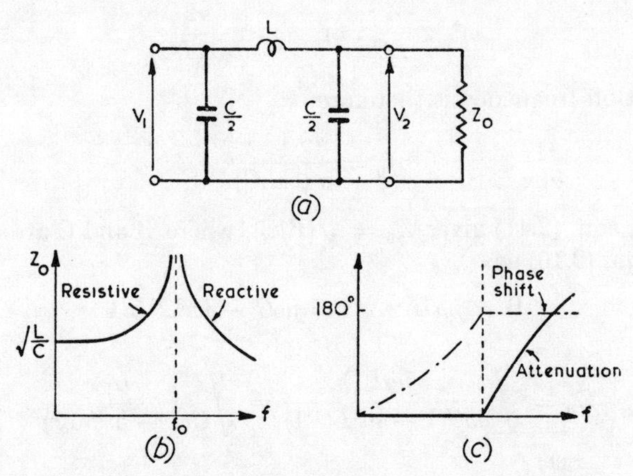

FIG. 2.15. PROTOTYPE π-SECTION LOW-PASS FILTER

elements (theoretically) in order to have zero dissipation. The ratio of total series reactance to total shunt susceptance per section is given the symbol k, and if this ratio is independent of frequency the section is termed a constant-k section. A filter is normally made up of a number of cascaded sections.

For the π-section shown in Fig. 2.15 (*a*), $k = \omega L/\omega C = L/C$, and this is therefore a constant-k filter section. Eqn. (2.10) gives

$$\mathrm{A} = \mathrm{D} = 1 + YZ = 1 - \omega^2 LC/2$$

From eqn. (2.56),

$$\gamma = \log_e [\mathrm{A} + \sqrt{(\mathrm{A}^2 - 1)}]$$

If $A^2 < 1$ this becomes

$$\begin{aligned}\gamma &= \log_e [A + j\sqrt{(1 - A^2)}] \\ &= \log_e 1/\tan^{-1} [\sqrt{(1 - A^2)}/A] \\ &= \log_e e^{j\beta} \quad \text{where } \beta = \tan^{-1} [\sqrt{(1 - A^2)}/A] \\ &= j\beta\end{aligned}$$

Since the attenuation coefficient, α, is zero, there will be no attenuation up to the frequency for which $A^2 = 1$, provided that the section is terminated in its characteristic impedance, Z_0. The pass band is given by the inequality

$$-1 < A < +1 \quad \text{or} \quad -2 < \frac{-\omega^2 LC}{2} < 0$$

Hence $$\omega^2 < \frac{4}{LC} \quad \text{or} \quad \omega < \frac{2}{\sqrt{(LC)}}$$

The cut-off frequency is therefore

$$f_0 = \frac{1}{\pi\sqrt{(LC)}} \quad . \quad . \quad . \quad (2.65)$$

Now, eqn. (2.41) gives $Z_0 = \sqrt{(B/C)}$, where B and C are derived from eqn. (2.10) as

$$B = j\omega L \qquad C = j\omega C - j\omega^3 C^2 L/4$$

Hence

$$Z_0 = \sqrt{\frac{j\omega L}{j\omega C(1 - \omega^2 LC/4)}} = \sqrt{\frac{L}{C(1 - \omega^2/\omega_0{}^2)}} \quad . \quad (2.66)$$

since $\omega_0{}^2 = 4/LC$.

When $\omega = \omega_0$ the characteristic impedance becomes infinite, while for $\omega \ll \omega_0$ it tends to $\sqrt{(L/C)}$. The value of $\sqrt{(L/C)}$ is called the *design impedance*, R_0. Between $\omega = 0$ and $\omega = \omega_0$, Z_0 is purely resistive, while for $\omega > \omega_0$, Z_0 is a pure reactance, tending to zero as the frequency increases. The variation of Z_0 with frequency is shown in Fig. 2.15 (*b*). This variation of Z_0 makes constant-k filters inconvenient to use in practice. Above ω_0 the propagation coefficient, γ, has a reference component, and so signals are attenuated between input and output.

The phase angle between input and output, for the characteristic-impedance termination, is

$$\beta = \tan^{-1} \frac{\sqrt{(1 - A^2)}}{A} = \tan^{-1} \frac{\sqrt{(\omega^2 LC - \omega^4 L^2 C^2/4)}}{1 - \omega^2 LC/2}$$

This varies from $\tan^{-1}(0/1)$, or 0°, at $\omega = 0$, to $\tan^{-1}(0/-1)$, or 180°, at $\omega = \omega_0 = 2/\sqrt{(LC)}$. Between $\omega = 0$ and $\omega = \omega_0$, β is positive, and the output *lags* behind the input.

Above ω_0, the value of $\sqrt{(A^2 - 1)}$ is a reference number and A is negative, so that $A + \sqrt{(A^2 - 1)}$ is always negative, say $-\xi$ and the propagation coefficient is

$$\begin{aligned}\gamma &= \log_e(-\xi) \\ &= \log_e \xi\underline{/180^\circ} = \log_e \xi e^{j\pi} \\ &= \log_e \xi + j\pi \qquad . \quad . \quad . \quad . \quad (2.67)\end{aligned}$$

The phase-change coefficient has a constant value of 180° independent of frequency. The response curve is shown in Fig. 2.15 (*c*).

Example 2.7. The prototype constant-k low-pass filter section shown in Fig. 2.15(*a*) is terminated in its design resistance R_0. Determine the insertion loss ratio and phase lag at (*a*) low frequencies, (*b*) 0·707 of the cut-off frequency, (*c*) cut-off frequency, (*d*) twice cut-off frequency.

From eqn. (2.30),

$$V_1 = (A + B/R_0)V_2$$

where $A = 1 - \omega^2 LC/2$ and $B = j\omega L$.

Hence $\quad \dfrac{V_1}{V_2} = 1 - \dfrac{\omega^2 LC}{2} + j\omega\sqrt{(LC)} \quad$ since $R_0 = \sqrt{(L/C)}$

(*a*) For $\omega \to 0$, $\left|\dfrac{V_1}{V_2}\right| \to 1$ and the phase change $\to 0$

(*b*) For $\omega = 0{\cdot}707\,\omega_0$, $\left|\dfrac{V_1}{V_2}\right| = |1 - 1 + j1{\cdot}41| = \underline{\underline{1{\cdot}41}}$

and the phase change is $\underline{\underline{90^\circ}}$

(*c*) For $\omega = \omega_0$, $\left|\dfrac{V_1}{V_2}\right| = |1 - 2 + j2| = \underline{\underline{\sqrt{5}}}$

and the phase change is $\tan^{-1}(2/-1) = \underline{\underline{116{\cdot}5^\circ}}$

(*d*) For $\omega = 2\omega_0$, $\left|\dfrac{V_1}{V_2}\right| = |1 - 8 + j4| = \underline{\underline{\sqrt{65}}}$

and the phase change is $\tan^{-1}(4/-7) = \underline{\underline{150{\cdot}3^\circ}}$

The results are plotted in Fig. 2.16, and demonstrate that the cut-off in this type of filter section is not sharp. One way of improving this is to terminate the section in another similar section, and so produce a ladder network.

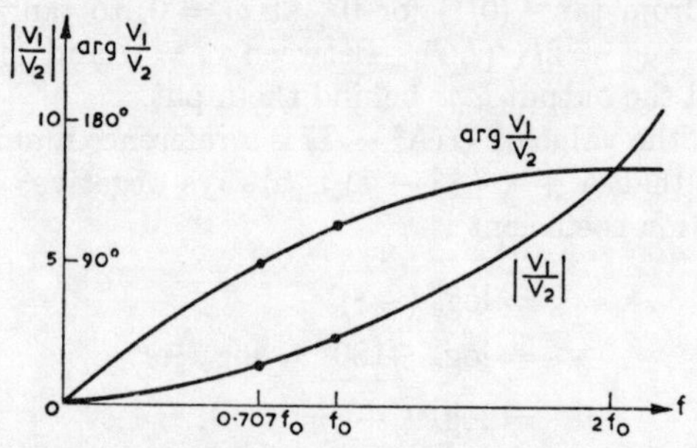

FIG. 2.16. CHARACTERISTICS OF CONSTANT-k π-SECTION TERMINATED IN R_0

2.17. The T-section constant-*k* Low-pass Filter

The T-section filter corresponding to the π-section, of Section 2.16 is shown in Fig. 2.17 (*a*). Eqn. (2.8) gives the value of A as

$$\mathrm{A} = 1 + YZ = 1 - \omega^2 LC/2$$

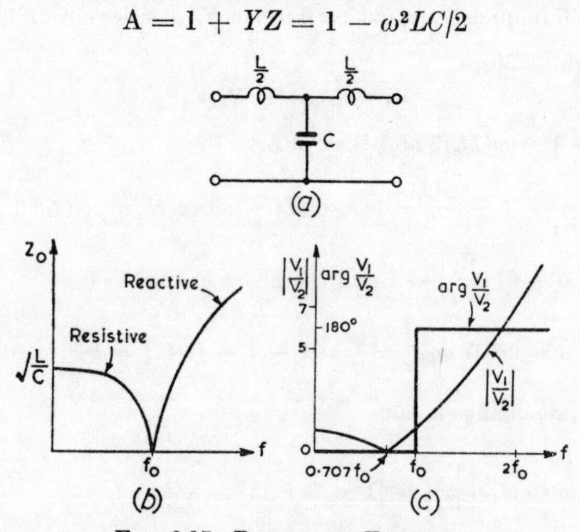

FIG. 2.17. PROTOTYPE T-SECTION

Since this is the same as for the π-section, the cut-off frequency and phase change are the same as those already derived. The constants B and C, from eqn. (2.8), are

$$\mathrm{B} = j\omega L - j\omega^3 \frac{L^2C}{4} \qquad \mathrm{C} = j\omega C$$

and hence

$$Z_0 = \sqrt{\frac{\mathrm{B}}{\mathrm{C}}} = \sqrt{\frac{L}{C}\left(1 - \frac{\omega^2 LC}{4}\right)} = \sqrt{\frac{L}{C}\left(1 - \frac{\omega^2}{{\omega_0}^2}\right)} \quad . \quad (2.68)$$

Thus Z_0 is equal to $\sqrt{(L/C)}$ at low frequencies, as for the π-section, but falls to zero at the cut-off frequency $1/\pi\sqrt{(LC)}$ and becomes reactive and tends to infinity as ω tends to infinity (Fig. 2.17 (*b*)).

Note that

$$L = \sqrt{(LC)}\sqrt{(L/C)} = \frac{2R_0}{\omega_0} \quad . \quad . \quad . \quad (2.69)$$

and

$$C = \frac{\sqrt{(LC)}}{\sqrt{(L/C)}} = \frac{2}{\omega_0 R_0} \quad . \quad . \quad . \quad (2.70)$$

This applies to both π- and T-sections.

Example 2.8. For the T-section shown in Fig. 2.17(*a*) determine the ratio of input to output voltage and transfer phase angle at zero frequency, 0·707 ω_0, ω_0 and $2\omega_0$, when the network is on open-circuit.

From eqn. (2.30),

$$\frac{V_1}{V_2} = \mathrm{A} + \frac{\mathrm{B}}{\infty} = \mathrm{A} = 1 - \omega^2 LC/2$$

(*a*) $\omega \to 0, \dfrac{V_1}{V_2} = 1$, and the phase change is zero

(*b*) $\omega = 0{\cdot}707\ \omega_0, \dfrac{V_1}{V_2} = 1 - 1 = 0$; hence $V_2 \to \infty$, and the phase shift is zero

(*c*) $\omega = \omega_0, \dfrac{V_1}{V_2} = 1 - 2 = -1$, and the phase change is 180°

(*d*) $\omega = 2\omega_0, \dfrac{V_1}{V_2} = 1 - 8 = -7$, and the phase change is 180°

The open-circuit characteristic is shown in Fig. 2.17(*c*).

2.18. Use of a Low-pass Filter as a Matching Device

Over a limited range of frequencies a low-pass constant-*k* filter section may be used as an impedance-matching element. Thus for the π-section shown in Fig. 2.18, eqn. (2.31) gives

$$Z_{in} = \frac{V_1}{I_1} = \frac{\mathrm{A}Z_L + \mathrm{B}}{\mathrm{C}Z_L + \mathrm{D}} = \frac{\mathrm{A}Z_L + \mathrm{B}}{(\mathrm{A}^2 - 1)Z_L/\mathrm{B} + \mathrm{A}}$$

since A = D (symmetrical network) and $\mathrm{C} = (\mathrm{A}^2 - 1)/\mathrm{B}$.

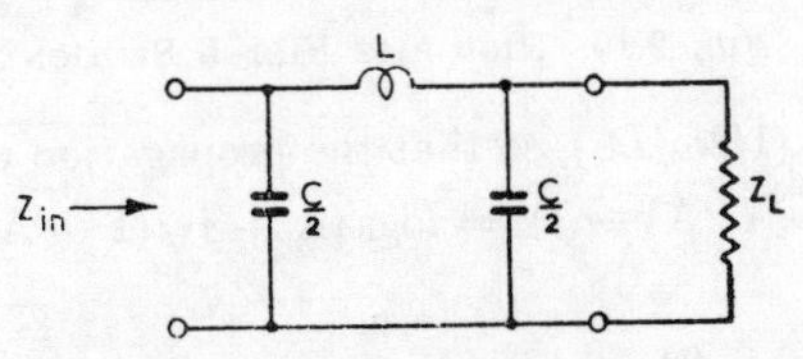

FIG. 2.18. ILLUSTRATING IMPEDANCE MATCHING

At frequencies near those for which A = 0 this reduces to

$$Z_{in} = \frac{B^2}{-Z_L} \quad . \quad . \quad . \quad (2.71)$$

For the π-section, $A = 1 - \omega^2 LC/2$ (eqn. (2.10)), and it follows that $A = 0$ around the frequencies for which $\omega^2 LC/2 = 1$, i.e. near

$$f = \frac{\sqrt{2}}{2\pi\sqrt{(LC)}} = \frac{f_0}{\sqrt{2}} \qquad \text{(from eqn. (2.65))}$$

Also, from eqn. (2.10), $B = j\omega L$, and hence, from eqn. (2.71),

$$Z_{in} = \frac{\omega^2 L^2}{Z_L} \quad . \quad . \quad . \quad (2.72)$$

This gives the desired matching of Z_L for resistive loads.

The components of the π-section are found from these equations. Thus

$$f_0 = \sqrt{2}f \quad . \quad . \quad . \quad . \quad (2.73)$$

From eqn. (2.72),

$$L = \frac{\sqrt{(Z_{in}Z_L)}}{\omega} \quad . \quad . \quad . \quad . \quad (2.74)$$

and, since $\omega^2 LC/2 = 1$,

$$C = \frac{2}{\omega^2 L} \quad . \quad . \quad . \quad . \quad (2.75)$$

The design impedance is therefore

$$R_0 = \sqrt{\frac{L}{C}} = \frac{L}{\sqrt{(LC)}} = \frac{\sqrt{(Z_{in}/Z_L)}}{\omega(\sqrt{2}/\omega)} = \sqrt{\frac{Z_{in}Z_L}{2}} \quad . \quad (2.76)$$

2.19. The Constant-*k* High-pass Filter

A high-pass filter has zero attenuation above the cut-off frequency. For the T-network shown in Fig. 2.19, $R_0 = \sqrt{(L/C)}$, and eqn. (2.8)

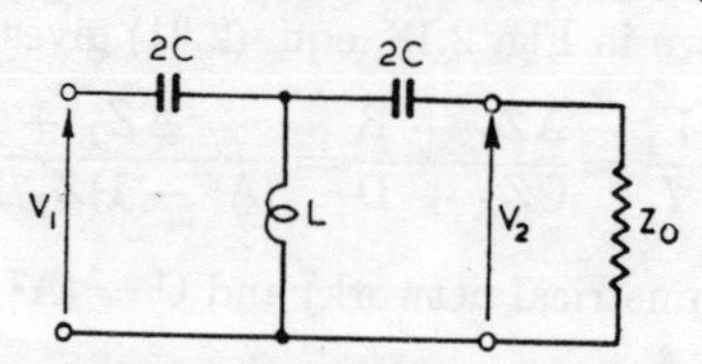

FIG. 2.19. HIGH-PASS FILTER SECTION

gives $A = 1 - (1/2\omega^2 LC)$, so that the propagation coefficient is

$$\gamma = \log_e [A + \sqrt{(A^2 - 1)}] = \log_e [A + j\sqrt{(1 - A^2)}] \quad \text{if } A^2 < 1$$
$$= j\beta$$

where $\beta = \tan^{-1} [\sqrt{(1 - A^2)}/A]$.

Now, if $A^2 < 1$,

$$-1 < A < 1 \quad \text{or} \quad -2 < -1/2\omega^2 LC < 0$$

There is no attenuation when $4 > 1/\omega^2 LC$, and the cut-off frequency occurs when $1/\omega^2 LC = 4$, or

$$\omega = \omega_0 = \frac{1}{2\sqrt{(LC)}} \quad \text{and} \quad f_0 = \frac{1}{4\pi\sqrt{(LC)}} \quad . \quad (2.77)$$

Below this frequency $1/\omega^2 LC > 4$ and hence $A^2 > 1$, which means that there will be attenuation. Also

$$L = \sqrt{(LC)}\sqrt{(L/C)} = \frac{R_0}{2\omega_0} \quad . \quad . \quad . \quad (2.78)$$

and

$$C = \frac{\sqrt{(LC)}}{\sqrt{(L/C)}} = \frac{1}{2\omega_0 R_0} \quad . \quad . \quad . \quad (2.79)$$

2.20. The Delay Line

An artificial *delay line* may be constructed of a ladder network of series inductors and shunt capacitors. This network has the property of delaying a signal which is applied to it. It is made up of cascaded T- or π-sections, adjacent impedances being lumped together in practice (Fig. 2.20).

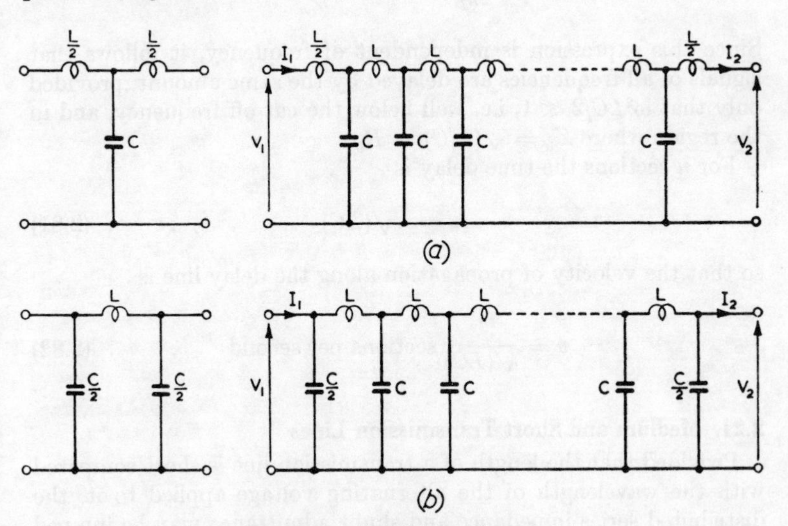

FIG. 2.20. THE DELAY LINE
(*a*) T-Section ladder network
(*b*) π-Section ladder network

The artificial delay line is obviously equivalent to a low-pass ladder filter, whose cut-off frequency is given by eqn. (2.65) as $f_0 = 1/\pi\sqrt{(LC)}$, and which has a phase lag per section of $\beta = \tan^{-1}[\sqrt{(1 - A^2)}/A]$ when the network is terminated in its characteristic impedance (this will vary with frequency as in the constant-k low-pass filter sections).

Hence

$$\tan\beta = \sqrt{\left[\frac{1}{A^2} - 1\right]} = \sqrt{\left[1\Big/\left(1 - \frac{\omega^2 LC}{2}\right)^2 - 1\right]}$$

$$\simeq \sqrt{(1 + \omega^2 LC - 1)} \quad \text{if} \quad \frac{\omega^2 LC}{2} \ll 1$$

$$= \omega\sqrt{(LC)}$$

But if $\omega^2 LC/2 \ll 1$, then $\omega\sqrt{(LC)}$ is also $\ll 1$ and $\tan\beta \simeq \beta$. Hence $\beta = \omega\sqrt{(LC)}$ radians per section.

Now, a phase lag of 2π radians at f cycles per second corresponds to a time delay of $1/f$ seconds, so that a phase lag of $\omega\sqrt{(LC)}$ radians corresponds to a time delay of

$$\tau = \frac{\omega\sqrt{(LC)}}{2\pi f} = \sqrt{(LC)} \text{ seconds} \quad . \quad . \quad (2.80)$$

Since this expression is independent of frequency, it follows that signals of all frequencies are delayed by the same amount, provided only that $\omega^2 LC/2 \ll 1$, i.e. well below the cut-off frequency, and in the region where $Z_0 = \sqrt{(L/C)} = R_0$.

For n sections the time delay is

$$\tau_n = n\sqrt{(LC)} \quad . \quad . \quad . \quad . \quad (2.81)$$

so that the velocity of propagation along the delay line is

$$v = \frac{1}{\sqrt{(LC)}} \text{ sections per second} \quad . \quad . \quad (2.82)$$

2.21. Medium and Short Transmission Lines

Provided that the length of a transmission line is short compared with the wavelength of the alternating voltage applied to it, the distributed series impedance and shunt admittance may be lumped together, and the line may be represented by a so-called nominal-T or nominal-π section, with reasonable accuracy.

Example 2.9. A 50-mile length of a high-voltage 3-phase 50-c/s transmission line has a conductor resistance of 0·3 Ω/mile, an inductance of 3·6 mH/mile and a shunt capacitance of 0·0318 μF/mile for each phase. Draw the nominal-T equivalent circuit and determine the ABCD parameters.

For a 50-mile length of line,

Series resistance/phase = 50 × 0·3 = 15 Ω
Series inductance/phase = 50 × 3·6 = 180 mH
Shunt capacitance/phase = 50 × 0·0318 = 1·59 μF

The nominal-T circuit is shown in Fig. 2.21, the total series inductance being split between the two series arms as shown.

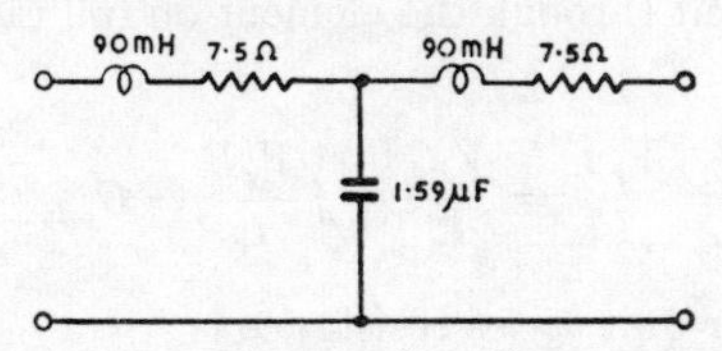

FIG. 2.21. NOMINAL T-CIRCUIT FOR ONE PHASE OF A THREE-PHASE LINE

From eqn. (2.8),

$$\begin{bmatrix} A & B \\ C & D \end{bmatrix} = \begin{bmatrix} (1 + YZ) & (2Z + YZ^2) \\ 1 & (1 + YZ) \end{bmatrix}$$

Hence
$$\begin{aligned} A = D &= 1 + j\omega C(R + j\omega L) \\ &= 1 + j\omega\, 1{\cdot}59 \times 10^{-6}(7{\cdot}5 + j\omega 0{\cdot}09) \\ &= 0{\cdot}986 + j0{\cdot}0037 \simeq \underline{\underline{0{\cdot}986}} \end{aligned}$$

$$\begin{aligned} B &= 2(R + j\omega L) + (R + j\omega L)^2 j\omega C \\ &= 2(7{\cdot}5 + j28{\cdot}2) + (56 + j423 - 796)\, j0{\cdot}0005 \\ &= \underline{\underline{14{\cdot}8 + j56.}} \end{aligned}$$

and
$$C = j\omega C = \underline{\underline{j0{\cdot}0005}}$$

Note that, since A is nearly equal to unity, and has a small quadrate component, the alternative evaluation of B from the expression $(A^2 - 1)/C$ must include the quadrate term in A.

2.22. The Per-unit System

In a supply system which has a rated voltage V_r and a rated current I_r, we may define any voltages in terms of the rated voltage, and any currents in terms of the rated current.

These are then called *per-unit* voltages and currents. Thus

$$V_{1pu} = \frac{V_1}{V_r} \quad \text{and} \quad I_{1pu} = \frac{I_1}{I_r} \qquad (2.83)$$

where V_1, I_1 are the actual voltage and current and V_r, I_r are the rated values.

The per-unit value of a series impedance element, $Z = R \pm jX$ is the voltage drop across the element on full load, expressed as a fraction of the rated voltage.

$$Z_{pu} = \frac{I_r Z}{V_r} = \frac{I_r R}{V_r} \pm j\,\frac{I_r X}{V_r} = R_{pu} \pm jX_{pu} \quad . \quad (2.84)$$

The per-unit value of a shunt admittance $Y = G \pm jB$ is the ratio of the current through the element on full rated voltage to the rated current.

$$Y_{pu} = \frac{V_r Y}{I_r} = \frac{V_r G}{I_r} \pm j\,\frac{V_r B}{I_r} = G_{pu} \pm jB_{pu} \quad . \quad (2.85)$$

Note that the per-unit series resistance is

$$R_{pu} = \frac{I_r R}{V_r} = \frac{I_r{}^2 R}{V_r I_r} = \frac{\text{Full-load power loss in } R}{\text{System volt-ampere rating}}$$

$$= \text{P.U. power loss} \quad . \quad . \quad . \quad . \quad . \quad (2.86)$$

Also, the per-unit shunt conductance is

$$G_{pu} = \frac{V_r G}{I_r} = \frac{V_r{}^2 G}{V_r I_r} = \frac{\text{Full-voltage power loss in } G}{\text{System volt-ampere rating}}$$

$$= \text{P.U. shunt power loss} \quad . \quad . \quad . \quad . \quad (2.87)$$

The linear parameters of a two-port network can also be expressed in per-unit values as follows. From eqn. (2.1),

$$V_1 = \text{A}V_2 + \text{B}I_2$$

Hence
$$\frac{V_1}{V_r} = \text{A}\,\frac{V_2}{V_r} + \frac{\text{B}}{V_r}\,I_r\,\frac{I_2}{I_r}$$

or
$$V_{1pu} = \text{A}V_{2pu} + \text{B}_{pu}I_{2pu} \quad . \quad . \quad (2.88)$$

where $\text{B}_{pu} = \text{B}I_r/V_r$.

Similarly,
$$I_1 = \text{C}V_2 + \text{D}I_2$$

Hence
$$\frac{I_1}{I_r} = \frac{\text{C}}{I_r}\,V_r\,\frac{V_2}{V_r} + \text{D}\,\frac{I_2}{I_r}$$

or
$$I_{1pu} = \text{C}_{pu}V_{2pu} + \text{D}I_{2pu} \quad . \quad . \quad (2.89)$$

where $\text{C}_{pu} = \text{C}V_r/I_r$.

Frequently, especially in networks which involve transformers, it is simpler to work in per-unit values than in actual values, since, for example, the per-unit impedance of a transformer is the same referred to either side.

Example 2.10. A 66-kVA 1-phase 3·3-kV/250-V transformer has a no-load current of 0·013 p.u. and a core loss of 0·005 p.u. The p.u. series impedance is 0·05 and the full-load copper loss is 0·03 p.u. Draw the equivalent π-circuit in p.u. values, and in actual values referred to the 3·3-kV side.

Let the no-load circuit be represented by a conductance G_0 in parallel with a susceptance B_0 (Fig. 2.22). Then, from eqn. (2.87),

$$G_{0\,pu} = \text{P.U. no-load loss} = 0{\cdot}005$$

But
$$Y_{pu} = \frac{V_r Y}{I_r} = \frac{I_0}{I_r}$$

where I_0 is the no-load current, and I_r the rated current.

Hence
$$Y_{pu} = I_{0\,pu} = G_{0\,pu} - jB_{0\,pu}$$

So that $B_{0\,pu} = \sqrt{(I_{0\,pu}{}^2 - G_{0\,pu}{}^2)} = \sqrt{(0{\cdot}013^2 - 0{\cdot}005^2)} = 0{\cdot}012$

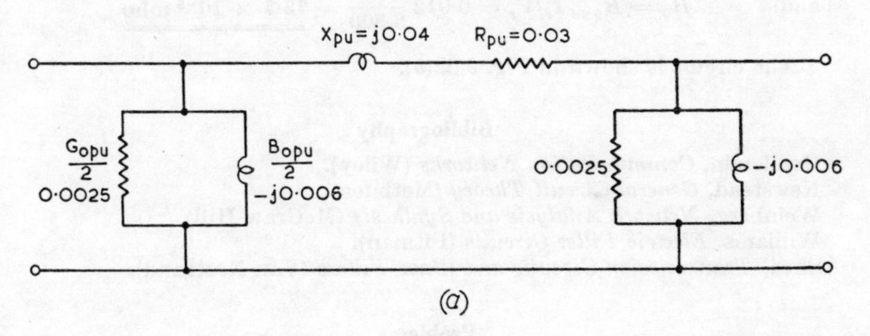

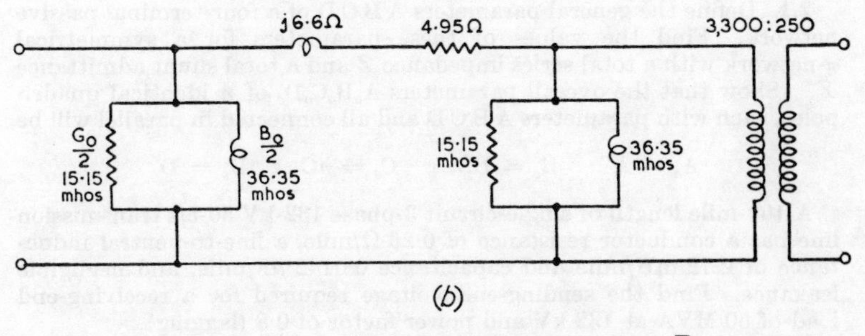

FIG. 2.22. PER-UNIT AND ACTUAL π-EQUIVALENTS OF A TRANSFORMER

If the series impedance is $Z = R + jX$, then, from eqn. (2.86),

$$R_{pu} = \text{p.u. copper loss} = 0{\cdot}03 \text{ p.u.}$$

Also $$Z_{pu} = R_{pu} + jX_{pu}$$

Hence $$0{\cdot}05 = |0{\cdot}03 + jX_{pu}|$$

or $$X_{pu} = \underline{\underline{0{\cdot}04}}$$

The equivalent π-circuit in p.u. values will have half the per-unit admittance as each shunt limb, as shown in Fig. 2.22(*a*).

The equivalent circuit in actual, values referred to the 3·3-kV side is obtained from

$$R = R_{pu}\frac{V_r}{I_r} = 0{\cdot}03\,\frac{3{,}300}{20} = \underline{\underline{4{\cdot}95\ \Omega}}$$

since $$I_r = \frac{66{,}000}{3{,}300} = 20 \text{ A}$$

Also $$X = X_{pu}\,V_r/I_r = 0{\cdot}04\,\frac{3{,}300}{20} = \underline{\underline{6{\cdot}6\ \Omega}}$$

$$G_0 = G_{0\,pu}I_r/V_r = 0{\cdot}005\,\frac{20}{3{,}300} = \underline{\underline{30{\cdot}3 \times 10^{-6} \text{ mho}}}$$

and $$B_0 = B_{0\,pu}\,I_r/V_r = 0{\cdot}012\,\frac{20}{3{,}300} = \underline{\underline{72{\cdot}7 \times 10^{-6} \text{ mho}}}$$

The circuit is shown in Fig. 2.22(*b*).

Bibliography

Guellemin, *Communication Networks* (Wiley).
Newstead, *General Circuit Theory* (Methuen).
Weinberg, *Network Analysis and Synthesis* (McGraw-Hill).
Williams, *Electric Filter Circuits* (Pitman).
Shea, *Transmission Circuits and Wave Filters* (Van Nostrand).

Problems

2.1. Define the general parameters A B C D of a four-terminal passive network. Find the values of these parameters for a symmetrical π-network with a total series impedance Z and a total shunt admittance Y. Show that the overall parameters $A_oB_oC_oD_o$ of n identical quadripoles, each with parameters A B C D and all connected in parallel will be

$$A_o = A \qquad B_o = B/n \qquad C_o = nC \qquad D_o = D$$

A 100-mile length of single-circuit 3-phase 132-kV 50-c/s transmission line has a conductor resistance of 0·26 Ω/mile, a line-to-neutral inductance of 2·12 mH/mile and capacitance 0·0142 μF/mile, and negligible leakance. Find the sending-end voltage required for a receiving-end load of 50 MVA at 132 kV and power factor of 0·8 (lagging).
(*Ans.* 152 kV) (*A.E.E.*, 1956)

2.2. Show that, for a symmetrical passive quadripole (or four-terminal impedance network), the input and output voltages and currents, V_1, I_1, and V_2, I_2 are related by the expressions

$$V_1 = AV_2 + BI_2 \text{ and } I_1 = CV_2 + DI_2, \text{ and that } AD - BC = 1$$

Draw the equivalent T-network in (*a*) per-unit terms and (*b*) in actual values, referred to the l.v. side, for an 11/33-kV star/delta-connected 10-MVA transformer with a no-load current of 0·04 p.u., a core loss of 0·005 p.u., a full-load I^2R loss of 0·015 p.u., and a p.u. impedance of 0·06. (*A.E.E.*, 1956)

(*Ans.* Series arms, 0·0075 + *j*0·03 p.u. each; shunt arm, 0·005 − *j*0·04 p.u.)

2.3. A transmission circuit is represented by a T-network in which each series impedance is $90\underline{/60^\circ}$ ohms and the shunt admittance is $1{\cdot}5 \times 10^{-3}\underline{/90^\circ}$ mho. Determine (*a*) the general circuit constants ABCD; and (*b*) the open-circuit driving-point impedance of the network. (*A.E.E.*, 1958)

(*Ans.* (*a*) $0{\cdot}885\underline{/4{\cdot}3^\circ}$, $169{\cdot}2\underline{/62^\circ}$, $1{\cdot}5 \times 10^{-3}\underline{/90^\circ}$; (*b*) $Z_{o.c.} = 588\underline{/-85{\cdot}5^\circ}$)

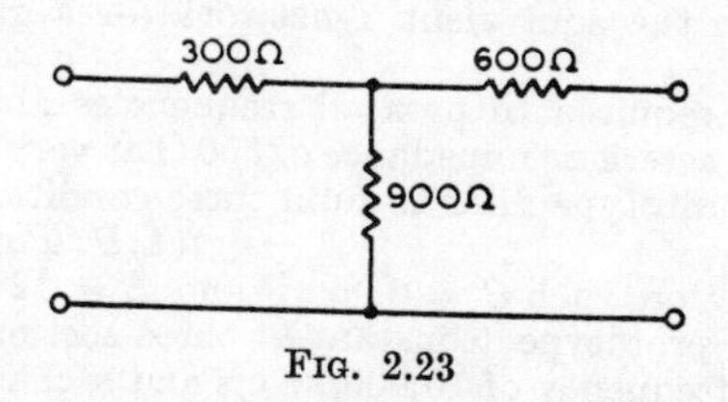

Fig. 2.23

2.4. Define the terms image impedance, characteristic impedance, and insertion loss for 4-terminal networks. Find the image impedances of the network shown in Fig. 2.23, and then calculate the insertion loss produced by the network when it is inserted between its image impedances. (*A.E.E.*, 1959)

(*Ans.* 890 Ω, 1,120 Ω, 8·2 dB)

2.5. A transmission circuit is represented by a symmetrical π-network in which the series impedance is $120\underline{/60^\circ}$ ohms, and each shunt admittance is $2{\cdot}5 \times 10^{-3}\underline{/90^\circ}$ ohms. Calculate (*a*) the values of the general circuit constants ABCD, and (*b*) the characteristic impedance of the circuit. (*A.E.E.*, 1960)

(*Ans.* $Z_0 = 165{\cdot}5\underline{/17^\circ}$ Ω; $A = D = 0{\cdot}775\underline{/11{\cdot}4^\circ}$; $B = 120\underline{/60^\circ}$; $C = 4{\cdot}4 \times 10^{-3}\underline{/94^\circ}$ mho)

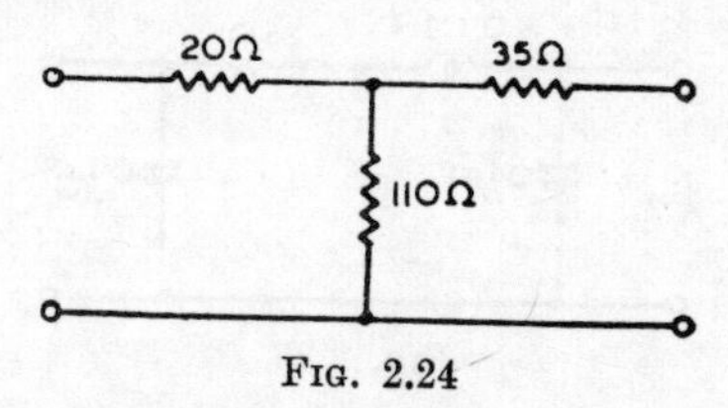

Fig. 2.24

2.6. Define the terms image impedance and iterative impedance as applied to 4-terminal networks. Calculate the iterative impedances for the network shown in Fig. 2.24. What is the insertion loss of the network when it is inserted between its iterative impedances? (*A.E.E.*, 1961)

(*Ans.* 75 Ω, 90 Ω; 6 dB)

2.7. Distinguish between image and iterative impedances referred to a 2-port network.

A four-terminal resistive network has input terminals A and B, and output terminals C and D. The resistances measured across AB when CD are first short-circuited then open-circuited are 720 Ω and 1,240 Ω. The resistance across CD with AB open-circuited is 910 Ω. Determine the equivalent T-network and the image impedances. Calculate the insertion loss when the network is inserted between its image impedances. (*A.E.E.*, 1962)

(*Ans.* 940 Ω, 695 Ω; 8·68 dB)

2.8. An impedance Z is removed from each of the four arms of a symmetrical lattice network, and placed in series with one of the input terminals and also with one of the output terminals of the network. Show that the resulting arrangement is electrically equivalent to the original lattice network. Hence indicate how this arrangement might be used to obtain the equivalent T-network for a given symmetrical lattice network. (*A.E.E.*, 1963)

2.9. A filter is required to pass all frequencies above $20{,}000/2\pi$ c/s and to have a characteristic impedance of 500 Ω at very high frequencies. Design a simple prototype filter to fulfil these conditions. (*L.U. Part III Tel.*, 1957)

(*Ans.* For T-section, each $C = 0{\cdot}05\ \mu$F and $L = 12{\cdot}5$ mH)

2.10. Design a prototype (constant-k) three-section high-pass filter to have a cut-off frequency of $10{,}000/2\pi$ c/s and a characteristic impedance of 500 Ω at a very high frequencies. (*L.U. Part III Tel.*, 1960)

(*Ans.* For T-section, each $C = 0{\cdot}1\ \mu$F and $L = 0{\cdot}025$ H)

2.11. State why attenuation is measured in decibels and why most attenuators are constructed of non-reactive resistors.

An attenuator is to be connected between a source and a load, each of which has a resistance of 500 Ω. One terminal of both source and load is earthed. The attenuator is required to give an attenuation of 5, 10 or 15 dB, controlled by switching. Make a diagram of a suitable arrangement and calculate the values required for the components. Explain the steps in your calculation.

What precautions would be necessary if the attenuator had to be satisfactory at frequencies up to 100 kc/s? (*L.U. Part III Tel.*, 1961)

(*Ans.* 5-dB T-section has each series resistance = 140·5 Ω, shunt resistance = 827 Ω)

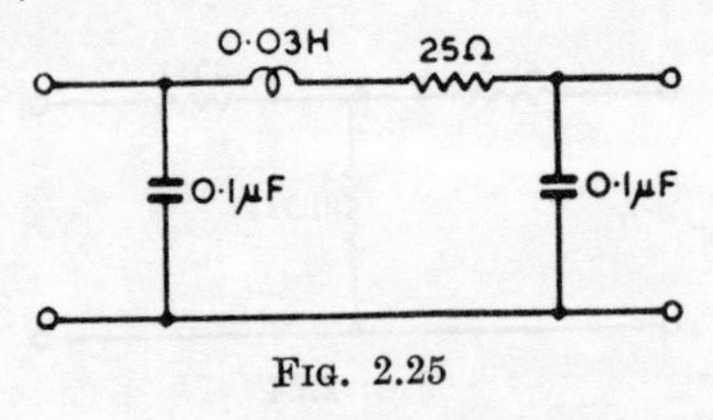

FIG. 2.25

2.12. Deduce the relationships between the impedances of a symmetrical π-network and those of the equivalent T-network.

For the circuit shown in Fig. 2.25 determine from first principles the impedances of the equivalent T-network for a frequency of 800 c/s.

If the resistance of one branch of the equivalent network is found to be negative, how can this effect be incorporated in the actual equivalent network? (*L.U. Part III Tel.*, 1962)

(*Ans.* Each series arm, $Z = 13 + j\omega \times 15{\cdot}6 \times 10^{-3}\ \Omega$; shunt arm $Y = -6{\cdot}3 \times 10^{-6} + j\omega \times 0{\cdot}193 \times 10^{-6}$ mho)

2.13. Each of the series arms of a symmetrical-T type low-pass filter section consists of an inductor of 18 mH having negligible resistance, while the shunt arm is a 0·1 μF capacitor. Indicate how the characteristic impedance varies with frequency, and calculate its values at 1 kc/s and 8 kc/s. Derive any formula used.

What are the limitations of such a filter section? Indicate how its performance can be improved. (*L.U. Part III Tel.*, 1963)

(*Ans.* 590 Ω, *j*678 Ω)

2.14. Explain what is meant by the iterative and image impedances of an unsymmetrical four-terminal network.

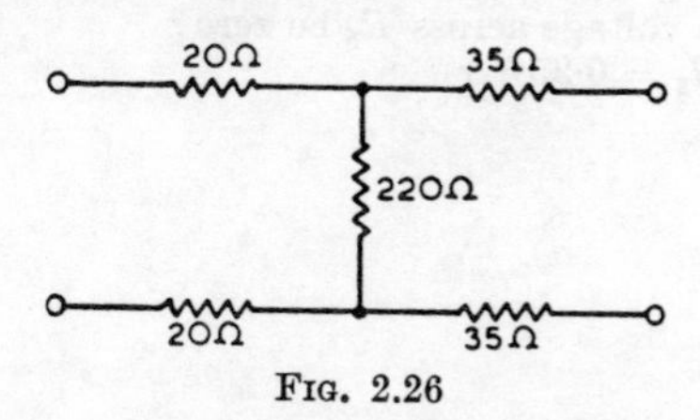

Fig. 2.26

Calculate (*a*) the iterative impedances of the network shown in Fig. 2.26, and (*b*) the iterative propagation coefficient of the network when connected between such impedances. (*A.E.E.*, 1963)

(*Ans.* 150 Ω, 180 Ω, 0·693 Np)

2.15. A short-circuited delay line, consisting of five identical π-sections, is used as the anode load of a pentode which has a mutual conductance of 5 mA/V. The series inductance per section is 100 μH, and the shunt capacitance is 100 pF. Across the input end of the line is a resistance equal to the characteristic impedance of the line. Sketch the output waveform of this system, and calculate its magnitude and duration when a 1-V positive rectangular pulse lasting 5 μsec is applied between the grid and cathode of the pentode. (*I.E.E. Eln.*, 1956)

(*Ans.* 2·5-V pulse, 1 μsec wide)

2.16. Show that a 3-phase system, having a line voltage V, total power P, and impedance per line Z, has the same regulation and efficiency as a single-phase system with a voltage V, a power P, and a loop impedance Z.

A 3-phase 50-c/s transmission line 100 miles long has a conductor resistance of 0·25 Ω/mile, a line-to-neutral inductance of 2·55 mH/mile, a capacitance of 0·0446 μF/mile, and negligible leakance. Find the sending-end voltage when supplying a receiving-end load of 60 MVA at 275 kV and power factor 0·8 (lagging). Draw a complexor diagram and derive any formula used in the calculation. (*A.E.E.*, 1959)

(*Ans.* 276 kV)

2.17. State the properties of an ideal transformer.

An ideal transformer has two windings each of N turns, and a further winding of nN turns as shown in Fig. 2.27. A generator connected to the input terminals supplies a current I. Derive an expression for the

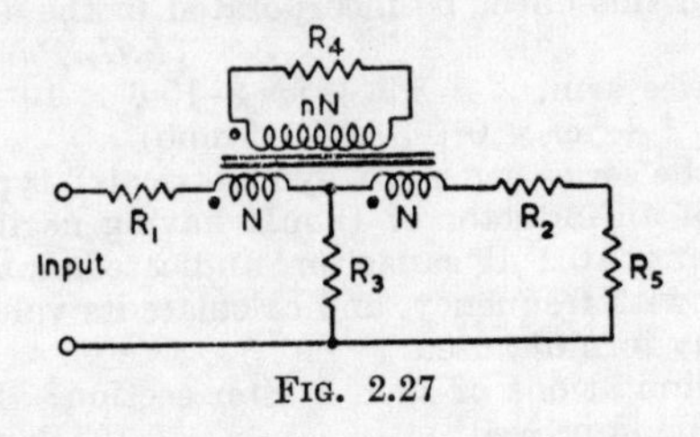

FIG. 2.27

resulting output voltage across R_5. If $n = 2$, $R_1 = R_2 = 100\ \Omega$ and $R_3 = R_4 = R_5 = 1{,}000\ \Omega$, calculate the change in voltage across R_5 for a current change of $0{\cdot}1\ A$ at the input terminals. Under what conditions will the voltage across R_5 be zero? (*A.E.E.*, 1964)

(*Ans.* $31{\cdot}9$ V; $R_3 = 0{\cdot}25R_4$)

CHAPTER 3

Complexor Locus Diagrams, and Frequency Response

It is often convenient to describe the behaviour of circuits or entire systems, when one of the parameters is varied, by means of a diagram showing the locus of the quantity considered as the parameter alters. If the quantity described is a complexor (i.e. if it may be represented by a complex number) the diagram may take one of two forms. In the first form the end point of the complexor is plotted on an *Argand diagram*. This end point will describe a locus as the required parameter is varied. When frequency is the variable such a locus is called a *Nyquist diagram*. Alternatively, the magnitude and phase of the complexor may be plotted to a base of the variable parameter. When frequency is the variable and the base scale is log frequency the diagrams are called *Bode diagrams*.

3.1. Complexor Equation of a Straight Line

Consider two constant complexors *$\boldsymbol{a}$ and $\boldsymbol{b}$ as shown in Fig. 3.1 (*a*). The equation of the line through the end point of $\boldsymbol{a}$ and parallel to $\boldsymbol{b}$ is

$$\boldsymbol{z} = \boldsymbol{a} + \boldsymbol{b}u \qquad (3.1)$$

where u is a scalar parameter whose value may range from $+\infty$ to $-\infty$. This follows from the normal rule for the addition of two complexors (i.e. they are placed end to end in order). When $u = 0$ then $\boldsymbol{z} = \boldsymbol{a}$. As u is varied, the end point of $\boldsymbol{z}$ moves in the direction of $\boldsymbol{b}$ from the end point of $\boldsymbol{a}$ by a distance $\pm|\boldsymbol{b}|u$. Thus point P ($u = 1$) gives $\boldsymbol{z} = \boldsymbol{a} + \boldsymbol{b}$, while at point Q ($u = 2$), $\boldsymbol{z} = \boldsymbol{a} + 2\boldsymbol{b}$. In this way a scale of u can be constructed along the line. Since $\boldsymbol{a}$ and $\boldsymbol{b}$ can be any two complexors, eqn. (3.1) represents the general equation of a straight line.

Important particular cases occur when the lines are parallel to the axes. Thus the line $\boldsymbol{z}_1$, parallel to the quadrate axis in Fig. 3.1 (*b*), is given by the equation $\boldsymbol{z}_1 = \boldsymbol{a}_1 + \boldsymbol{b}_1 u$. If the complexors

* **Bold type** is used in this chapter to distinguish complexors from scalar quantities.

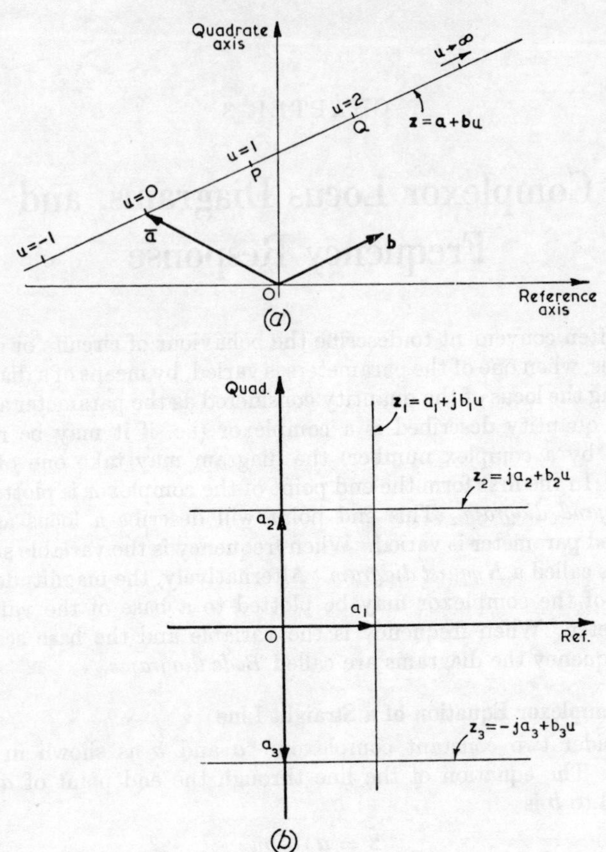

FIG. 3.1. LINEAR LOCI

$\boldsymbol{a}_1$ and $\boldsymbol{b}_1$ are represented in rectangular form, then $\boldsymbol{a}_1 = a_1$ (i.e. a complexor in the direction of the reference axis) and $\boldsymbol{b}_1 = jb_1$ (a complexor in the direction of the positive quadrate axis). The value of a_1 determines the point at which the line cuts the reference axis, and the equation for z_1 becomes

$$z_1 = a_1 + jb_1u \quad . \quad . \quad . \quad . \quad (3.2)$$

where a_1 and b_1 are constants, and u is the variable.

In the same way, the line z_2 parallel to and above the reference axis, is given by

$$z_2 = ja_2 + b_2u \quad . \quad . \quad . \quad . \quad (3.3.)$$

and the line z_3, parallel to but below the reference axis, by

$$z_3 = -ja_3 + b_3u \qquad . \quad . \quad . \ (3.4)$$

These results may be applied directly in electrical circuit theory to the plotting of the impedance loci for simple series circuits or the admittance loci for simple parallel circuits, as one circuit parameter is varied. Equal scales must be used for both axes. In the impedance diagram, resistance is plotted in the reference direction and reactance in the quadrate direction, while in the admittance diagram conductance is plotted in the reference direction and susceptance in the quadrate direction. A scale indicating changes in the variable can be constructed along the locus.

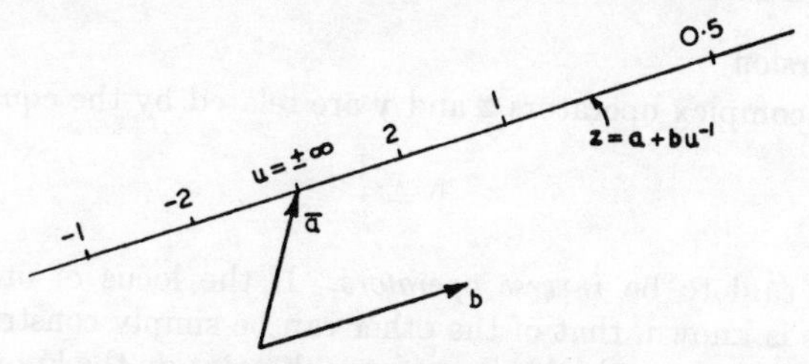

FIG. 3.2. RECIPROCAL SCALE

A further common type of linear locus is given by the equation

$$\boldsymbol{z} = \boldsymbol{a} + \boldsymbol{b}u^{-1} \qquad . \quad . \quad . \quad . \ (3.5)$$

This represents the line through the end point of $\boldsymbol{a}$ and parallel to $\boldsymbol{b}$, but with a reciprocal scale of u, so that when $u = \infty$, $\boldsymbol{z} = \boldsymbol{a}$ (Fig. 3.2.).

3.2. Complexor Equation of a Circle

A circle is defined in terms of its centre and radius. The complexor equation for a circle must consist of the sum of the complexor from the origin to the centre and a radius complexor. Thus in Fig. 3.3 the complexor, $\boldsymbol{z}$, from the origin to any point P on the circle is given by

$$\boldsymbol{z} = \boldsymbol{a} + Re^{j\theta} \qquad . \quad . \quad . \quad . \ (3.6)$$

where the centre lies at the end point of $\boldsymbol{a}$ and $Re^{j\theta}$ is the radius complexor which makes an angle θ with the reference direction. The angle θ is the variable parameter. Alternative forms are

$$\left.\begin{aligned} \boldsymbol{z} &= \boldsymbol{a} + R\underline{/\theta} \\ \text{and} \qquad \boldsymbol{z} &= \boldsymbol{a} + R\cos\theta + jR\sin\theta \end{aligned}\right\} \qquad . \quad . \ (3.6a)$$

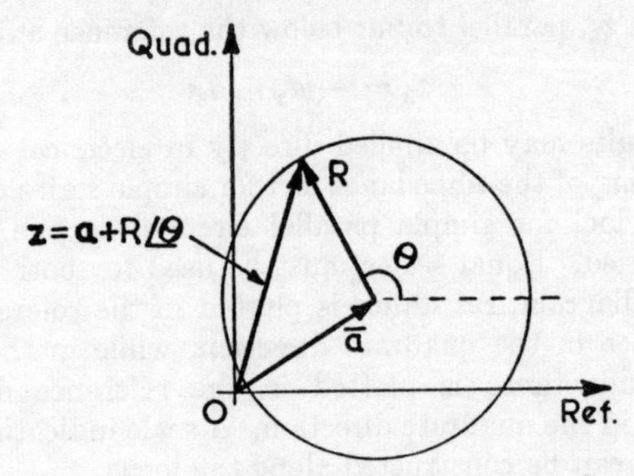

FIG. 3.3. COMPLEXOR REPRESENTATION OF A CIRCLE

3.3. Inversion

If two complex operators z and y are related by the equation

$$y = \frac{1}{z} \quad . \quad . \quad . \quad . \quad (3.7)$$

they are said to be *inverse operators.* If the locus of one of the operators is known, that of the other can be simply constructed by plotting the reciprocal complexor for each point on the known locus. For linear loci the inverse is always a circle, provided only that the linear locus does not pass through the origin. Some simple geometrical inversions will now be considered.

(a) *Inversion of a straight line through the origin.* The equation of a line whose slope is the same as that of a complexor $\boldsymbol{b}$, and which passes through the origin, is

$$z = bu$$

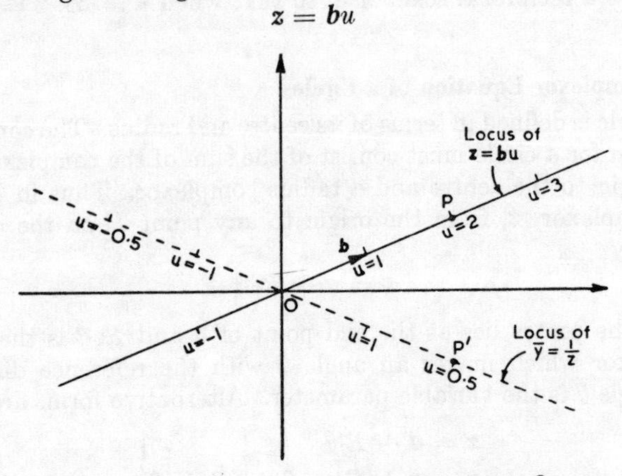

FIG. 3.4. INVERSION OF A LINE THROUGH THE ORIGIN

The inverse of this is $\boldsymbol{y} = 1/\boldsymbol{z} = 1/\boldsymbol{b}u$, and if $\boldsymbol{b}$ is written in polar form as $\boldsymbol{b} = b\underline{/\theta}$, the expression for $\boldsymbol{y}$ becomes

$$\boldsymbol{y} = \left(\frac{1}{b}\underline{/-\theta}\right) u^{-1} \quad . \quad . \quad . \quad (3.8)$$

This is also a line through the origin, but with a slope which is the negative of the slope of $\boldsymbol{z}$, and with a reciprocal scale (Fig. 3.4). Thus, if $\boldsymbol{b} = 1\underline{/30°}$, say, then when $u = 2$, $\boldsymbol{z} = 2\underline{/30°}$ and $\boldsymbol{y} = 0{\cdot}5\underline{/-30°}$ (points P and P' on the diagram).

(*b*) *Inversion of a straight line not passing through the origin.* Let SS' in Fig. 3.5 be the locus representing the complex quantity

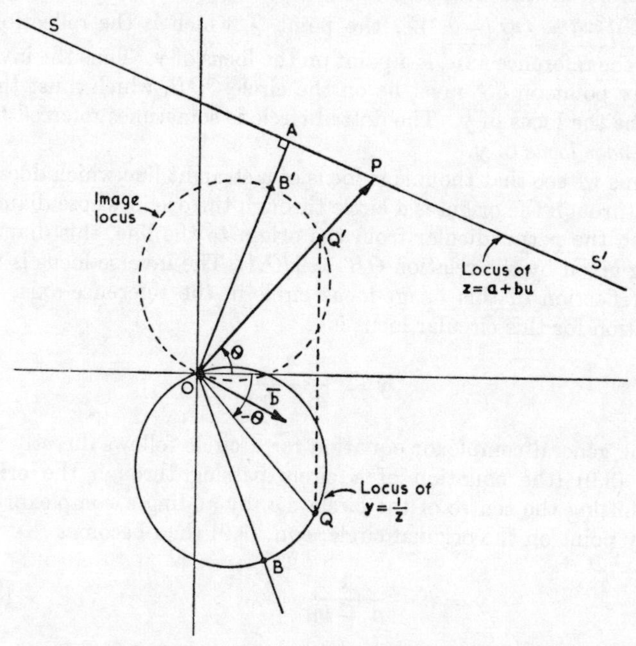

Fig. 3.5. Inversion of a Line not through the Origin

$\boldsymbol{z} = \boldsymbol{a} + \boldsymbol{b}u$, and let P be any point on SS'. Then $\boldsymbol{z}_P = OP\underline{/\theta}$. We wish to find the locus of $\boldsymbol{y} = 1/\boldsymbol{z}$.

Draw the line OA perpendicular to SS', and choose a point B' on OA, or OA produced, so that

$$OA \,.\, OB' = 1$$

On OB' as diameter construct a circle to cut OP, or OP produced, at Q'. Then angle $OQ'B'$ is a right angle and triangles $OB'Q'$ and OAP are similar,

so that
$$\frac{OP}{OB'} = \frac{OA}{OQ'}$$

Hence
$$OP\,.\,OQ' = OA\,.\,OB'$$

But, since $OA.OB' = 1$, it follows that $OP.OQ' = 1$, so that Q' is the inverse of P and the dotted circle must therefore be the locus of $|\mathbf{y}| = 1/|\mathbf{z}|$.

Now consider the full circle which is the image of the dotted circle in the reference axis. Since $\mathbf{z}_P = OP/\underline{\theta}$, then $\mathbf{y}_P = (1/OP)/\underline{-\theta} = OQ'/\underline{-\theta}$, i.e. the point Q, which is the reflection of Q' in the reference axis, is a point on the locus of $\mathbf{y}$. Thus the inverse of any point on SS' must lie on the circle OQB, which must therefore be the locus of $\mathbf{y}$. The dotted circle is sometimes referred to as the *image locus* of $\mathbf{y}$.

Thus we see that the image locus of a straight line which does not pass through the origin is a circle through the origin whose diameter lies on the perpendicular from the origin to the line, this diameter being given by the relation $OB' = 1/OA$. The inverse locus is then the reflection of this image-locus circle in the reference axis. The equation for this circular locus is

$$\mathbf{y} = \frac{1}{\mathbf{a} + \mathbf{b}u} \qquad (3.9)$$

The general complexor equation for a circle follows directly from eqn. (3.9) (the equation of a circle passing through the origin), by shifting the centre of the circle, i.e. by adding a complexor $\mathbf{e}$ to every point on the original circle, eqn. (3.9) then becomes

$$\mathbf{y} = \mathbf{e} + \frac{1}{\mathbf{a} + \mathbf{b}u} \qquad (3.10)$$

$$= \frac{\mathbf{a}.\mathbf{e} + 1 + \mathbf{b}.\mathbf{e}u}{\mathbf{a} + \mathbf{b}u} = \frac{\mathbf{c} + \mathbf{d}u}{\mathbf{a} + \mathbf{b}u} \qquad (3.10a)$$

where $\mathbf{c} = \mathbf{ae} + 1$ and $\mathbf{d} = \mathbf{be}$.

In circuit analysis a common linear locus is one which lies parallel to one of the co-ordinate axes. Four possibilities are shown in Fig. 3.6. At (*a*) the linear $\mathbf{z}$-locus is

$$\mathbf{z} = ja + bu$$

The image locus of z is therefore the dotted circle on OB' as diameter, and with $OB' = 1/OA$, while the inverse locus is the full-line circle OQB, which is the reflection of the dotted circle in the reference axis. For $u = 0$, $z = ja$, and hence $y = 1/ja = -j/a = OB$. For $u \to \infty$, $z \to \infty$ and $y \to 0$. For any point P on the

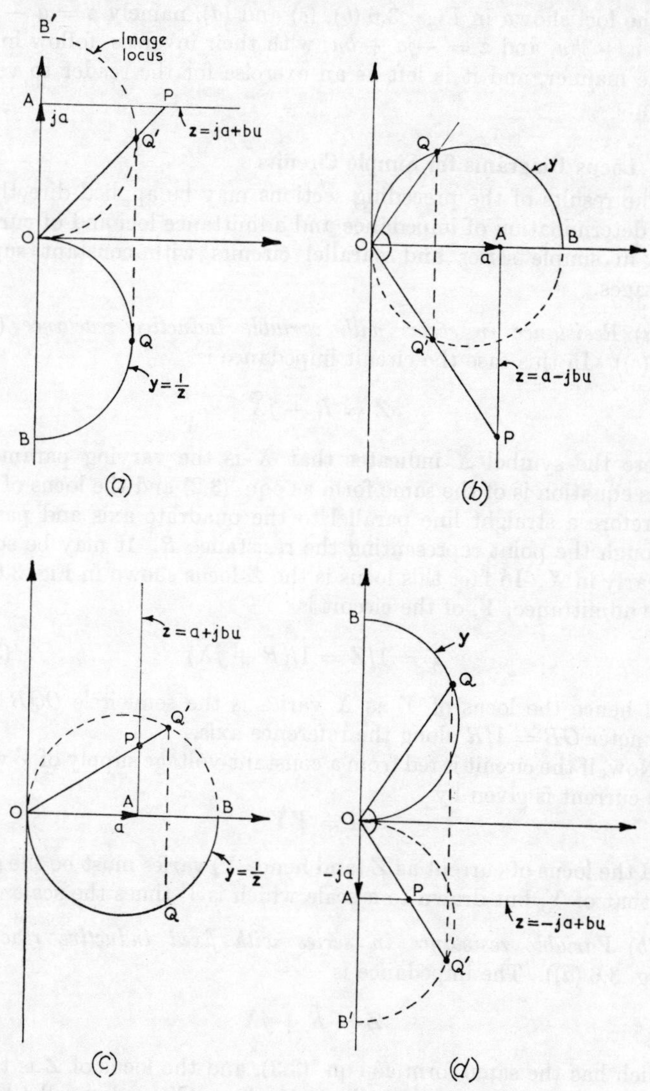

FIG. 3.6. INVERSION OF LINES PARALLEL TO THE CO-ORDINATE AXIS

z-locus, the corresponding point on the image locus is Q', where the line OP cuts the dotted circle. The inverse of P is Q, which is the reflection of Q' in the reference axis. It should be noted that, since we have considered only a half-line, we obtain a semicircle as the inverse locus.

The loci shown in Figs. 3.6 (*b*), (*c*) and (*d*), namely $z = a - jbu$, $z = a + jbu$, and $z = -ja + bu$, with their inverses, follow in the same manner, and it is left as an exercise for the reader to verify them.

3.4. Locus Diagrams for Simple Circuits

The results of the preceding sections may be applied directly to the determination of impedance and admittance loci and of current loci in simple series and parallel circuits with constant supply voltages.

(*a*) *Resistance in series with variable inductive reactance* (Fig. 3.6 (*c*)). In this case the circuit impedance is

$$\mathbf{Z} = R + j\tilde{X}$$

where the symbol $\tilde{X}$ indicates that X is the varying parameter. This equation is of the same form as eqn. (3.2) and the locus of $\mathbf{Z}$ is therefore a straight line parallel to the quadrate axis and passing through the point representing the resistance R. It may be scaled linearly in X. In fact this locus is the Z-locus shown in Fig. 3.6 (*c*). The admittance, $\mathbf{Y}$, of the circuit is

$$\mathbf{Y} = 1/\mathbf{Z} = 1/(R + j\tilde{X}) \quad . \quad . \quad . \quad (3.11)$$

and hence the locus of Y as X varies is the semicircle OQB on a diameter $OB = 1/R$ along the reference axis.

Now, if the circuit is fed from a constant-voltage supply of V volts, the current is given by

$$\mathbf{I} = V\mathbf{Y}$$

and the locus of current as $\mathbf{Z}$ (and hence $\mathbf{Y}$) varies must be the same as that of $\mathbf{Y}$, but drawn to a scale which is V times the scale of $\mathbf{Y}$.

(*b*) *Variable resistance in series with fixed inductive reactance* (Fig. 3.6 (*a*)). The impedance is

$$\mathbf{Z} = \tilde{R} + jX$$

which has the same form as eqn. (3.3), and the locus of $\mathbf{Z}$ is therefore a straight line through the point $(0 + jX)$ and parallel to the reference axis, as illustrated in Fig. 3.6 (*a*). The locus of admittance, $\boldsymbol{Y}$, is thus the semicircle on $OB = 1/jX$ as diameter, as shown, and this (to a different scale) is also the locus of current from a constant-voltage supply.

(*c*) *Resistance in series with variable capacitive reactance* (Fig. 3.6 (*b*)). The circuit impedance is

$$\boldsymbol{Z} = R - j\tilde{X}$$

and the locus of $\boldsymbol{Z}$ in the impedance diagram is AP in Fig. 3.6 (*b*), giving an admittance and current locus as shown in the semicircle OQB. It should be noted that in this case $X = 1/\omega C$, and hence if either the frequency, f, or the capacitance, C, is the variable, the impedance locus will have a reciprocal scale in either f or C.

(*d*) *Variable resistance in series with fixed capacitive reactance* (Fig. 3.6 (*d*)). In this case we may write

$$\boldsymbol{Z} = -jX + \tilde{R}$$

The impedance locus will have the form shown by the straight line AP in Fig. 3.6 (*d*), and the current and admittance locus is the semicircle OQB.

(*e*) *Resistance in parallel with variable capacitive reactance* (Fig. 3.6 (*c*)). For the parallel connexion of elements we may write the circuit admittance as

$$\boldsymbol{Y} = G + jB$$

The locus of $\boldsymbol{Y}$ as B varies is thus a straight line parallel to the quadrate axis and passing through the point representing the conductance, G. It is the line AP in Fig. 3.6 (*c*). The impedance of this circuit is

$$\boldsymbol{Z} = 1/\boldsymbol{Y} = 1/(G + j\tilde{B})$$

and the locus of $\boldsymbol{Z}$ as B varies is therefore the semicircle OQB.

In the case of parallel circuits fed by a constant current, I, the voltage across the circuits is given by

$$\boldsymbol{V} = \boldsymbol{IZ}$$

so that the locus of $\boldsymbol{V}$ will have the same form as the $\boldsymbol{Z}$-locus, but will be to a different scale.

It is left to the reader to equate the other forms of two-element parallel circuits with the loci shown in Fig. 3.6.

Example 3.1. A 0·0318-μF capacitor is connected in series with a 5-kΩ resistor across a 10-V supply whose frequency varies between 500 c/s and 2·5 kc/s. Plot the impedance, admittance and current loci. Determine from these loci (*a*) the maximum and minimum values of current, and (*b*) the frequency and the current when the circuit phase angle is 40°.

The circuit is basically a fixed resistance in series with a variable capacitive reactance, but the limitations on the frequency set upper and lower limits to the value of the impedance, $\boldsymbol{Z}$. For the general circuit,

$$\boldsymbol{Z} = R - j\tilde{X}$$

and the locus of $\boldsymbol{Z}$ is a straight line through the point R, and parallel to the negative quadrate axis (line AF in Fig. 3.7). In the problem,

$$\boldsymbol{Z} = 5{,}000 - j/2\pi f \times 0{\cdot}0318 \times 10^{-6} = 5{,}000 - j\,5 \times 10^{6} \times f^{-1}$$

The locus of $\boldsymbol{Z}$ extends from $f^{-1} = 1/500 = 2 \times 10^{-3}$ up to $f^{-1} = 1/2{,}500 = 0{\cdot}4 \times 10^{-3}$, i.e. from $\boldsymbol{Z} = 5{,}000 - j10{,}000$ up to $\boldsymbol{Z} = 5{,}000 - j2{,}000$. A suitable scale is chosen for the impedance diagram, and this is then drawn as the line D_1E_1. Note that we could construct a linear scale in f^{-1} on this locus by making use of the fact that, when f^{-1} changes by 2×10^{-3}, the reactance changes by 10,000, so that the scale distance representing 10 kΩ will also represent 2×10^{-3} seconds.

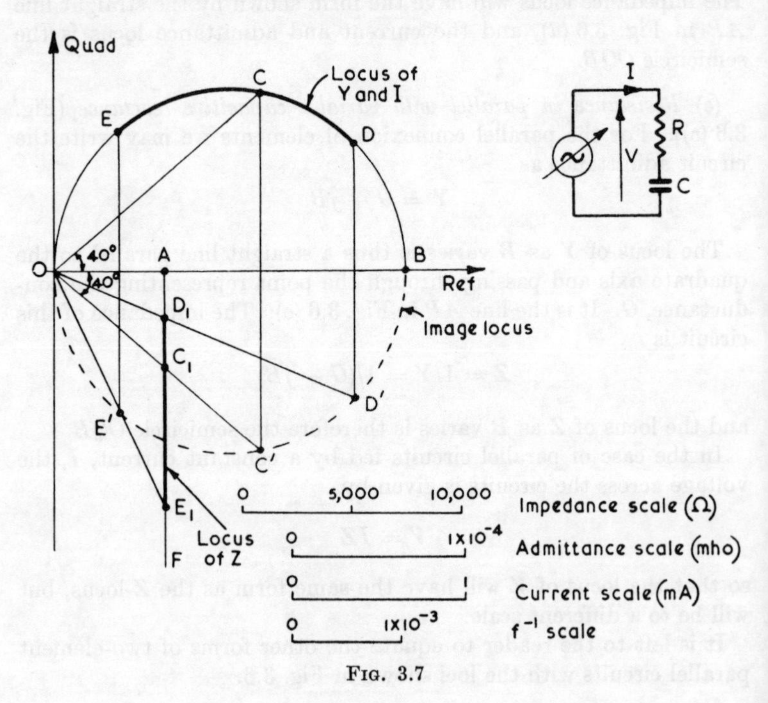

Fig. 3.7

From our previous work the admittance locus for this circuit is a semicircle with diameter $OB = 1/OA$ on the reference axis and lying above this axis. A suitable scale must be chosen to give a semicircle of reasonable dimensions as illustrated in Fig. 3.7. Thus $OB = 1/5{,}000 = 2 \times 10^{-4}$ mho. The extent of the actual admittance locus is from D to E, where D is the inverse of D_1 and E is the inverse of E_1, the construction being indicated on the diagram. The current locus will be the same as the admittance locus, but in this case the distance OB will represent $10 \times 2 \times 10^{-4}$ A or 2 mA.

By measurement from the diagram,

(*a*) Maximum current $= OD = \underline{\underline{1{\cdot}83 \text{ mA}}}$

Minimum current $= OE = \underline{\underline{0{\cdot}87 \text{ mA}}}$

(*b*) When the phase angle is 40° the current is $OC = \underline{\underline{1{\cdot}53 \text{ mA}}}$

The impedance at this point is OC_1, so that the distance AC_1 in f^{-1} units is $0{\cdot}85 \times 10^{-3}$, giving $f = \underline{\underline{1{\cdot}18 \text{ kc/s}}}$.

3.5. Resonant Circuits

The locus diagrams of series and parallel resonant circuits provide useful information about the behaviour of the circuits, particularly near resonance.

(*a*) *Series-resonant circuit.* The impedance of an LRC series circuit is

$$\mathbf{Z} = R + j\omega L + \frac{1}{j\omega C} \quad . \quad . \quad . \quad (3.12)$$

If R is constant and the reactance is varied, the locus of $\mathbf{Z}$ is a straight line. The reactance may be varied in three ways—by variation in L, C or ω. In each case resonance occurs when $\omega L = 1/\omega C$, and under this condition $\mathbf{Z} = R$. Variation of L gives a linear scale of L on the $\mathbf{Z}$-locus, variation in C gives a reciprocal scale of C, while variation in the frequency gives a non-uniform scale of ω. In each case the admittance locus is circular and of diameter $1/R$ mho. The locus of current from a constant-voltage supply is therefore also circular. From this current locus the current resonance curve can be readily constructed.

Fig. 3.8 shows the impedance locus (line Q_1AP_1) for a series-resonant circuit supplied at constant voltage and frequency, in which L is varied from zero towards infinity. The admittance locus is the circular locus shown. When $L = 0$, $\mathbf{Z} = R - j/\omega C$, and this gives point Q_1 on the impedance locus and Q on the admittance locus. At resonance the corresponding points are A and B. At a point through resonance, say P_1, the corresponding point on the admittance locus is P. Hence the complete admittance or current locus is $QBPO$. The current when $L = 0$ is proportional to OQ, when $L = 1/\omega C$ to OB, etc., and since a linear scale of L can be con-

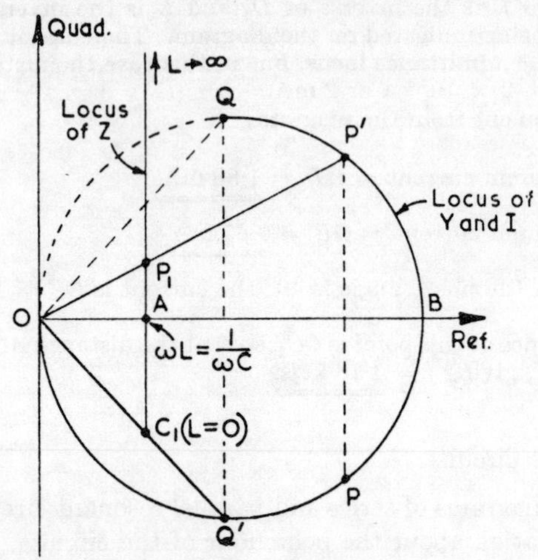

Fig. 3.8. Locus Diagram for a Series-resonant Circuit with Variable Inductance

structed along the Z-locus, we may readily plot the resonance curve of current to a base of L by measurement from the diagram.

(*b*) *Parallel-resonant circuit.* Consider the parallel circuit of Fig. 3.9 (*a*). The admittance of the circuit is

$$Y = Y_C + Y_L = j\omega C + \frac{1}{R + j\omega L} \quad . \quad . \quad (3.13)$$

If the frequency, f, is the variable, then both Y_C and Y_L will vary, and the locus of Y will be obtained from the loci of Y_C and Y_L by adding these quantities vectorially for each value of f. From this locus combined with its scale of frequency the resonance curve of the circuit can be drawn.

Example 3.2. A parallel-resonant circuit has a coil of resistance 50 Ω and inductance 0·01 H. The capacitor has a value of 0·5 μF. Plot the locus of current from a 1-V supply for which ω varies from 0 to 30,000 rad/sec. Hence draw the resonance curve.

The locus of $Z_L = R + j\omega L$ will be the line AS (Fig. 3.9(*b*)). When $\omega = 10{,}000$ rad/sec, $\omega L = 100$, so that, if a scale of 50 Ω = 1 unit is chosen, then a distance of 2 units along AS corresponds to $\omega = 10{,}000$.

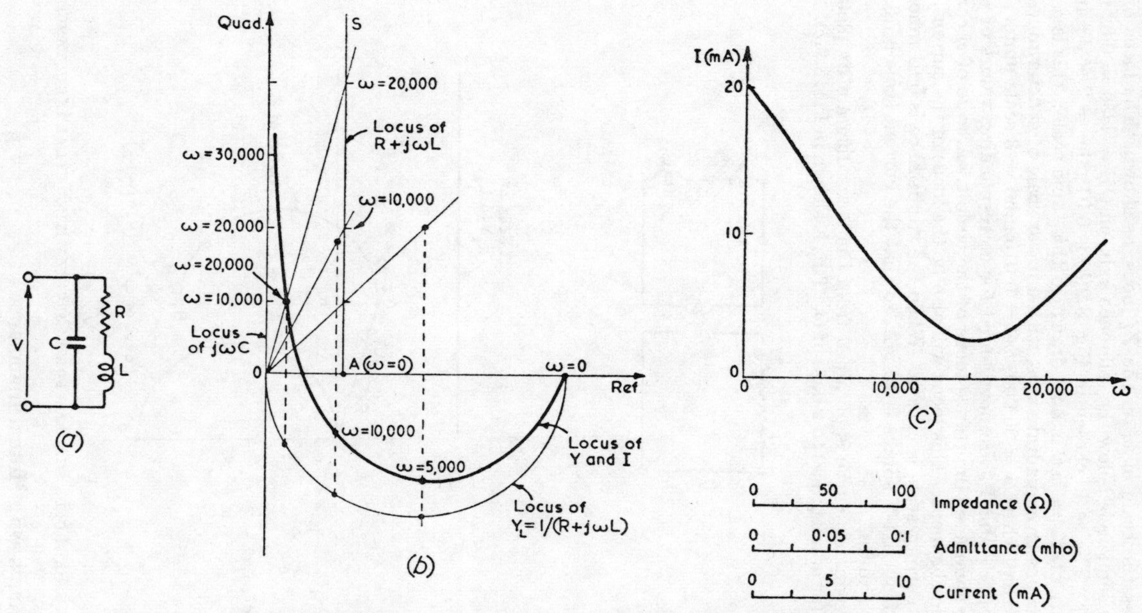

Fig. 3.9. Frequency Response of a Parallel-resonant Circuit

This fixes the scale of ω on the $\boldsymbol{Z}_L$ locus, as shown. The locus of $\boldsymbol{Y}_L$ will be a semicircle below the reference axis, on $OB = 1/50 = 0{\cdot}02$ mho diameter. The scale chosen in Fig. 3.9(*b*) is 0·01 mho = 2 scale units. The locus of $\boldsymbol{Y}_C = j\omega C$ is now drawn to this same scale: this is a line along the positive quadrate axis, with a linear scale of ω determined by the fact that when $\omega = 30{,}000$, $\omega C = 0{\cdot}015$ mho = 3 scale units. The values of $\boldsymbol{Y}_L$ and $\boldsymbol{Y}_C$ are now added for corresponding values of ω to give the complete admittance locus of the circuit. A scale of ω can be placed on the locus. This locus will also be the locus of the current, if a suitable current scale is chosen. When the admittance is 0·01 mho the current from a 1-V source is 10 mA, so that the current scale is 10 mA = 2 scale units.

The resonance curve is obtained from the locus diagram by plotting current (or admittance) to a base of ω. This is shown in Fig. 3.9(*c*).

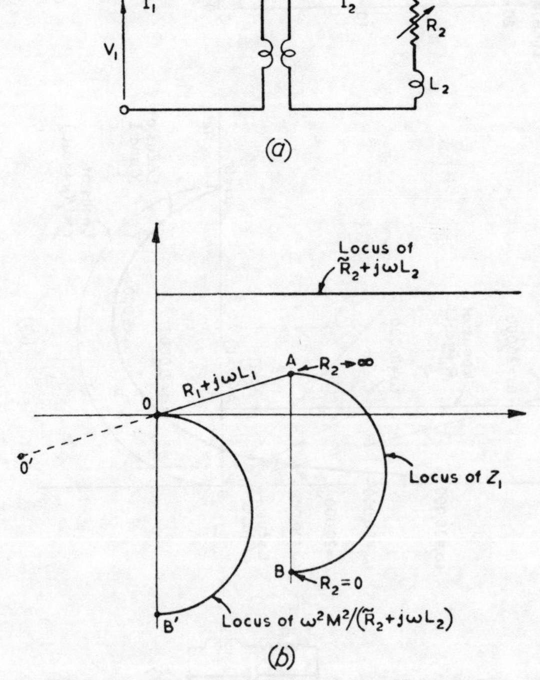

FIG. 3.10. LOCUS DIAGRAM FOR A CIRCUIT WITH MUTUAL INDUCTANCE

3.6. Circuits with Mutual Inductance

So far, we have considered circuits which have a constant of inversion of unity (e.g. $\boldsymbol{Z}.\boldsymbol{Y} = 1$). If the constant of inversion differs from unity, as in a circuit which includes a mutual coupling, we can take this into account by an appropriate change of scale.

For example, consider the circuit shown in Fig. 3.10 (*a*). The circuit equations are

$$V_1 = I_1(R_1 + j\omega L_1) - j\omega M I_2$$

and

$$0 = -j\omega M I_1 + (R_2 + j\omega L_2)I_2$$

Eliminating I_2 yields the familiar expression

$$V_1 = I_1\left[(R_1 + j\omega L_1) + \frac{\omega^2 M^2}{(R_2 + j\omega L_2)}\right] \quad . \quad (3.14)$$

The locus of input impedance (V_1/I_1) as R_2, say, varies will be made up of the constant primary impedance plus the inverse of ($\tilde{R}_2 + j\omega L_2$) multiplied by the constant of inversion $\omega^2 M^2$. This is illustrated in Fig. 3.10 (*b*). Note that in this instance the scale of the inverse locus must be the scale which corresponds to the primary impedance. The required impedance locus is the semicircle on AB. It can be shown that the inverse of this semicircle (i.e. the locus of input admittance, and therefore input current) is also semicircular. The reader is invited to verify this either by analysis or by a point-by-point graphical inversion of the curve AB. This point-by-point inversion is carried out for any point P, say, on the original locus by drawing the line OP, and on it marking off the point Q' such that $OP \, . \, OQ' = 1$. The point Q' is then reflected in the reference axis to give a point Q on the required inverse locus. The scale for the inverse locus may then be altered, if necessary, to give a suitable diagram.

An alternative is to consider that the semicircle OB' is the required locus and that the additional term Z_1 causes a shift of the origin to O' where $OO' = -R_1 - j\omega L_1$.

3.7. Two-port Network Loci

In Chapter 2 we saw that the operation of any linear, passive, bilateral two-port network can be represented in terms of a set of

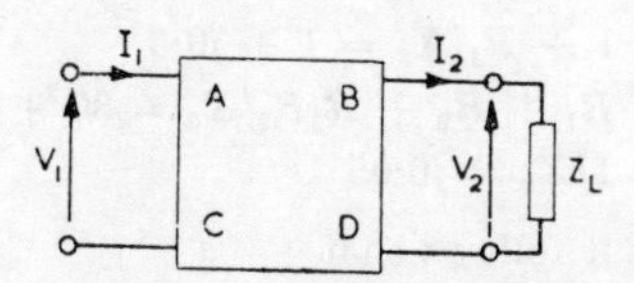

Fig. 3.11. The Loaded Two-port Network

four parameters. In terms of the **ABCD** constants we can write down the relationship between input and output (Fig. 3.11) as

$$V_1 = \mathbf{A}V_2 + \mathbf{B}I_2$$

and

$$I_1 = \mathbf{C}V_2 + \mathbf{D}I_2$$

The input admittance is

$$Y_{in} = \frac{I_1}{V_1} = \frac{\mathbf{C}V_2 + \mathbf{D}I_2}{\mathbf{A}V_2 + \mathbf{B}I_2}$$

For a load impedance, Z_L, the output voltage and current are related by $V_2 = I_2Z_L$, and hence

$$Y_{in} = \frac{\mathbf{C}Z_L + \mathbf{D}}{\mathbf{A}Z_L + \mathbf{B}} = \frac{\frac{\mathbf{C}}{\mathbf{A}}\left(\mathbf{A}Z_L + \mathbf{B} + \frac{\mathbf{DA}}{\mathbf{C}} - \mathbf{B}\right)}{\mathbf{A}Z_L + \mathbf{B}}$$

$$= \frac{\mathbf{C}}{\mathbf{A}} + \frac{\frac{\mathbf{C}}{\mathbf{A}}\left(\frac{\mathbf{AD} - \mathbf{BC}}{\mathbf{C}}\right)}{\mathbf{A}Z_L + \mathbf{B}}$$

Hence $$Y_{in} = \frac{\mathbf{C}}{\mathbf{A}} + \frac{1}{\mathbf{A}^2Z_L + \mathbf{AB}}, \quad \text{since } \mathbf{AD} - \mathbf{BC} = 1 \quad . \quad (3.15)$$

Now $\mathbf{C}/\mathbf{A}$ is a constant of the two-port network, and if the load impedance has either a resistive or reactive variable term, then the locus of $\mathbf{A}^2Z_L + \mathbf{AB}$ is a straight line, and the locus of $1/(\mathbf{A}^2Z_L + \mathbf{AB})$ is a circle. It follows that the locus diagram for any two-*terminal* network (irrespective of how complicated it may be) containing only one variable may be found by using only one inversion. This is achieved by rearranging the network so that the variable element is the load on a two-*port* network, and then applying the above method.

Example 3.3. Determine the locus of input admittance and current for the circuit shown in Fig. 3.12(*a*) as the capacitance C_x varies between zero and infinity. The applied voltage is 10 V.

The circuit may be rearranged as shown at (*b*) to give C_x as the load on a two-port network, whose auxiliary parameters are given by eqns. (2.7) as

$$\mathbf{A} = 1 + R_1/X_3 = 1 + j0{\cdot}5$$
$$\mathbf{B} = R_1 + R_2 + R_1R_2/X_3 = 30 + j10$$
$$\mathbf{C} = 1/X_3 = j0{\cdot}05$$

From these values it follows that

$$\mathbf{C}/\mathbf{A} = 0{\cdot}02 + j0{\cdot}04; \quad \mathbf{AB} = 25 + j25; \quad \text{and } \mathbf{A}^2 = 0{\cdot}75 + j1;$$

hence, from eqn. (3.15),

$$Y_{in} = 0{\cdot}02 + j0{\cdot}04 + 1/[-j(0{\cdot}75 + j1)\,\tilde{X} + 25 + j25]$$

where $\tilde{X} = 1/\omega C_x$ varies from zero when $C_x = \infty$ to infinity when $C_x = 0$. The locus of $[-j(0{\cdot}75 + j1)\tilde{X} + 25 + j25]$ is the straight line through

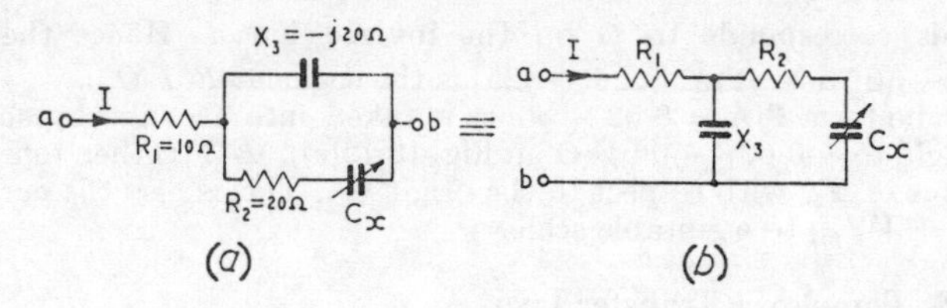

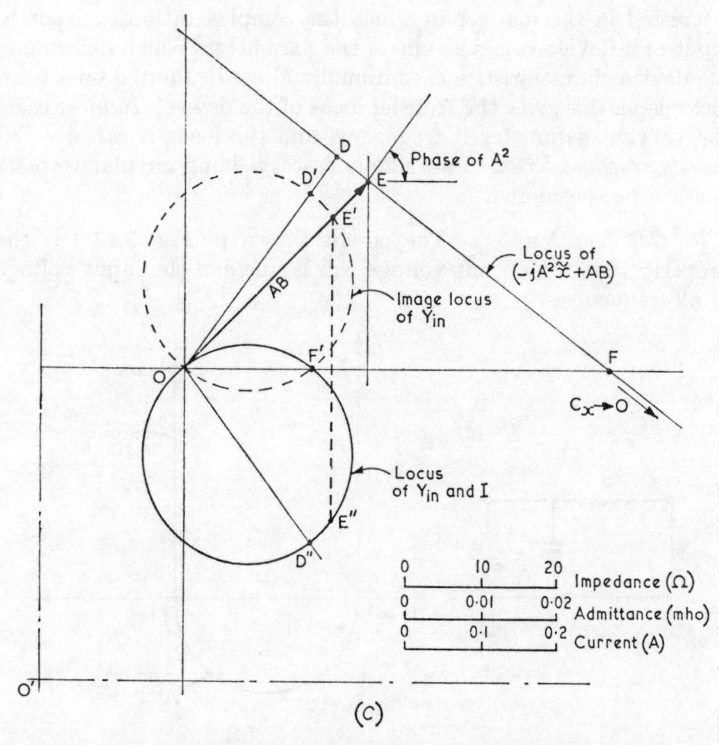

FIG. 3.12. LOCUS DIAGRAM FOR A TWO-PORT NETWORK WITH ONE VARIABLE

the end point of **AB** and making an angle of $-90°$ with the direction of $\mathbf{A}^2$. This is the line EF in Fig. 3.12(c). The image locus of this line is a circle (drawn to a suitable scale) through O and with its diameter along the perpendicular from O to EF. Since the perpendicular OD is by measurement 35 Ω, the circle diameter is $1/35 = 0{\cdot}0286$ mho. The inverse of EF is the reflection of this image circle (shown dotted) in the reference axis.

When $C_x = \infty$, $X = 0$. This corresponds to point E on the linear locus, E' on the image, and E'' on the inverse. When $C_x = 0$, $X = \infty$.

This corresponds to O on the inverse locus. Hence the locus of $1/[-j(0{\cdot}75 + j1)\tilde{X} + 25 + j25]$ is the segment $E''F'O$.

The term $\mathbf{C}/\mathbf{A} = 0{\cdot}02 + j0{\cdot}04$ is taken into account by shifting the origin to $-0{\cdot}02 - j0{\cdot}04$ (O' in Fig. 13.12(c)). $E''F'O$ then represents the locus of $\boldsymbol{Y}_{in}$ with respect to the origin O'. This is also the current locus ($\boldsymbol{I} = \boldsymbol{V}\boldsymbol{Y}_{in}$) to a suitable scale.

3.8. Complexor Transfer Loci

Very frequently (for example in servomechanism theory) we are interested in the manner in which the complex ratio of output to input of a device varies as one of the parameters which determines the device characteristics is continually altered. Plotted on a complex plane, this gives the transfer locus of the device. In most cases the varying parameter is frequency, and the locus is called a *frequency response locus*. Two simple cases yielding circular transfer loci will be considered.

(*a*) *CR Lag Network*. The circuit shown in Fig. 3.13 has the property that the output voltage will lag behind the input voltage at all frequencies.

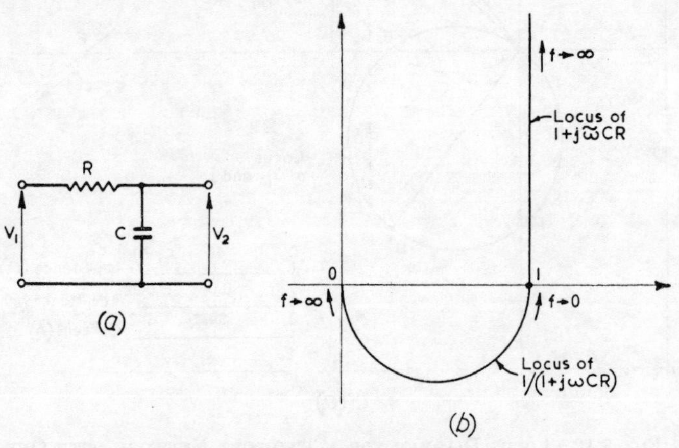

Fig. 3.13. Simple Lag Network

With suitable choice of C and R the circuit is also well known as a simple form of integrating network. The complex ratio of output to input is

$$G(j\omega) = \frac{\boldsymbol{V}_2}{\boldsymbol{V}_1} = \frac{1/j\omega C}{R + 1/j\omega C} = \frac{1}{1 + j\omega CR} \quad . \ (3.16)$$

Since, with frequency as the variable parameter, $1 + j\omega CR$ will give a straight-line locus through the point $(1 + j0)$, the locus of

$1/(1 + j\omega CR)$ is the semicircle shown in Fig. 3.13 (*b*). The diameter of the semicircle is obviously unity. The point $(1 + j0)$ corresponds to zero frequency, while the origin corresponds to infinite frequency.

(*b*) *CR Lead Network.* By interchanging the positions of *C* and *R* in the previous circuit, a network is produced whose output voltage leads the input voltage for all values of frequency, Fig. 3.14 (*a*).

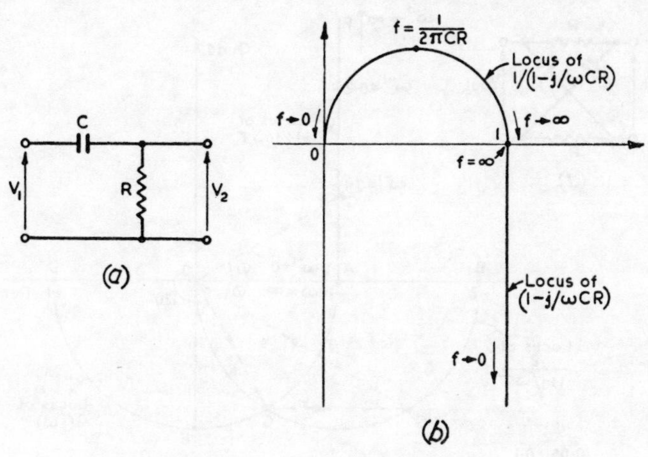

FIG. 3.14. SIMPLE LEAD NETWORK

Again with suitable values of *C* and *R* this is a simple form of differentiating network. The complex ratio of output to input is,

$$G(j\omega) = \frac{V_2}{V_1} = \frac{R}{R + 1/j\omega C} = \frac{1}{1 - j/\omega CR} \quad . \quad (3.17)$$

The locus of $(1 - j/\omega CR)$ as the frequency varies is the straight line parallel to the negative quadrate axis and passing through the point $(1 + j0)$. Hence the locus of $1/(1 - j/\omega CR)$ is the semicircle of diameter unity shown in Fig. 3.14 (*b*). The point $(1 + j0)$ corresponds to infinite frequency, while the locus passes through the origin at zero frequency. When $f = 1/2\pi CR$, the phase angle is 45° leading.

Example 3.4. Determine the locus of the complex voltage transfer function $G(j\omega)$ for the lattice circuit shown in Fig. 3.15(*a*) as the frequency varies. If $C = 1\ \mu$F and $R = 150$ kΩ, find the frequency at which the phase shift is (*a*) −90°, (*b*) −120°.

From the circuit we can write $V_{out} = \dfrac{V_{in}(1/j\omega C)}{R + 1/j\omega C} - \dfrac{V_{in}R}{R + 1/j\omega C}$

Hence $$G(j\omega) = \frac{V_{out}}{V_{in}} = \frac{1}{(R + 1/j\omega C)}\left(\frac{1}{j\omega C} - R\right)$$

$$= \frac{1 - j\omega CR}{1 + j\omega CR} = \frac{1 - j\omega\tau}{1 + j\omega\tau} \qquad \text{(i)}$$

$$= 1 - \frac{2}{1 - j/\omega\tau} = 1 + \frac{2}{-1 + j/\omega\tau} \qquad \text{(ii)}$$

where $\tau = CR$.

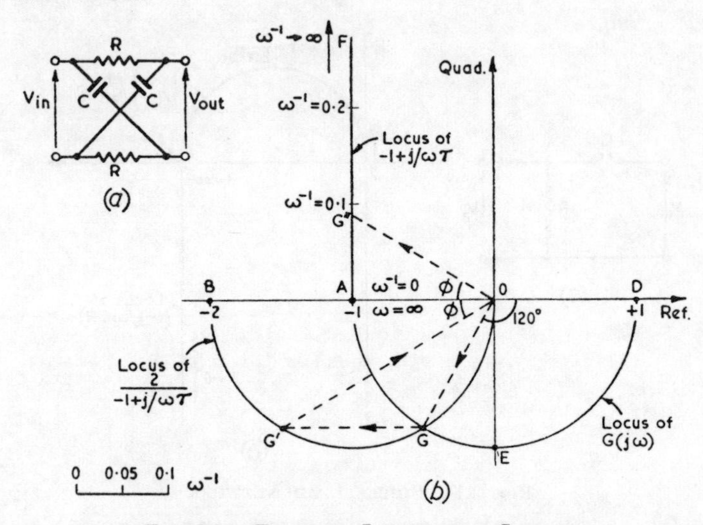

FIG. 3.15. TRANSFER LOCUS FOR A LATTICE

With ω as the variable, eqn. (ii) has the same form as eqn. (3.10) and therefore represents a circle. The diameter and location of this circle are found as follows. The locus of $(-1 + j/\omega\tau)$ is a straight line from the point $(-1 + j0)$ and parallel to the positive quadrate axis (line AF in Fig. 3.15(*b*)). The scale of frequency is an inverse one, so that a linear scale of ω^{-1} can be drawn along AF. With the values given for C and R one unit of this scale represents $\omega^{-1} = CR = 0{\cdot}15$. The point A corresponds to $\omega^{-1} = 0$ $(\omega = \infty)$.

The locus of $2/(-1 + j/\omega\tau)$ is therefore the semicircle on $OB = 2$ as diameter, and lying below the negative reference axis; and the locus of $G(j\omega)$ is the same semicircle moved one unit to the right (semicircle AED). Notice that the radius of this semicircle is unity, so that the lattice is an all-pass network, with a non-minimum phase characteristic. This means that there is no value of ω (apart from zero) for which the phase lag of the output has a minimum value, since the lag increases with ω right up to $-180°$.

From eqn. (i) it can readily be seen that when $\omega = 1/CR$ the phase angle of $G(j\omega)$ is $-90°$. Hence, for $90°$ phase shift,

$$f = \omega/2\pi = 6{\cdot}67/2\pi = \underline{\underline{1{\cdot}06 \text{ c/s}}}$$

For a phase angle of $-120°$, the corresponding point on the $G(j\omega)$ locus is G. This in turn corresponds to G' and G'' on the other loci, as shown in the diagram. The distance AG'' represents $\omega^{-1} = 0{\cdot}085$, by measurement and hence

$$f = 1/2\pi\omega^{-1} = \underline{\underline{1{\cdot}87 \text{ c/s}}}$$

3.9. Multiplication of Complexor Loci

When elements of a complete system are connected in cascade, the overall transfer locus is the product of the individual transfer loci. Thus, if the individual loci are known, the overall locus must be found by a point-by-point method for any given parameter value.

Suppose, for example, that the transfer function of two units in cascade is of the form

$$G(j\omega) = \frac{1}{j\omega\tau(1 + j\omega\tau)} \quad . \quad . \quad . \quad (3.18)$$

Now, as f varies, the locus of $1/(1 + j\omega\tau)$ is the semicircle shown in Fig. 3.16, while the locus of $1/j\omega\tau$ lies along the negative quadrate

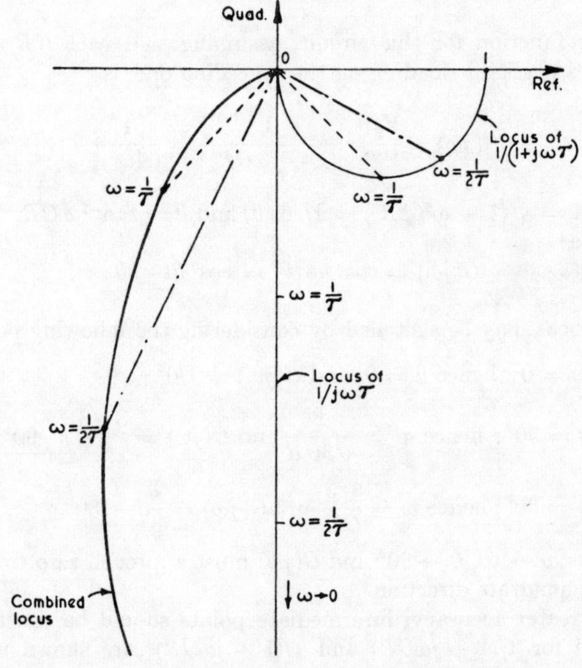

FIG. 3.16. MULTIPLICATION OF TWO COMPLEXOR LOCI

axis. The points on each locus corresponding to angular frequencies of 0, $1/2\tau$ and $1/\tau$ are shown. The point on the combined locus at any frequency is obtained by multiplying together the magnitudes of the components at this frequency and adding the phase angles.

A further example which is of some importance in *RC* oscillator theory is the triple-lag circuit shown in Fig. 3.17 (*a*). The overall

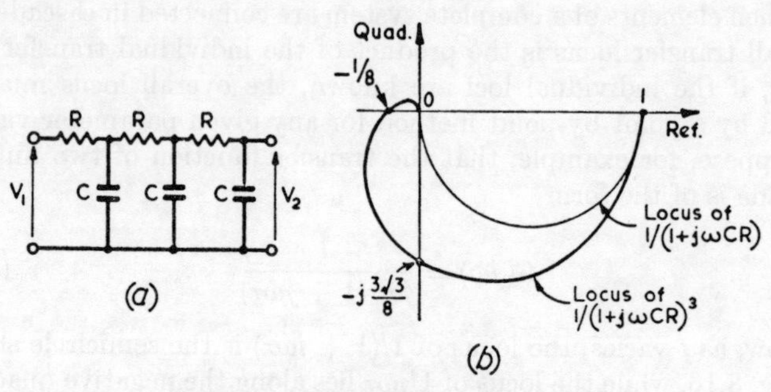

Fig. 3.17. The Triple-lag Circuit

transfer function for this circuit, assuming that each *CR* section produces negligible loading on the preceding one, is

$$G(j\omega) = \frac{V_2}{V_1} = \frac{1}{(1 + j\omega CR)^3} = \frac{1}{(Ae^{j\theta})^3}$$

where $A = \sqrt{(1 + \omega^2C^2R^2)} = 1/\cos\theta$, and $\theta = \tan^{-1}\omega CR$.

Hence

$$G(j\omega) = \cos^3\theta e^{-j3\theta} = \cos^3\theta\underline{/-3\theta} \quad . \quad . \quad (3.19)$$

The locus may be sketched by considering the following points—

(i) $\omega = 0$; hence $\theta = 0$, and $G(j\omega) = 1\underline{/0°}$

(ii) $\theta = 30°$; hence $\omega = \dfrac{1}{\sqrt{3}CR}$ and $G(j\omega) = \dfrac{3\sqrt{3}}{8}\underline{/-90°}$

(iii) $\theta = 60°$; hence $\omega = \dfrac{\sqrt{3}}{CR}$ and $G(j\omega) = \dfrac{1}{8}\underline{/-180°}$

(iv) as $\omega \to \infty$, $\theta \to 90°$ and $G(j\omega)$ must approach zero from the positive quadrate direction.

For greater accuracy, intermediate points should be considered. The loci for $1/(1 + j\omega CR)$ and $1/(1 + j\omega CR)^3$ are shown in Fig. 3.17 (*b*).

3.10. Phase-advance Circuit (Transitional Lead)

In many cases the overall response of a closed-loop control system may be improved by including a phase-advance circuit such

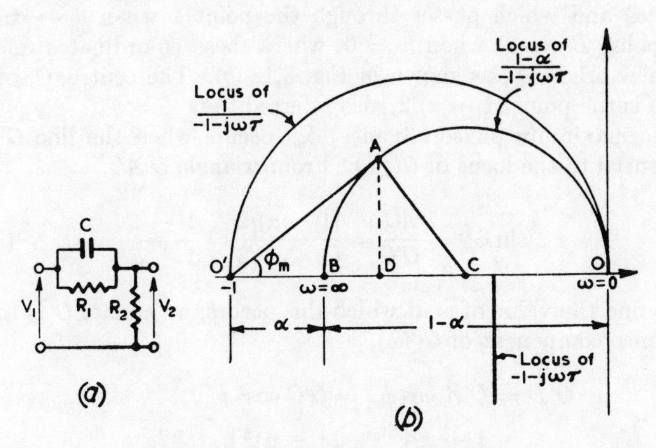

FIG. 3.18. PHASE-ADVANCE CIRCUIT

as is shown in Fig. 3.18 (a). The transfer function of this circuit is

$$G(j\omega) = \frac{V_2}{V_1} = \frac{R_2}{R_2 + \dfrac{R_1(1/j\omega C)}{R_1 + 1/j\omega C}} = \frac{R_2(1 + j\omega CR_1)}{R_1 + R_2 + j\omega CR_1R_2}$$

$$= \frac{R_2}{R_1 + R_2}\,\frac{1 + j\omega CR_1}{1 + j\omega CR_1R_2/(R_1 + R_2)}$$

$$= \alpha\,\frac{1 + j\omega\tau}{1 + j\omega\alpha\tau} \qquad . \quad . \quad . \quad (3.20)$$

where $\alpha = R_2/(R_1 + R_2)$ (i.e. $\alpha < 1$) and $\tau = CR_1$.

Hence $\qquad G(j\omega) = \dfrac{\alpha + j\omega\alpha\tau}{1 + j\omega\alpha\tau} = \dfrac{1 + j\omega\alpha\tau - (1 - \alpha)}{1 + j\omega\alpha\tau}$

$$= 1 + (1 - \alpha)\,\frac{1}{(-1 - j\omega\alpha\tau)} \qquad . \quad . \quad (3.20a)$$

From Section 3.3, the locus of $1/(-1 - j\omega\alpha\tau)$ as ω varies is a semicircle with diameter 1 on the reference axis and above it, as shown in Fig. 3.18 (b). The locus of $(1 - \alpha)/(-1 - j\omega\alpha\tau)$ must also be a semicircle, but the diameter is now reduced to $(1 - \alpha)$. When

$\omega = 0$ this circle passes through the point B $(= -1 + \alpha + j0)$ and when $\omega \to \infty$ it passes through the origin. To obtain the locus of $G(j\omega)$ the origin is shifted to O' $(= -1 + j0)$. The locus of $G(j\omega)$ is thus a semicircle *above* the reference axis, whose diameter is $(1 - \alpha)$ and which passes through the point 1 when $\omega \to \infty$ and the point B $(= \alpha)$ when $\omega = 0$, where these co-ordinates refer to the new origin O', as shown in Fig. 3.18 (*b*). The centre, C, of the circle is the point $(1 + \alpha)/2$, also referred to O'.

The maximum phase advance, ϕ_m, occurs when the line $O'A$ is tangential to the locus of $G(j\omega)$. From triangle $O'AC$,

$$\sin \phi_m = \frac{AC}{O'C} = \frac{(1 - \alpha)/2}{(1 + \alpha)/2} = \frac{1 - \alpha}{1 + \alpha} \quad . \quad . \quad (3.21)$$

To find the value of ω at which this occurs, we equate $O'D$ to the reference component of $G(j\omega)$,

$$O'D = O'A \cos \phi_m = O'C \cos^2 \phi_m$$

$$= \frac{1 + \alpha}{2}\left[1 - \left(\frac{1 - \alpha}{1 + \alpha}\right)^2\right] = \frac{2\alpha}{1 + \alpha} \quad . \quad . \quad (3.22)$$

The reference component of $G(j\omega)$ is the reference part of

$$\frac{\alpha(1 + j\omega\tau)}{1 + j\omega\alpha\tau} = \mathcal{R}\alpha\,\frac{1 + j\omega\tau}{1 + j\omega\alpha\tau} = \alpha\,\frac{1 + \omega^2\alpha\tau^2}{1 + \omega^2\alpha\tau^2} \quad (3.23)$$

Hence, equating (3.22) and (3.23),

$$\frac{2}{1 + \alpha} = \frac{1 + \omega^2\alpha\tau^2}{1 + \omega^2\alpha^2\tau^2}$$

which simplifies to $\omega^2\tau^2 = 1/\alpha$, or

$$\omega = \frac{1}{\tau\sqrt{\alpha}} \quad . \quad . \quad . \quad . \quad (3.24)$$

3.11. Phase-retard Circuit (Transitional Lag)

The transfer function of the phase-retard circuit shown in Fig. 3.19 is

$$G(j\omega) = \frac{V_2}{V_1} = \frac{R_2 + 1/j\omega C}{R_1 + R_2 + 1/j\omega C} = \frac{1 + j\omega CR_2}{1 + j\omega C(R_1 + R_2)} = \frac{1 + j\omega\tau}{1 + j\omega\beta\tau}$$

$$. \quad . \quad . \quad (3.25)$$

where $\tau = CR_2$, and $\beta = (R_1 + R_2)/R_2$ (i.e. $\beta > 1$).

Hence

$$G(j\omega) = \frac{1}{\beta}\left(\frac{\beta + j\omega\beta\tau}{1 + j\omega\beta\tau}\right)$$

$$= \frac{1}{\beta}\left(1 + \frac{\beta - 1}{1 + j\omega\beta\tau}\right) \quad . \quad . \quad (3.25a)$$

The locus of $(\beta - 1)/(1 + j\omega\beta\tau)$ is a semicircle of diameter $\beta - 1$ on the reference axis. The circle passes through the origin and lies below the reference axis. The locus of $1 + (\beta - 1)/(1 + j\omega\beta\tau)$ is

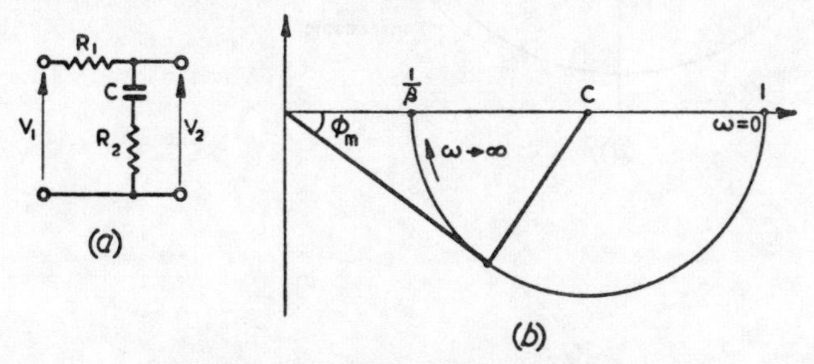

FIG. 3.19. PHASE-RETARD CIRCUIT

the same semicircle shifted to the right by one unit (i.e. the origin is shifted to the left by one unit); it has the same diameter and passes through the points 1 and β (with respect to the new origin) on the reference axis. Hence the locus of $G(j\omega)$ is this same semicircle reduced in scale by $1/\beta$, so that the extremities of the diameter are $1/\beta$ and 1, as shown in Fig. 3.19 (*b*).

The maximum phase retardation may be worked out in the same way as for the phase-advance circuit, and is given by

$$\sin \phi_m = \frac{1 - 1/\beta}{1 + 1/\beta} = \frac{\beta - 1}{\beta + 1} \quad . \quad . \quad . \quad (3.26)$$

By considering the reference component of $G(j\omega)$ at maximum phase retardation the frequency at which this maximum occurs is found to be

$$f = \frac{1}{2\pi\tau\sqrt{\beta}} \quad . \quad . \quad . \quad . \quad (3.27)$$

3.12. The Smith Circle Diagram

A further important example of the locus diagram technique occurs in relation to a diagram which is known as the *Smith circle diagram.*

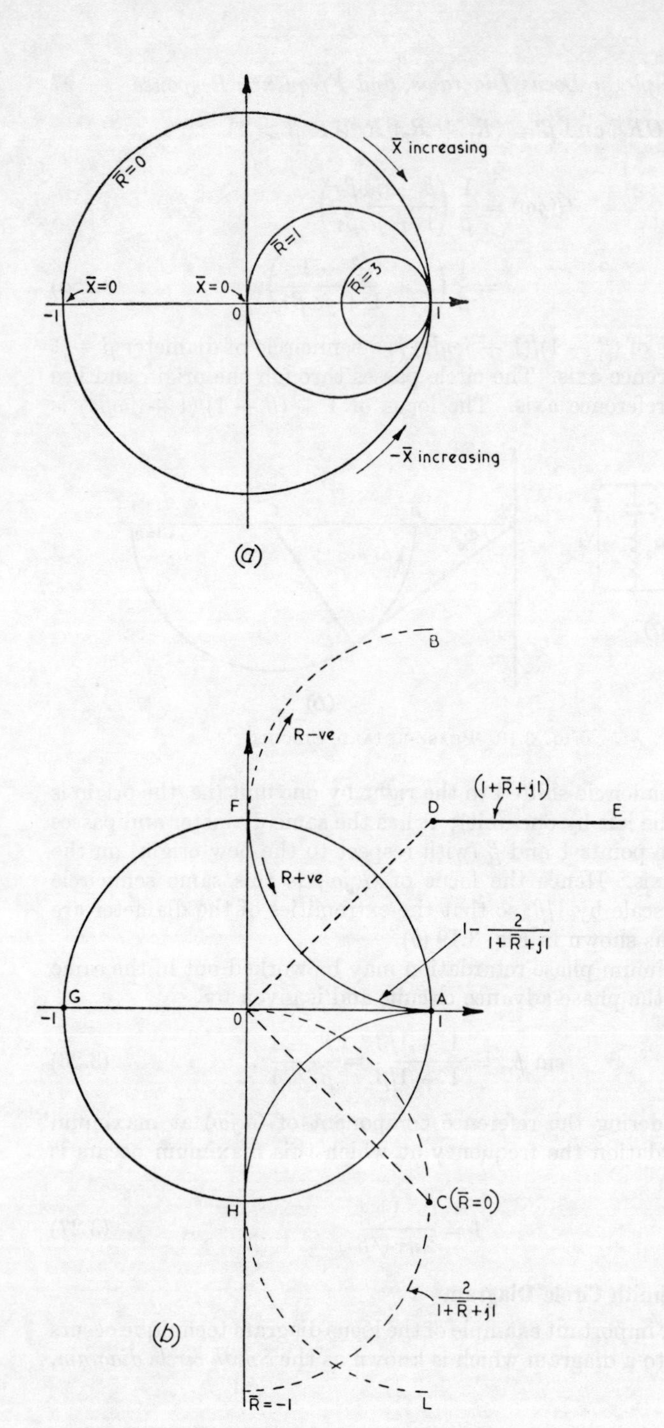

FIG. 3.20. CONSTRUCTION OF THE SMITH CHART

In Chapter 7 we define a quantity called the reflection coefficient, ρ, of a transmission line by the relation

$$\rho = \frac{Z_L/Z_0 - 1}{Z_L/Z_0 + 1} = \frac{Z_{LN} - 1}{Z_{LN} + 1}. \quad . \quad . \quad (3.28)$$

where Z_L is a variable, and is the load impedance at the end of the line, Z_0 is a constant (complex) called the characteristic impedance, and $Z_{LN} = Z_L/Z_0$ is the *normalized load impedance.* The Smith chart gives the locus of ρ for all possible variations in the complex quantity Z_{LN}. It is convenient to consider separately (*a*) the case when the reference component of Z_{LN} is constant, and the quadrate component varies between $\pm\infty$, and (*b*) the case when the quadrate component of Z_{LN} is constant and the reference (or resistive) component varies between zero and ∞.

Let us first consider case (*a*). We can rewrite eqn. (3.28) as

$$\rho = \frac{Z_{LN} + 1}{Z_{LN} + 1} - \frac{2}{Z_{LN} + 1} = 1 - \frac{2}{Z_{LN} + 1} \quad . \quad (3.29)$$

and since $Z_{LN} = R_N + jX_N$, where R_N and X_N are the normalized series resistance and reactance of the load, we may write

$$\rho = 1 - \frac{2}{1 + R_N + jX_N} = 1 - \frac{2/(1 + R_N)}{1 + jX_N/(1 + R_N)} \quad . \quad (3.30)$$

As X_N varies the locus of $2/(1 + R_N + jX_N)$ is a circle of diameter $2/(1 + R_N)$ on the reference axis (a complete circle, since X_N can be positive or negative), and passing through the origin. Comparison with eqn. (3.20) for the phase advance circuit shows that the locus of ρ is a circle of diameter $2/(1 + R_N)$ passing through the points 1 and $[1 - 2/(1 + R_N)]$ on the reference axis. For $R_N = 0$ the diameter is 2, for $R_N = 1$ the diameter is 1, for $R_N = 3$ the diameter is $\frac{1}{2}$, etc., and these circles are shown in Fig. 3.20 (*a*). For X_N positive the locus lies above the reference axis and for X_N negative it lies below the axis. As $X_N \to \infty$, $\rho \to 1$, and when $X_N = 0$, $\rho = (1 - 2/(1 + R_N))$. Hence, for a constant resistance, ρ varies along a constant-resistance circle from $(1 - 2/(1 + R_N))$ to 1, as X_N increases from zero to $\pm\infty$.

Now consider the case where the normalized load reactance, X_N, is constant and R_N is varied. In eqn. (3.30) the locus of the second term may be compared with the loci for the inverse of the lines $z = ja + bu$ and $z = -ja + bu$ of Section 3.3. Suppose, for instance, that $X_N = +1$. The locus of $(1 + R_N + j1)$ is shown as the line

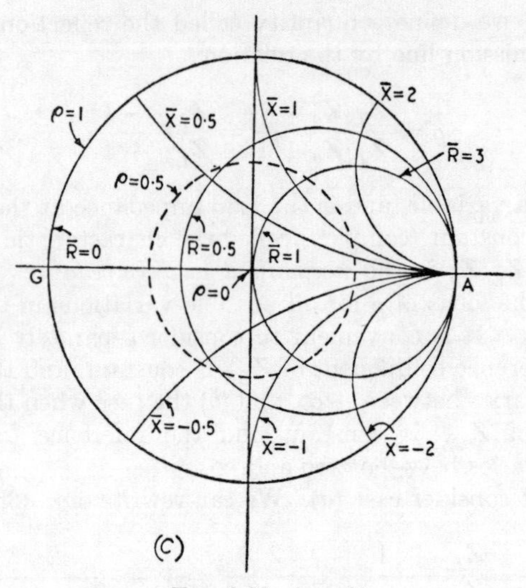

FIG. 3.20. (*contd.*)

DE in Fig. 3.20 (*b*). The locus of $2/(1 + R_N + j1)$ will be the semicircle *OC* of diameter 2 on the negative quadrate axis. The lower portion of this semicircle corresponds to negative values of *R*. The locus of ρ is obtained by subtracting this semicircle from unity (subtract the reference terms and reverse the sign of the quadrate term). This gives a semicircle of diameter 2 on the line *AB* as diameter, and passing through the point *A*. Obviously this locus cannot extend beyond the $R_N = 0$ circle, since this would imply negative values of R_N. In the same way, for any other value of $+X_N$ the locus of ρ will be a semicircle of diameter $2/X_N$ along *AB* and passing through point *A*. For negative values of X_N the loci will lie below the reference axis, so that the whole family of loci are as indicated in Fig. 3.20 (*c*).

When $X_N = 0$, the locus of $(1 + R_N)$ lies along the reference axis, and the locus of $1 - 2/(1 + R_N)$ is the line *AG* extending from $G(-1)$ when $R_N = 0$ to A $(+1)$ when $R_N \to \infty$. Hence a scale of R_N may be constructed along *AG*.

Circles with centre *O* give the loci for constant values of ρ, the outside circle corresponding to $\rho = 1$. The circle for $\rho = 0{\cdot}5$ is indicated in Fig. 3.20 (*c*) by the dotted line.

3.13. Gain and Phase Characteristics

An alternative method of describing the transfer characteristics of a circuit or a system is to plot the gain (i.e. the logarithmic ratio of output to input magnitudes) and the phase shift to a logarithmic base of frequency. The diagrams are called *Bode diagrams.* The Bode diagram method has the advantage that, for non-interacting elements, the overall characteristic is obtained by *adding* the individual element gains and phase shifts, whereas on the complex polar diagram the complexor loci must be multiplied together in order to obtain the combined locus. A further advantage is that in most cases the gain characteristic may be represented by straight lines which are asymptotic to the actual graph. These linear asymptotes are very readily found. Let us consider some straightforward cases to illustrate the concept.

(*a*) *Lag Circuit.* For the simple CR lag network shown in Fig. 3.21 (*a*) we have already obtained the complex voltage transfer ratio (eqn. (3.16)).

$$G(j\omega) = \frac{V_2}{V_1} = \frac{1}{1 + j\omega CR} = \frac{1}{1 + j\omega\tau}$$

from eqn. (3.16), with $\tau = CR$. Hence

$$G = 20 \log_{10} |V_2/V_1| = 20 \log_{10} 1 - 20 \log_{10} \sqrt{(1 + \omega^2\tau^2)}$$

$$= -10 \log_{10} (1 + \omega^2\tau^2) \text{ decibels} \quad . \quad . \quad . \quad . \quad (3.31)$$

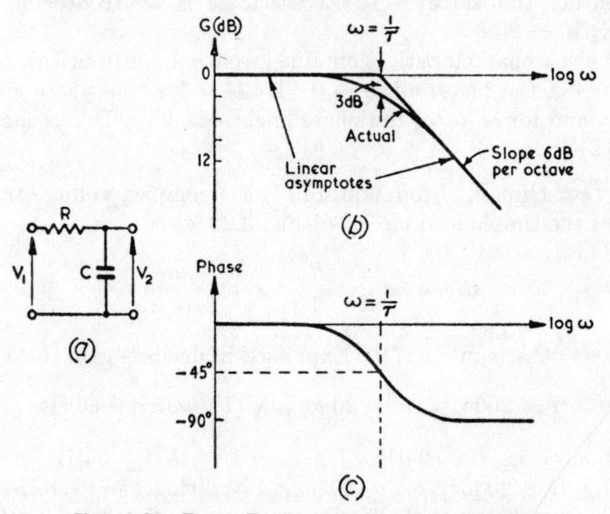

FIG. 3.21. BODE DIAGRAM FOR A LAG CIRCUIT

For $\omega \ll 1/\tau$, this gives $G = 0$ dB; for $\omega = 1/\tau$ it gives $G = -10 \log_{10} 2 \simeq -3$ dB; and for $\omega \gg 1/\tau$, G becomes $-20 \log_{10} \omega\tau = -20 \log_{10} \tau - 20 \log_{10} \omega$ decibels. This latter expression represents a straight line from the point $\omega = 1/\tau$ on the horizontal axis and with a slope of -6 dB/octave. This is seen by considering two angular frequencies ω_1 and ω_2 which satisfy the condition $\omega \gg 1/\tau$ and where $\omega_2 = 2\omega_1$ (i.e. ω_2 is one octave higher than ω_1).

Then

$$G_1 = -20 \log_{10} \tau - 20 \log_{10} \omega_1$$

and

$$\begin{aligned} G_2 &= -20 \log_{10} \tau - 20 \log_{10} 2\omega_1 \\ &= -20 \log_{10} \tau - 20 \log_{10} \omega_1 - 20 \log_{10} 2 \\ &= (G_1 - 6) \text{ decibels} \end{aligned}$$

In words, the gain drops 6 dB for each doubling of frequency. The gain curve is shown in Fig. 3.21 (*b*). The frequency $f_c = 1/2\pi\tau$ at which the gain is 3 dB down is called the *corner frequency.* The linear asymptotes are, as shown, the horizontal line to the corner frequency, and the line falling at 6 dB/octave from the corner frequency. The error is a maximum of 3 dB at the corner frequency (also called the *break point* of the curve). An alternative way of designating the slope is to say that it is 20 dB/decade, since $20 \log_{10} 10 = 20$.

The phase characteristic is obtained from eqn. (3.16) with $CR = \tau$. For $\omega \to 0$, the phase angle $\to 0$; for $\omega = 1/\tau$, the phase angle is $-45°$; and for $\omega \to \infty$, the phase angle $\to -90°$. This is shown in Fig. 3.21 (*c*).

(*b*) *Lead Circuit.* From eqn. (3.17) the complex voltage transfer ratio of the simple lead circuit of Fig. 3.22 (*a*) is

$$G(j\omega) = \frac{R}{R + 1/j\omega C} = \frac{j\omega\tau}{1 + j\omega\tau}$$

where $\tau = CR$ (eqn. (3.17)). Expressed in decibels gain this gives

$$G = 20 \log_{10} \omega\tau - 20 \log_{10} \sqrt{(1 + \omega^2\tau^2)} \text{ decibels} \qquad (3.32)$$

For $\omega \to \infty$, $G = 0$ dB; for $\omega = 1/\tau$, $G = -3$ dB; and for $\omega \ll 1/\tau$, $G \simeq 20 \log_{10} \omega\tau - 20 \log_{10} 1 = 20 \log_{10} \omega\tau$ decibels.

As before, doubling the frequency causes a change of 6 dB in G, and hence the equation represents a straight line *rising* at 6 dB/octave from a large negative value (since $\omega\tau \ll 1$) towards the point $\omega = 1/\tau$. The curve and its linear asymptotes are shown in Fig. 3.22 (*b*).

The phase characteristic is shown in Fig. 3.22 (*c*). For $\omega \to 0$, the

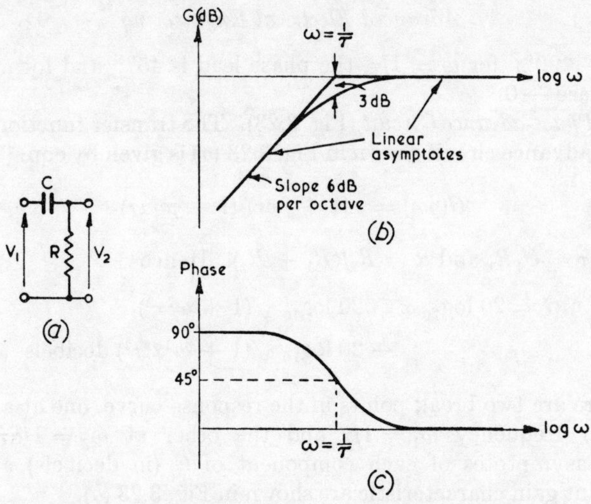

FIG. 3.22. BODE DIAGRAM FOR THE SIMPLE LEAD CIRCUIT

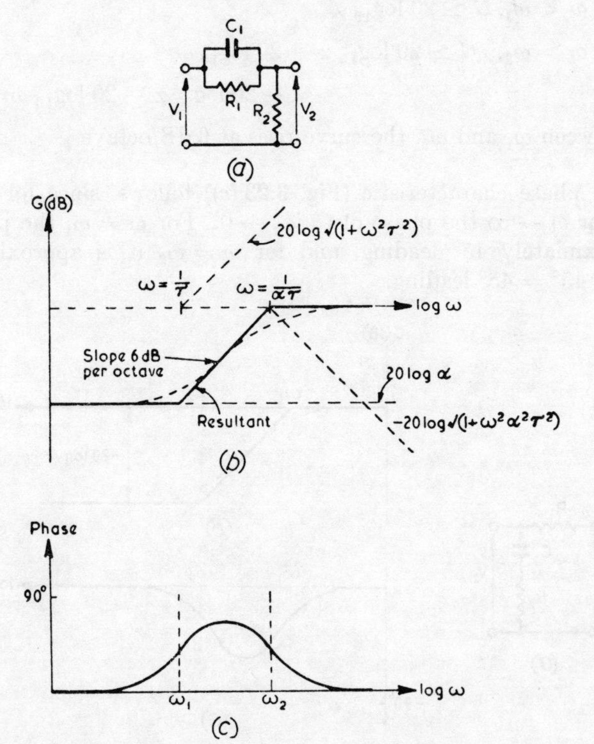

FIG. 3.23. BODE DIAGRAM FOR A PHASE-ADVANCE CIRCUIT

phase $\rightarrow 90°$; for $\omega = 1/\tau$, the phase lead is 45°; and for $\omega \rightarrow \infty$, the phase $\rightarrow 0$.

(*c*) *Phase-advance Circuit* (Fig. 3.23). The transfer function of the phase-advance circuit shown in Fig. 3.23 (*a*) is given by eqn. (3.20) as

$$G(j\omega) = \alpha(1 + j\omega\tau)/(1 + j\omega\alpha\tau)$$

where $\tau = C_1R_1$ and $\alpha = R_2/(R_1 + R_2)$. Hence

$$G = 20 \log_{10} \alpha + 20 \log_{10} \sqrt{(1 + \omega^2\tau^2)} - 20 \log_{10} \sqrt{(1 + \omega^2\alpha^2\tau^2)} \text{ decibels} \quad . \quad (3.35)$$

There are two break points in the response curve, one at a corner angular frequency $\omega_1 = 1/\tau$ and the other at $\omega_2 = 1/\alpha\tau$. The linear asymptotes of each component of G (in decibels) and the resultant gain characteristic are shown in Fig. 3.23 (*b*).

For $\omega < \omega_1$, $G \simeq 20 \log_{10} \alpha$

For $\omega > \omega_2$, $G \simeq 20 \log_{10} \alpha + 20 \log_{10} \omega\tau - 20 \log_{10} \alpha - 20 \log_{10} \omega\tau = 0$

Between ω_1 and ω_2, the curve rises at 6 dB/octave

The phase characteristic (Fig. 3.23 (*c*)) follows, since for $\omega \rightarrow 0$ and for $\omega \rightarrow \infty$ the phase of $G(j\omega) \rightarrow 0$. For $\omega = \omega_1$ the phase is approximately 45° leading, and for $\omega = \omega_2$ it is approximately $90° - 45° = 45°$ leading.

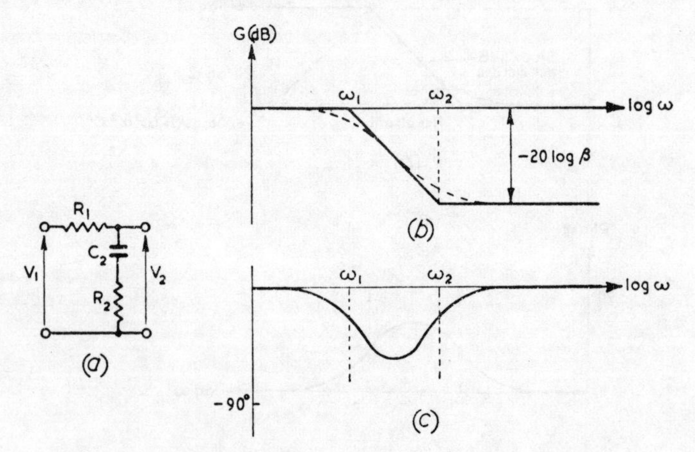

Fig. 3.24. Bode Diagram for a Phase-retard Circuit

(*d*) *Phase-retard Circuit* (Fig. 3.24). The transfer function of the phase-retard circuit shown in Fig. 3.24 (*a*) is given by eqn. (3.25) as

$$G(j\omega) = \frac{1 + j\omega\tau}{1 + j\omega\beta\tau}$$

where $\tau = C_2R_2$, and $\beta = (R_1 + R_2)/R_2$. Hence

$$G = 20 \log_{10} \sqrt{(1 + \omega^2\tau^2)} - 20 \log_{10} \sqrt{(1 + \omega^2\beta^2\tau^2)} \text{ decibels}$$

As in the previous case, there are two break points, at $\omega_1 = 1/\beta\tau$ and $\omega_2 = 1/\tau$. It is left as an exercise for the reader to verify that the resultant gain and phase characteristics are those shown in Figs. 3.24 (*b*) and (*c*).

Bibliography

Hammond, *Feedback Theory* (English Universities Press).

Problems

3.1. A series circuit consists of an inductance of 3·18 mH and a resistance whose value can be varied from zero to infinity. Draw to scale the impedance, admittance and current loci for the circuit when it is fed from a 100-V 50-c/s supply. Determine the impedance, admittance, conductance, susceptance and current when the resistance is 2 Ω. What value of resistance will make the power absorbed a maximum?

(*Ans.* Z = 2·24 Ω, Y = 0·46 mho, G = 0·392 mho, B = 0·21 mho; resistance = 1 Ω for maximum power.)

3.2. A resistance of 10 Ω is connected in series with an inductive reactance (whose value can be varied between 2 and 20 Ω) to a 250-V constant-frequency supply. Draw the admittance, impedance and current loci, and find the current and power factor when the reactance is 5 Ω.

(*Ans.* 22·3 A, 0·894.)

3.3. A fixed capacitive reactance of 10 Ω is connected in series with a variable resistance. Across this combination is connected a fixed coil of 20 Ω resistance and 50 Ω reactance, the combination being supplied from a constant-frequency 200-V supply. Draw the admittance and current loci as the resistance varies. From these determine (*a*) the current when the power factor is 0·95 lagging, (*b*) the maximum current and the corresponding resistance, (*c*) the maximum power absorbed and the corresponding current and resistance.

(*Ans.* (*a*) 6·5 A; (*b*) 16·5 A, 1·5 Ω; (*c*) 2·2 kW, 13 A, 10 Ω.)

3.4. A variable capacitance and a resistance of 300 Ω are connected in series across a 240-V 50-c/s supply. Draw the complexor locus of impedance and current as the capacitance changes from 5 μF to 30 μF. From the diagram find (*a*) the capacitance to give a current of 0·7 A, and (*b*) the current when the capacitance is 10 μF. (*L.U.*)

(*Ans.* 19·2 μF, 0·55 A.)

3.5. A pure inductance L is connected in series with a variable resistance R. Across the combination is a capacitance C. Sketch the loci for total admittance and current when the circuit is fed from a constant-frequency supply of V volts, the resistance is varied from zero to infinity, and $\omega C = 1/2\omega L$. For this condition show that the total current is constant at a value of $V\omega C$ as R varies. Verify the result analytically.

3.6. The approximate equivalent circuit of an induction motor is shown in Fig. 3.25. If $V = 300$ V, $X_0 = 30\ \Omega$, $R_0 = 600\ \Omega$, $R_2 = 0{\cdot}4\ \Omega$ and $X_2 = 4{\cdot}2\ \Omega$, draw the complexor locus of I_1 as s varies from 0 to 1. From this determine the value of input current and power factor when $s = 0{\cdot}05$.

(*Ans.* 43·8 A, 0·73 lagging.)

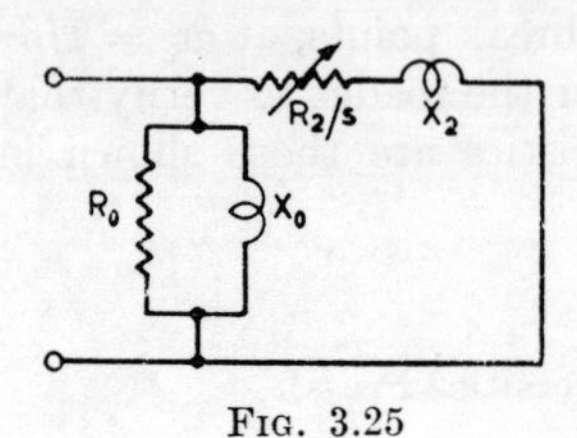

FIG. 3.25

FIG. 3.26

3.7. For the circuit shown in Fig. 3.26 draw the loci for impedance and current as R_L is varied. From this determine the total current when R_L is (i) open-circuited, (ii) short-circuited and (iii) 5 Ω. Also find the value of R_L for (*a*) maximum power and (*b*) maximum power factor.

(*Ans.* (i) 0·097 A, (ii) 0·158 A, (iii) 0·119 A; (*a*) 15 Ω; (*b*) 8·3 Ω.)

3.8. A constant-frequency constant-current source feeds a coil of inductive reactance 5 Ω and resistance 10 Ω in series with which is connected a parallel combination of a 10-Ω resistor and a variable capacitor. Draw the loci for impedance and total voltage when the source supplies a constant current of 10 A and the capacitance is varied. Find the total voltage and power factor when the capacitive reactance is 40 Ω.

(*Ans.* 196 V, 0·99 lagging.)

CHAPTER 4

Harmonics and Non-linear Elements

If a sinusoidal voltage is applied to a device which has a non-linear current/voltage relationship, the resulting current will not be sinusoidal in waveform. It can be shown that such a distorted waveform will be made up of the sum of components whose frequencies are harmonically related to that of the fundamental. Iron-cored coils and transformers, rectifiers and electronic power amplifiers are some of the devices which will introduce harmonics into a system.

At power frequencies the presence of harmonics is undesirable for several reasons, including (*a*) increased iron losses, (*b*) circuit resonance may occur at a harmonic frequency, giving undesirably large harmonic currents, (*c*) rectifier-type instruments no longer indicate true r.m.s. values, (*d*) large third-harmonic neutral currents can exist in three-phase systems, and (*e*) increased possibility of interference with communication circuits.

In communication networks, where non-sinusoidal waveforms may often be encountered, it is important to be able to determine the harmonic content in a particular signal waveform in order to assess the bandwidth required for its accurate transmission.

4.1. R.M.S. Value of a Complex Wave

Non-sinusoidal a.c. waveforms are often referred to as *complex waves*. The instantaneous value of such a complex current wave is given by

$$i = I_0 + I_{1m} \sin (\omega t + \phi_1) + I_{2m} \sin (2\omega t + \phi_2) + \ldots + I_{nm} \sin (n\omega t + \phi_n) \quad (4.1)$$

where I_0 represents a d.c. term, I_{nm} represents the peak value, and ϕ_n the phase of the nth harmonic.

By definition, the r.m.s. value of any alternating current is

$$I = \sqrt{(\text{Average value of } i^2)}$$

Now,

$$i^2 = I_0{}^2 + I_{1m}{}^2 \sin^2 (\omega t + \phi_1) + \ldots + I_{nm}{}^2 \sin^2 (n\omega t + \phi_n) + \ldots + 2I_{1m}I_{2m} \sin (\omega t + \phi_1) \sin (2\omega t + \phi_2) + \ldots + 2I_0 I_{1m} \sin (\omega t + \phi_1) + \ldots \quad (4.2)$$

The right-hand side of this equation is seen to consist of three types of term, namely (*a*) self-products of terms of the same frequency, (*b*) cross-products involving two terms of differing frequencies, and (*c*) products of the d.c. component with each harmonic.

For the general self-product, the average value over one cycle is given by

$$\frac{1}{2\pi}\int_0^{2\pi} I_{nm}{}^2 \sin^2 (n\omega t + \phi_n)\, \mathrm{d}(\omega t)$$

which readily simplifies to $I_{nm}{}^2/2$ by expressing $\sin^2 (n\omega t + \phi_n)$ in the form $\frac{1}{2}[1 - \cos 2(n\omega t + \phi_n)]$.*

The average value of $I_0{}^2$ is, of course, $I_0{}^2$.

The average values of the cross-products and the products of the d.c. component with each harmonic are zero, since

$$\int_0^{2\pi} \sin (p\omega t + \phi_p) \sin (q\omega t + \phi_q)\, \mathrm{d}(\omega t)$$

$$= \int_0^{2\pi} \tfrac{1}{2}\{\cos[(p-q)\omega t + \phi_p - \phi_q] - \cos[(p+q)\omega t + \phi_p + \phi_q]\}\, \mathrm{d}(\omega t)$$

$$= 0 \qquad (4.3)$$

and

$$\int_0^{2\pi} \sin (p\omega t + \phi_p)\, \mathrm{d}(\omega t) = 0 \qquad (4.4)$$

We may therefore write

$$\text{Average value of } i^2 = I_0{}^2 + \frac{I_{1m}{}^2}{2} + \frac{I_{2m}{}^2}{2} + \ldots$$

so that

$$I = \sqrt{\left(I_0{}^2 + \frac{I_{1m}{}^2}{2} + \frac{I_{2m}{}^2}{2} + \ldots\right)} \qquad (4.5)$$

By using r.m.s. values for each harmonic component, this equation may be simplified to

$$I = \sqrt{(I_0{}^2 + I_1{}^2 + I_2{}^2 + \ldots)} \qquad (4.6)$$

where $I_1 = I_{1m}/\sqrt{2}$, etc.

Exactly similar expressions will apply for voltage waves.

Eqn. (4.6) may be represented graphically in the form of a series of right-angled triangles, as shown in Fig. 4.1. Inspection of this

* *See* Shepherd, Morton and Spence, *Higher Electrical Engineering* (Section 5.3) (Pitman).

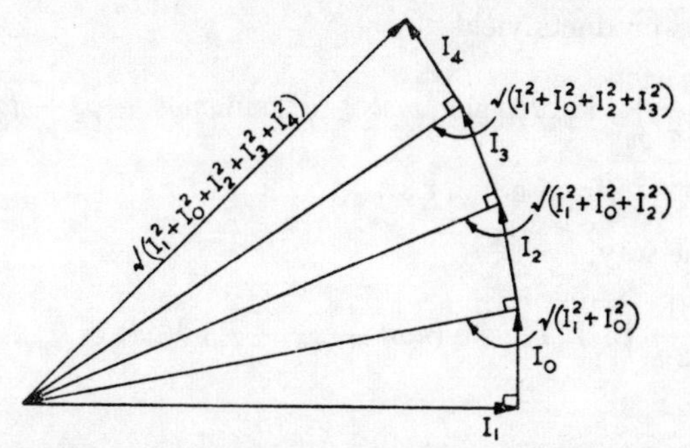

FIG. 4.1. GRAPHICAL REPRESENTATION OF R.M.S. VALUE OF A COMPLEX WAVE

diagram shows that the r.m.s. value of a complex wave will not be much larger than the largest component, provided that this component is some three times larger than the next largest.

4.2. Power and Power Factor

If a complex voltage

$$v = V_0 + V_{1m} \sin \omega t + V_{2m} \sin (2\omega t + \psi_2) + \ldots$$

is applied to a circuit, and the current flowing is

$$i = I_0 + I_{1m} \sin (\omega t - \phi_1) + I_{2m} \sin (2\omega t + \psi_2 - \phi_2) + \ldots$$

the instantaneous power is $p = vi$ watts, and the average power is

$$P = \frac{1}{2\pi}\int_0^{2\pi} vi\, \mathrm{d}(\omega t)$$

The product vi is seen to involve self-products, cross-products and products of the d.c. components with each harmonic, in the same way as the expression for i^2 in the previous section.

The average value over one cycle of the general term for the self-products gives

$$\begin{aligned}
P_n &= \frac{1}{2\pi}\int_0^{2\pi} V_{nm} I_{nm} \sin (n\omega t + \psi_n) \sin (n\omega t + \psi_n - \phi_n)\, \mathrm{d}(\omega t) \\
&= \frac{V_{nm} I_{nm}}{2\pi}\int_0^{2\pi} \tfrac{1}{2}\,[\cos \phi_n - \cos (2n\omega t + 2\psi_n - \phi_n)]\, \mathrm{d}(\omega t) \\
&= \frac{V_{nm} I_{nm}}{2\pi\,.\,2}\,[2\pi \cos \phi_n - 0] \\
&= V_n I_n \cos \phi_n \quad . \quad . \quad . \quad . \quad . \quad . \quad . \quad (4.7)
\end{aligned}$$

where V_n and I_n are r.m.s. values.

The cross-products yield

$$P_{pq} = \frac{1}{2\pi}\int_0^{2\pi} V_{pm}I_{qm}\sin(p\omega t + \psi_p)\sin(q\omega t + \psi_2 - \phi_q)\,\mathrm{d}(\omega t)$$

$$= 0 \qquad \text{(from eqn. (4.3))}$$

In the same way,

$$P_{0m} = \frac{1}{2\pi}\int_0^{2\pi} V_0 I_{nm}\sin(n\omega t + \psi_n - \phi_n)\,\mathrm{d}(\omega t) = 0$$

and

$$P_{n0} = \frac{1}{2\pi}\int_0^{2\pi} V_{nm} I_0 \sin(n\omega t + \psi_n)\,\mathrm{d}(\omega t) = 0$$

The total power is thus

$$P = V_0 I_0 + V_1 I_1 \cos\phi_1 + V_2 I_2 \cos\phi_2 + \ldots \qquad (4.8)$$

= Sum of separate powers for each harmonic component alone

The above derivation gives the theory underlying the use of an electrodynamic wattmeter for the analysis of the harmonic content of a complex current wave. The complex current is passed through the current coil of the wattmeter, and the voltage coil is fed separately at a constant voltage from a variable-frequency sinusoidal oscillator. The oscillator is tuned to the fundamental and to each harmonic in turn. The maximum indication as the oscillator is tuned through each frequency will then correspond to the product $V_n I_n$ (i.e. unity power factor at that harmonic), which is proportional to I_n if V is kept constant (*see* Section 4.6).

If R_s is the effective series resistance of the circuit through which a complex current of r.m.s. value I is passed, then from the definition of an r.m.s. current,

$$P = I^2 R_s \text{ watts} \qquad (4.9)$$

In the same way, if R_p is the effective parallel resistance of the circuit and the r.m.s. value of the complex voltage is V, the power is

$$P = \frac{V^2}{R_p} \text{ watts} \qquad (4.10)$$

The overall power factor of a single-phase network when harmonics are present is defined as

$$\text{P.F.} = \frac{\text{Total power}}{VI} \qquad (4.11)$$

where V and I are the r.m.s. values of the complex voltage and current. The power factor cannot be expressed in terms of a phase angle, and hence the terms lag and lead cannot be applied.

4.3. Effect of Harmonics in Single-phase Circuits

The effect of harmonic voltages or currents when applied to linear circuits can most easily be evaluated by considering each harmonic separately, and then combining the results to find overall r.m.s. values and total power dissipation. Since inductive reactance at the nth harmonic ($X_n = 2\pi nfL$) is n times the value at the fundamental ($X_1 = 2\pi fL$), an inductive circuit will have a smaller harmonic content in its current wave than in its voltage wave. On the other hand, since capacitive reactance decreases with frequency (at nth harmonic, $X_{nc} = 1/n2\pi fC$) capacitive circuit currents will have a higher harmonic content than the applied voltages.

It may happen that, in a circuit containing both inductance and capacitance, series or parallel resonance occurs at one of the harmonic frequencies. This is known as selective or *harmonic resonance.* Series harmonic resonance is most undesirable in power supply systems, since it may result in large harmonic currents.

Example 4.1. An r.m.s. current of 5 A, which has a third-harmonic content, is passed through a coil having a resistance of 1 Ω and an inductance of 10 mH. The r.m.s. voltage across the coil is 20 V. Calculate the magnitudes of the fundamental and harmonic components of current if the fundamental frequency is $300/2\pi$ cycles per second. Also find the power dissipated. (*L.U. Part II*)

At fundamental frequency, $\omega = 300$, so that

$$X_1 = \omega L = 300 \times 10^{-2} = 3\ \Omega$$

Impedance at fundamental frequency is

$$Z_1 = 1 + j3 = 3{\cdot}16\underline{/73{\cdot}3^\circ}\ \Omega$$

If V_1 is the fundamental voltage across the coil,

$$I_1 = \frac{V_1}{Z_1} \quad \text{or} \quad |V_1| = 3{\cdot}16 I_1 \qquad . \quad . \quad \text{(i)}$$

At third-harmonic frequency, the inductive reactance is

$$X_3 = 3\omega L = 9\ \Omega$$

Corresponding impedance is

$$Z_3 = 1 + j9 = 9{\cdot}05\underline{/83{\cdot}7^\circ}\ \Omega$$

Also

$$I_3 = \frac{V_3}{Z_3} \quad \text{or} \quad |V_3| = 9{\cdot}05 I_3 \qquad . \quad . \quad \text{(ii)}$$

The r.m.s. current is 5A, so that, from eqn. (4.6),

$$5 = \sqrt{(I_1^2 + I_3^2)} \quad . \quad . \quad . \quad \text{(iii)}$$

Similarly

$$20 = \sqrt{(V_1^2 + V_3^2)} \quad . \quad . \quad . \quad \text{(iv)}$$

Substituting from eqns. (i) and (ii) in eqn. (iv),

$$20 = \sqrt{[(3{\cdot}16I_1)^2 + (9{\cdot}05I_3)^2]} \quad . \quad . \quad . \quad \text{(v)}$$

Hence, from eqn. (iii), $25 = I_1^2 + I_3^2$ or $I_1^2 = 25 - I_3^2$
and, from eqn. (v), $400 = 10I_1^2 + 82I_3^2 = 250 - 10I_3^2 + 82I_3^2$.

Hence $\underline{I_3 = 1{\cdot}44 \text{ A}}$

and thus $I_1 = \sqrt{(25 - 1{\cdot}44^2)} = \underline{\underline{4{\cdot}8 \text{ A}}}$

Note that I_1 lags behind V_1 by 73·3°, while I_3 lags behind V_3 by 83·7°. In this case the power dissipated is simply $I^2R = 25 \times 1 = \underline{\underline{25 \text{ W}}}$.

Example 4.2. Explain what is meant by harmonic resonance in a.c. circuits.

A current having an instantaneous value of $2(\sin \omega t + \sin 3\,\omega t)$ amperes is passed through a circuit which consists of a coil of resistance R and inductance L in series with a capacitor C. Derive an expression for the value of ω at which the r.m.s. circuit voltage is a minimum. Determine this voltage if the coil has inductance 0·1 H and resistance 150 Ω and the capacitance is 10 μF. Determine also the circuit voltage at the fundamental resonant frequency. (*A.E.E.*, 1961)

The impedance of the circuit at the fundamental frequency is

$$Z_1 = R + j(\omega L - 1/\omega C)$$

and its magnitude is $\sqrt{[R^2 + (\omega L - 1/\omega C)^2]}$.

Hence the peak fundamental voltage magnitude is

$$V_{1m} = |I_{1m}Z_1| = 2\sqrt{[R^2 + (\omega L - 1/\omega C)^2]}$$

In the same way, the magnitude of the peak third-harmonic voltage is

$$V_{3m} = |I_{3m}Z_3| = 2\sqrt{[R^2 + (3\omega L - 1/3\omega C)^2]}$$

Hence the r.m.s. circuit voltage is

$$V = \sqrt{\left(\frac{V_{1m}^2}{2} + \frac{V_{3m}^2}{2}\right)}$$

giving $V^2 = 2[R^2 + (\omega L - 1/\omega C)^2 + R^2 + (3\omega L - 1/3\omega C)^2]$. (i)

If V, and hence V^2, has a minimum value as ω varies, then

$$\frac{dV^2}{d\omega} = 0 = 2[2(\omega L - 1/\omega C)(L + 1/\omega^2 C) + 2\,(3\omega L - 1/3\omega C)(3L + 1/3\omega^2 C)]$$

Hence, $\omega^2L^2 - 1/\omega^2C^2 + 9\omega^2L^2 - 1/9\omega^2C^2 = 0$

so that $$10\omega^2L^2 = 10/9\omega^2C^2$$

and $$\omega = {}^4\sqrt{(1/9\omega L^2C^2)} = \frac{1}{\sqrt{(3LC)}}$$

The value of V^2 at this value of ω is, from eqn. (i),

$$V^2 = 2\left[2R^2 + \omega^2L^2\left(1 - \frac{1}{\omega^2LC}\right)^2 + 9\omega^2L^2\left(1 - \frac{1}{9\omega^2LC}\right)^2\right]$$

$$= 2\left[2R^2 + \frac{L^2}{3LC}\left(1 - \frac{3LC}{LC}\right)^2 + \frac{9L^2}{3LC}\left(1 - \frac{3LC}{9LC}\right)^2\right]$$

$$= 2\left(2R^2 + \frac{8L}{3C}\right)$$

Substituting the given values,

$$V^2 = 2(45{,}000 + 26{,}670) = 143{,}340$$

so that V = 378 V = minimum circuit voltage.

At the fundamental resonant frequency, $\omega L = 1/\omega C$, and hence from eqn. (i),

$$V^2 = 2[2R^2 + (3\omega L - 1/3\omega C)^2]$$
$$= 2(45{,}000 + 71{,}200)$$

Hence V = 482 V at the fundamental resonant frequency.

In reactive circuits the percentage harmonic content of the applied voltage will differ from that of the current because reactance varies with frequency. Errors will arise if the impedance of such a circuit is taken to be the quotient of r.m.s. voltage and current, as the following example will show.

Example 4.3. A current given by

$$i = 2 \sin \omega t + 0{\cdot}5 \sin 3\omega t + 0{\cdot}2 \sin 5\omega t$$

passes through a coil of resistance 10 Ω and inductance 31·8 mH. The fundamental frequency is 50 c/s. Deduce an expression for the applied voltage. Determine the error (as a fraction of the impedance at 50 c/s) if the impedance of the circuit is taken as the quotient of r.m.s. voltage and current.

For the fundamental, $Z_1 = 10 + j10 = 14{\cdot}14\underline{/45^\circ}$, so that

$$V_{1m} = I_{1m}Z_1 = 28{\cdot}28\underline{/45^\circ}$$

For the third harmonic, $Z_3 = 10 + j30 = 31{\cdot}6\underline{/71{\cdot}5^\circ}$, so that

$$V_{3m} = I_{3m}Z_3 = 15{\cdot}8\underline{/71{\cdot}5^\circ}$$

For the fifth harmonic, $Z_5 = 10 + j50 = 51\underline{/78{\cdot}7°}$, so that

$$V_{5m} = I_{5m}Z_5 = 10{\cdot}2\underline{/78{\cdot}7°}$$

The expression for the applied voltage follows as

$$v = 28{\cdot}28\sin(\omega t + 45°) + 15{\cdot}8\sin(3\omega t + 71{\cdot}5°) + 10{\cdot}2\sin(5\omega t + 78{\cdot}7°)$$

To find the apparent impedance with the applied complex voltage we obtain the r.m.s. values of the voltage and current. Thus

$$I = \sqrt{\left(\frac{4}{2} + \frac{0{\cdot}25}{2} + \frac{0{\cdot}04}{2}\right)} = 1{\cdot}47\ \text{A}$$

and
$$V = \sqrt{\left(\frac{28{\cdot}28^2}{2} + \frac{15{\cdot}8^2}{2} + \frac{10{\cdot}2^2}{2}\right)} = 24\ \text{V}$$

so that the apparent impedance is

$$Z_{app} = \frac{V}{I} = \frac{24}{1{\cdot}47} = 16{\cdot}4\ \Omega$$

The impedance at the fundamental frequency is $Z_1 = 14{\cdot}14\ \Omega$, so that the fractional error is

$$\frac{16{\cdot}4 - 14{\cdot}14}{14{\cdot}14} = \underline{\underline{0{\cdot}16}}$$

Note that while the current has 25 per cent third and 10 per cent fifth harmonic, the voltage wave has 56 per cent third and 36 per cent fifth harmonic.

4.4. Mathematical Relations for Repetitive Waveforms

Many waveforms which are met in practice can be represented by simple mathematical expressions and the magnitudes and relative phases of their harmonic components can conveniently be found by Fourier analysis. Numerical methods are used to analyse waveforms for which simple mathematical expressions cannot be obtained. In this section some of the more common simple waveforms will be examined.

One form of the Fourier series expansion of a periodic function of time, $f(t)$, of period T (i.e. $f(t) = f(t + T)$) is

$$f(t) = a_0 + \sum_{n=1}^{\infty} (a_n \cos n\omega t + b_n \sin n\omega t) \quad . \quad (4.12)$$

provided that $f(t)$ has a finite number of discontinuities and that

$$\int_{-T/2}^{T/2} |f(t)|\ dt$$

exists (conditions which are inevitably met in electrical engineering practice), where*

$$a_0 = \frac{1}{T}\int_{-T/2}^{T/2} f(t)\,\mathrm{d}t \qquad (4.13)$$

$$a_n = \frac{2}{T}\int_{-T/2}^{T/2} f(t)\cos n\omega t\,\mathrm{d}t \qquad (4.14)$$

$$b_n = \frac{2}{T}\int_{-T/2}^{T/2} f(t)\sin n\omega t\,\mathrm{d}t \qquad (4.15)$$

and

$$\omega = 2\pi/T = 2\pi \times \text{Fundamental frequency}$$

Note that if eqns. (4.12) and (4.1) are alternative expressions for the same time function, then $I_0 = a_0$, $I_{nm} = \sqrt{(a_n^2 + b_n^2)}$, and $\tan\phi_n = a_n/b_n$.

Examination of a complex waveform will often reveal much about the harmonic content, as the following considerations will show.

(i) *D.C. component.* Eqn. (4.13) shows that there will be a d.c. term unless the average value of $f(t)$ over one cycle is zero, i.e. unless the area above the time axis is equal to that below it.

(ii) *Even functions.* An even function is defined as one for which $f(t) = f(-t)$, i.e. it is symmetrical about the chosen zero of the time axis. This is true of eqn. (4.12) only if it contains no sine terms. A typical wave with no sine terms is shown in Fig. 4.2 (*a*).

(iii) *Odd functions.* Odd functions are those for which $f(t) = -f(-t)$. This is true of eqn. (4.12) only if there are no cosine terms and no d.c. component (Fig. 4.2 (*b*)).

(iv) $f(t) = f(t + T/2)$. This represents a wave which repeats after half a cycle of the fundamental. The condition may be expressed as $f(\omega t) = f(\omega t + \pi)$, since $\omega T/2 = \pi$, and hence even harmonics only can be present. Thus

$$a_1 = a_3 = a_5 = \ldots = b_1 = b_3 = b_5 = \ldots = 0$$

A typical wave of this form is shown with a d.c. component in Fig. 4.2 (*c*).

(v) $f(t) = -f(t + T/2)$. This represents a wave for which the positive and negative half-cycles are identical in shape, and hence since $f(\omega t) = -f(\omega t + \pi)$, there can be no even harmonics or d.c. term. Only odd harmonics are present (Fig. 4.2 (*d*)).

* *See* Pedoe, *Advanced National Certificate Mathematics*, Vol. II (E.U.P.), or other mathematics textbooks.

A suitable choice of the zero time position will often simplify a Fourier analysis, by enabling one or more of the above five conditions to be fulfilled. Sometimes alternative expressions are possible; for example, the wave shown in Fig. 4.2 (e) may be analysed as a cosine

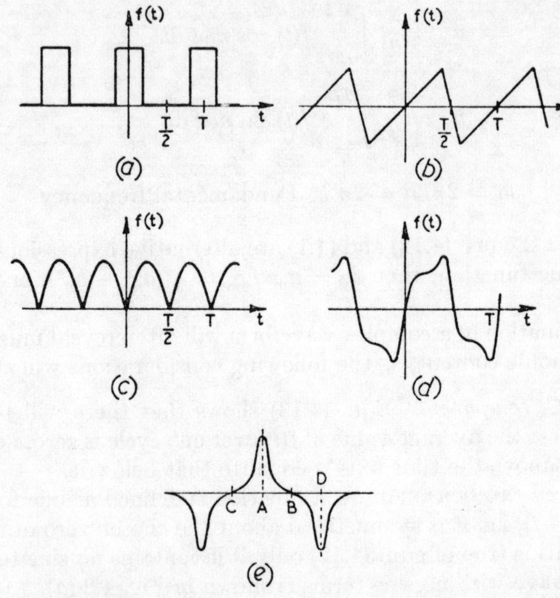

FIG. 4.2. TYPICAL WAVEFORMS AND THEIR HARMONIC COMPONENTS

(a) No sine terms
(b) No cosine terms
(c) D.C. and even harmonics only
(d) Odd harmonics only
(e) Waveforms which may be analysed as a series of odd harmonics

series of odd harmonics if the time zero is taken at A or D, or as a sine series if the time zero is taken at C or B.

Example 4.4. Determine the frequency spectrum of a positive-going periodic pulse train of period T, pulse width τ, and pulse height h.

If the zero of time is chosen at the centre of the pulse as shown in Fig. 4.3(a), the Fourier expansion will contain no sine terms, since $f(t)$ will then be an even function of time.

The d.c. component is evaluated from eqn. (4.13) as

$$a_0 = \frac{1}{T}\int_{-T/2}^{T/2} f(t)\,\mathrm{d}t = \frac{1}{T}\int_{-\tau/2}^{\tau/2} h\,\mathrm{d}t = \frac{h\tau}{T}$$

since $f(t)$ is zero except between $-\tau/2$ and $\tau/2$ in one cycle.

The harmonic amplitudes are obtained from eqn. (4.14) as

$$a_n = \frac{2}{T}\int_{-\tau/2}^{\tau/2} h \cos n\omega_1 t \, dt \qquad \text{where} \qquad \omega_1 = 2\pi/T$$

$$= \frac{2h}{Tn\omega_1}\left[\sin\left(\frac{n\omega_1\tau}{2}\right) - \sin\left(-\frac{n\omega_1\tau}{2}\right)\right] = \frac{2h\tau}{T}\frac{\sin(n\omega_1\tau/2)}{n\omega_1\tau/2}$$

Hence
$$f(t) = \frac{h\tau}{T} + \frac{2h\tau}{T}\sum_{n=1}^{\infty}\frac{\sin(n\omega_1\tau/2)}{n\omega_1\tau/2}\cos n\omega t$$

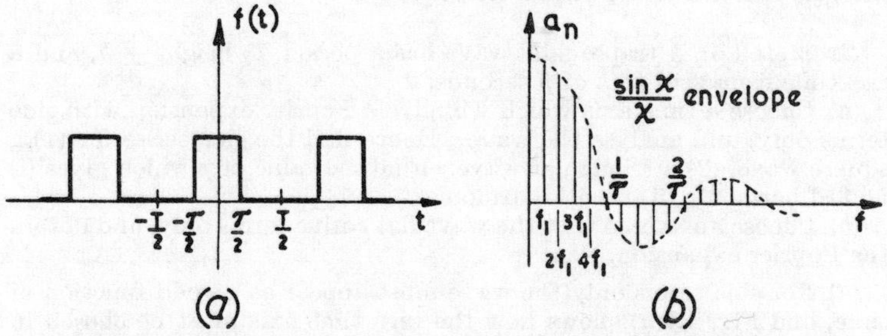

FIG. 4.3. PULSE TRAIN AND ITS LINE FREQUENCY SPECTRUM

From the last equation in Example 4.4 the following important points may be deduced.

1. The magnitudes of the harmonic components vary according to a sin $(n\omega_1\tau/2)/(n\omega_1\tau/2)$ envelope distribution (i.e. a $(\sin x)/x$ distribution), which has zero values when $n\omega_1\tau/2 = k\pi$, i.e. when $f = nf_1 = k/\tau$, where $f_1 = 1/T$, and $k = 1, 2, 3 \ldots$ The points at which zeros occur are thus $1/\tau$, $2/\tau$, $3/\tau$, etc., which do not necessarily correspond to integral values of n. These points depend only on the pulse width, τ, and are independent of the pulse-repetition frequency, $1/T$.
2. The frequency of each component harmonic, $n\omega_1/2\pi$, depends on the pulse-repetition frequency only, and not on the pulse width.
3. An approximation to the bandwidth required for such a signal is from zero to $1/\tau$ cycles per second.

The line frequency spectrum sketched in Fig. 4.3 (*b*) gives the magnitude of each harmonic component. This frequency spectrum is usually drawn with magnitudes only plotted against frequency.

If the mark/space ratio of a pulse train is $1/m$, then $\tau = T/(1 + m)$, and zeros will occur at frequencies of $f = k(1 + m)/T$. Thus for a 1:1 mark/space ratio the zeros will occur at frequencies of $k2/T$, i.e. at all even-harmonic frequencies. The spectrum will contain only odd harmonics. This agrees with consideration (v) on page 115. If the mark/space ratio is 1:2, zeros will occur when $f = 3k/T$, i.e. there will be no third or integral multiples of the third harmonic (no triplen harmonics).

A final point to notice is that, if the period is increased until only a single pulse remains ($T \to \infty$), the frequency separation of the harmonics tends to zero. The discrete frequency line spectrum becomes a continuous frequency spectrum, given by a Fourier integral instead of a Fourier series.

Example 4.5. A trapezoidal wave has a period T, height $\pm h$, and a rise time from zero to h of p seconds.

(*a*) Choose a time axis which will give a Fourier expansion with sine terms only, and analyse the wave. Hence find the sine series for (1) a square wave, (2) a triangular wave. Find the value of p which gives (i) no 3rd harmonic (ii) no 5th harmonic.

(*b*) Choose an axis so that the wave has cosine terms only, and obtain the Fourier expansion.

(*a*) For sine terms only the wave must appear as an odd function of time, and Fig. 4.4(*a*) shows how the zero time axis must be chosen in order that this shall occur.

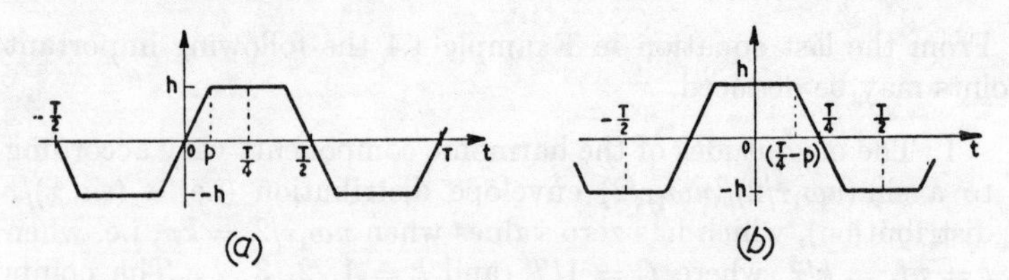

FIG. 4.4. THE TRAPEZOIDAL WAVE: CHOICE OF TIME ZERO

(*a*) Odd function of time
(*b*) Even function of time

Examination of the waveform shows that $f(t) = -f(t + T/2)$, and hence only odd harmonics are present. The mathematical representation of the trapezoid over one cycle from $-T/2$ to $T/2$ is

$$f(t) = -\frac{h}{p}(t + T/2) \qquad -T/2 < t < (-T/2 + p)$$

$$f(t) = -h \qquad (-T/2 + p) < t < -p$$

$$f(t) = \frac{h}{p}t \qquad -p < t < p$$

$$(t) = h \qquad p < t < (T/2 - p)$$

$$f(t) = -\frac{h}{p}(t - T/2) \qquad (T/2 - p) < t < T/2$$

In the Fourier expansion $a_0 = a_n = 0$. From eqn. (4.15),

$$b_n = \frac{2}{T}\int_{-T/2}^{T/2} f(t) \sin n\omega t \, dt$$

$$= \frac{2}{T} 4 \int_0^{T/4} f(t) \sin n\omega t \, dt \qquad \text{since } f(t) \sin n\omega t \text{ is the same shape in each quarter-cycle}$$

$$= \frac{8}{T}\left\{\int_0^p \frac{ht}{p} \sin n\omega t \, dt + \int_p^{T/4} h \sin n\omega t \, dt\right\}$$

$$= \frac{8h}{T}\left\{\left[\left(\frac{-t}{pn\omega}\right)\cos n\omega t\right]_0^p + \int_0^p \frac{\cos n\omega t}{pn\omega} dt - \left[\frac{\cos n\omega t}{n\omega}\right]_p^{T/4}\right\}$$

$$= \frac{4h}{\pi n} \frac{\sin n\omega p}{n\omega p}$$

Hence
$$f(t) = \frac{4h}{\pi} \sum_{n=2q+1}^{\infty} \frac{1}{n} \frac{\sin n\omega p}{n\omega p} \sin n\omega t$$

for $q = 0, 1, 2$, i.e. for all odd integral values of n.

(1) For a square wave $p \to 0$, so that $\dfrac{\sin n\omega p}{nwp} \to 1$

and thus

$$f(t) = \frac{4h}{\pi} \sum \frac{1}{n} \sin n\omega t = \frac{4h}{\pi}\left(\sin \omega t + \frac{1}{3} \sin 3\omega t + \ldots\right)$$

(2) For a triangular wave $p = T/4$, and hence $\dfrac{\sin n\omega p}{n\omega p} = \dfrac{\sin n\pi/2}{n\pi/2}$

so that
$$f(t) = \frac{8h}{\pi^2} \sum \frac{\sin n\pi/2}{n^2} \sin n\omega t$$
$$= \frac{8h}{\pi^2}\left(\sin \omega t - \frac{1}{9} \sin 3\omega t + \frac{1}{25} \sin 5\omega t - \ldots\right)$$

Note that these results may be obtained by direct analysis of square or triangular waves.

For the trapezoidal wave, the third harmonic will be absent if $\sin 3\omega p = 0$, provided that $p \neq 0$ (since $\dfrac{\sin n\omega p}{nwp}$ is then 1);

i.e. if
$$3\omega p = \pi, 2\pi, \ldots$$

or
$$p = \frac{T}{6}, \frac{2T}{6}, \ldots \text{ etc.}$$

Only the first solution is physically possible.

The fifth harmonic will be absent if

$$\sin 5\omega p = 0 \quad \text{and} \quad p \neq 0 \quad \text{i.e.} \quad p = \frac{T}{10} \quad \text{or} \quad \frac{T}{5} \text{ etc.}$$

(*b*) If the zero of the time axis is chosen as shown in Fig 4.4(*b*), the expression for $f(t)$ becomes an even function of time, and hence the expansion contains cosine terms only. Inspection of the waveform reveals that again only odd harmonics will be present, with no d.c. term. It is left for the reader to verify that the expression for a_n is

$$(4h/n\pi)\sin(n\pi/2)\frac{\sin n\omega p}{n\omega p}$$

and hence that the cosine series for a rectangular wave is

$$f(t)=\frac{4h}{\pi}\left(\cos\omega t-\frac{1}{3}\cos 3\omega t+\frac{1}{5}\cos 5\,\omega t-\ldots\right)$$

and for a triangular wave is

$$f(t)=\frac{8h}{\pi^2}\left(\cos\omega t+\frac{1}{9}\cos 3\,\omega t+\frac{1}{25}\cos 5\omega t+\ldots\right)$$

Example 4.6. Find the frequency spectrum of a half-wave rectified sine wave of peak value V_m.

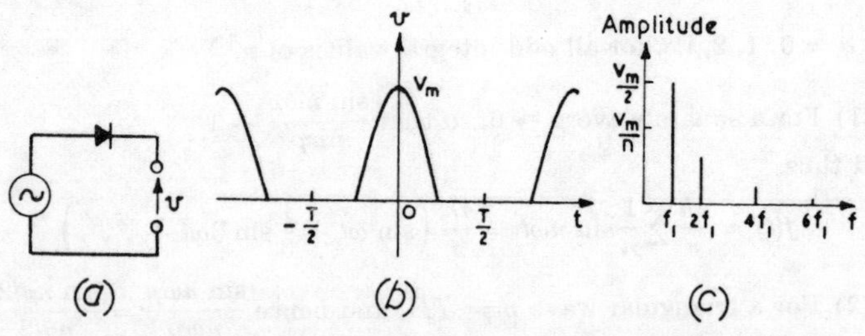

FIG. 4.5. SPECTRUM OF A HALF-WAVE RECTIFIED SINE WAVE

If the origin is chosen to coincide with a peak of the output voltage wave (Fig. 4.5(*b*)) the function will be even and there will be no sine terms in the Fourier expansion. From eqn. (4.13),

$$a_0=\frac{1}{T}\int_{-T/2}^{T/2}f(t)\,dt=\frac{1}{T}\int_{-T/4}^{T/4}V_m\cos(2\pi t/T)\,dt=\frac{V_m}{\pi}$$

since in the interval from $-T/2$ to $T/2$ the function $f(t)$ is zero except between $-T/4$ and $T/4$.

For the nth harmonic,

$$a_n=\frac{2}{T}\int_{-T/4}^{T/4}V_m\cos\frac{2\pi t}{T}\cos\frac{n2\pi t}{T}\,dt$$

$$=\frac{V_m}{T}\int_{-T/4}^{T/4}\left[\cos\frac{(n-1)2\pi t}{T}+\cos\frac{(n+1)2\pi t}{T}\right]dt$$

$$=\frac{V_m}{\pi(n-1)}\sin(n-1)\pi/2+\frac{V_m}{\pi(n+1)}\sin(n+1)\,\pi/2$$

If $n=1$, $$a_1=\underset{(n-1)\to 0}{\text{Limit}}\left[\frac{V_m}{\pi}\frac{\sin(n-1)\pi/2}{n-1}\right]=\frac{1}{2}V_m$$

For all other odd values of n, $a_n = 0$

For $n = 2$, $$a_2 = \frac{V_m}{\pi} - \frac{V_m}{3\pi} = \frac{2}{3}\frac{V_m}{\pi}$$

For $n = 4$, $$a_4 = -\frac{V_m}{3\pi} + \frac{V_m}{5\pi} = -\frac{2}{15}\frac{V_m}{\pi}$$

Hence $$f(t) = \frac{V_m}{\pi} + \frac{1}{2}V_m \cos \omega t + \frac{2}{3}\frac{V_m}{\pi}\cos 2\omega t - \frac{2}{15}\frac{V_m}{\pi}\cos 4\omega + \ldots$$

The frequency spectrum is shown in Fig. 4.5(c).

4.5. Form Factor of a Complex Wave

In power-frequency networks, calculations are generally made in terms of r.m.s. values, and these true r.m.s. values are measured by means of moving-iron or electrodynamic instruments. Rectifier instruments, on the other hand, have deflexions proportional to the average value of the rectified current over one cycle, and if these instruments are to be scaled in r.m.s. values, a constant ratio between mean and r.m.s. must be assumed. This leads to errors if the actual ratio differs from the assumed one—the actual errors are generally well below 10 per cent.

The waveforms produced by a.c. machines will usually have identical positive and negative half-cycles; hence if harmonics are present these will only be odd ones. For this case it is convenient to define the form factor, k_f, of an a.c. wave as the ratio of the r.m.s. value of the wave to its mean value taken over a half-period beginning at a zero point.

For a pure sine wave of peak value V_m,

$$k_f = \frac{V_m/\sqrt{2}}{(2/\pi)V_m} = 1{\cdot}11$$

Rectifier instruments are scaled with this assumed form factor, and hence, for a wave with a form factor k_f',

$$\text{True r.m.s. reading} = \frac{k_f' \times \text{Rectifier instrument reading}}{1{\cdot}11}$$

It follows that the form factor of a complex wave may be found by comparing the readings on a true r.m.s. indicating instrument and on a rectifier instrument. Thus

$$k_f = 1{\cdot}11 \times \frac{\text{R.M.S. instrument reading}}{\text{Rectifier instrument reading}} \quad . \quad (4.16)$$

If the wave contains d.c. and second-harmonic terms, this equation may still be used to determine a form factor for the wave, but k_f must now be defined as the ratio of r.m.s. to full-wave rectified mean over one cycle (rectifier instruments normally incorporate full-wave rectifiers). Note that the first definition of k_f cannot generally include cases where direct currents or even harmonics are present, since then positive- and negative-going parts of the wave will only exceptionally have the same mean value, and zeros will not normally occur at half-cycle intervals. In some simple cases, however, general expressions for the form factor may be deduced.

(*a*) *Sine series.* Consider the voltage

$$v = V_{1m} \sin \omega t \pm V_{3m} \sin 3\omega t \pm V_{5m} \sin 5\omega t \pm \ldots$$

where $\omega = 2\pi/T$. Zeros occur at $t = 0$ and $t = T/2$. The half-cycle mean is therefore

$$\begin{aligned} V_{av} &= \frac{1}{T/2}\int_0^{T/2} v \, dt \\ &= \frac{2}{T}\left(V_{1m}\int_0^{T/2} \sin \omega t \, dt \pm V_{3m}\int_0^{T/2} \sin 3\omega t \, dt \pm \ldots\right) \\ &= \frac{2}{T}\left(\frac{V_{1m}T}{\pi} \pm \frac{V_{3m}T}{3\pi} \pm \frac{V_{5m}T}{5\pi} \pm \ldots\right) \end{aligned}$$

and hence

$$k_f = \frac{(1/\sqrt{2})\sqrt{(V_{1m}{}^2 + V_{3m}{}^2 + \ldots)}}{(2/\pi)\left(V_{1m} \pm \dfrac{V_{3m}}{3} \pm \dfrac{V_{5m}}{5} \pm \ldots\right)} \quad . \quad (4.17)$$

(*b*) *Cosine series.* Consider the cosine series

$$v = V_{1m} \cos \omega t \pm V_{3m} \cos 3\omega t \pm V_{5m} \cos 5\omega t \pm \ldots$$

Zeros occur at $\pm T/4$ and positive and negative half-cycles are symmetrical. Hence

$$\begin{aligned} V_{av} &= \frac{2}{T}\int_{-T/4}^{T/4} (V_{1m} \cos \omega t \pm V_{3m} \cos 3\omega t \pm V_{5m} \cos 5\omega t \pm \ldots) \, dt \\ &= \frac{2}{\pi}\left(V_{1m} \mp \frac{V_{3m}}{3} \pm \frac{V_{5m}}{5} \mp \ldots\right) \end{aligned}$$

and

$$k_f = \frac{(1/\sqrt{2})\sqrt{(V_{1m}{}^2 + V_{3m}{}^2 + \ldots)}}{(2/\pi)\left(V_{1m} \mp \dfrac{V_{3m}}{3} \pm \dfrac{V_{5m}}{5} \mp \ldots\right)} \quad . \quad . \quad . \quad . \quad (4.18)$$

Example 4.7. Define the term "form factor" and discuss one example of its significance in electrical measurements. Derive a general expression for the form factor of a complex wave which is symmetrical about its maximum value. Hence calculate the form factor of an alternating voltage which is represented by

$$v = 100 \sin 314t + 24 \sin 942t + 10 \sin 1{,}570t$$

(*A.E.E.*, 1958)

A wave which is symmetrical about its maximum value and for which the time zero coincides with this maximum is represented by an even time function, i.e. by a Fourier expansion containing cosine terms only. Hence the theory given for the cosine series applies, and k_f is given by eqn. (4.18).

For a wave given by

$$v = 100 \sin 314t + 24 \sin 3\omega t + 10 \sin 5\omega t$$

the time zero coincides with a voltage zero. If the time scale is moved by a quarter-cycle of the fundamental, the time zero will coincide with the maximum of the fundamental. The expression for v then becomes

$$\begin{aligned} v &= 100 \sin (\omega t + \pi/2) + 24 \sin 3(\omega t + \pi/2) + 10 \sin 5(\omega t + \pi/2) \\ &= 100 \cos \omega t - 24 \cos 3\omega t + 10 \cos 5\omega t \end{aligned}$$

Taking into account the negative sign of the third-harmonic term, eqn. (4.18) gives the half-cycle average as

$$V_{av} = \frac{2}{\pi}\left(100 + \frac{24}{3} + \frac{10}{5}\right) = 70{\cdot}2 \text{ V}$$

The r.m.s. value is

$$V = (1/\sqrt{2})\sqrt{(100^2 + 24^2 + 10^2)} = 72{\cdot}7 \text{ V}$$

Hence $$k_f = \frac{72{\cdot}7}{70{\cdot}2} = \underline{\underline{1{\cdot}036}}$$

An example of the application of form-factor variation is found in the separation of iron losses into hysteresis and eddy-current components. Below the knee of the magnetization curve, the hysteresis loss for a maximum flux density B_m and a frequency f is given by

$$P_h = k_h f B_m{}^n \text{ watts}$$

where k_h is a constant for a given material, and n is also constant at about 1·6.

The eddy-current loss depends on the resistivity of the material, and on the induced e.m.f. in the iron. The average value of this e.m.f., V_{av}, will depend on the average flux linked per second, and since $4B_mA$ webers are changed per cycle, where A is the iron cross-sectional area,

$$V_{av} = 4B_mAf$$

Hence the r.m.s. induced e.m.f. is

$$V = k_f \times 4 B_m A f$$

and the eddy-current loss is

$$P_e \ (\propto V^2) = k_e k_f^2 B_m^2 f^2 \text{ watts}$$

where k_e is a constant which depends on the dimensions and composition of the material.

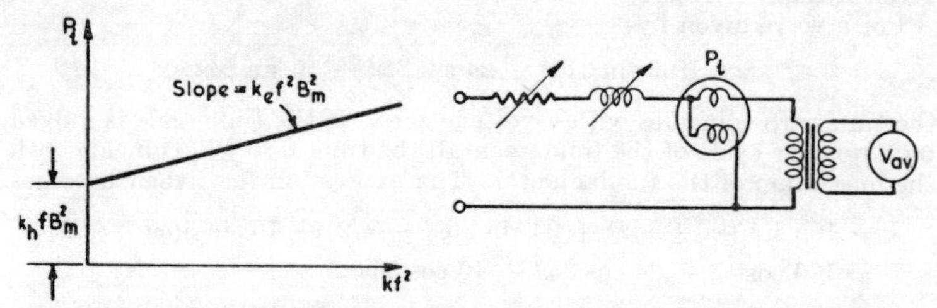

FIG. 4.6. IRON-LOSS SEPARATION BY VARIATION OF FORM FACTOR

The total iron loss, P_i, is then

$$P_i = P_h + P_e = k_h f B_m^n + k_e k_f^2 f^2 B_m^2$$

Hence if the total iron loss is measured for constant f and B_m (i.e. constant V_{av} as measured on a rectifier voltmeter) while the form factor of the flux is varied by varying the flux waveform by means of a variable series inductance and resistance, and if this total loss is plotted to a base of k_f^2, the linear graph which results enables the loss components to be separated (Fig. 4.6).

4.6. Superposition of Non-harmonically Related Currents

If currents of differing frequencies are applied to linear circuits, the principle of superposition applies, and the effect of each current may be calculated independently. If the combined currents are measured on a true r.m.s. indicating instrument, then, provided that the frequency separation is not small, the instrument will record the square root of the sum of the squares of each individual current. If, however, the frequency separation is small, so that the period of the difference frequency is greater than the response time of the instrument, the indication will fluctuate about the r.m.s. value.

Suppose that currents $i_1 = I_{1m} \sin \omega t$ and $i_2 = I_{2m} \sin (\omega + \delta\omega)t$

are passed through a true r.m.s. indicating ammeter. The square of the instantaneous current will be

$$
\begin{aligned}
i^2 &= (i_1 + i_2)^2 \\
&= I_{1m}^{\ 2} \sin^2 \omega t + 2I_{1m}I_{2m} \sin \omega t \sin (\omega + \delta\omega)t \\
&\qquad\qquad\qquad\qquad + I_{2m}^{\ 2} \sin^2 (\omega + \delta\omega)t \\
&= \frac{I_{1m}^{\ 2}}{2}(1 - \cos 2\omega t) + I_{1m}I_{2m}[\cos \delta\omega t - \cos (2\omega + \delta\omega)t] \\
&\qquad\qquad\qquad\qquad + \frac{I_{2m}^{\ 2}}{2}[1 - \cos 2(\omega + \delta\omega)t]
\end{aligned}
$$

If the response of the instrument is such that it cannot follow variations at angular frequencies of ω or $\omega + \delta\omega$, the instantaneous deflexion, θ, will be

$$\theta \propto I^2 = \frac{I_{1m}^{\ 2}}{2} + I_{1m}I_{2m} \cos \delta\omega t + \frac{I_{2m}^{\ 2}}{2}$$

If $\delta\omega$ is also large compared to the angular frequency which the instrument can follow, this becomes, simply,

$$\theta \propto \frac{I_{1m}^{\ 2}}{2} + \frac{I_{2m}^{\ 2}}{2}$$

and the indication can be calibrated in true r.m.s. values. If, however, the instrument can follow variations at the angular frequency $\delta\omega$, the deflexion will vary between values proportional to

$$\frac{I_{1m}^{\ 2}}{2} + \frac{I_{2m}^{\ 2}}{2} + I_{1m}I_{2m} \quad \text{and} \quad \frac{I_{1m}^{\ 2}}{2} + \frac{I_{2m}^{\ 2}}{2} - I_{1m}I_{2m}$$

at a frequency of $\delta\omega/2\pi$. This is called the *beat frequency*. Obviously if $\delta\omega = 0$ there is no beat-frequency variation.

A similar situation exists if the current and voltage coils of an electrodynamic wattmeter are fed from sources of slightly different frequency. As one frequency is varied through the other, the pointer will oscillate at the beat frequency. As the voltage coil frequency is swept through the harmonic of the current, the instantaneous torque will be

$$
\begin{aligned}
\text{Torque} &\propto V_m \sin \omega_v t \times I_{nm} \sin (n\omega t + \phi_n) \\
&\propto V_m I_{nm} \tfrac{1}{2}[\cos (n\omega t - \omega_v t + \phi_n) - \cos (n\omega t + \omega_v t + \phi_n)]
\end{aligned}
$$

The second term in this proportionality has a period which is too small to produce any response from the instrument. If ω_v is nearly equal to $n\omega$, the pointer will oscillate at the beat frequency and the peak indication will be proportional to I_{nm}. The mean torques produced by the interaction of the voltage coil flux with all the other harmonics of the current will be zero. Note that, if $\omega_v = n\omega$, an error is possible due to the possible phase difference between V and I_n—it is the maximum of the beat frequency oscillation which gives a true indication of relative harmonic amplitude.

4.7. Non-linear Circuit-elements

A non-linear circuit-element is one where the impedance varies with the applied voltage. If a sinusoidal voltage is applied to such an element, the current wave will contain harmonics. Non-linear

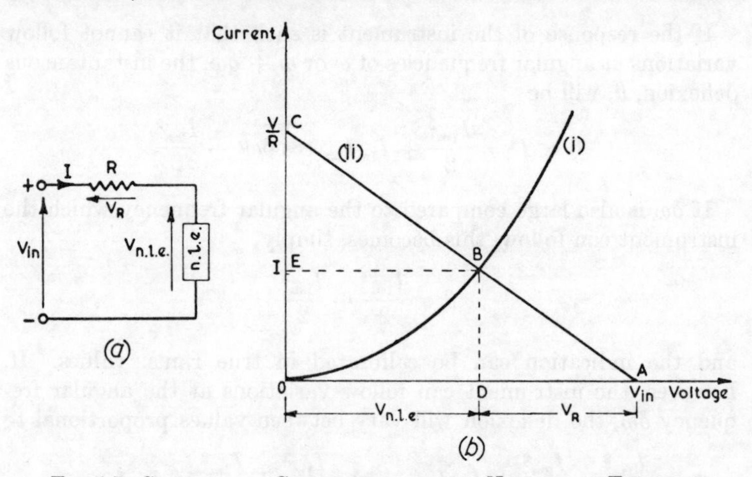

FIG. 4.7. CURRENT IN A CIRCUIT CONTAINING A NON-LINEAR ELEMENT (n.l.e.)

resistors, capacitors and inductors can be made, and indeed all iron-cored coils are non-linear. Certain materials, such as ceramics containing silicon carbide (e.g. Thyrite), have non-linear resistance. The depletion-layer capacitance of a *p-n* semiconductor junction diode varies with voltage. Applications of such devices include stabilization and protection of voltage or current sources (non-linear resistors), frequency modulation and parametric amplifier techniques (non-linear capacitors) and control systems (non-linear inductors).

With non-linear resistors a relation of the form $i = kv^n$ may normally be assumed to exist between current and voltage. The question arises: What is the resultant current when a voltage is applied across a non-linear element (n.l.e.) in series with a linear element, or across two different non-linear elements in series? Consider the circuit of Fig. 4.7 (*a*), consisting of a resistor R in series with an n.l.e. whose I/V characteristic is shown as curve (i) in Fig. 4.7 (*b*). From A ($OA = V_{in}$) a *load line AC* (ii) is drawn to the point, C where $I = V/R$ (i.e. the same construction as that used to draw the load line of a transistor or valve amplifier circuit). The point B where curve (i) cuts line (ii) is the d.c. operating point of the circuit. The resultant current is $I = OE$ and the voltage across the resistor is AD, while that across the n.l.e. is OD.

If two n.l.e.s are connected in series the construction is similar, but this time the characteristic of the second n.l.e. is drawn from a

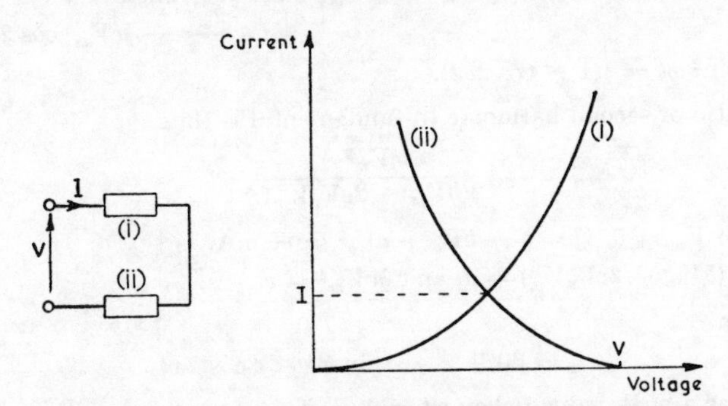

FIG. 4.8. TWO NON-LINEAR ELEMENTS IN SERIES

point on the voltage axis representing the supply voltage but in the reverse direction (i.e. towards zero). The intersection of the two curves gives the operating current (Fig. 4.8).

When a varying voltage is applied to a resistor and an n.l.e. in series, the instantaneous current is found by drawing the resistance line from points on the voltage axis representing various instantaneous voltages. If these currents are plotted on a time axis the waveform of the current can be found. Alternatively, if the equation of the n.l.e. is known, the analytical expression for the current may be simply derived.

Frequently circuits containing active elements have a non-linear current/voltage relation. This applies particularly to power amplifiers where the relation may often be expressed as a power series. The distortion of a signal due to this effect is termed *non-linear distortion.*

Example 4.8. The dynamic I_a/V_g characteristic of a triode with a resistive load is given by

$$i_a = a + bv_g + cv_g^2$$

Deduce the ratio of second-harmonic component in the output to the fundamental.

If the valve has a steady no-signal anode current of 60 mA and the application of a sinusoidal grid voltage causes the anode current to vary between 105 and 25 mA, calculate the percentage second harmonic in the output current. (*A.E.E.*, 1957)

Assume that the grid signal is given by a steady voltage (the grid bias voltage) of $-V_g$ upon which is superimposed a sinusoidal signal $V_m \sin \omega t$. Then

$$\begin{aligned} i_a &= a + b(-V_g + V_m \sin \omega t) + c(-V_g + V_m \sin \omega t)^2 \\ &= a - bV_g + cV_g^2 + (bV_m - 2cV_gV_m) \sin \omega t + cV_m^2 \sin^2 \omega t \\ &= a - bV_g + cV_g^2 + \tfrac{1}{2}cV_m^2 + (bV_m - 2cV_gV_m) \sin \omega t \\ &\qquad - \tfrac{1}{2}cV_m^2 \cos 2\omega t \end{aligned}$$

since $\sin^2 \omega t = \frac{1}{2}(1 - \cos 2\omega t)$.

The ratio of second harmonic to fundamental is thus

$$\frac{cV_m^2}{2(bV_m - 2cV_gV_m)}$$

With $V_m = 0$, $i_a = a - bV_g + cV_g^2 = 60$ mA.

Let $(bV_m - 2cV_gV_m) = d$, and $\frac{1}{2}cV_m^2 = e$.

Then

$$i_a = 60 + e + d \sin \omega t - e \cos 2\omega t$$

This has a peak value (when $\omega t = 90°$) of

$$I_{max} = 60 + d + 2e$$

and a minimum value (when $\omega t = 270°$) of

$$I_{min} = 60 - d + 2e$$

Hence $$I_{max} + I_{min} = 120 + 4e = 130$$

since $I_{max} = 105$ and $I_{min} = 25$; and

$$I_{max} - I_{min} = 30 = 2d$$

This gives $e = 2{\cdot}5$ and $d = 40$, so that the percentage second harmonic is

$$\frac{2{\cdot}5}{40} \times 100 = \underline{\underline{6{\cdot}25 \text{ per cent}}}$$

Bibliography

Shepherd, Morton, Spence, *Higher Electrical Engineering* (Pitman).
Goldman, *Frequency Analysis, Modulation, and Noise* (McGraw-Hill).
Shea, *Transmission Networks and Wave Filters* (Van Nostrand).
Golding and Widdis, *Electrical Measurements* (Pitman).

Problems

4.1. Derive the theory for the use of electrodynamic instruments to ascertain the harmonic content of a non-sinusoidal current waveform. Give a circuit diagram and indicate how relative magnitudes of the harmonics are determined. (*I.E.E. Meas.*, 1956)

4.2. A rectifying device produces a square-topped wave of current of equal conducting and non-conducting periods of 10 msec. The value during the conducting period is 10 A. Determine the r.m.s. values of the two harmonic components having frequencies nearest to 500 c/s.

(*Ans.* 0·5 A; 0·41 A.) (*I.E.E. Meas.*, 1957)

4.3. Show from first principles that the r.m.s. value of a complex current or voltage is obtained from the square root of the sum of the mean square values of all the components.

A 50-c/s current which is represented by

$$i = 20 \sin 314t + 8 \sin 942t + 5 \sin 1570t$$

is passed through a coil of negligibly small resistance. The current in the coil and the voltage across it are then measured by electrodynamic instruments. Determine the indicated current. If the measured terminal voltage is 40 V find the percentage error in the calculated value of the inductance when its reactance is found directly from the readings of the instruments mentioned above. (*A.E.E.*, 1957)

(*Ans.* 15·65 A, 44·8 per cent.)

4.4. Define the term "form factor" and give an example of its use in electrical measurements.

Derive a general expression for the form factor of a complex wave containing only odd-order harmonics. Hence calculate the form factor of the alternating current represented by

$$i = 2{\cdot}5 \sin 157t + 0{\cdot}7 \sin 471t + 0{\cdot}4 \sin 785t$$

(*Ans.* 1·038.) (*A.E.E.*, 1960)

4.5. Explain why an inspection of a complex waveform enables the presence of (*a*) even harmonics and (*b*) a steady component to be determined. An r.m.s. current of 7 A containing a third-harmonic component flows in a coil which has an inductance of 25 mH and a resistance of 6 Ω. The fundamental frequency is 50 c/s and the r.m.s. voltage across the coil is 80 V. Calculate the magnitudes of the fundamental and harmonic components of the current. (*A.E.E.*, 1959)

(*Ans.* 6·75 A, 1·81 A.)

4.6. An e.m.f. given by

$$e = 100 \sin \omega t + 40 \sin \left(3\omega t - \frac{\pi}{6}\right) + 10 \sin \left(5\omega t - \frac{\pi}{3}\right) \quad \text{volts}$$

is applied to a series circuit having a resistance of 100 Ω, an inductance of 40·6 mH and a capacitance of 10 μF. Derive an expression for the current in the circuit. Also find the r.m.s. value of the current and the power dissipated in the circuit. Take $\omega = 314$ rad/sec.

(*Ans.* 0·329 A, 10·8 W.) (*L.U. Part II Theory*, 1960)

4.7. The voltage applied to the primary winding of a small unloaded single-phase transformer is given by

$$v = 200 \sin \omega t - 100 \cos 3\omega t \quad \text{volts}$$

The uniform cross-section of iron in the transformer core is 10 cm², the mean length of magnetic path is 40 cm, the primary winding has 500 turns, and the frequency of the fundamental component of the applied voltage is 50 c/s. The resistance and leakage reactance of the winding are to be neglected.

Derive an expression for the core flux density in relation to time. Sketch the waveforms of the applied voltage and flux density in their correct relative phase positions, and estimate approximately the peak flux density in the core.

If the magnetic field intensity in amperes per metre acting on the core material due to the magnetizing current, neglecting higher harmonics, is given by

$$H = 150 \sin(\omega t - 60°) - 60 \cos(3\omega t - 15°)$$

calculate the total iron loss in the core. (*A.E.E.*, 1962)
(*Ans.* 1·36 Wb/m², 8·34 W.)

4.8. A coil having resistance and inductance carries a known periodic non-sinusoidal current. Derive an expression for the factor by which the impedance (referred to fundamental frequency) is apparently increased when its value is taken as the quotient of the r.m.s. voltage and the r.m.s. current. Assume the current to contain a fundamental and a third-harmonic component.

Determine the percentage error in measuring the impedance of a coil in this way when the current contains a 20 per cent fifth harmonic and a 10 per cent seventh harmonic in addition to the fundamental. The coil has a reactance/resistance ratio of 4/1 at the fundamental frequency.
(*Ans.* 51·2 per cent.) (*A.E.E.*, 1962)

4.9. A non-sinusoidal periodic voltage is applied to a circuit and causes a non-sinusoidal periodic current to flow. Develop an expression for the power dissipated in the circuit.

An uncompensated electrodynamic wattmeter reads 250 W when direct currents of 1·0 A and 0·05 A flow in its current and voltage coils, respectively. Calculate the reading of the wattmeter when a current of

$$10 \sin(314t + 15°) + 5 \sin 628t + 4 \sin 1{,}256t \quad \text{amperes}$$

flows in the current coil and the voltage applied to the voltage-coil circuit is

$$400 \cos(314t - 30°) + 150 \sin(628t + 15°) + 100 \sin 942t \quad \text{volts}$$

The inductance of the voltage-coil circuit is negligible. What is the resistance of the voltage-coil circuit? (*A.E.E.*, 1963)
(*Ans.* 1,779 W, 5,000 Ω.)

4.10. Explain what is meant by the term "non-linear distortion" when applied to an amplifier.

The dynamic I_a/V_g characteristic of a triode with resistive load may be represented by

$$I_a = a + bV_g + cV_g^2$$

Deduce the ratio of the second-harmonic component in the output to the fundamental.

If the valve has a steady no-signal anode current of 60 mA and the application of a sinusoidal grid voltage causes the anode current to vary between 105 and 25 mA calculate the percentage second harmonic in the output current. (*A.E.E.*, 1957)
(*Ans.* 6·25 per cent.)

4.11. Discuss the cause and effects of non-linearity in amplifiers. The output-current/input-voltage characteristic of an amplifier with a resistive load may be represented by the relation

$$i = a + bV + cV^2 \quad \text{milliamperes}$$

Deduce the ratio of the second-harmonic component in the output to the fundamental.

The application of a sinusoidal input voltage to an amplifier causes the output current to vary between the limits of 180 per cent and 45 per cent of the value with no input voltage. Determine the percentage second harmonic in the output current. (*A.E.E.*, 1959)

(*Ans.* 9·2 per cent.)

4.12. A 10-Ω fixed resistor is connected in series with another resistor for which the current/voltage relationship is

$$i = 0{\cdot}002\, v^3$$

where i is in amperes and v is in volts. A sinusoidal voltage of 5 V r.m.s. in series with a direct voltage of 10 V is applied to the combination. Determine, graphically or otherwise, (*a*) the maximum current and (*b*) the voltage across each resistor at maximum current.

(*Ans.* 0·93 A, 7·77 V, 9·3 V.) (*A.E.E.*, 1963)

4.13. Discuss the difficulties that arise in calculations on circuits containing elements which have non-linear current/voltage characteristics.

A 12-Ω resistor is connected in series with a resistor for which the current/voltage relation is

$$i = (5 \times 10^{-2})v + (4 \times 10^{-4})v^2 + (1{\cdot}5 \times 10^{-6})v^3$$

where i is in amperes and v in volts.

If a sinusoidal voltage of 100 V r.m.s. is applied to the combination, determine the two peak values of current. Indicate a procedure by which the current waveform may be obtained. (*A.E.E.*, 1962)

(*Ans.* 5·7 A, 2·6 A.)

4.14. The characteristics of a triode may be represented over the range where v_a is positive and $i_a > 0$ by

$$i_a = 5 \times 10^{-4}\,(v_a + 40\, v_g)^2$$

where i_a is in milliamperes and v_a and v_g are in volts. The input signal $(-10 + 8 \sin \omega t)$ volts is applied between grid and cathode. The anode–cathode voltage is maintained constant at 250 V. Calculate the ratio of the peak to the mean anode current, and sketch to a common base of time the grid-voltage and anode-current waveforms. *Hint.* Find the part of the cycle during which conduction takes place. Outside this range i_a is zero. The mean anode current is found by integrating over the conduction period. (*I.E.E. App. El.*, 1963)

(*Ans.* 5·75.)

4.15. Two alternating currents of frequencies f_1 and f_2 and of equal maximum values are superimposed and passed simultaneously through a hot-wire ammeter. The ammeter has been accurately calibrated by means of direct current. What will it now read? If one frequency can be varied above and below the other, describe and explain the effects observed as the variation is carried out. (*I.E.E. Meas.*, 1954)

4.16. A rectifying device conducts as follows in the forward direction—

V (volts)	0	2·4	4·2	6·3	7·8	8·7	9·6
I (amperes)	0	0·05	0·1	0·2	0·3	0·4	0·6

In the reverse direction conduction is negligible.

The rectifier is connected in series with a resistive load of 15 Ω and a sinusoidal supply of 12 V (r.m.s.). What will be the reading on a high-resistance moving-coil voltmeter connected across the load, and what will be average power developed in the load. (*I.E.E. Eln.*, 1957)
(*Ans.* 2·1 V; 0·81 W.)

CHAPTER 5

Transient Analysis of Networks

WHENEVER the conditions in a physical system in which there is an energy-storage element are changed, some time elapses before the system attains a new steady state. During the change from one steady state to another the system is said to be subject to a transient disturbance. It is the purpose of this chapter to analyse the transient disturbances which can occur in electrical networks including inductive and capacitive elements in which energy is stored. The analysis will be carried out by the method of *Laplace transformation*, which is a development from the work of the brilliant British electrical engineer, Oliver Heaviside. In essence this method transforms the time differential equations which are always associated with transient analysis into algebraic equations. After algebraic manipulation, inverse transforms are used to obtain the time-varying solution.

Similar methods may be employed for the transient analysis of other physical systems containing energy-storage elements, as the chapter on servomechanisms will show.

5.1. The Laplace Transformation

We define the *Laplace transform*, $\bar{f}(s)$, of a function of time, $f(t)$, by the following equation—

$$\bar{f}(s) = \int_0^\infty f(t)\mathrm{e}^{-st}\,\mathrm{d}t \quad . \quad . \quad . \quad (5.1)$$

where s is a parameter (which will generally be complex). The restriction on the value of s is that the integral of eqn. (5.1) must be convergent to zero for $t \to \infty$. If no such value of s exists, the function of time is not transformable. Sometimes $\bar{f}(s)$ is written as $\mathscr{L}[f(t)]$. The quantity $f(t)$ may be considered as the inverse transform of $\bar{f}(s)$, this being written as $f(t) = \mathscr{L}^{-1}[\bar{f}(s)]$.

If $f(t)$ is given, and is not too complicated a function of time, it is usually quite easy to determine $\bar{f}(s)$ from eqn. (5.1). However, if $\bar{f}(s)$ is given, the corresponding function of time is

$$f(t) = \mathscr{L}^{-1}[\bar{f}(s)] = \frac{1}{2\pi j}\int_{\alpha-j\infty}^{\alpha+j\infty} \bar{f}(s)\mathrm{e}^{st}\,\mathrm{d}s \quad . \quad . \quad (5.2)$$

This equation is stated without proof, and is best evaluated as a contour integral. In order to avoid this rather difficult process we shall build a table of Laplace transform pairs (Appendix, p. 507), from which, by inspection, the value of $f(t)$ may be read for the more common values of $\bar{f}(s)$.

Generally the s in $\bar{f}(s)$ will be omitted for brevity. In all the following derivations it is implied that $f(t) = 0$ for $t < 0$. The derivations refer to the Laplace transform pairs of the Appendix.

(1) $f(t) =$ *constant*, A

Then the Laplace transform is

$$\mathscr{L}[A] = \bar{f}_1 = \int_0^\infty A\mathrm{e}^{-st}\,\mathrm{d}t = -\frac{A}{s}\,\mathrm{e}^{-st}\Big]_0^\infty = \frac{A}{s}$$

provided that at $t = \infty$ the integral converges to zero. This gives the transform of a time function which has a value of zero for $t < 0$ and of A for $t \geqslant 0$, i.e. a step function.

(2) $f(t) = \mathrm{e}^{at}$

Then

$$\mathscr{L}[\mathrm{e}^{at}] = \bar{f}_2 = \int_0^\infty \mathrm{e}^{at}\mathrm{e}^{-st}\,\mathrm{d}t = \int_0^\infty \mathrm{e}^{-(s-a)t}\,\mathrm{d}t$$

$$= -\frac{1}{(s-a)}\,\mathrm{e}^{-(s-a)t}\Big]_0^\infty = \frac{1}{s-a}$$

provided that the reference part of s is greater than the reference part of a.

It follows directly that $\mathscr{L}[\mathrm{e}^{-at}] = \dfrac{1}{s+a}$

(3) $f(t) = t$

Then

$$\mathscr{L}[t] = \bar{f}_3 = \int_0^\infty t\,\mathrm{e}^{-st}\,\mathrm{d}t$$

$$= -\frac{t}{s}\,\mathrm{e}^{-st}\Big]_0^\infty - \int_0^\infty -\frac{\mathrm{e}^{-st}}{s}\,\mathrm{d}t \quad \text{(integrating by parts)}$$

$$= \frac{1}{s^2}$$

(Note that $t\mathrm{e}^{-st} = 0$ for $t = \infty$, and for $t = 0$.)

(4) $f(t) = t^n$

Then

$$\mathscr{L}[t^n] = \bar{f}_4 = \int_0^\infty t^n e^{-st}\, dt = -\frac{t^n}{s} e^{-st}\Big]_0^\infty - \int_0^\infty -\frac{nt^{n-1}}{s} e^{-st}\, dt$$

$$= 0 + \frac{nt^{n-1}}{-s^2} e^{-st}\Big]_0^\infty - \int_0^\infty \frac{n(n-1)t^{n-2}}{-s^2} e^{-st}\, dt$$

$$= 0 + 0 + 0 + \ldots - \int_0^\infty \frac{n!}{-s^n} e^{-st}\, dt$$

$$= \frac{n!}{s^{n+1}}$$

(5) $f(t) = \dfrac{t^{n-1}}{(n-1)!}$

It follows from the previous transform that the transform of t^{n-1} is

$$\mathscr{L}[t^{n-1}] = \int_0^\infty t^{n-1} e^{-st}\, dt = \frac{(n-1)!}{s^n}$$

Hence

$$\mathscr{L}\left[\frac{t^{n-1}}{(n-1)!}\right] = \frac{1}{s^n} = \bar{f}_5$$

(6) $f(t) = \sin \omega t$

From Euler's formula, $\sin \omega t = \dfrac{1}{2j}(e^{j\omega t} - e^{-j\omega t})$. Hence

$$\mathscr{L}[\sin \omega t] = \bar{f}_6 = \frac{1}{2j}\int_0^\infty (e^{j\omega t} - e^{-j\omega t})e^{-st}\, dt$$

$$= \frac{1}{2j}\left(\frac{1}{(s-j\omega)} - \frac{1}{(s+j\omega)}\right)$$

$$= \frac{\omega}{s^2 + \omega^2}$$

(7) $f(t) = \cos \omega t$

$$\mathscr{L}[\cos \omega t] = \bar{f}_7 = \frac{1}{2}\int_0^\infty (e^{j\omega t} + e^{-j\omega t})e^{-st}\, dt$$

$$= \frac{1}{2}\left(\frac{1}{s-j\omega} + \frac{1}{s+j\omega}\right) = \frac{s}{s^2 + \omega^2}$$

(8) $f(t) = \sinh at$

Since $\sinh at = \frac{1}{2}(e^{at} - e^{-at})$, we may write

$$\mathscr{L}[\sinh at] = \bar{f}_8 = \frac{1}{2}\int_0^\infty (e^{at} - e^{-at})e^{-st}\,dt$$

$$= \frac{1}{2}\left(\frac{1}{s-a} - \frac{1}{s+a}\right) = \frac{a}{s^2 - a^2}$$

(9) $f(t) = \cosh at$

In the same way

$$\mathscr{L}[\cosh at] = \bar{f}_9 = \frac{1}{2}\int_0^\infty (e^{at} + e^{-at})e^{-st}\,dt = \frac{s}{s^2 - a^2}$$

In some cases more complicated Laplace transform pairs can be obtained by differentiating with respect to s under the integral sign, as shown by the following.

(10) $f(t) = t \sin \omega t$

From transform pair 6,

$$\mathscr{L}[\sin \omega t] = \int_0^\infty \sin \omega t\, e^{-st}\,dt = \frac{\omega}{s^2 + \omega^2}$$

Differentiating both sides with respect to s yields

$$\int_0^\infty \frac{\partial}{\partial s}(\sin \omega t\, e^{-st})\,dt = \int_0^\infty -t \sin \omega t\, e^{-st}\,dt$$

$$= \frac{\partial}{\partial s}\frac{\omega}{(s^2 + \omega^2)} = \frac{-2\omega s}{(s^2 + \omega^2)^2}$$

Hence

$$\mathscr{L}[t \sin \omega t] = \bar{f}_{10} = \frac{2\omega s}{(s^2 + \omega^2)^2}$$

The next two examples show how transform pairs may be built up from already known pairs.

(11) $f(t) = (1 - \cos \omega t)$

$$\bar{f}_{11} = \mathscr{L}[1 - \cos \omega t] = \mathscr{L}[1] - \mathscr{L}[\cos \omega t]$$

$$= \frac{1}{s} - \frac{s}{s^2 + \omega^2} = \frac{\omega^2}{s(s^2 + \omega^2)}$$

(12) $f(t) = \sin(\omega t + \phi)$

$$\bar{f}_{12} = \mathscr{L}[\sin(\omega t + \phi)] = \mathscr{L}[\sin \omega t \cos \phi + \cos \omega t \sin \phi]$$

$$= \frac{\omega \cos \phi}{s^2 + \omega^2} + \frac{s \sin \phi}{s^2 + \omega^2}$$

Theorem 1

An important theorem gives the Laplace transforms of exponentially decaying functions. The theorem states that if $\bar{f}(s)$ is the transform of $f(t)$, then $\bar{f}(s+a)$ is the transform of $e^{-at}f(t)$. The proof is immediately obvious since if

$$\bar{f}(s) = \int_0^\infty f(t)e^{-st}\,dt$$

then

$$\bar{f}(s+a) = \int_0^\infty f(t)e^{-(s+a)t}\,dt = \int_0^\infty e^{-at}f(t)e^{-st}\,dt$$

$$= \mathscr{L}[e^{-at}f(t)] \quad . \quad . \quad . \quad . \quad . \quad (5.3)$$

The following three examples illustrate the use of the above theorem.

(13) $f(t) = e^{-at}\sin\omega t$

$$\mathscr{L}[e^{-at}\sin\omega t] = \bar{f}_{13}(s) = \bar{f}_6(s+a) = \frac{\omega}{(s+a)^2+\omega^2}$$

(14) $f(t) = e^{-at}\cos\omega t$

$$\mathscr{L}[e^{-at}\cos\omega t] = \bar{f}_{14}(s) = \bar{f}_7(s+a) = \frac{s+a}{(s+a)^2+\omega^2}$$

(15) $f(t) = e^{-at}\left(\cos\omega t - \frac{a}{\omega}\sin\omega t\right)$

$$\mathscr{L}\left[e^{-at}\left(\cos\omega t - \frac{a}{\omega}\sin\omega t\right)\right] = \bar{f}_{15}(s) = \bar{f}_7(s+a) - \frac{a}{\omega}\bar{f}_6(s+a)$$

$$= \frac{s+a}{(s+a)^2+\omega^2} - \frac{a}{(s+a)^2+\omega^2}$$

$$= \frac{s}{(s+a)^2+\omega^2}$$

(16) *Unit Impulse, δ*

The unit impulse, or δ-function, is defined as a pulse which is completed in an infinitesimal time Δt (where $\Delta t \to 0$), and which has a magnitude such that the area under the curve of the impulse waveform remains unity as $\Delta t \to 0$, i.e.

$$\int_0^\infty (\text{impulse})\,dt = 1 = \int_0^{\Delta t} (\text{impulse})\,dt \quad . \quad . \quad (5.4)$$

since the impulse ends at time Δt.

Suppose that

$$\mathscr{L}[\delta] = \bar{f}_{16} = \int_0^\infty (\text{impulse})\, e^{-st}\, dt$$

Since the impulse duration tends to zero, it may be assumed that $e^{-st} = 1$ for the Δt seconds, and that thereafter the integral has zero value.

Hence

$$\bar{f}_{16} = \int_0^{\Delta t} (\text{impulse})\, 1\, dt = 1 \quad (\text{from eqn. } (5.4))$$

(17) *Delayed function*

Fig. 5.1 shows a step function which has zero value up to some time T, and thereafter has a value A. The transform of this function is

$$\bar{f}_{17} = \int_0^\infty f(t) e^{-st}\, dt = \int_T^\infty A e^{-st}\, dt = \frac{e^{-sT}}{s} A$$

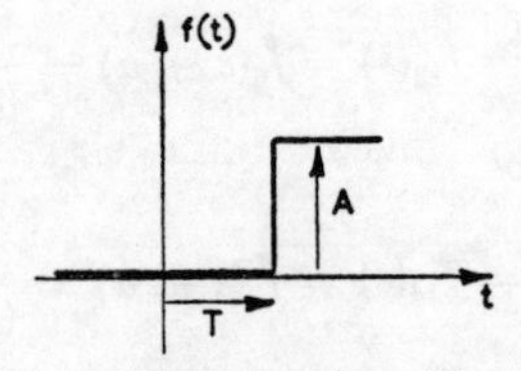

FIG. 5.1. DELAYED STEP FUNCTION

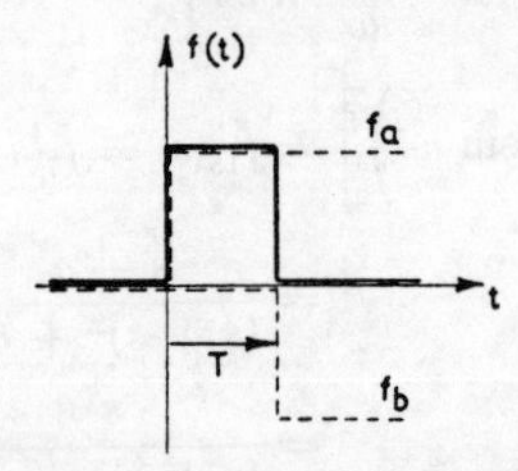

FIG. 5.2. A PULSE FORMED BY TWO STEP FUNCTIONS

(18) *Pulse of height A and width T*

A single pulse of height A and width T can be made up of a step function, f_a, occurring at $t = 0$ from which is subtracted an equal step function, f_b, which occurs at $t = T$ (Fig. 5.2). Hence

$$\mathscr{L}[\text{pulse}] = \bar{f}_{18} = \bar{f}_a - \bar{f}_b = \int_0^\infty A e^{-st}\, dt - \int_T^\infty A e^{-st}\, dt$$

$$= \frac{A}{s} - \frac{A e^{-sT}}{s} = \frac{A}{s}(1 - e^{-sT})$$

5.2. Transform of a Derivative

Suppose that some function of time $f(t)$ has a first derivative $f'(t)$. Then

$$\mathscr{L}[f'(t)] = \int_0^\infty f'(t)e^{-st}\,dt$$

$$= \Big[e^{-st}f(t)\Big]_0^\infty - \int_0^\infty -se^{-st}f(t)\,dt \quad \text{(integrating by parts)}$$

If $f(t)$ is Laplace transformable, s must be chosen in such a way that $f(t)e^{-st} \to 0$ as $t \to \infty$, and hence the upper limit of the first term on the right-hand side is zero, and we obtain

$$\begin{aligned}\mathscr{L}[f'(t)] &= -f(0) + s\int_0^\infty f(t)e^{-st}\,dt \\ &= s\,\mathscr{L}[f(t)] - f(0) \\ &= s\bar{f}(s) - f(0) \qquad . \quad . \quad (5.5)\end{aligned}$$

where $f(0)$ is the value of $f(t)$ as t approaches zero from the direction of $t > 0$.

In the same way, by substituting $f''(t)$ for $f'(t)$ we obtain

$$\begin{aligned}\mathscr{L}[f''(t)] &= s\,\mathscr{L}[f'(t)] - f'(0) \\ &= s^2\,\mathscr{L}[f(t)] - sf(0) - f'(0) \qquad . \quad . \quad (5.6)\end{aligned}$$

where $f'(0)$ is the value of $f'(t)$ at $t = 0$.

Continuing this process we can write the general relation

$$\mathscr{L}[f^n(t)] = s^n\bar{f}(s) - s^{n-1}f(0) - s^{n-2}f'(0) - s^{n-3}f''(0) - \ldots - f^{n-1}(0) \qquad . \quad . \quad . \quad (5.7)$$

These results show how derivatives of any order are transformed into algebraic expressions by the Laplace transformation.

5.3. Impedance Functions

Consider a linear resistor R through which there is a current, $i(t)$, which is some function of time. The voltage across the resistor is

$$v(t) = Ri(t)$$

Taking Laplace transforms,

$$\bar{v}(s) = R\bar{i}(s)$$

or simply

$$\bar{v} = R\bar{i} \qquad . \quad . \quad . \quad . \quad (5.8)$$

If the same current flowed through an inductor L the voltage across it would be

$$v(t) = L\frac{di(t)}{dt}$$

Hence

$$\bar{v} = L(s\bar{i} - i(0)) \qquad \text{(from eqn. (5.5))}$$

$$= sL\bar{i} \quad \text{(assuming zero initial current)} \quad . \quad . \quad (5.9)$$

In the same way, if the p.d. across a capacitor C is $v(t)$ (i.e. a time-varying voltage), the current is

$$i(t) = C\frac{dv(t)}{dt}$$

Hence

$$\bar{i} = C[s\bar{v} - v(0)]$$

$$= sC\bar{v} \quad \text{(assuming zero initial voltage across } C\text{)} \quad . \quad (5.10)$$

and so

$$\bar{v} = \frac{\bar{i}}{sC} \quad . \quad . \quad . \quad . \quad (5.11)$$

We may extend this to the case where the current $i(t)$ flows through R, L and C in series, when the transform of the overall voltage will be

$$\bar{v} = \left(R + sL + \frac{1}{sC}\right)\bar{i}$$

where it is assumed that there is zero initial current through L and zero initial voltage across C. The term in brackets is the *impedance function*, $\bar{z}$, of the circuit or the impedance to transient current.

It is apparent that this argument can be extended to admittance functions, and to any circuit with zero initial conditions. The impedance or admittance functions will be similar to the normal complex impedances or admittances of the circuit but with the constant s replacing $j\omega$ (and hence s^2 replacing $(j\omega)^2 = -\omega^2$, and so on).

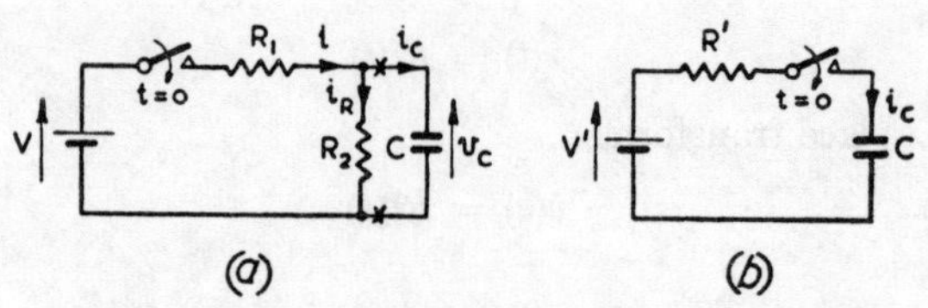

Fig. 5.3

Example 5.1. Find expressions for the current through, and the voltage across, the capacitor and the overall current when the switch in the circuit shown in Fig. 5.3(*a*) is suddenly closed. Evaluate these if $R_1 = 20$ kΩ, $R_2 = 20$ kΩ and $C = 0{\cdot}1$ μF, and the supply is a direct voltage of 20 V.

Thévenin's theorem may be used to simplify the circuit at (*a*) to that at (*b*). Thus the voltage across XX is $VR_2/(R_1 + R_2) = V'$, and the impedance looking into XX with the supply short-circuited is $R_1R_2/(R_1 + R_2) = R'$.

The impedance function for Fig. 5.3(*b*) is

$$\bar{z} = R' + \frac{1}{sC}$$

Hence

$$\bar{i}_c = \frac{\bar{v}}{R' + \dfrac{1}{sC}}$$

$\bar{v}$ is called the *driving function*, and is the transform of V'. Transform 1 gives

$$\bar{v} = \frac{V'}{s}$$

Hence

$$\bar{i}_c = \frac{V'}{s\left(R' + \dfrac{1}{sC}\right)} = \frac{V'}{R'\left(s + \dfrac{1}{CR'}\right)}$$

From Transform 2,

$$i_c = \frac{V'}{R'}\,\mathrm{e}^{-t/CR'} = \frac{VR_2}{(R_1 + R_2)}\,\frac{1}{R_1R_2/(R_1 + R_2)}\,\mathrm{e}^{-(R_1+R_2)t/CR_1R_2}$$

$$= \frac{V}{R_1}\,\mathrm{e}^{-(R_1+R_2)t/CR_1R_2}$$

From eqn. (5.11),

$$\bar{v}_c = \frac{\bar{i}_c}{sC} = \frac{V'}{sCR'\left(s + \dfrac{1}{CR'}\right)}$$

Each term in the denominator appears separately in the table of transforms which we have derived, but the combined expression does not. This situation is dealt with by splitting into partial fractions, so that we obtain

$$\bar{v}_c = \frac{V'}{CR'}\left(\frac{1}{s} - \frac{1}{(s + 1/CR')}\right)CR'$$

From Transforms 1 and 2,

$$v_c = V'(1 - \mathrm{e}^{-t/CR'})$$

$$= \frac{VR_2}{(R_1 + R_2)}(1 - \mathrm{e}^{-(R_1+R_2)t/CR_1R_2})$$

From Fig. 5.3(*a*),

$$i = i_R + i_c = \frac{v_c}{R_2} + i_c$$

$$= \frac{V}{(R_1 + R_2)}\left(1 - e^{-(R_1+R_2)t/CR_1R_2}\right) + \frac{V}{R_1} e^{-(R_1+R_2)t/CR_1R}$$

Inserting the given numerical values yields

$$i_c = 1e^{-1{,}000t} \text{ milliampere}$$

$$v_c = 10(1 - e^{-1{,}000t}) \text{ volts}$$

and

$$i = 0{\cdot}5\,(1 - e^{-1{,}000t}) + 1\,e^{-1{,}000t}$$

$$= 0{\cdot}5 + 0{\cdot}5e^{-1{,}000t} \text{ milliampere}$$

Example 5.2. A voltage given by $v = V_m \sin(\omega t + \phi)$ is suddenly applied to a circuit consisting of a resistance R in series with an inductance L. Derive an expression for the resulting current, and determine the value of ϕ which will give (*a*) zero transient, (*b*) maximum transient, assuming zero initial conditions.

The impedance function for the circuit is

$$\bar{z} = R + sL$$

so that the transform of current is

$$\bar{i} = \frac{\bar{v}}{R + sL}$$

where $\bar{v} = \mathscr{L}[V_m \sin(\omega t + \phi)] = V_m \dfrac{\omega\cos\phi + s\sin\phi}{s^2 + \omega^2}$ (Transform 12)

Hence $\bar{i} = V_m \dfrac{\omega\cos\phi + s\sin\phi}{L(s^2 + \omega^2)(R/L + s)}$

Splitting into partial fractions,

$$\bar{i} = \frac{V_m}{L}\left(\frac{ms + n}{s^2 + \omega^2} + \frac{l}{s + R/L}\right)$$

where m, n and l have to be determined. For all values of s,

$$\omega\cos\phi + s\sin\phi = (ms + n)(s + R/L) + l(s^2 + \omega^2)$$

Equating coefficients of s^2 gives $0 = m + l$, so that

$$m = -l$$

Equating coefficients of s, $\sin\phi = n + \dfrac{R}{L}m$

Putting $s = 0$, $\qquad \omega\cos\phi = n\dfrac{R}{L} + l\omega^2$

Solving these three equations,

$$l = \frac{-L}{\sqrt{(R^2 + \omega^2L^2)}}\sin(\phi - \theta) = -m$$

and, after considerable manipulation,

$$n = \frac{\omega L}{\sqrt{(R^2 + \omega^2 L^2)}} \cos (\phi - \theta)$$

where $\theta = \tan^{-1} (\omega L/R)$.

Thus

$$\bar{i} = \frac{V_m/L}{\sqrt{(R^2 + \omega^2 L^2)}} \left[\frac{sL \sin (\phi - \theta) + \omega L \cos (\phi - \theta)}{s^2 + \omega^2} - \frac{L \sin (\phi - \theta)}{s + R/L} \right]$$

From Transforms 12 and 2,

$$i = \frac{V_m}{\sqrt{(R^2 + \omega^2 L^2)}} [\sin (\omega t + \phi - \theta) - e^{-Rt/L} \sin (\phi - \theta)]$$

The first term in the square brackets represents the steady-state solution, which could have been deduced from straightforward circuit theory, and the second term represents the transient solution, since it has an exponential decay factor.

When the Laplace transform of an unknown quantity shows that there is a steady-state part in the solution, it will often be simpler to solve only the transient part using Laplace transforms, and to write down the steady-state part from conventional circuit theory. In this example such an approach would have avoided the necessity of solving for the constants m and n.

The transient will be zero when $\sin (\phi - \theta) = 0$, i.e. when $\phi = \theta$, and will be a maximum when $\sin (\phi - \theta) = \pm 1$, i.e. when $\phi - \theta = \pm 90°$.

5.4. Double-energy Series Circuit

A double-energy circuit is one which contains two separate energy-storage elements. In an *RLC* series circuit, which is a

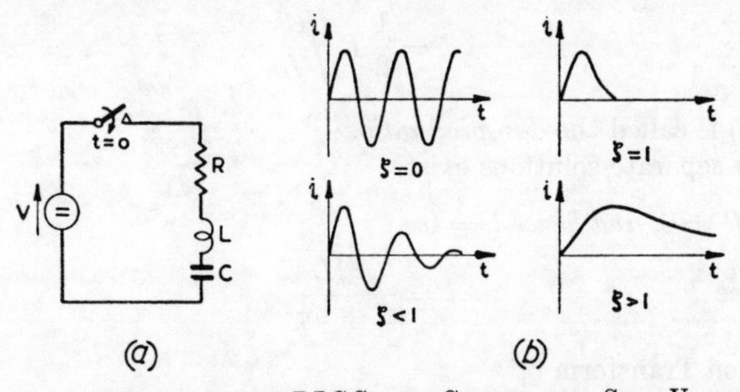

Fig. 5.4. Response of an *RLC* Series Circuit to a Step Voltage

typical double-energy circuit, the current response to a step voltage depends on the values of the elements. Consider the circuit shown in Fig. 5.4 (*a*). The impedance function is

$$\bar{z} = \left(R + sL + \frac{1}{sC} \right)$$

and so we may write

$$\bar{i} = \frac{\bar{v}}{sL + R + \dfrac{1}{sC}}$$

For a suddenly-applied direct voltage, V, the driving function is

$$\bar{v} = \frac{V}{s} \qquad \text{(Transform 1)}$$

Hence

$$\bar{i} = \frac{V}{sL\left(s + \dfrac{R}{L} + \dfrac{1}{sLC}\right)} = \frac{V}{L\left(s^2 + \dfrac{R}{L}s + \dfrac{1}{LC}\right)}$$

This expression may be represented in a standard form as

$$\bar{i} = \frac{V/L}{s^2 + 2\zeta\omega_n s + \omega_n^2} \qquad . \quad . \quad . \quad (5.12)$$

where ω_n is the undamped natural frequency, given by

$$\omega_n = \frac{1}{\sqrt{(LC)}} \qquad . \quad . \quad . \quad . \quad (5.13)$$

and $2\zeta\omega_n = R/L$, so that

$$\zeta = \frac{1}{2} R \sqrt{\frac{C}{L}} \qquad . \quad . \quad . \quad . \quad (5.14)$$

ζ (zeta) is called the *damping ratio.*

Four separate solutions exist.

(*a*) *$R = 0$, and hence $\zeta = 0$*

Hence
$$\bar{i} = \frac{V/L}{s^2 + \omega_n^2}$$

and from Transform 6,

$$i = \frac{V}{\omega_n L} \sin \omega_n t \qquad . \quad . \quad . \quad . \quad (5.15)$$

The current is an undamped sinusoidal oscillation of frequency $f_n = \omega_n/2\pi = 1/2\pi\sqrt{(LC)}$.

(*b*) *$\zeta < 1$, so that $R < 2\sqrt{(L/C)}$ (underdamped)*

By completing the squares in the denominator, the expression for current (eqn. (5.12)) may be written as

$$\bar{i} = \frac{V/L}{(s + \zeta\omega_n)^2 + (\omega_n^{\,2} - \zeta^2\omega_n^{\,2})}$$

Now, Transform 13 is of the form $\omega/(s^2 + \omega^2)$, and so to put the expression for $\bar{i}$ in this form we multiply and divide by $\omega_n\sqrt{(1 - \zeta^2)}$, thus obtaining

$$\bar{i} = \frac{(V/L)\omega_n\sqrt{(1 - \zeta^2)}}{\omega_n\sqrt{(1 - \zeta^2)}[(s + \zeta\omega_n)^2 + \omega_n^{\,2}(1 - \zeta^2)]}$$

Since $\omega_n^{\,2}(1 - \zeta^2)$ will be positive if $\zeta < 1$, we can use Transform 13 to obtain i:

$$i = \frac{V}{\omega L}\,\mathrm{e}^{-\zeta\omega_n t} \sin \omega t \qquad . \quad . \quad . \ (5.16)$$

where $\omega = \omega_n\sqrt{(1 - \zeta^2)}$.

(*c*) $\zeta = 1$, *or* $R = 2\sqrt{(L/C)}$ (*critical damping*)

The transform of current in this case reduces to

$$\bar{i} = \frac{V/L}{(s + \zeta\omega_n)^2}$$

Applying Theorem 1 and Transform 3,

$$i = \frac{V}{L}\,t\mathrm{e}^{-\zeta\omega_n t} \qquad . \quad . \quad . \quad . \ (5.17)$$

(*d*) $\zeta > 1$, *so that* $R > 2\sqrt{(L/C}$ (*overdamped*).

If $\zeta > 1$ then $\omega_n^{\,2}(1 - \zeta^2)$ is negative, and the transform of current may be written as

$$\bar{i} = \frac{(V/L)\omega_n\sqrt{(\zeta^2 - 1)}}{\omega_n\sqrt{(\zeta^2 - 1)}[(s + \zeta\omega_n)^2 - \omega_n^{\,2}(\zeta^2 - 1)]}$$

Using Transform 8 and Theorem 1, the current is

$$i = \frac{V}{aL}\,\mathrm{e}^{-\zeta\omega_n t} \sinh at \qquad . \quad . \quad . \ (5.18)$$

where $a = \omega_n\sqrt{(\zeta^2 - 1)}$.

These four solutions are illustrated in Fig. 5.4 (*b*).

Example 5.3. An RLC series circuit has $R = 10\ \Omega$, $L = 10$ mH and $C = 1\ \mu$F. Find an expression for the voltage across L after a direct voltage of 100 V is suddenly applied, assuming zero initial conditions.

The transform of circuit current is given by eqn. (5.12) as

$$\bar{\imath} = \frac{V/L}{s^2 + 2\zeta\omega_n s + \omega_n^2}$$

The transform of voltage across L is thus

$$\bar{v}_L = sL\bar{\imath} = \frac{sV}{s^2 + 2\zeta\omega_n s + \omega_n^2}$$

Now from eqn. (5.14),

$$\zeta = \tfrac{1}{2} \times 10\sqrt{(10^{-6}/10^{-2})} = 0{\cdot}05$$

and so the circuit is underdamped. Also

$$\omega_n = \frac{1}{\sqrt{(LC)}} = 10^4$$

so that $\zeta\omega_n = 500$, and hence, completing the squares,

$$\bar{v}_L = \frac{100s}{(s + \zeta\omega_n)^2 + \omega_n^2(1 - \zeta^2)}$$

$$= \frac{100s}{(s + 500)^2 + 10^8(0{\cdot}9975)}$$

From Transform 15 this gives

$$v_L = 100e^{-500t}\left(\cos 9{,}980t - \frac{500}{9{,}980}\sin 9{,}980t\right)$$

$$= 100e^{-500t}(\cos 9{,}980t - 0{\cdot}05 \sin 9{,}980t)$$

$$= \underline{\underline{101e^{-500t}\cos(9{,}980t + 3°)}}$$

The solution is a damped sinusoidal oscillation.

5.5. Transfer Functions

The term *transfer function* is applied to networks which have a pair of input and a pair of output terminals, i.e. two-port networks.

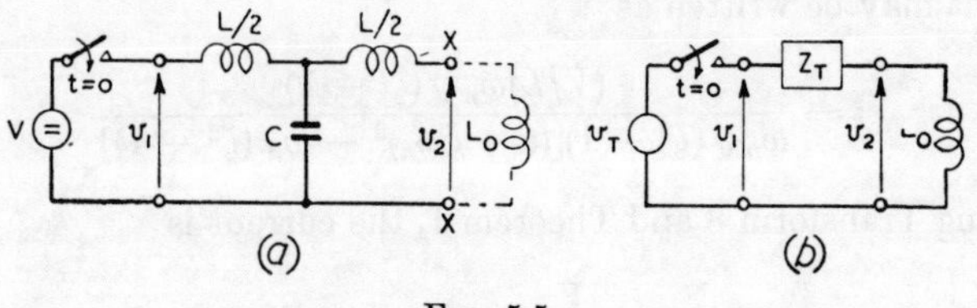

FIG. 5.5

The transfer function of a two-port network is the ratio of output to input (expressed in Laplace form) for a given loading. Thus the open-circuit voltage-transfer function is $\bar{v}_{out}/\bar{v}_{in}$ with the output open-circuited, and the short-circuit current-transfer ratio is $\bar{i}_{out}/\bar{i}_{in}$ with the output short-circuited.

Example 5.4. For a filter circuit shown in Fig. 5.5(a), determine (a) the open-circuit voltage-transfer function, (b) the voltage-transfer function when the filter is terminated in an inductance L_o, (c) the output voltage when the input is a step voltage, V, and the termination is L_o.

(a) On open-circuit, the input impedance function is $sL/2 + 1/sC$ (from Section 5.3).

Hence
$$\bar{v}_{1oc} = \bar{i}_{oc}\left(\frac{sL}{2} + \frac{1}{sC}\right)$$

so that
$$\bar{v}_{2\,oc} = \bar{i}_{oc}\frac{1}{sC} = \frac{\bar{v}_{1oc}}{sC\left(\dfrac{sL}{2} + \dfrac{1}{sC}\right)}$$

and
$$\frac{\bar{v}_{2\,oc}}{\bar{v}_{1\,oc}} = \frac{1}{s^2LC/2 + 1}$$

This is the open-circuit voltage-transfer function.

(b) With the load L_o connected, the circuit may be simplified by using Thévenin's theorem. Thus, if the circuit is broken at XX the transform of the voltage, v_T, appearing across the break is

$$\bar{v}_T = \bar{v}_{2\,oc} = \frac{\bar{v}_1}{s^2LC/2 + 1}$$

The impedance function looking into the break is

$$\bar{z}_T = \frac{sL}{2} + \frac{(sL/2)(1/sC)}{sL/2 + 1/sC} = \frac{sL}{2} + \frac{sL/2}{1 + s^2LC/2}$$

The equivalent circuit is shown in Fig. 5.5(b).
The Laplace transform of the current is

$$\bar{i} = \frac{\bar{v}_T}{\bar{z}_T + sL_0} = \frac{\bar{v}_1}{(s^2LC/2 + 1)\left[s(L/2 + L_0) + \dfrac{sL/2}{1 + s^2LC/2}\right]}$$

$$= \frac{\bar{v}_1}{s\left[\dfrac{s^2LC(L/2 + L_0)}{2} + (L + L_0)\right]}$$

Hence
$$\bar{v}_2 = \bar{i}sL_0 = \frac{2\bar{v}_1L_0}{LC(L/2 + L_0)\left[s^2 + \dfrac{2(L + L_0)}{LC(L/2 + L_0)}\right]}$$

and
$$\frac{\bar{v}_2}{\bar{v}_1} = \frac{2L_0}{LC(L/2 + L_0)\left[s^2 + \dfrac{2(L + L_0)}{LC(L/2 + L_0)}\right]}$$

(*c*) If the input is a step voltage, then $\bar{v}_1 = V/s$ and $\bar{v}_2$ is given by

$$\bar{v}_2 = \frac{2VL_0}{sLC(L/2 + L_0)\left[s^2 + \dfrac{2(L + L_0)}{LC(L/2 + L_0)}\right]}$$

$$= \frac{VL_0\omega^2}{(L + L_0)\, s(s^2 + \omega^2)}$$

where $\omega^2 = \dfrac{2(L + L_0)}{LC(L/2 + L_0)}$.

From Transform 11,

$$v_2 = \frac{VL_0}{L + L_0}(1 - \cos \omega t)$$

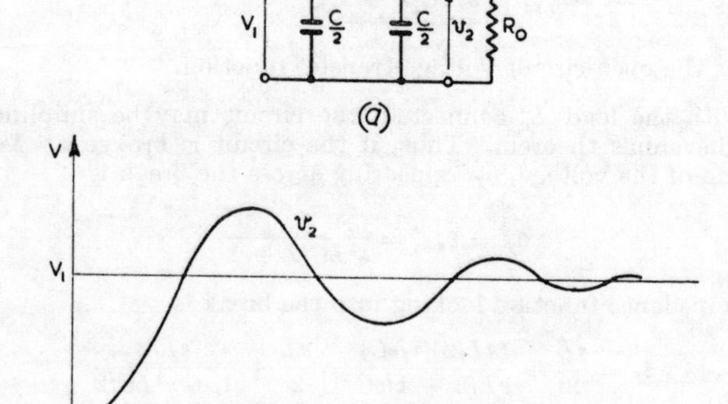

FIG. 5.6. RESPONSE OF A π-SECTION FILTER

Example 5.5. A π-section filter has a series inductance L and shunt capacitances $C/2$ each. Determine the transient response to a step input voltage when the filter is terminated in its design resistance R_0. The circuit is shown in Fig. 5.6(*a*).

From eqn. (2.10), the linear parameters A and B of the network are

$$\text{A} = 1 - \omega^2 LC/2 \qquad \text{and} \qquad \text{B} = j\omega L$$

Hence, from eqn. (2.30) and since $R_o = \sqrt{(L/C)}$,

$$V_1 = (\text{A} + \text{B}/R_o)V_2 = [1 - \omega^2 LC/2 + j\omega L\sqrt{(C/L)}]V_2$$

so that

$$V_2 = \frac{V_1}{-\omega^2 LC/2 + j\omega\sqrt{(LC)} + 1}$$

Taking Laplace transforms, and assuming zero initial conditions.

$$\bar{V}_2 = \frac{\bar{V}_1}{s^2 LC/2 + s\sqrt{(LC)} + 1}$$

$$\bar{V}_2 = \frac{V_1}{s\dfrac{LC}{2}\left[s^2 + \dfrac{2}{\sqrt{(LC)}}s + \dfrac{2}{LC}\right]} \qquad \text{since } \bar{V}_1 = V_1/s$$

$$= \frac{2V_1}{LC}\left[\frac{1}{s} - \frac{s + 2/\sqrt{(LC)}}{s^2 + \dfrac{2}{\sqrt{(LC)}}s + \dfrac{2}{LC}}\right]\frac{LC}{2}$$

$$= V_1\left\{\frac{1}{s} - \frac{s + 1/\sqrt{(LC)}}{[s + 1/\sqrt{(LC)}]^2 + 1/LC} - \frac{1/\sqrt{(LC)}}{[s + 1/\sqrt{(LC)}]^2 + 1/LC}\right\}$$

Hence

$$V_2 = V_1\left\{1 - e^{-t/\sqrt{(LC)}}\cos\left[\frac{1}{\sqrt{(LC)}}t\right] - e^{-t/\sqrt{(LC)}}\sin\left[\frac{1}{\sqrt{(LC)}}t\right]\right\}$$

$$= V_1\left\{1 - e^{-t/\sqrt{(LC)}}\sqrt{2}\cos\left[\frac{t}{\sqrt{(LC)}} - 45^\circ\right]\right\}$$

This function is shown in Fig. 5.6(*b*). This type of response is known as *ringing*, the frequency of the "ring" being $1/2\pi\sqrt{(LC)}$.

5.6. Transform of an Integral

Suppose that a function of time $f(t)$ has a Laplace transform $\bar{f}(s)$ given by eqn. (5.1) as

$$\bar{f}(s) = \int_0^\infty f(t)e^{-st}\,dt$$

The transform of the definite integral $\int_0^t f(t)\,dt$ will be

$$\int_0^\infty \left[\int_0^t f(t)\,dt\right]e^{-st}\,dt$$

$$= \left[-\frac{1}{s}e^{-st}\int_0^t f(t)\,dt\right]_0^\infty - \int_0^\infty -\frac{1}{s}e^{-st}f(t)\,dt \qquad . \quad (5.19)$$

i.e.

$$\mathscr{L}\left[\int_0^t f(t)dt\right] = 0 + \frac{1}{s}\bar{f}(s) \quad . \quad . \quad . \quad (5.20)$$

This shows that, if $\bar{f}(s)$ is the transform of $f(t)$, $(1/s)\bar{f}(s)$ is the transform of the definite integral $\int_0^t f(t)\,dt$. The first term on the right-hand side of eqn. (5.19) is zero at the lower limit ($t = 0$), since the integral $\int_0^t f(t)\,dt$ is then $\int_0^0 f(t)\,dt$, which is zero. It is zero at the upper limit from the definition of the Laplace transform (Section 5.1). If the indefinite integral were used, $\int f(t)\,dt$ would not be zero for $t = 0$, and the transform would be

$$\mathscr{L}[\int f(t)\,dt] = \frac{-1}{s}\left[\int f(t)\,dt\right]_{t=0} + \frac{1}{s}\bar{f}(s)$$

where $\bar{f}(s) = \mathscr{L}[f(t)]$. In practice, the indefinite integral is not important.

5.7. Circuits with Initial Charges and Currents

The circuits dealt with so far have been specifically confined to those with zero initial charges and currents. This is the condition in the majority of practical circuits. Suppose, however, that a capacitor, C, has an initial charge which gives rise to an initial p.d. of V_C across its terminals. At a later instant, t, the p.d., v, across the terminals is

$$v(t) = V_C + \frac{1}{C}\int_0^t i(t)\,dt \quad . \quad . \quad . \quad (5.21)$$

where $i(t)$ is the charging current.

Taking Laplace transforms,

$$\bar{v} = \int_0^\infty \left[V_C + \frac{1}{C}\int_0^t i(t)\,dt\right]e^{-st}\,dt$$

$$= \frac{V_C}{s} + \frac{\bar{i}}{sC} \quad . \quad . \quad . \quad . \quad . \quad . \quad (5.22)$$

This is the transform for a circuit consisting of a source of direct voltage, V_C, in series with an uncharged capacitor, C, at time $t = 0$.

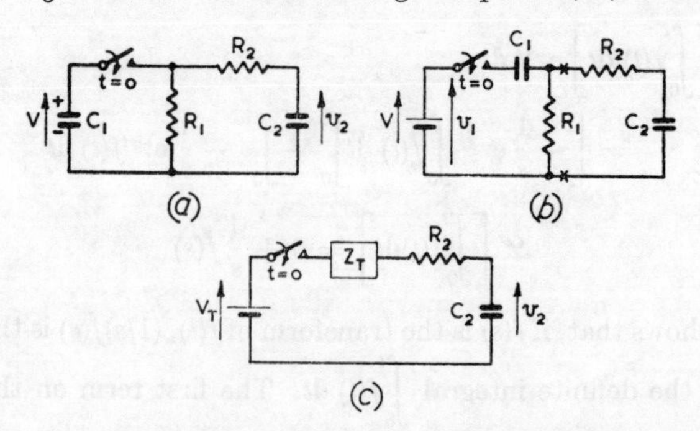

Fig. 5.7. Circuit with an Initially Charged Capacitor

Example 5.6. The capacitor C_1 in Fig. 5.7(a) is initially charged to V volts. At $t = 0$ the switch is closed. Derive an expression for the voltage across capacitor C_2.

In Fig. 5.7(b) the charged capacitor is replaced by a source, V, in series with an uncharged capacitor C_1. This circuit is further simplified by Thévenin's theorem, by breaking it at XX. The transform of the voltage, v_T, across the break is

$$\bar{v}_T = \frac{\bar{v}_1}{R_1 + 1/sC_1} R_1 = \frac{\bar{v}_1 sC_1R_1}{1 + sC_1R_1}$$

The impedance, Z_T, looking into the break is

$$Z_T = \frac{R_1/j\omega C_1}{R_1 + 1/j\omega C_1} = \frac{R_1}{1 + j\omega C_1R_1}$$

Hence the impedance function is

$$\bar{z}_T = \frac{R_1}{1 + sC_1R_1}$$

The overall impedance function is

$$\bar{z} = \frac{R_1}{1 + sC_1R_1} + R_2 + \frac{1}{sC_2}$$

and the transformed circuit current is then $\bar{\imath} = \bar{v}_T/\bar{z}$, while the transformed voltage across C_2 is

$$\bar{v}_2 = \frac{\bar{\imath}}{sC_2} = \frac{\bar{v}_1 sC_1R_1}{sC_2(1 + sC_1R_1)} \frac{1}{\left(\dfrac{R_1}{1 + sC_1R_1} + R_2 + \dfrac{1}{sC_2}\right)}$$

$$= \frac{\bar{v}_1 C_1R_1}{C_2R_1 + C_2R_2 + C_1R_1 + sC_1C_2R_1R_2 + 1/s}$$

Now, $\bar{v}_1$ is the transform of V, i.e. $\bar{v}_1 = V/s$, so that

$$\bar{v}_2 = \frac{VC_1R_1}{C_1C_2R_1R_2\left[s^2 + \dfrac{s(C_2R_1 + C_2R_2 + C_1R_1)}{C_1C_2R_1R_2} + \dfrac{1}{C_1C_2R_1R_2}\right]}$$

$$= \frac{V}{C_2R_2(s^2 + 2\zeta\omega_n s + \omega_n^2)}$$

where $\omega_n = 1/\sqrt{(C_1C_2R_1R_2)}$ and $\zeta = \dfrac{C_2R_1 + C_2R_2 + C_1R_1}{2\sqrt{(C_1C_2R_1R_2)}}$

It can be shown that ζ must be greater than unity, so that, from the results of Section 5.4(d), we may write

$$v_2 = \frac{V}{aC_2R_2} e^{-\zeta\omega_n t} \sinh at$$

where $a = \omega_n\sqrt{(\zeta^2 - 1)}$.

Expanding sinh at as $\frac{1}{2}(e^{at} - e^{-at})$,

$$v_2 = \frac{V}{2aC_2R_2}\left[e^{(a-\zeta\omega_n)t} - e^{-(a+\zeta\omega_n)t}\right]$$

This represents a voltage which starts from zero, rises to a maximum, and falls to zero as $t \to \infty$, as illustrated in Fig. 5.4(*b*) for $\zeta > 1$.

Now consider an inductor carrying an initial current, I_0, connected to a circuit at $t = 0$. At a later instant, t, the inductor current will be given by

$$i = I_0 + \frac{1}{L}\int_0^t v(t)\,dt$$

where $v(t)$ is the voltage across the inductor.

From Transform 1 and eqn. (5.20),

$$\bar{i} = \frac{I_0}{s} + \frac{\bar{v}}{sL} \quad . \quad . \quad . \quad (5.23)$$

Thus the current can be assumed to be the same as that through an inductor L (which initially carries no current) in parallel with a constant-current source I_0 which is connected in such a way as to give the correct direction of external current at $t = 0$ and is switched on at $t = 0$.

Example 5.7. An inductor L of series resistance r is connected in parallel with a resistor R across a d.c. supply of V volts.

(*a*) Deduce an expression for the current in the inductor when the supply is suddenly disconnected.

(*b*) If the resistor R is replaced by a capacitor C, such that $r\sqrt{(C/L)} = 2$, determine the inductor current when the supply is suddenly disconnected.

(*a*) The circuit is shown in Fig. 5.8(*a*). The equivalent circuit when the switch is opened is shown at (*b*). From the previous section, the circuit when the switch is opened is equivalent to an inductor L across which is connected (at $t = 0$) a constant-current source, $I_0 = V/r$. The actual inductor current is the current entering terminal A in this equivalent circuit, *not* the current through the equivalent inductance L.

The transform of voltage across L in Fig. 5.8(*b*) is $\bar{v}_L = \bar{i}_A(R + r)$.

Hence the transform of current through L is $\bar{i}_L = \bar{v}_L/sL$.

The initial current through L in the circuit of Fig. 5.8(*a*) is $I_0 = V/r$.

Now, $I_0 = i_L + i_A$, in Fig. 5.8(*b*), and the transform is

$$\frac{I_0}{s} = \bar{i}_L + \bar{i}_A = \frac{\bar{i}_A(R + r)}{sL} + \bar{i}_A$$

$$= \frac{\bar{i}_A(R + r + sL)}{sL}$$

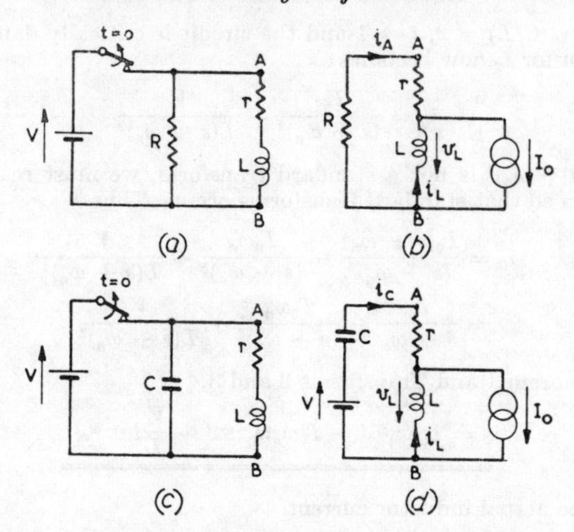

FIG. 5.8. SWITCHING PROBLEM WITH INITIAL CONDITIONS

This gives

$$\bar{i}_A = \frac{I_0}{s}\frac{sL}{R + r + sL} = \frac{V/r}{s + (R + r)/L}$$

From Transform 2,
$$i_A = \frac{V}{r}\,e^{-(R+r)t/L}$$

(*b*) When the inductor is shunted by a capacitor C, then at $t = 0$, when the switch is opened, there will be a current $I_0 = V/r$ through the coil, and the capacitor will be charged to V volts (Fig. 5.8(*c*)). The equivalent circuit when the switch is opened is shown at (*d*).

The transform of voltage across L in Fig. 5.8(*d*) is

$$\bar{v}_L = -\frac{V}{s} + \bar{i}_C\left(r + \frac{1}{sC}\right) \quad . \quad . \quad . \quad \text{(i)}$$

But $\bar{i}_L = \bar{v}_L/sL$, and $I_0 = i_L + i_C$, so that

$$\frac{I_0}{s} = \bar{i}_L + \bar{i}_C = \frac{\bar{v}_L}{sL} + \bar{i}_C \quad . \quad . \quad . \quad \text{(ii)}$$

Eqns. (i) and (ii) give
$$\bar{i}_C = \frac{(I_0L + V/s)}{r + 1/sC + sL}$$

$$= \frac{I_0 s}{s^2 + \frac{r}{L}s + \frac{1}{LC}} + \frac{V}{L\left(s^2 + \frac{r}{L}s + \frac{1}{LC}\right)}$$

$$= \frac{I_0 s}{s^2 + 2\zeta\omega_n s + \omega_n^2} + \frac{V}{L(s^2 + 2\zeta\omega_n s + \omega_n^2)}$$

where $\omega_n = 1/\sqrt{(LC)}$ and $\zeta = \frac{1}{2}r\sqrt{(C/L)}$.

Since $r\sqrt{(C/L)} = 2$, $\zeta = 1$ and the circuit is critically damped. The expression for $\bar{i}_C$ now becomes

$$\bar{i}_C = \frac{I_0 s}{(s+\omega_n)^2} + \frac{V}{L(s+\omega_n)^2}$$

Since $s/(s+\omega_n)^2$ is not a standard transform, we must rearrange the expression so that standard transforms occur. Thus

$$\bar{i}_C = \frac{I_0(s+\omega_n)}{(s+\omega_n)^2} - \frac{I_0\omega_n}{(s+\omega_n)^2} + \frac{V}{L(s+\omega_n)^2}$$

$$= \frac{I_0}{s+\omega_n} - \frac{I_0\omega_n}{(s+\omega_n)^2} + \frac{V}{L(s+\omega_n)^2}$$

From Theorem 1 and Transforms 2 and 3,

$$i_C = I_0 \mathrm{e}^{-\omega_n t} - I_0\omega_n t\mathrm{e}^{-\omega_n t} + \frac{V}{L} t\mathrm{e}^{-\omega_n t}$$

This is the actual inductor current.

5.8. Transients in Mutually-coupled Circuits

The impedance function of a circuit which contains a mutual coupling is obtained in a manner similar to that already described for circuits containing R, L and C. Thus the e.m.f., v_2, induced in one winding of two coupled coils is related to the current, i_1, in the other winding by the expression

$$v_2 = M\frac{\mathrm{d}i_1}{\mathrm{d}t}$$

Taking Laplace transforms, and assuming zero initial current,

$$\bar{v}_2 = sM\bar{i}_1 \quad . \quad . \quad . \quad . \quad (5.24)$$

Similarly,

$$\bar{v}_1 = sM\bar{i}_2 \quad . \quad . \quad . \quad . \quad (5.25)$$

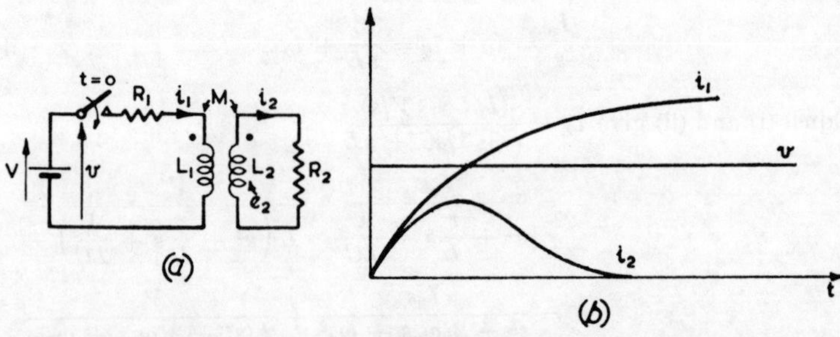

Fig. 5.9. Application of a Direct Voltage to a Mutually Coupled Circuit

Example 5.8. Derive expressions for the primary and secondary current in the mutually coupled circuit shown in Fig. 5.9(a) when a direct voltage, $V = 10$ V, is suddenly applied, and $R_1 = R_2 = 100\ \Omega$, $L_1 = L_2 = 100$ mH, and $M = 60$ mH.

The circuit equations are

$$v = (R_1 + j\omega L_1)i_1 - j\omega M i_2$$

and

$$j\omega M i_1 = (R_2 + j\omega L_2)i_2$$

Using the results of Section 5.3 and eqns. (5.24) and (5.25),

$$\bar{v} = (R_1 + sL_1)\bar{i}_1 - sM\bar{i}_2 \quad . \quad . \quad . \quad \text{(i)}$$

and

$$sM\bar{i}_1 = (R_2 + sL_2)\bar{i}_2. \quad . \quad . \quad . \quad \text{(ii)}$$

Substituting for $\bar{i}_1$ in eqn. (i) gives

$$\bar{v} = \frac{(R_1 + sL_1)(R_2 + sL_2)\bar{i}_2}{sM} - sM\bar{i}_2$$

so that

$$\bar{i}_2 = \frac{sM\bar{v}}{R_1R_2 + s(L_1R_2 + L_2R_1) + s^2(L_1L_2 - M^2)} \quad . \quad \text{(iii)}$$

Since the applied voltage is a step, $\bar{v} = V/s$, and hence

$$\bar{i}_2 = \frac{MV}{(L_1L_2 - M^2)\left(s^2 + s\dfrac{L_1R_2 + L_2R_1}{L_1L_2 - M^2} + \dfrac{R_1R_2}{L_1L_2 - M_2}\right)} \quad \text{(iv)}$$

This is of the same form as eqn. (5.12), and the solution will depend on the numerical relationship between the circuit elements. It should be noted that, for the special case of perfect coupling, $L_1L_2 = M^2$ and eqn. (iii) reduces to

$$\bar{i}_2 = \frac{sM\bar{v}}{R_1R_2 + s(L_1R_2 + L_2R_1)}$$

For the numerical values given, eqn. (iv) may be rewritten as

$$\bar{i}_2 = \frac{MV}{(L_1L_2 - M^2)(s^2 + 2\zeta\omega_n s + \omega_n^2)}$$

$$= \frac{0{\cdot}06 \times 10}{64 \times 10^{-4}\,(s^2 + 3{,}125s + 1{,}250^2)}$$

where $\omega_n = \sqrt{[R_1R_2/(L_1L_2 - M^2)]} = 1{,}250$, and $2\zeta\omega_n = 20/64 \times 10^{-4} = 3{,}126$, so that $\zeta = 1{\cdot}25$, and the circuit is overdamped. The solution is

$$i_2 = \frac{0{\cdot}6}{64 \times 10^{-4}} \frac{e^{-\zeta\omega_n t} \sinh at}{a} \quad \text{(from Section 5.4}(d))$$

where $a = \omega_n\sqrt{(\zeta^2 - 1)} = 1{,}250\sqrt{(0{\cdot}56)} = 937$, and $\zeta\omega_n = 1{,}563$.

Hence

$$i_2 = 0{\cdot}1e^{-1{,}563t}\{\tfrac{1}{2}(e^{937t} - e^{-937t})\}$$

$$= 0{\cdot}05\,(e^{-626t} - e^{-2{,}500t})$$

The form of this current is shown in Fig. 5.9(b).

The peak secondary current occurs when

$$\frac{d}{dt}(e^{-626t} - e^{-2,500t}) = 0$$

i.e. when $-626e^{-626t} + 2{,}500e^{-2,500t} = 0$, or $e^{1,874t} = 2{,}500/626$, or

$$\underline{\underline{t = 0{\cdot}74 \text{ msec}}}$$

The transform for the primary current is found by substituting from eqn. (ii) in the derived expression for $\bar{i}_2$. Thus

$$\bar{i}_1 = \frac{(R_2 + sL_2)}{sM} \frac{MV}{(L_1L_2 - M^2)(s^2 + 2\zeta\omega_n s + \omega_n^2)}$$

$$= \frac{V}{L_1L_2 - M^2}\left[\frac{R_2}{s(s^2 + 2\zeta\omega_n^2 + \omega_n^2)} + \frac{L_2}{(s^2 + 2\zeta\omega_n s + \omega_n^2)}\right]$$

$$= \frac{V}{L_1L_2 - M^2}\left[\frac{R_2}{\omega_n^2}\frac{1}{s} - \frac{R_2}{\omega_n^2}\frac{s + 2\zeta\omega_n}{s^2 + 2\zeta\omega_n s + \omega_n^2} + \frac{L_2}{s^2 + 2\zeta\omega_n s + \omega_n^2}\right]$$

$$= V\left[\frac{1}{R_1 s} - \frac{1}{R_1}\frac{s + 2\zeta\omega_n}{s^2 + 2\zeta\omega_n s + \omega_n^2} + \frac{L_2}{(L_1L_2 - M^2)(s^2 + 2\zeta\omega_n s + \omega_n^2)}\right]$$

since $\omega_n = \sqrt{[R_1R_2/(L_1L_2 - M^2)]}$.

Substituting numerical values, we obtain

$$\bar{i}_1 = 0{\cdot}1\left[\frac{1}{s} - \frac{s + 3{,}126}{s^2 + 3{,}126s + 1{,}250^2} + \frac{1{,}563}{s^2 + 3{,}126s + 1{,}250^2}\right]$$

$$= 0{\cdot}1\left[\frac{1}{s} - \frac{s + 1{,}563}{(s + 1{,}563)^2 + (1{,}250^2 - 1{,}563^2)}\right]$$

$$= 0{\cdot}1\left[\frac{1}{s} - \frac{s + 1{,}563}{(s + 1{,}563)^2 - 937^2}\right]$$

Using Theorem 1 and Transforms 1 and 9,

$$i_1 = 0{\cdot}1\,(1 - e^{-1,563t}\cosh 937t)$$

Finally, expressing the hyperbolic function in exponential form,

$$i_1 = 0{\cdot}1\,[1 - \tfrac{1}{2}e^{-1,563t}(e^{937t} + e^{-937t})]$$

$$= \underline{\underline{0{\cdot}1\,(1 - \tfrac{1}{2}e^{-626t} - \tfrac{1}{2}e^{-2,500t})}}$$

This has no peak value, and represents a current which rises to a value $V/R_1 = 0{\cdot}1$ (Fig. 5.9(*b*)).

A simplification is obtained if perfect coupling exists between L_1 and L_2, since then $L_1L_2 = M^2$.

The problem of what happens in the secondary circuit of a mutual inductor when the primary current supplied from a d.c. source it suddenly quenched is solved by considering the flux in the core as the instant of quenching. Suppose that the steady primary current, $I_1 = V/R_1$, in the primary coil of Fig. 5.10 (*a*) is suddenly quenched (i.e. reduced instantaneously to zero). Since the secondary is a closed winding it may be assumed that the flux linking it remains

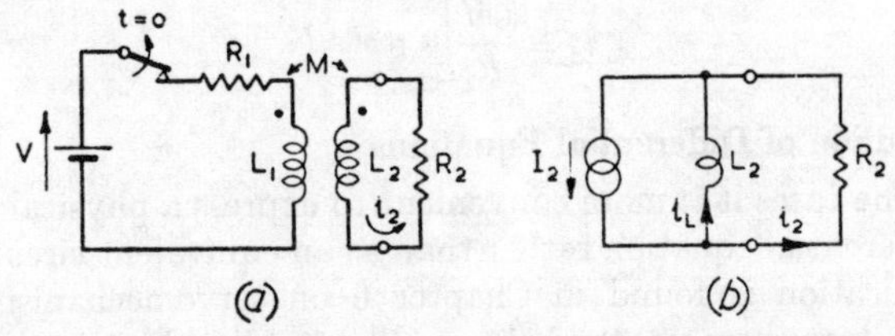

FIG. 5.10. QUENCHING A MUTUALLY COUPLED CIRCUIT

unchanged at the instant of quenching, or in other words that a secondary current flows of such a value that

$$MI_1 = L_2 I_2 \quad \text{at } t = 0$$

MI_1 is the flux linkage produced by the primary which links the secondary, and $L_2 I_2$ is the flux linkage produced by the secondary current I_2. Perfect coupling is assumed.

Hence
$$I_2 = \frac{MI_1}{L_2} \quad \text{at } t = 0$$
$$= \frac{MV}{L_2 R_1}$$

The problem now reduces to the case of an inductor, L_2, which carries an initial current I_2 and is suddenly connected to a resistor R_2. Using the results of Section 5.7, the equivalent secondary circuit is as shown in Fig. 5.10 (*b*), from which

$$I_2 = i_2 + i_L$$

Hence

$$\bar{I}_2 = \frac{I_2}{s} = \frac{MV}{sL_2R_1} = \bar{i}_2 + \bar{i}_L \quad \text{(taking Laplace transforms)}$$

Now
$$i_2 R_2 = L_2 \frac{di_L}{dt}$$

so that
$$\bar{i}_2 = \frac{sL_2}{R_2}\bar{i}_L$$

and therefore
$$\frac{MV}{sL_2R_1} = \bar{i}_2\left(1 + \frac{R_2}{sL_2}\right)$$

or

$$i_2 = \frac{MV}{R_1(R_2 + sL_2)} = \frac{MV}{R_1L_2(s + R_2/L_2)}$$

Taking inverse transforms (Transform 2),

$$i_2 = \frac{MV}{R_1L_2} e^{-R_2t/L_2} \qquad (5.26)$$

5.9. Solution of Differential Equations

In some cases it is more convenient to express a physical problem as a differential equation rather than as an equivalent circuit. Such an application is found in Chapter 6 on servomechanisms. The Laplace transform method is readily applicable to such cases, provided that the equations are linear and have constant coefficients. Consider, for instance, the equation

$$L\frac{d^2i}{dt^2} + R\frac{di}{dt} + \frac{1}{C}i = f(t) \qquad (5.27)$$

Making use of the results of Section 5.2, this equation transforms to

$$s^2L\bar{i} - sLi(0) - Li'(0) + sR\bar{i} - Ri(0) + \frac{\bar{i}}{C} = \bar{f}(s)$$

where $i(0)$ is the value of i at $t = 0$, and $i'(0)$ is the value of di/dt at $t = 0$. Hence

$$\left(s^2L + sR + \frac{1}{C}\right)\bar{i} = \bar{f}(s) + Li'(0) + (R + sL)i(0)$$

and so

$$i = f_1(t) + i'(0)f_2(t) + i(0)f_3(t) \qquad (5.28)$$

where $f_1(t) = \mathscr{L}^{-1}\left[\frac{\bar{f}(s)}{s^2L + sR + 1/C}\right]$

$$f_2(t) = \mathscr{L}^{-1}\left[\frac{L}{s^2L + sR + 1/C}\right]$$

$$f_3(t) = \mathscr{L}^{-1}\left[\frac{R + sL}{s^2L + sR + 1/C}\right]$$

It will be seen that the solution for i contains a term $f_1(t)$ which is independent of the initial conditions of the problem. This represents the particular integral. The two other terms depend on initial conditions and constitute the complementary function. It is seen that the Laplace transform method evaluates the particular integral and the complementary function at the same time.

Example 5.9. A mass, M, of 0·5 kg is suspended from a spring which has a constant of 50 N per metre, and the system is subject to viscous damping of 5 N per metre per second. If the mass is initially displaced by 0·1 m and then released, determine the expression for its resulting motion.

The equation of motion of the system is

$$m\frac{d^2x}{dt^2} + k_f\frac{dx}{dt} + kx = f(t)$$

where m is the mass, k_f the damping coefficient, k the spring constant and $f(t)$ the driving function. Taking Laplace transforms,

$$s^2m\bar{x} - sm\,x(0) - mx'(0) + sk_f\bar{x} - k_f\,x(0) + k\bar{x} = \bar{f}(s)$$

Now, at $t = 0$, $x = 0{\cdot}1 = x(0)$ and $x'(0) = 0$. Also, since there is no driving function, $\bar{f}(s) = 0$. Hence

$$(ms^2 + k_fs + k)\bar{x} = (sm + k_f)0{\cdot}1$$

and

$$\bar{x} = \frac{0{\cdot}1(sm + k_f)}{\left(s^2 + \dfrac{k_f}{m}s + \dfrac{k}{m}\right)m}$$

$$= \frac{0{\cdot}1(s + 10)}{s^2 + 10s + 100} = \frac{0{\cdot}1(s + 10)}{(s + 5)^2 + 100 - 25}$$

$$= 0{\cdot}1\left[\frac{s + 5}{(s + 5)^2 + 75} + \frac{5}{(s + 5)^2 + 75}\right]$$

Hence $x = 0{\cdot}1\,e^{-5t}\left(\cos\sqrt{(75)}t + \dfrac{5}{\sqrt{75}}\sin\sqrt{(75)}t\right)$

(Transforms 13(a) and 14)

$$= 0{\cdot}115\,e^{-5t}\cos\left[\sqrt{(75)}t - 30°\right]\text{ metres}$$

The resultant motion is seen to be a damped sinusoidal oscillation.

5.10. Initial and Final Value Theorems

There are occasionally cases where we wish to find the values of a function only at the beginning and end of a transient disturbance. If the Laplace transform of the function is known, and if, further, the function and its first derivative tend to definite values as $t \to \infty$, then the initial and final values may be obtained direct from the transformed equation without the necessity of determining the time-domain expression.

Theorem 2

If $\bar{f}(s)$ is the Laplace transform of $f(t)$, the initial value of $f(t)$ is

$$f(0) = \lim_{s\to\infty} s\bar{f}(s) \qquad . \quad . \quad . \quad . \quad (5.29)$$

This is proved from the expression for the transform of a derivative thus:

$$\mathscr{L}\left[\frac{\mathrm{d}}{\mathrm{d}t}f(t)\right] = \int_0^\infty \frac{\mathrm{d}\,f(t)}{\mathrm{d}t}\,\mathrm{e}^{-st}\,\mathrm{d}t = s\bar{f}(s) - f(0) \text{ (eqn. (5.5))}$$

Now as $s \to \infty$, $\mathrm{e}^{-st} \to 0$, and the integral must also tend to zero provided that $\mathrm{d}\,f(t)/\mathrm{d}t$ is finite. Hence

$$0 = \lim_{s\to\infty} (s\bar{f}(s) - f(0))$$

and so

$$f(0) = \lim_{s\to\infty} s\bar{f}(s)$$

since $f(0)$ is independent of s.

Theorem 3

If $\bar{f}(s)$ is the Laplace transform of $f(t)$, then the value of $f(t)$ as $t \to \infty$ is

$$f(t)_{t\to\infty} = \lim_{s\to 0} s\bar{f}(s) \quad . \quad . \quad . \quad (5.30)$$

Again consider eqn. (5.5) and let $s \to 0$. Then $\mathrm{e}^{-st} \to 1$, and we may write

$$\int_0^\infty \frac{\mathrm{d}}{\mathrm{d}t}f(t)\,\mathrm{d}t = \lim_{s\to 0} (s\bar{f}(s) - f(0))$$

Hence

$$f(t)]_0^\infty = -f(0) + \lim_{s\to 0} s\bar{f}(s)$$

and

$$f(\infty) - f(0) = -f(0) + \lim_{s\to 0} s\bar{f}(s)$$

which proves the theorem. Note that neither theorem applies to sinusoidal functions, which have no definite value as $t \to \infty$.

Example 5.10. The complex-domain expression for the angular position θ_o of the output shaft of a certain servomechanism is

$$\bar{\theta}_o = \frac{\omega_n^2\bar{\theta}_i}{s^2 + 2\zeta\omega_n s + \omega_n^2}$$

where $\bar{\theta}_i$ is the transform of the input shaft position (driving function). If the driving function is a ramp input $\theta_i = 10t$, determine the final angular velocity of the output shaft.

The angular velocity of the output shaft is $\mathrm{d}\theta_o/\mathrm{d}t = \omega_o$, and its transform is $s\bar{\theta}_o$ (assuming zero initial conditions).

Hence
$$\bar{\omega}_o = s\bar{\theta}_o = \frac{s\omega_n^2\bar{\theta}_i}{s^2 + 2\zeta\omega_n s + \omega_n^2}$$

Also

$$\bar{\theta}_i = \frac{10}{s^2} \qquad \text{(Transform 3)}$$

so that

$$\bar{\omega}_o = \frac{10\omega_n^2}{s(s^2 + 2\zeta\omega_n s + \omega_n^2)}$$

Also, from eqn. (5.30),

$$\omega_0(\infty) = \lim_{s\to 0} s\bar{f}(s)$$

$$= \lim_{s\to 0} \frac{10\omega_n^2}{s^2 + 2\zeta\omega_n s + w_n^2} = \underline{\underline{10}}$$

Bibliography

Holbrook, *Laplace Transforms for Electrical Engineers* (Pergamon Press).
Jaeger, *The Laplace Transformation* (Methuen).

Problems

5.1. A primary coil, of inductance 0·9 H and resistance 30 Ω, is tightly coupled (with a coupling coefficient $k = 1$) to a secondary coil of inductance 0·1 H and resistance 10 Ω. The secondary coil is short-circuited.

(*a*) At a time $t = 0$ the primary coil is suddenly connected to a 100-V d.c. supply of negligible internal impedance. Find an expression for the secondary current, and roughly sketch its time variation.

(*b*) After steady-state current has been established in the primary, the primary circuit is suddenly opened and its current instantaneously quenched. Find the secondary current resulting from this operation. Briefly state the reasons for any difference in the currents obtained in (*a*) and (*b*). (*A.E.E.*, 1956)

(*Ans.* $2{\cdot}5e^{-25t}$; $10e^{-100t}$).

5.2. At the instant of spark-gap breakdown the circuit of a surge generator is equivalent to a 0·01-μF capacitor charged to 200 kV, connected to a 5,250-Ω resistor through an inductance of 1·0 mH. The test specimen, whose impedance may be assumed infinite, is connected across a 4,000-Ω section of the resistor. Sketch, approximately to scale, the surge voltage impressed across the test terminals during the first few microseconds following the breakdown of the gap. What is the initial rate of rise of the test voltage, and how can it be adjusted?

(*Ans.* 800 kV/μsec.) (*A.E.E.*, 1956)

5.3. The primary and secondary windings of a transformer each have an inductance of 1 H and a resistance of 40 Ω, and the mutual inductance between the windings is 0·6 H. The secondary winding is short-circuited and the primary winding is suddenly connected to a 40-V d.c. supply of negligible impedance. Derive expressions for the primary and secondary currents, and calculate the time at which the secondary current has its maximum value. Indicate by a rough sketch the variation with time of the currents. (*A.E.E.*, 1958)

(*Ans.* $i_1 = 1 - 0{\cdot}5e^{-100t} - 0{\cdot}5e^{-25t}$; $i_2 = 0{\cdot}5(e^{-25t} - e^{-100t})$; 18·45 msec.)

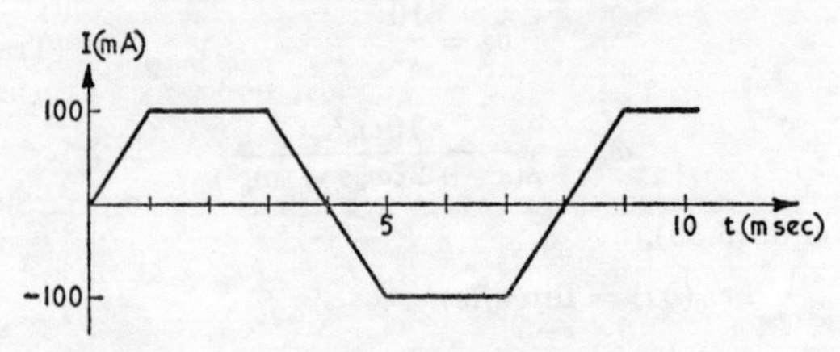

Fig. 5.11

5.4. The alternating current whose waveform is shown in Fig. 5.11 is made to flow, from instant $t = 0$, through a circuit comprising a 250-Ω resistor, a 5-μF capacitor and a 0·5-H pure inductor, all in series. Draw curves to scale to show the voltages that will be developed across each of these circuit-elements, and the overall voltage across the circuit. Find the maximum rate of energy supply to the circuit. (N.B. A Fourier analysis is unnecessary.) (*A.E.E.*, 1956)
(*Ans.* 8·5 W.)

5.5. The capacitor C_1 in Fig. 5.12 is charged to 1,000 V and then discharged by the sudden closing of the switch S. Derive an expression for the subsequent variation with time of the voltage across C_2 (which is initially uncharged) if $C_1 = 0{\cdot}02\ \mu$F, $C_2 = 0{\cdot}0025\ \mu$F, $R_1 = 10{,}000\ \Omega$ and $R_2 = 2{,}000\ \Omega$. (*A.E.E.*, 1957)
(*Ans.* 905[exp $(-4{\cdot}5 \times 10^3 t)$ − exp $(-2{\cdot}3 \times 10^5 t)$].)

5.6. If a direct voltage is suddenly applied to the circuit of Fig. 5.13,

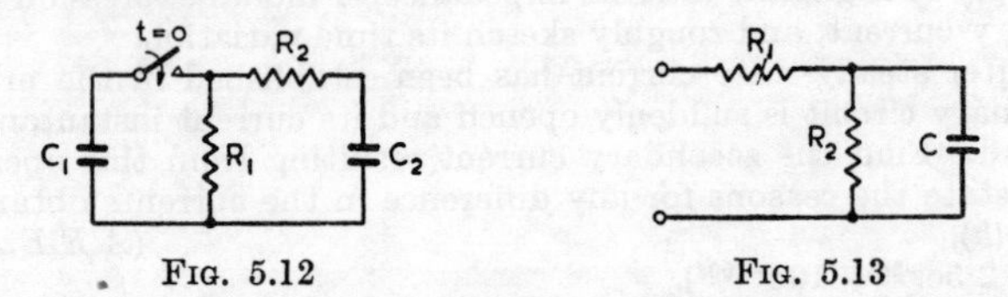

Fig. 5.12 Fig. 5.13

derive an expression for the subsequent variation with time of the voltage across C (initially uncharged).

If $R_1 = R_2 = 100$ kΩ, $C = 0{\cdot}5\ \mu$F and $V = 2{\cdot}5$ kV, determine the initial rate of change of current in R_1. Indicate by a rough sketch how the current in each element of the circuit changes with time.
(*Ans.* −0·5 A/sec.) (*A.E.E.*, 1959)

5.7. A direct voltage from a source of zero impedance is suddenly applied across the terminals of the circuit of Fig. 5.14. Derive an expression for the subsequent variation with time of the voltage across L_1.

If $C = 0{\cdot}02\ \mu$F, $L = 50$ mH and $L_1 = 10$ mH, and the applied voltage is 11 kV, calculate the maximum value of the voltage across L_1. Determine also the time in which this value is first attained.
(*Ans.* 2 kV; 73 μsec.) (*A.E.E.*, 1960)

5.8. A 50-μF capacitor, charged to 300 V, is suddenly connected to a coil of inductance 2·0 H and resistance 500 Ω. Derive expressions for, and draw rough sketches of, (*a*) the capacitor voltage, and (*b*) the current in the circuit subsequent to the connexion. Calculate also the time at which the current reaches its maximum value. (*A.E.E.*, 1961)
(*Ans.* 9·25 msec.)

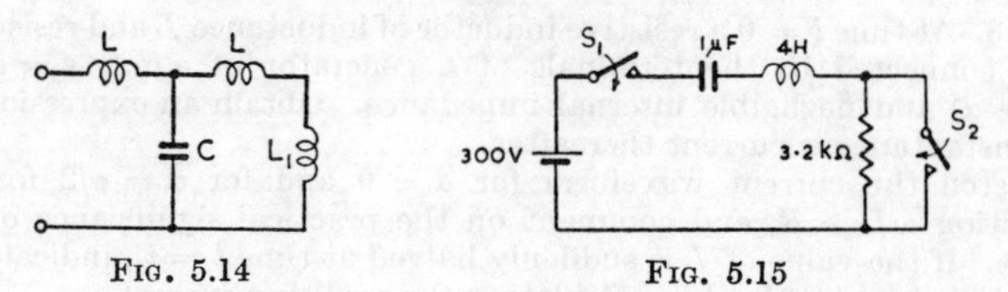

FIG. 5.14 FIG. 5.15

5.9. In the circuit shown in Fig. 5.15, the switch S_1 is closed at time $t = 0$. At this time S_2 is open and the capacitor is uncharged. Derive an expression for the current at a time t seconds after S_1 is closed, S_2 remaining open.

If S_2 is subsequently closed at time $t = \pi/600$ second, find the rate of change of current immediately *after* S_2 is closed, and the frequency of this current. (*L.U. Part III Theory*, 1962)

(*Ans.* $0{\cdot}25e^{-400t} \sin 300t$, 6,170 A/sec, 79·5 c/s.)

5.10. A transformer consists of primary and secondary windings of inductance L_1 and L_2, respectively, with mutual inductance M. The windings have negligible resistance. A capacitor C is connected across the secondary winding, and a voltage $V_0e^{-\gamma t}$ is applied to the primary circuit at time $t = 0$. Show that the secondary current, i, at any time t must satisfy a differential equation of the form

$$\frac{d^2i}{dt^2} + a^2i = k \exp(-\gamma t)$$

and determine a and k in terms of L_1, L_2, M, V_0, C and γ. The general solution of this equation is of the form

$$i = A \cos at + B \sin at + \frac{k}{\gamma^2 + a^2} \exp - (\gamma t)$$

Determine A and B in terms of a, k and γ, assuming that the capacitor is initially uncharged. (*L.U. Part III Theory*, 1961)

(*Ans.* $a = L_1/C(L_1L_2 - M^2)$; $k = -MV_0\gamma/(L_1L_2 - M^2)$; $A = -k/(\gamma^2 + a^2)$; $B = -ka/\gamma(\gamma^2 + a^2)$.)

5.11. A coil having inductance L and resistance R is connected in parallel with a capacitor C. The initial charge and current are zero. At $t = 0$ a constant current, I, is fed into this parallel combination. Assuming that $R^2 < 4L/C$, show that the current flowing through the coil can be represented by

$$i = I\left[1 - \frac{e^{-\alpha t}}{\beta}(\alpha \sin \beta t + \beta \cos \beta t)\right]$$

where $\alpha = R/2L$ and $\beta = \left[\frac{1}{LC} - \frac{R^2}{4L^2}\right]^{1/2}$.

(*L.U. Part III Theory*, 1960)

5.12. A circuit consists of two resistors, each of R ohms, in series, and an uncharged capacitor of C farads is connected across one of the resistors. If this circuit is connected across a charged capacitor also of C farads, derive an expression for the time required for the p.d. across the first capacitor to reach its maximum value. Sketch graphs representing the variation with time of the voltages across the capacitors.

(*Ans.* $0{\cdot}865CR$ sec.) (*L.U. Part III Theory*, 1959)

5.13. At time $t = 0$ a resistive inductor of inductance L and resistance R is connected to the terminals of a generator of e.m.f. $e = E \sin(\omega t + \alpha)$ and negligible internal impedance. Obtain an expression for the instantaneous current thereafter.

Sketch the current waveform for $\alpha = 0$ and for $\alpha = \pi/2$ for the condition $\omega L \gg R$, and comment on the practical significance of the result. If the value of L is suddenly halved at time $t = t_1$, indicate the essential steps involved in calculating the resulting current.

(*L.U. Part III Theory*, 1963)

5.14. Two coils, A and B, have mutual inductance between them, coil A having a self-inductance of 1 H. When the two coils are connected in series, it is found that the total inductance is either 2·3 H or 0·7 H, depending upon the method of connexion.

The series connexion is now removed and a p.d. given by

$$v = 100 \sin 314t \qquad \text{volts}$$

is switched across A at an instant 10 msec after the instant of zero voltage, B being on open-circuit. Assuming that the coils have no resistance, find (*a*) the maximum value of the current in A, (*b*) the value of the e.m.f. across B 12 msec after the switch was closed, and (*c*) the current in A at this instant.

Sketch the waveform of the primary current.

Deduce an expression (without evaluating it) for the current if the coil A has resistance. (*L.U. Part III Theory*, 1957)

(*Ans.* 0·637 A, 23·6 V, −0·575 A.)

5.15. An oscillatory circuit consists of an inductor of inductance L and resistance r connected in parallel with a capacitance C. A negative resistance, $-R$, is connected in parallel with the capacitor. Derive an expression for the frequency of the damped oscillations of the circuit, and determine the value of the negative resistance required for continuous oscillations to be maintained. (*L.U. Part III Tel.*, 1957.)

(*Ans.* $-R = -L/Cr$.)

5.16. A capacitor of capacitance C, charged to a voltage V, is discharged at $t = 0$ through a resistance R in series with a leaky capacitor of capacitance C_2 and leakage conductance G initially uncharged. Derive an expression for the voltage that will develop across C_2.

(*L.U. Part III Tel.*, 1957)

(*Ans.*

$$v = \frac{V}{\omega_n RC_2\sqrt{(\zeta^2 - 1)}} e^{-\omega_n \zeta t} \sinh \omega_n\sqrt{(\zeta^2 - 1)}t$$

where $\omega_n^2 = G/RCC_2$, and $2\zeta\omega_n = 1/RC_2 + G/C_2 + 1/RC$.)

CHAPTER 6

Feedback Systems

WHERE accurate control of some desired quantity is required (such as temperature, flow, speed or position), it is necessary to compare the controlled quantity with some reference value, and to adjust the system in such a way that any difference between actual and desired output is reduced. This is called a *closed-loop control system*, since the processes of measurement, comparison with the desired value, and subsequent correction (if any error is found), may be thought of as forming a closed loop. The term *servomechanism* is often applied to such a system, and this may be defined as an error-actuated control system with power amplification. The control action may be simply proportional to the error (proportional control), but may also include terms depending on the time integral of the error (reset), or the time derivative of the error (rate action), or both (three-term control). Control systems may be pneumatically, hydraulically or electrically operated. This chapter will deal with basic principles only, and will use electrical operation as the example. Some simple systems employ on-off control—e.g. the control of water temperature by an immersion heater. This type of control causes cycling of the output.

When accurate control is not required, a simple *open-loop system* may be used, where there is no feedback of information from the output to the input.

6.1. Block diagrams—Stability

Feedback systems are frequently represented in terms of block or dependence diagrams, each block of the system having a transfer function which relates the output to the input. These transfer functions will normally consist of a numerical factor, K, and a frequency- or time-dependent factor $G(j\omega)$. If Laplace transforms are used the factor $G(j\omega)$ becomes a function of s, $G(s)$. For example, the simple lag circuit (Fig. 6.1) has a transfer function as follows

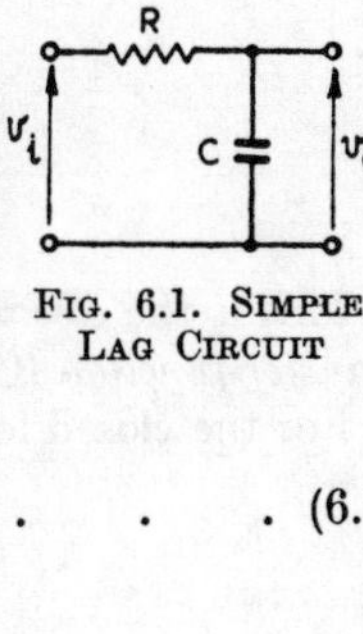

FIG. 6.1. SIMPLE LAG CIRCUIT

$$KG(j\omega) = \frac{v_o}{v_i} = \frac{1}{1 + j\omega CR} \quad . \quad . \quad . \quad (6.1)$$

or

$$KG(s) = \frac{\bar{v}_o}{\bar{v}_i} = \frac{1}{1 + sCR} \quad . \quad . \quad (6.2)$$

In this instance the ratio of output to input is non-dimensional, but generally the transfer function will have dimensions; for example, when the block represents the mechanical output of a motor for a given electrical input.

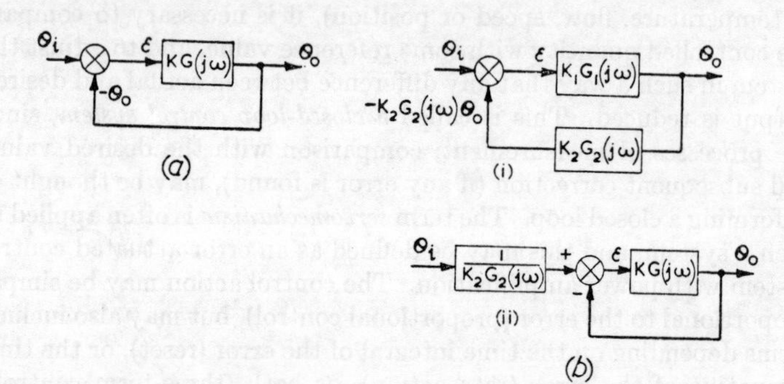

FIG. 6.2. SIMPLE CLOSED LOOPS

The simplest form of feedback circuit is shown in Fig. 6.2 (*a*). The circular symbol represents an algebraic summation. For this network the error, ε, is

$$\varepsilon = \theta_i - \theta_o$$

and

$$\theta_o = KG(j\omega)\,\varepsilon = KG(j\omega)\,(\theta_i - \theta_o)$$

Hence

$$\frac{\theta_o}{\theta_i} = \frac{KG(j\omega)}{1 + KG(j\omega)} \quad . \quad . \quad . \quad . \quad (6.3)$$

or

$$\frac{\bar{\theta}_o}{\bar{\theta}_i} = \frac{KG(s)}{1 + KG(s)} \quad . \quad . \quad . \quad (6.3 \text{ bis})$$

This is the *closed-loop transfer function* (C.L.T.F.). The *open-loop transfer function* (O.L.T.F.) is simply $KG(j\omega)$.

For the closed loop shown in Fig. 6.2 (*b*) (i),

$$\varepsilon = \theta_i - K_2G_2(j\omega)\,\theta_o$$

and

$$\theta_o = K_1G_1(j\omega)\,\varepsilon = K_1G_1(j\omega)\,\theta_i - K_1G_1(j\omega)\,K_2G_2(j\omega)\,\theta_o$$

Hence

$$\frac{\theta_o}{\theta_i} = \frac{K_1G_1(j\omega)}{1 + K_1G_1(j\omega)\,K_2G_2(j\omega)}$$

$$= \frac{1}{K_2G_2(j\omega)}\frac{KG(j\omega)}{1 + KG(j\omega)} \quad . \quad . \quad . \quad (6.4)$$

where $KG(j\omega) = K_1G_1(j\omega)\,K_2G_2(j\omega)$.

This may be represented by the block (ii), which has a simple unity feedback loop. In the same way, more complicated blocks can be simplified to cascaded blocks with unity feedback loops.

In the above examples the feedback is negative (the feedback component is subtracted from the input). In active networks or systems under certain conditions this feedback may become positive, and in this case care must be taken to prevent the system becoming unstable. From eqn. (6.3), instability will occur if $KG(j\omega) = -1$, since then θ_o/θ_i becomes indeterminate. If the O.L.T.F., $KG(j\omega)$, is drawn as a polar plot on the complex plane as ω varies (Nyquist plot), instability will arise if the locus passes through or encloses the point $(-1, 0)$. This is a simplified expression of Nyquist's criterion for stability. If the locus encloses the point $(-1, 0)$, the condition that $KG(j\omega) = -1$ is possible, and the system will oscillate at the particular frequency for which this occurs. If the locus does not enclose the point $(-1, 0)$, instability cannot occur (Figs. 6.3 (*a*) and (*b*)).

A special case arises where the locus may have negative real values greater than -1 but the locus does not enclose the point $(-1, 0)$. In this case the system is said to be *conditionally stable*, since, for the given gain, K, the condition $KG(j\omega) = -1$ does not exist (Fig. 6.3 (*c*)). If the gain, K, falls, however, the point $(-1, 0)$ may be enclosed and the system will become unstable.

In system design care is taken to prevent instability, and two parameters, the gain margin and the phase margin, are used as design criteria. The *gain margin* is the factor by which K must be changed in order that the locus shall pass through the point $(-1, 0)$. In Fig. 6.3 (*a*) the locus has a negative real value of 0·5, and hence the gain margin is 2. In a conditionally stable system such as is shown in Fig. 6.3 (*c*) the gain margin would be smaller than unity. The *phase margin* is the difference between 180° and the angle of

$G(j\omega)$ at that frequency for which $|KG(j\omega)| = 1$ (angle ϕ in Fig. 6.3 (a)). As a rule of thumb, systems may be considered adequately stable if the gain margin is greater than 3 and the phase margin is greater than 35°.

Example 6.1. The open-loop transfer function of a servo depends on the angular frequency, ω, according to the expression

$$KG(j\omega) = \frac{K(9 - \omega^2 + 2j\omega)}{3 - \omega^2 + j\omega(4 - \omega^2)}$$

where K is a real positive number. Determine the angular frequencies for which the transfer function is real. Sketch the locus of the function, and hence find the values of K for which the system is (a) stable, (b) unstable, (c) conditionally stable. (*L.U. Part III Elect.*, 1960)

Rationalizing the denominator,

$$KG(j\omega) = K\left[\frac{(9 - \omega^2 + 2j\omega)(3 - \omega^2 - j\omega(4 - \omega^2))}{(3 - \omega^2)^2 + \omega^2(4 - \omega^2)^2}\right]$$

$$= K\left[\frac{(9 - \omega^2)(3 - \omega^2) + 2\omega^2(4 - \omega^2) + j(2w(3 - \omega^2) - \omega(9 - \omega)(4 - \omega^2))}{(3 - \omega^2)^2 + \omega^2(4 - \omega^2)^2}\right]$$

This will have no quadrate term when

$$2\omega(3 - \omega^2) - \omega(9 - \omega^2)(4 - \omega^2) = 0$$

i.e. when $\omega = 0$, or when

$$2(3 - \omega^2) = (9 - \omega^2)(4 - \omega^2)$$

This gives

$$\omega^4 - 11\omega^2 + 30 = 0$$

or

$$(\omega^2 - 6)(\omega^2 - 5) = 0$$

Hence the real angular frequencies for a reference (real) value of $KG(j\omega)$ are given by

$$\omega_1 = \sqrt{6} = 2{\cdot}45 \qquad \omega_2 = \sqrt{5} = 2{\cdot}24 \qquad \text{and} \qquad \omega = 0$$

In the same way, there is no reference term when

$$(9 - \omega^2)(3 - \omega^2) + 2\omega^2(4 - \omega^2) = 0$$

i.e. when

$$\omega^4 + 4\omega^2 - 27 = 0$$

This has a positive root for $\omega^2 = 3{\cdot}56$ or $\omega = 1{\cdot}89$.

Substituting appropriate values of ω in the expression for $KG(j\omega)$, the following table can be drawn up—

ω	0	1	1·89	1·95	2	2·24	2·45	∞
$KG(j\omega)$	$3K$	$(1{\cdot}7 - j1{\cdot}54)K$	$-j6{\cdot}7K$	$-3{\cdot}34 - j8{\cdot}5)K$	$(-5 - j4)K$	$-2K$	$-K$	$-j0$

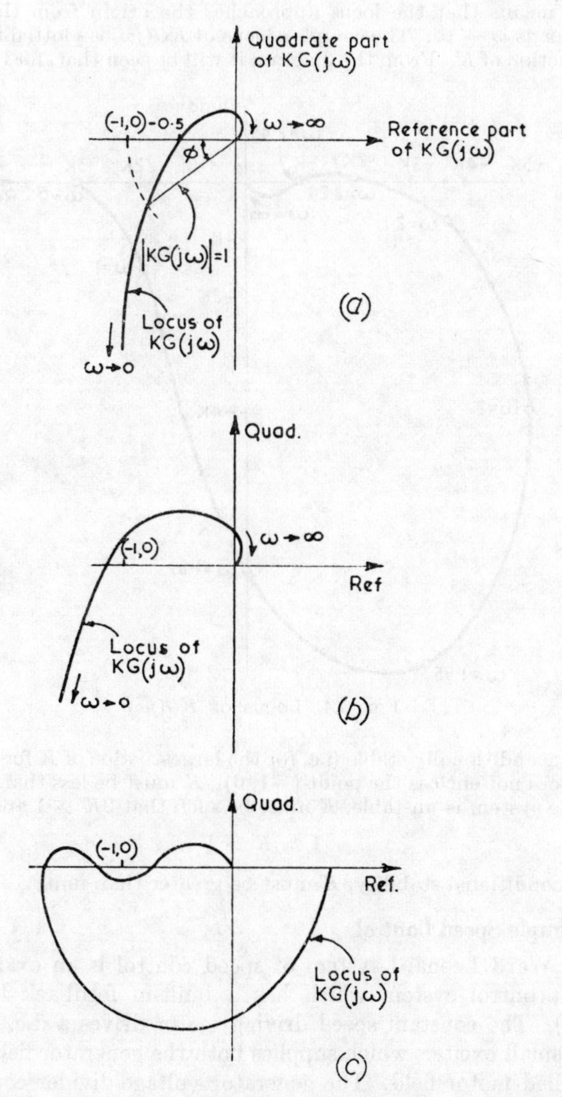

FIG. 6.3. ILLUSTRATING THE NYQUIST CRITERION OF STABILITY

For the case where $\omega \to \infty$, only the highest-order terms in ω need be considered. Thus

$$KG(j\omega)_{\omega\to\infty} \to \frac{K(-\omega^2)}{-j\omega^3} \to \frac{-jK}{\omega} \to -j0$$

This means that the locus approaches the origin from the minus-j direction as $\omega \to \infty$. The complex locus of $KG(j\omega)$ is plotted in Fig. 6.4 as a function of K. From the diagram it will be seen that, for the system

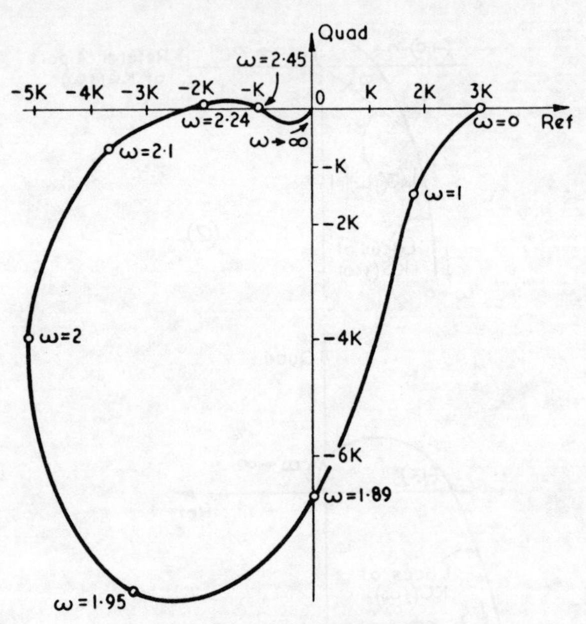

Fig. 6.4. Locus of $KG(j\omega)$

to be unconditionally stable (i.e. for the largest value of K for which the locus does not enclose the point $(-1, 0)$), K must be less than 0·5.

If the system is unstable, K must be such that $2K > 1$ and $K < 1$,

i.e. $$1 > K > \tfrac{1}{2}$$

For conditional stability, K must be greater than unity.

6.2. Simple Speed Control

The Ward Leonard system of speed control is an example of a simple control system which has a built-in feedback loop (Fig. 6.5 (*a*)). The constant-speed driving motor drives a d.c. generator and a small exciter, which supplies both the generator field and the controlled motor field. The generator voltage-divider controls the speed of the controlled motor.

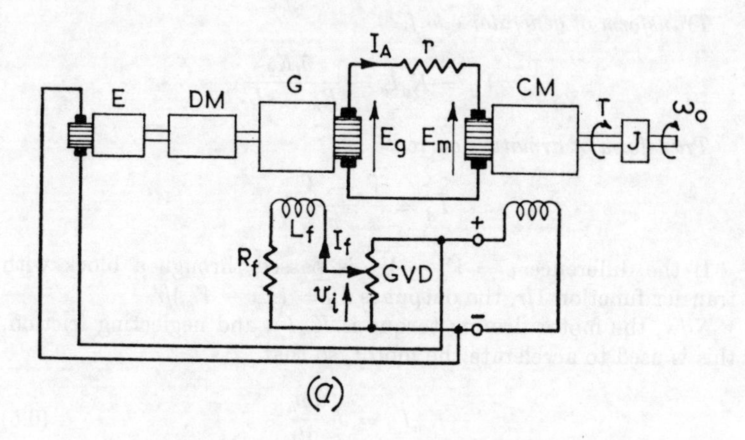

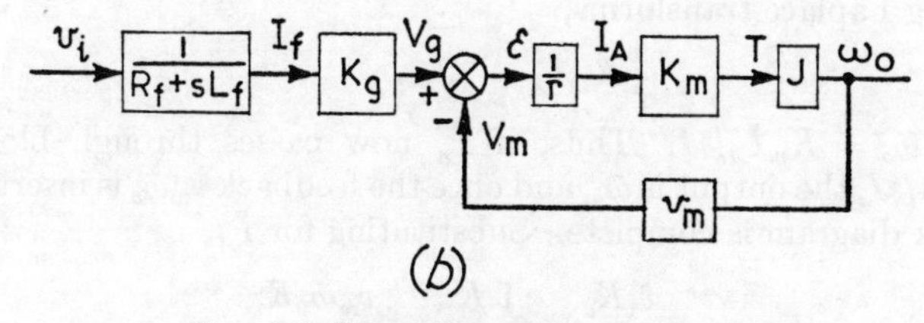

Fig. 6.5. Ward Leonard Speed Control

CM. Controlled motor
DM. Driving motor
E. Exciter
G. Generator
GVD. Generator voltage-divider

Let V_g = Generator e.m.f.
V_m = Motor e.m.f.
$v_m = V_m/\omega_o$ = Motor e.m.f. per unit shaft speed (rad/sec)
I_A = Armature current
J = Moment of inertia of motor and load
K_g = Generator e.m.f. per unit of field current
K_m = Motor torque per unit of armature current
L_f = Inductance of generator field
R_f = Resistance of generator field
r = Resistance of armature circuit
ω_o = Speed of output shaft (rad/sec)

Friction will be neglected.

The block diagram of the system (Fig. 6.5 (*b*)) is built up as follows, Laplace transforms being used for the transfer functions.

Transform of generator e.m.f.

$$\bar{V}_g = K_g \bar{I}_g = \frac{\bar{v}_i K_g}{R_f + sL_f}$$

Transform of armature current

$$\bar{I}_A = \frac{\bar{V}_g - \bar{V}_m}{r}$$

If the difference $\varepsilon = V_g - V_m$ is passed through a block with transfer function $1/r$, the output is $I_A = (V_g - V_m)/r$.

Now, the motor driving torque is $K_m I_A$, and neglecting friction, this is used to accelerate the motor, so that

$$K_m I_A = J \frac{\mathrm{d}\omega_o}{\mathrm{d}t} \quad . \quad . \quad . \quad . \quad (6.5)$$

or, taking Laplace transforms,

$$K_m \bar{I}_A = sJ\bar{\omega}_o$$

so that $\bar{\omega}_o = K_m \bar{I}_A / sJ$. Thus, if $\bar{I}_A$ now passes through blocks K_m and $1/sJ$, the output is $\bar{\omega}_o$, and once the feedback $v_m \bar{\omega}_o$ is inserted, the block diagram is complete. Substituting for $\bar{I}_A$,

$$\bar{\omega}_o = \frac{\bar{v}_1 K_g}{(R_f + sL_f)} \frac{1}{r} \frac{K_m}{sJ} - \frac{v_m \bar{\omega}_0 K_m}{rsJ}$$

so that

$$\bar{\omega}_o = \frac{K_g K_m \bar{v}_1}{(R_f + sL_f) rsJ \left(1 + \dfrac{v_m K_m}{rsJ}\right)}$$

$$= \frac{K_g K_m \bar{v}_1}{L_f rJ (s + R_f/L_f)(s + v_m K_m / rJ)} \quad . \quad . \quad (6.6)$$

$$= \frac{K \bar{v}_1}{(s + 1/\tau_f)(s + 1/\tau_m)} \quad . \quad . \quad . \quad . \quad (6.6a)$$

where $K = K_g K_m / L_f rJ$, $\tau_f = L_f / R_f$, and $\tau_m = rJ / v_m K_m$.

τ_f and τ_m are called the *time-constants* of the system.

From eqn. (6.6*a*), the response of the system to changes of v_1 can readily be found. If, for example, v_1 is a sudden change of V volts, and the system constants are such that $K = 10^3$, $\tau_f = 0{\cdot}01$, $\tau_m = 0{\cdot}04$, then $\bar{v}_1 = V/s$ and eqn. (6.6*a*) becomes

$$\bar{\omega}_o = \frac{10^3 V}{s(s + 100)(s + 25)}$$

The right-hand side of this equation must be split into partial fractions to give

$$\bar{\omega}_o = \frac{10^3 V}{7{,}500}\left[\frac{3}{s} + \frac{1}{s+100} - \frac{4}{s+25}\right]$$

Then, from Transforms 1 (a) and 2,

$$\omega_o = 0{\cdot}133\,V(3 + \mathrm{e}^{-100t} - 4\mathrm{e}^{-25t})$$

This represents an exponential rise to a final speed change of $0{\cdot}4\,V$ rad/sec.

The frequency response of the system can be found by substituting $j\omega$ for s in eqn. (6.6a), and plotting the output speed, ω_o, in a complex plane as a function of ω.

In the above the inductance of the armature circuits has been neglected, as has any friction torque or load torque. If a load torque, T_L, is included, eqn. (6.5) becomes

$$K_m I_A = \frac{J\,\mathrm{d}\omega_o}{\mathrm{d}t} + T_L$$

Hence

$$sJ\bar{\omega}_o = K_m \bar{I}_A - \bar{T}_L$$

and a solution can be obtained as before. Thus

$$\bar{\omega}_o = \frac{K_g K_m \bar{v}_1}{sJr(R_f + sL_f)} - \frac{e_m K_m \bar{\omega}_o}{sJr} - \frac{\bar{T}_L}{sJ}$$

Therefore

$$\bar{\omega}_o = \frac{K\bar{v}_1}{(s + 1/\tau_f)(s + 1/\tau_m)} - \frac{\bar{T}_L}{J(s + 1/\tau_m)}$$

If the input voltage change is zero, and a sudden step load is applied ($\bar{T}_L = T_L/s$), then

$$\bar{\omega}_o = \frac{T_L}{sJ(s + 1/\tau_m)} = \frac{T_L \tau_m}{J}\left(\frac{1}{s} - \frac{1}{s + 1/\tau_m}\right)$$

so that

$$\omega_o = \frac{T_L \tau_m}{J}\,(1 - \mathrm{e}^{-t/\tau_m})$$

6.3. The Summing Junction

The block diagrams of closed-loop systems show junctions at which signals may be added or subtracted. These are *summing junctions*. The simplest form of summing junction is shown in Fig. 6.6.

By Millman's theorem,

$$V_{0'E} = \varepsilon = \frac{V_1/R_1 + V_2/R_2 + V_3/R_3}{1/R_1 + 1/R_2 + 1/R_3 + 1/R}$$
$$= K_1V_1 + K_2V_2 + K_3V_3 \qquad . \quad . \quad (6.7)$$

Hence ε is the sum of the voltages at the free ends of the resistors R_1, R_2, R_3, taken in ratios which depend on the values of the resistors and of the summing resistor, R. If R is the input resistance of an amplifier, it will normally be much larger than R_1, R_2 or R_3, and ε will be given by

$$\varepsilon = V_1\frac{R_p}{R_1} + V_2\frac{R_p}{R_2} + V_3\frac{R_p}{R_3}$$

where R_p is the equivalent resistance of R_1, R_2, R_3 in parallel (since $1/R$ will be negligible compared with $1/R_1$, etc.).

The addition of voltages at a summing junction is sometimes called *star adding.*

6.4. Measurement of Shaft Error

In a *remote position control* (R.P.C.) servo, it is necessary to measure the difference between the actual angular position of the output

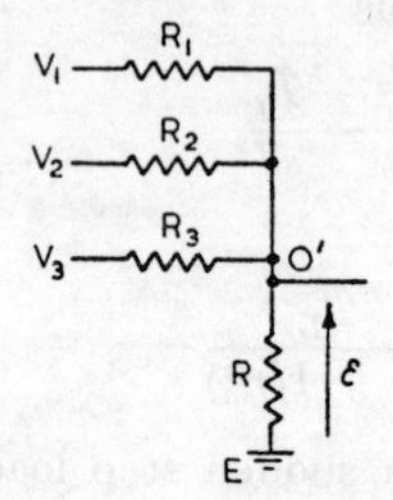

FIG. 6.6. SUMMING JUNCTION

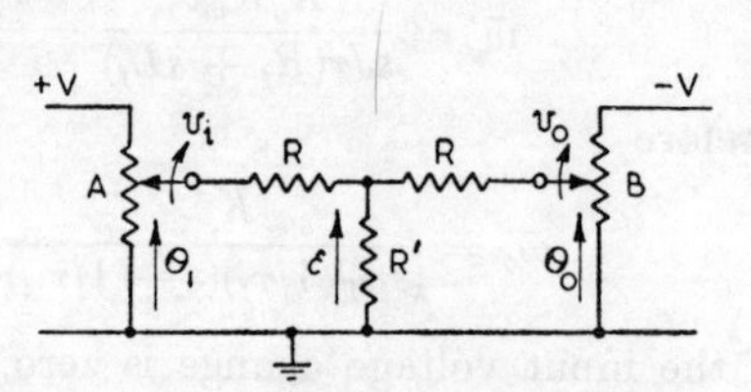

FIG. 6.7. ERROR DETECTION BY D.C. VOLTAGE DIVIDERS

shaft and the desired reference position, and to express this difference, or "error," as a voltage. Two methods of achieving this will be described.

(*a*) *Voltage dividers as shaft error detectors.* D.C. voltage dividers give a convenient means of shaft error detection. Consider the two voltage dividers A and B in Fig. 6.7. If R is very large compared with the voltage-divider resistance,

$$v_i = V\frac{\theta_i}{\theta_m} \quad \text{and} \quad v_o = -V\frac{\theta_o}{\theta_m}$$

where θ_i and θ_o are the shaft rotations of the voltage dividers, and θ_m is the maximum rotation. If also $R' \gg R$, the error voltage is

$$\varepsilon = \frac{v_i/R + v_o/R}{1/R + 1/R} = \frac{1}{2}(v_i + v_o)$$

$$= \frac{1}{2}V\frac{\theta_i}{\theta_m} - \frac{1}{2}V\frac{\theta_o}{\theta_m} = K_p(\theta_i - \theta_o). \quad (6.8)$$

where K_p is the voltage-divider constant, or error voltage per radian of shaft displacement. Only if $\theta_i = \theta_o$ is the error voltage, ε, zero. The polarity of the error voltage depends on whether $\theta_i > \theta_o$ or $\theta_i < \theta_o$.

Normally the voltage dividers have a 300° maximum rotation, and are used with a supply voltage not exceeding 20 V. In some cases special multi-turn helical voltage dividers are used for shaft rotations of more than 300°, or alternatively a step-down gear may be employed.

(*b*) *Synchros.* These are single-phase a.c. machines, whose rotors carry windings which are fed through slip-rings, and whose stators have three star-connected windings which are spaced 120 electrical degrees apart. For shaft error detection a transmitter and receiver are required, the main differences between the two being the provision of a damping device on the receiver, and the fact that the rotor of the transmitter normally has a concentrated winding, while the receiver has a distributed rotor winding.

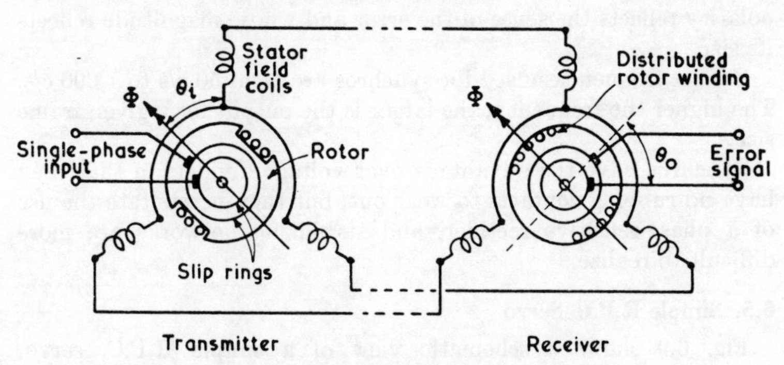

Fig. 6.8. Synchro Shaft-error Detection

Fig. 6.8 shows a typical error-measuring system. The rotor of the transmitter is fed from a single-phase supply. The magnitudes of the e.m.f.s induced in the stator windings will depend on the angular position of the rotor. They will, of course, all be in time phase.

The currents flowing between the two stators produce back-e.m.f.s in the receiver which are approximately equal to the transmitter e.m.f.s, so that a flux is set up in the air-gap of the receiver which will be in the same space relation to the receiver stator coils as the transmitter flux is to the transmitter stator coils. If the shaft position is measured from a reference axis in the receiver which is 90° different from that in the transmitter, then for a peak flux Φ_m the r.m.s. value of the e.m.f. induced in the receiver rotor is

$$V_r = 4{\cdot}44 f N_R \Phi_m \cos(90° + \theta_i - \theta_o)$$
$$= V_m \sin(\theta_o - \theta_i) \qquad (6.9)$$

If $\theta_o > \theta_i$, the receiver rotor e.m.f. is in phase with the transmitter rotor supply; if $\theta_o < \theta_i$, the receiver output is in antiphase with the transmitter supply; while if $\theta_o = \theta_i$ the receiver output is zero (coincidence).

Also, for small displacements, the error voltage is

$$\varepsilon = V_r = V_m(\theta_o - \theta_i) \qquad (6.10)$$

i.e. the a.c. error voltage is proportional to the shaft error. This voltage is in the form of a modulated carrier, whose phase depends on the sense of the error and whose magnitude depends on the size of the error.

If the a.c. error voltage, ε, is applied to a phase-sensitive rectifier (P.S.R.), the output will be a direct current or voltage whose polarity reflects the sense of the error and whose magnitude reflects its size.

Carrier frequencies used for synchros are from 50 c/s to 1,000 c/s. The higher the frequency, the larger is the output for a given frame size.

Synchros have the advantage over voltage dividers in that they have no rubbing contacts to wear out, but they necessitate the use of a phase-sensitive rectifier, and stabilizing networks are more difficult to realize.

6.5. Simple R.P.C. Servo

Fig. 6.9 shows a schematic view of a simple R.P.C. servo, with viscous friction damping. Error measurement using d.c. voltage dividers is shown. The error voltage signal is fed to a d.c. amplifier which supplies the centre-tapped split field winding of the servo-motor from a push-pull output stage. With no input to the amplifier the two halves of the field winding carry equal currents in opposite senses so that there is no net motor torque. When an error signal exists, the current in one half of the field winding increases while that in the other half decreases. The particular half in which the current increases depends upon the polarity of the error voltage, and hence on the sense of the angular shaft error. The armature of the motor is fed from a constant-current supply, so that the magnitude and direction of the output torque depend on the differential field current, and hence on the size and direction of the error. The motor torque is always arranged to be in such a direction as to reduce any angular shaft error.

Let K_m = Motor torque constant (N-m/field mA)
K_A = Amplifier transconductance (mA/V)
K_p = Voltage-divider constant (V/rad error)
J = Moment of inertia of motor and shaft (kg-m²)
F = Viscous friction constant (N-m per rad/sec)
ε = Error voltage

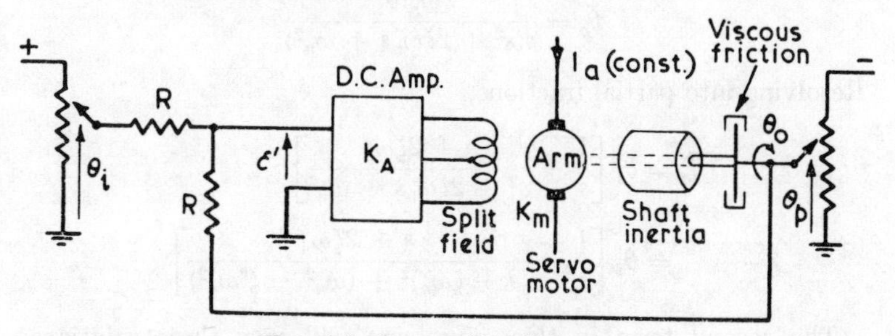

Fig. 6.9. Simple R.P.C. Servo

Then, for equilibrium, the opposing torque due to inertia and friction is equal to the motor driving torque. Hence

$$J\frac{d^2\theta_o}{dt^2} + F\frac{d\theta_o}{dt} = K_m K_A \varepsilon = K_m K_A K_p(\theta_i - \theta_o)$$

neglecting any load torque. Rearranging gives

$$J\frac{d^2\theta_o}{dt^2} + F\frac{d\theta_o}{dt} + K\theta_o = K\theta_i$$

where $K = K_m K_A K_p$; whence

$$\frac{d^2\theta_o}{dt^2} + \frac{F}{J}\frac{d\theta_o}{dt} + \frac{K}{J}\theta_o = \frac{K}{J}\theta_i \qquad (6.11)$$

This is conventionally written

$$\frac{d^2\theta_o}{dt^2} + 2\zeta\omega_n\frac{d\theta_o}{dt} + \omega_n^2\theta_o = \omega_n^2\theta_i \qquad (6.11a)$$

where ω_n, the undamped natural angular frequency, is given by

$$\omega_n = \sqrt{\frac{K}{J}} \quad . \quad . \quad . \quad (6.12)$$

and ζ, the damping ratio, by

$$\zeta = \frac{F}{2\sqrt{(JK)}} \quad . \quad . \quad . \quad (6.13)$$

since $2\zeta\omega_n = F/J$. The product $\zeta\omega_n$ is called the *damping constant.*

Taking Laplace transforms of eqn. (6.11*a*),

$$\bar{\theta}_o = \frac{\omega_n{}^2\bar{\theta}_i}{s^2 + 2\zeta\omega_n s + \omega_n{}^2}$$

For a step input, $\bar{\theta}_i = \theta_i/s$, and hence

$$\bar{\theta}_o = \frac{\omega_n{}^2\theta_i}{s(s^2 + 2\zeta\omega_n s + \omega_n{}^2)}$$

Resolving into partial fractions,

$$\bar{\theta}_o = \theta_i\left[\frac{1}{s} - \frac{s + 2\zeta\omega_n}{s^2 + 2\zeta\omega_n s + \omega_n{}^2}\right]$$

$$= \theta_i\left[\frac{1}{s} - \frac{s + 2\zeta\omega_n}{(s + \zeta\omega_n)^2 + (\omega_n{}^2 - \zeta^2\omega_n{}^2)}\right]$$

The second term in this expression will give three solutions, which are

$$\left.\begin{array}{ll} (a)\ \textit{Oscillatory} & \zeta < 1 \\ (b)\ \textit{Overdamped} & \zeta > 1 \\ (c)\ \textit{Critically damped} & \zeta = 1 \end{array}\right\} \quad . \quad . \quad . \quad (6.14)$$

For critical damping the friction constant is F_{crit}, given by

$$F_{crit}/2\sqrt{(JK)} = 1$$

Hence for any other value of F,

$$\zeta = \frac{F}{F_{crit}} = \frac{\text{Actual damping}}{\text{Critical damping}} \quad . \quad . \quad (6.15)$$

It is usual for the damping to be adjusted to give just less than critical damping, since this gives a fast response with not too much "overshoot."

If there is no damping ($F = 0$), the output will contain a term corresponding to continuous undamped oscillation.

(*a*) *Oscillatory solution.* In this case $\zeta < 1$; hence

$$\bar{\theta}_o = \theta_i \left[\frac{1}{s} - \frac{s + \zeta\omega_n}{(s + \zeta\omega_n)^2 + \omega_n{}^2(1 - \zeta^2)} - \frac{\zeta\omega_n}{(s + \zeta\omega_n)^2 + \omega_n{}^2(1 - \zeta^2)}\right]$$

and Transforms 1, 24 (*a*) and 23 (*a*) give

$$\theta_o = \theta_i \left\{1 - e^{-\zeta\omega_n t}\left[\cos \omega_n \sqrt{(1 - \zeta^2)}t + \frac{\zeta}{\sqrt{(1 - \zeta^2)}} \sin \omega_n \sqrt{(1 - \zeta^2)}t\right]\right\} \quad . \quad (6.16)$$

(*b*) *Overdamped solution.* In this case $\zeta > 1$; hence

$$\bar{\theta}_o = \theta_i \left[\frac{1}{s} - \frac{s + \zeta\omega_n}{(s + \zeta\omega_n)^2 - \omega_n{}^2(\zeta^2 - 1)} - \frac{\zeta\omega_n}{(s + \zeta\omega_n)^2 - \omega_n{}^2(\zeta^2 - 1)}\right]$$

which gives

$$\theta_o = \theta_i \left[1 - e^{-\zeta\omega_n t}\left\{\cosh \omega_n \sqrt{(\zeta^2 - 1)}t + \frac{\zeta}{\sqrt{(\zeta^2 - 1)}} \sinh \omega_n \sqrt{(\zeta^2 - 1)}t\right\}\right] \quad . \quad (6.17)$$

(Transforms 1, 24 (*b*) and 23 (*b*).)

(*c*) *Critically damped solution.* In this case $\zeta = 1$; hence

$$\bar{\theta}_o = \theta_i \left[\frac{1}{s} - \frac{s + \omega_n}{(s + \omega_n)^2} - \frac{\omega_n}{(s^2 + \omega_n)^2}\right]$$

giving

$$\theta_o = \theta_i[1 - e^{-\omega_n t} - \omega_n t\, e^{-\omega_n t}] \quad . \quad . \quad (6.18)$$

(Transforms 1, 24 (*c*) and 23 (*c*).)

6.6. Stabilization by Velocity Feedback

In practice, viscous-friction damping is not used since it involves a large power loss. One alternative is to have a second feedback loop which feeds the amplifier with a negative signal proportional to the output shaft velocity. This is velocity feedback. The schematic diagram for an R.P.C. servo with velocity feedback is shown in Fig. 6.10. Synchros are shown as the error-detecting devices. The output of the receiver synchro is fed to the phase-sensitive rectifier, and then to the servo amplifier. A d.c. tacho-generator is connected to the motor shaft, giving an output proportional to shaft speed.

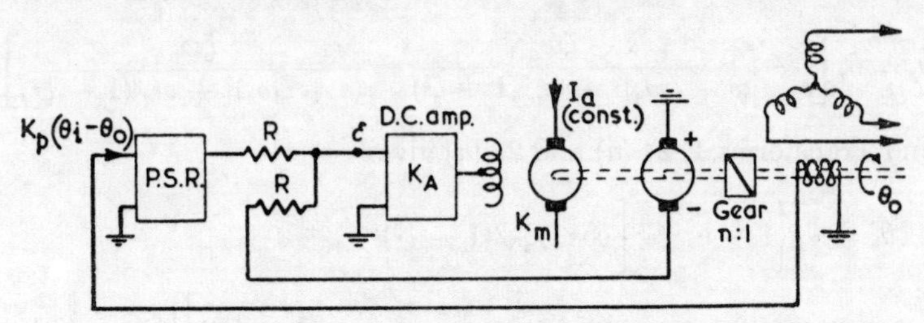

FIG. 6.10. R.P.C. SERVO WITH SYNCHRO ERROR DETECTION AND VELOCITY FEEDBACK

The actual output shaft is shown connected through an $n:1$ reduction gear.

Let J_m = Moment of inertia of motor
J_o = Moment of inertia of output shaft and load
J = Moment of inertia referred to output shaft
$= n^2 J_m + J_o$
K_s = Voltage input to d.c. amplifier per radian error of output shaft (V/rad)
K_t = Voltage input to amplifier per unit of output shaft velocity (V per rad/sec)
n = Gear ratio

Since the opposing torque is equal to the driving torque at the output shaft, friction and load torques being neglected,

$$J\frac{d^2\theta_o}{dt^2} = nK_mK_A\varepsilon$$

$$= nK_mK_A\left(K_s(\theta_i - \theta_o) - K_t\frac{d\theta_o}{dt}\right)$$

whence

$$\frac{d^2\theta_o}{dt^2} + \frac{nK_mK_AK_t}{J}\frac{d\theta_o}{dt} + \frac{nK_mK_AK_s}{J}\theta_o = \frac{nK_mK_AK_s}{J}\theta_i \quad . \quad (6.19)$$

which is of the same form as eqn. (6.11), and so has the same solutions. The advantage is that frictional power loss is minimized, so that full motor torque may be used for accelerating the load.

Example 6.2. For the servo shown in Fig. 6.10, the moment of inertia of the motor is 10^{-6} kg-m², and that of the load is negligible. The synchros and phase-sensitive rectifier give 0·3 V at the amplifier per degree error in the position of the output shaft, and the tacho-generator gives 1 V at the amplifier input for 3,000 rev/min of motor shaft. The motor torque constant is $2{\cdot}5 \times 10^{-4}$ N-m/mA, and the gear ratio is 100:1.

Determine the gain of the amplifier required to give critical damping, and the undamped natural angular frequency. (*R.R.E.*)

The differential equation of operation is eqn. (6.19). Comparison with eqn. (6.11*a*) gives

$$\omega_n{}^2 = \frac{nK_mK_AK_s}{J} \quad \text{and} \quad 2\zeta\omega_n = \frac{nK_mK_AK_T}{J}$$

Hence $$\omega_n{}^2 = \frac{100 \times 2{\cdot}5 \times 10^{-4} \times (0{\cdot}3 \times 57{\cdot}3)}{10^{-6} \times 10^4} K_A = 43\ K_A \qquad \text{(i)}$$

since K_s must be in volts per radian, and $J = n^2J_m$.

Also $$2\zeta\omega_n = \frac{100 \times 2{\cdot}5 \times 10^{-4} \times 60}{10^{-6} \times 10^4 \times 30 \times 2\pi} K_A$$

since $K_t = \dfrac{60}{2\pi(3{,}000/100)}$ volts per unit angular velocity (rad/sec) of the output shaft.

For critical damping, $\zeta = 1$, so that

$$2\omega_n = 0{\cdot}796\ K_A \qquad \text{(ii)}$$

Eliminating ω_n from eqns (i) and (ii),

$$\underline{\underline{K_A = 167 \text{ mA/V}}}$$

Hence, from eqn. (ii), $$\underline{\underline{\omega_n = 66{\cdot}3 \text{ rad/sec}}}$$

6.7. Phase-advance Stabilization

Damping of a servo may also be achieved by use of a passive phase-advance circuit either in the feedback loop or in the amplifier

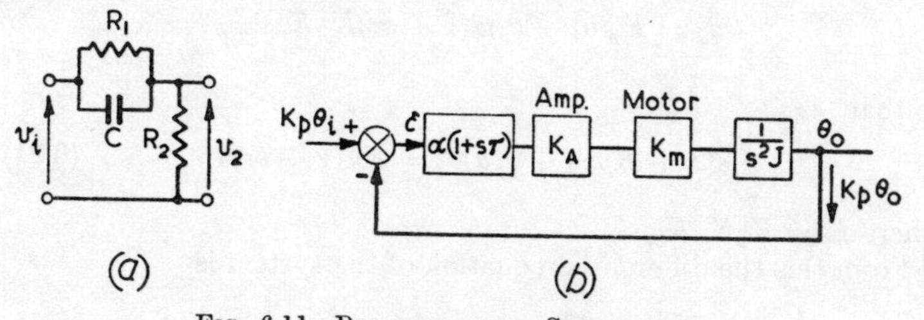

FIG. 6.11. PHASE-ADVANCE STABILIZATION

input. For the phase-advance circuit of Fig. 6.11 (*a*),

$$V_2 = \frac{V_1R_2(1 + j\omega CR_1)}{R_1 + R_2 + j\omega CR_1R_2}$$

so that

$$\frac{V_2}{V_1} = \frac{R_2}{R_1 + R_2} \frac{1 + j\omega CR_1}{1 + j\omega CR_1R_2/(R_1 + R_2)} = \alpha \frac{1 + j\omega\tau}{1 + j\omega\alpha\tau}$$

where $\alpha = R_2/(R_1 + R_2)$ and $\tau = CR_1$.

In Laplace transform notation,

$$\frac{\bar{V}_2}{\bar{V}_1} = \frac{\alpha(1 + s\tau)}{1 + s\alpha\tau}$$

$$\simeq \alpha(1 + s\tau) \quad . \quad . \quad . \quad (6.20)$$

provided that $\alpha s\tau \ll 1$.

The block diagram for a simple R.P.C. servo with phase-advance stabilization in the error path is shown in Fig. 6.11 (*b*). The block $1/s^2J$ is obtained from the differential equation of the servo, since, neglecting friction,

$$\text{Developed torque} = J\frac{d^2\theta_o}{dt^2}$$

Hence, taking Laplace transforms,

$$s^2J\bar{\theta}_o = K_mK_A \times \text{Amplifier input}$$

or

$$\bar{\theta}_o = \frac{1}{s^2J}K_mK_A \times \text{Amplifier input}$$

Each of the term/ $1/s^2J$, K_m and K_A can be represented by blocks in the forward loop. The overall response of the system is

$$\bar{\theta}_o = K_p(\bar{\theta}_i - \bar{\theta}_o)\alpha(1 + s\tau)K_AK_m\frac{1}{s^2J}$$

so that

$$(s^2J + K\alpha\tau s + K\alpha)\bar{\theta}_o = K\alpha(1 + s\tau)\bar{\theta}_i \quad . \quad . \quad (6.21)$$

where $K = K_pK_AK_m$.

From this the differential equation of the system is

$$J\frac{d^2\theta_o}{dt^2} + K\alpha\tau\frac{d\theta_o}{dt} + K\alpha\theta_o = K\alpha\left(\theta_i + \tau\frac{d\theta_i}{dt}\right) \quad . \quad (6.22)$$

Normally α is about 0·1 and the gain of the amplifier must be increased to compensate for this.

For a step input, $\bar{\theta}_i = \theta_i/s$, and eqn. (6.21) yields

$$\bar{\theta}_o = \frac{\theta_i K\alpha(1 + s\tau)}{sJ\left(s^2 + \dfrac{K\alpha\tau}{J}s + \dfrac{K\alpha}{J}\right)}$$

$$= \theta_i \left[\frac{1}{s} - \frac{s}{s^2 + \dfrac{K\alpha\tau}{J}s + \dfrac{K\alpha}{J}}\right]$$

$$= \theta_i \left[\frac{1}{s} - \frac{s}{s^2 + 2\zeta\omega_n s + \omega_n^2}\right] \quad . \quad . \quad (6.23)$$

where $\omega_n{}^2 = K\alpha/J$, and $2\zeta\omega_n = K\alpha\tau/J$.

As before, the solution depends on the damping factor, being oscillatory if $\zeta < 1$, overdamped if $\zeta > 1$ and critically damped if $\zeta = 1$. Fig. 6.12 shows these responses.

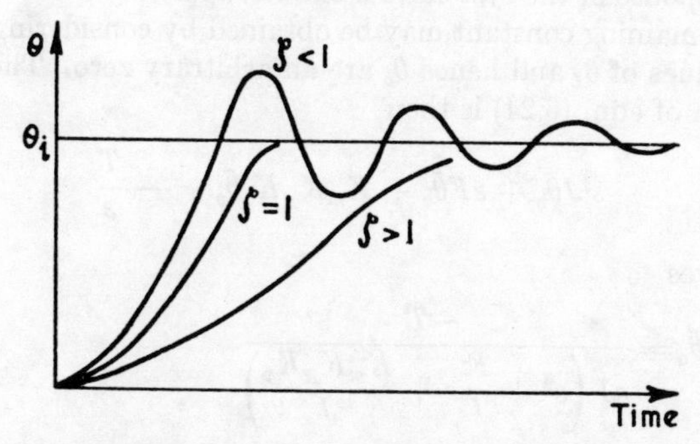

FIG. 6.12. RESPONSE OF PHASE-ADVANCE-STABILIZED SERVO TO A STEP INPUT

6.8. Errors in an R.P.C. Servo

Two causes of error in an R.P.C. servo will be considered. Firstly, if there is a constant load torque on the output shaft, then in the simple system of Fig. 6.9 there must be a constant torque from the motor, and this is possible only if there is an input to the motor—in other words, an error signal must be present at the amplifier input. Secondly, if the input is given a constant velocity, we shall show that the output shaft lags behind the input shaft position by a constant amount. This is called the *velocity lag*. Correction networks are used to eliminate these errors.

(*a*) *Load Torque Error.* Consider the simple R.P.C. servo of Fig. 6.9, driving a load which has a constant torque T. The differential equation of the system is

$$J\frac{d^2\theta_o}{dt^2} + F\frac{d\theta_o}{dt} + T = K_m K_A K_p(\theta_i - \theta_o) = K_m K_A K_p \varepsilon \quad . \quad (6.24)$$

where $\varepsilon = \theta_i - \theta_o$.

Under steady-state conditions there is no acceleration or velocity in the output shaft, and hence $d^2\theta_o/dt^2 = 0$ and $d\theta_o/dt = 0$, so that we may write

$$\varepsilon = \frac{T}{K_m K_A K_p} \quad . \quad . \quad . \quad . \quad (6.25)$$

The load-torque error can be made small by increasing the amplifier gain, K_A, but care must be taken in a practical system that instability does not arise. A very-high-gain amplifier also increases the problems of discrimination against noise.

The response of the system to a sudden application of load torque with θ_i remaining constant may be obtained by considering that the initial values of θ_i and hence θ_o are an arbitrary zero. The Laplace transform of eqn. (6.24) is then

$$s^2 J\bar{\theta}_o + sF\bar{\theta}_o + K_m K_A K_p \bar{\theta}_o = -\frac{T}{s}$$

which gives

$$\bar{\theta}_o = \frac{-T}{sJ\left(s^2 + \frac{F}{J}s + \frac{K_m K_A K_p}{J}\right)}$$

$$= \frac{-T}{K_m K_A K_p}\left[\frac{1}{s} - \frac{s + F/J}{s^2 + (F/J)s + K_m K_A K_p/J}\right]$$

$$= \frac{-T}{K_m K_A K_p}\left[\frac{1}{s} - \frac{s + 2\zeta\omega_n}{s^2 + 2\zeta\omega_n + \omega_n^2}\right]$$

where $2\zeta\omega_n = F/J$, and $\omega_n^2 = K_m K_A K_p/J$.

The solution consists of a step of size $-T/K_m K_A K_p$ plus a transient term, which is overdamped if $\zeta > 1$, underdamped if $\zeta < 1$, and critically damped if $\zeta = 1$. Note that, as we should expect, the step displacement is equal to the expression for angular error (eqn. (6.25)).

(*b*) *Velocity Lag.* Assume that the load torque is negligible, but that the input shaft is given a constant angular velocity, ω_i. When steady conditions have been reached the acceleration of the output shaft will be zero; and the output and input shafts will have the same velocity, ω_i, which can be seen as follows. The differential equation is

$$J\frac{d^2\theta_o}{dt^2} + F\frac{d\theta_o}{dt} + K\theta_o = K\theta_i = K\omega_i t \quad . \quad . \quad (6.26)$$

Taking Laplace transforms of this equation,

$$s^2J\bar{\theta}_o + sF\bar{\theta}_o + K\bar{\theta}_o = \frac{K\omega_i}{s^2}$$

or

$$\bar{\theta}_o = \frac{K\omega_i}{s^2(s^2J + sF + K)}$$

so that

$$\bar{\omega}_o = s\bar{\theta}_o = \frac{K\omega_i}{s(s^2J + sF + K)}$$

and, by the final value theorem, ω_o as $t \to \infty = s\bar{\omega}_o$ as $s \to 0$, so that $\omega_o = \omega_i$ in the steady state. We may therefore write the equation of motion (putting $d^2\theta_o/dt^2$ as zero in eqn. (6.26)) as

$$F\frac{d\theta_o}{dt} = K(\theta_i - \theta_o)$$

or

$$F\omega_o = F\omega_i = K\varepsilon$$

so that

$$\varepsilon = \frac{F\omega_i}{K} = \frac{F\omega_i}{K_m K_A K_p} \quad . \quad . \quad . \quad (6.27)$$

This is called the *velocity lag* or *velocity error*. It may be reduced by increasing the amplifier gain, K_A.

An alternative which reduces the error to zero in both the above cases is to introduce a device before the amplifier input which will give a signal proportional to the error *and* to the integral of the error. Only when the error is zero will the integral term stop increasing. Hence the motor torque will continue to increase until the error is zero. This integral control, or reset, is difficult to achieve electrically owing to the long time-constants required, but is a relatively simple process in pneumatic or hydraulic control systems.

If phase advance stabilization is used, there is no velocity error. This is seen, since if $\omega_o = \omega_i$, and $d^2\theta_o/dt^2 = 0$ under steady-state conditions, eqn. (6.22) becomes

$$K\alpha\tau\omega_o + K\alpha\theta_o = K\alpha\theta_i + K\alpha\tau\omega_i$$

so that $\theta_i - \theta_o = 0$.

In practice, a small amount of viscous damping will be present, and there will therefore be a small velocity error, the steady-state equation being

$$(F + K\alpha\tau)\omega_o + K\alpha\theta_o = K\alpha\theta_i + K\alpha\tau\omega_i$$

In passing it may be noted that eqn. (6.26) and all previous differential equations describing the action of servos can be expressed in terms of the error $\varepsilon = (\theta_i - \theta_o)$. Thus eqn. (6.26) becomes

$$J\frac{d^2(\theta_i - \varepsilon)}{dt^2} + F\frac{d(\theta_i - \varepsilon)}{dt} = k\varepsilon$$

or
$$J\frac{d^2\varepsilon}{dt^2} + F\frac{d\varepsilon}{dt} + k\varepsilon = J\frac{d^2\theta_i}{dt^2} + F\frac{d\theta_i}{dt}$$

which can be solved to give ε in terms of θ_i.

Example 6.3. An R.P.C. servomechanism with viscous friction damping uses synchros for error detection. The synchro output is 1 V per degree of error, the motor torque constant referred to the load shaft is 5×10^{-2}N per milliampere of field current, the moment of inertia of motor and load is 0·8 kg-m^2, and the amplifier transconductance is 400 mA/V. Determine (*a*) the viscous friction coefficient in newton-metres per radian per second to give a damping ratio of 0·6, (*b*) the undamped natural frequency, and (*c*) the steady-state velocity error when the input shaft rotates continuously at 12 rev/min. (*R.R.E.*)

(*b*) The equation of motion is

$$J\frac{d^2\theta_o}{dt^2} + F\frac{d\theta_o}{dt} = K_s K_A K_m(\theta_i - \theta_o)$$

where, from eqn. (6.12), $\omega_n = \sqrt{(K_s K_A K_m/J)}$.

K_s must be expressed in volts per radian.

Thus $\omega_n = \sqrt{(57{\cdot}3 \times 400 \times 5 \times 10^{-2}/0{\cdot}8)} = 37{\cdot}9$ rad/sec

and the undamped natural frequency is $37{\cdot}9/2\pi = \underline{\underline{6{\cdot}03 \text{ c/s}}}$

(*a*) Also, from eqn. (6.13) for critical damping ($\zeta = 1$),

$$2\zeta\omega_n = \frac{F}{J}$$

so that $F_{crit} = 2\omega_n J = 60{\cdot}6$. Hence, for a damping ratio of 0·6,

$$F = \zeta F_{crit} = 0{\cdot}6 \times 60{\cdot}6 = \underline{\underline{36{\cdot}36}} \text{ N-m per rad/sec.}$$

(*c*) From eqn. (6.27),

$$\text{Velocity lag} = \frac{F\omega_i}{K_s K_A K_m} = \frac{36{\cdot}36 \times 12 \times 2\pi}{60 \times 57{\cdot}3 \times 400 \times 5 \times 10^{-2}} \text{ radian}$$

$$= \underline{\underline{2{\cdot}28^\circ}}$$

Example 6.4. The angular position of an aerial is controlled by a closed-loop automatic control system to follow an input wheel. The input wheel is maintained in sinusoidal oscillation through $\pm 34°$ with an angular frequency $\omega = 1$ rad/s. The moving part of the aerial system, which is critically damped, has a moment of inertia of 200 kg-m^2 and a viscous frictional torque of 1,600 N-m per rad/s.

Calculate the amplitude of swing of the aerial and the time-lag between the aerial and the input wheel.

If the frictional torque were made negligibily small, the stiffness of the system being unaltered, what whould then be the amplitude of swing and the time-lag? (*A.E.E.*, 1964)

The differential equation representing the motion of the aerial is

$$J\frac{d^2\theta_0}{dt^2} + F\frac{d\theta_0}{dt} + K_s\theta_0 = K_s\theta_i$$

where θ_o is the instantaneous output angular position, θ_i is the input wheel angular position, and K_s respresents the stiffness of the system. Taking Laplace transforms,

$$\bar{\theta}_0 = \frac{K_s\bar{\theta}_i}{J\left(s^2 + \dfrac{F}{J}s + \dfrac{K_s}{J}\right)} \quad . \quad . \quad . \quad \text{(i)}$$

Since the system is critically damped, eqn. (6.13) gives

$$\frac{F}{2\sqrt{(JK_s)}} = 1$$

so that the undamped natural angular frequency is

$$\omega_n = \frac{F}{2J} = 4 \text{ rad/s}$$

The motion of the input wheel is represented by

$$\theta_i° = 34 \sin t$$

Hence

$$\bar{\theta}_i = \frac{34}{s^2 + 1}$$

and eqn. (i) becomes

$$\bar{\theta}_0 = \frac{16 \times 34}{(s^2+1)(s^2+8s+16)}$$

$$= 16 \times 34\left[\frac{as+b}{s^2+1} + \frac{ds+e}{s^2+8s+16}\right]$$

where

$$d = -a \quad . \quad . \quad . \quad . \quad \text{(ii)}$$

$$8a + b + e = 0 \quad . \quad . \quad . \quad . \quad \text{(iii)}$$

$$8b + 16a + d = 8b + 15a = 0 \quad . \quad . \quad . \quad \text{(iv)}$$

and

$$16b + e = 1 \quad . \quad . \quad . \quad . \quad \text{(v)}$$

The problem asks for the steady-state solution only, so that it is only necessary to find the constants a and b, since the second term in the equation for $\bar{\theta}_0$ involves a decaying exponential.

From eqns. (v) and (iii),

$$8a + 1 - 15b = 0 \quad . \quad . \quad . \quad . \quad \text{(vi)}$$

From eqns. (iv) and (vi),

$$a = \frac{-8}{289} \quad \text{and} \quad b = \frac{15}{289}$$

so that the steady-state part of the solution is

$$\bar{\theta}_{0\,s.s.} = 16 \times 34 \frac{\frac{-8}{289}s + \frac{15}{289}}{s^2 + 1}$$

$$= \frac{16 \times 34}{289} \sqrt{(8^2 + 15^2)} \frac{-s \sin \phi + 1 \cos \phi}{s^2 + 1}$$

where $\tan \phi = 8/15$, or $\phi = 28°$.

It follows from Transform 12 that

$$\theta_{0\,s.s.} = 32 \sin (t - 28°)$$

The amplitude of the aerial oscillation is $\underline{\underline{\pm\, 32°}}$

The time lag corresponds to the phase lag of 28° and is $28 \times (\pi/180)$ $= \underline{\underline{0{\cdot}49 \text{ sec}}}$

With negligible frictional torque, the equation for $\bar{\theta}_0$ becomes

$$\bar{\theta}_0 \simeq \frac{16 \times 34}{(s^2 + 1)(s^2 + 16)} = \frac{16 \times 34}{15} \left[\frac{1}{s^2 + 1} - \frac{1}{s^2 + 16} \right]$$

so that $\theta_0 \simeq 36{\cdot}2° (\sin t - \frac{1}{4} \sin 4t)$

The amplitude of the fundamental would be 36·2° and the time lag would be zero. There would, however, be an additional oscillation of four times the fundamental frequency. In a practical system with even very little damping this additional term would rapidly disappear to give the steady-state solution

$$\theta_{0\,s.s.} = \underline{\underline{36{\cdot}2° \sin t}}$$

6.9. The Velodyne

Accurate control of the speed of small motors may be achieved by the arrangement shown in Fig. 6.13. The motor field is fed by the error amplifier, and drives the output shaft at a speed which depends on the input voltage, v. A tacho-generator on the output shaft supplies the feedback voltage. This is called the *velodyne speed*

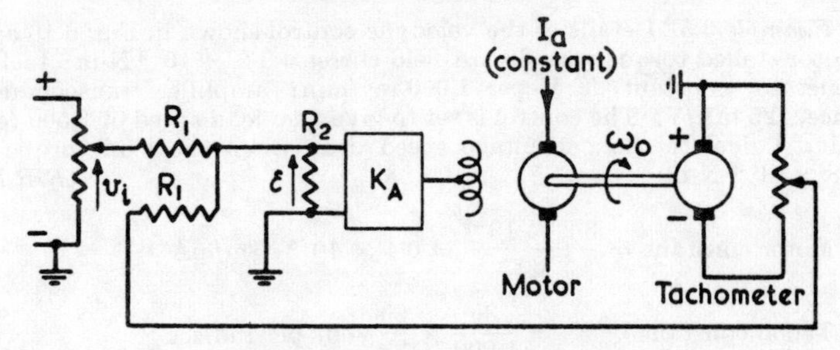

FIG. 6.13. VELODYNE SPEED CONTROL

control. If the motor torque constant is K_m newton-metres per milliampere, then, neglecting friction and loading,

$$J \frac{d\omega_o}{dt} = K_A K_m \varepsilon$$

where K_A is the amplifier constant and ε the error signal.

If $R_2 \gg R_1$ then

$$\varepsilon = \tfrac{1}{2}(v_i - K_t\omega_o) \quad . \quad . \quad . \quad (6.28)$$

where K_t is the tachometer constant in volts per radian per second. When a steady output speed is attained, $d\omega_o/dt = 0$ and hence the error, ε, is zero; then

$$v_i = K_t\omega_o \quad \text{or} \quad \omega_o = \frac{v_i}{K_t} \quad . \quad . \quad (6.29)$$

Effect of Loading. Consider a constant load torque, T_L, on the output shaft. The equation of motion is

$$J \frac{d\omega_o}{dt} + T_L = K_A K_m \varepsilon$$

For steady output speed, $d\omega_o/dt = 0$ and hence

$$\varepsilon = \frac{T_L}{K_A K_m} \quad . \quad . \quad . \quad . \quad (6.30)$$

This is the error signal required to supply the load torque, and represents a difference between the desired speed (as set by the input voltage-divider) and the output speed. Since the error or "droop" is independent of speed, its effect will be proportionately worse the lower the desired speed. Droop may be eliminated by the use of integral feedback, and will be small if K_A is large enough (but increasing K_A may lead to stability troubles).

Example 6.5. Details of the velodyne control shown in Fig. 6.13 are: motor stalled torque with 80 mA field current, $3{\cdot}2 \times 10^{-2}$ N-m; tacho-generator constant, 30 V per 1,000 rev/min; amplifier transconductance, 125 mA/V. The control is set to give a no-load speed of 1,500 rev/min. Calculate the percentage speed drop when the load torque is $1{\cdot}8 \times 10^{-2}$ N-m. (*R.R.E.*)

$$\text{Motor constant} = \frac{3{\cdot}2 \times 10^{-2}}{80} = 0{\cdot}4 \times 10^{-3}\ \text{N-m/mA}$$

$$\text{Tacho constant} = K_t = \frac{30}{1{,}000} \times \frac{60}{2\pi}\ \text{volts per rad/sec}$$

Under no-load conditions, the error is zero, and hence

$$v_i = K_t\omega_o$$

$$= \frac{30}{1{,}000} \times \frac{60}{2\pi} \times \frac{1{,}500}{60} \times 2\pi = \underline{\underline{45\ \text{V}}}$$

On load, the error voltage given by eqn. (6.30) is

$$\varepsilon = \frac{T_L}{K_A K_m} = \frac{1{\cdot}8 \times 10^{-2}}{125 \times 0{\cdot}4 \times 10^{-3}} = 0{\cdot}36\ \text{V}$$

If the on-load speed is ω_o', then, from eqn. (6.28),

$$0{\cdot}36 = \tfrac{1}{2}\,(v_i - K_t\omega_o')$$

so that

$$\omega_o' = \frac{45 - 0{\cdot}72}{30 \times 60/(1{,}000 \times 2\pi)}\ \text{rad/sec}$$

$$= \frac{44 \times 28 \times 1{,}000}{30} = \underline{\underline{1{,}476\ \text{rev/min}}}$$

Effect of Friction. If there is viscous friction present, the equation of motion can be written as

$$J\,\frac{d\omega_0}{dt} + F\omega_0 = K(\omega_i - \omega_0)$$

where ω_i is the desired or "set" input speed, K is the motor torque per unit error, and F is the friction torque per rad/sec.

Taking Laplace transforms,

$$\bar{\omega}_0 = \frac{K\bar{\omega}_i}{J\left(s + \dfrac{F + K}{J}\right)}$$

For a step input, $\bar{\omega}_i = \omega_i/s$, and hence

$$\bar{\omega}_0 = \frac{K\omega_i}{Js\left(s + \dfrac{F + K}{J}\right)}$$

From the final-value theorem, the steady-state output velocity is

$$\omega_{0\,s.s.} = \lim_{s\to 0} s\bar{\omega}_0 = \frac{K\omega_i}{(F+K)}$$

and the speed error is

$$\omega_i - \omega_{0\,s.s.} = \frac{F\omega_i}{F+K} \qquad . \qquad . \qquad (6.31)$$

6.10. The Operational Amplifier

The feedback principle may be employed with electronic amplifiers to make them perform mathematical operations—in particular, sign inversion, summation and integration. These *operational amplifiers* form the basic building bricks of electronic analogue computers. Fig. 6.14 (*a*), page 193, is the schematic of such an amplifier. The triangular block represents a d.c. amplifier of extremely high voltage gain ($-A$). The input impedance of the d.c. amplifier is R_{in}.

(*a*) *Overall Gain.* For the circuit shown in Fig. 6.14 (*a*),

$$i_1 = \frac{v_i - v_E}{Z_1} \quad \text{and} \quad i_2 = \frac{v_E - v_o}{Z_2} \qquad . \qquad . \qquad (6.32)$$

Also

$$v_o = -Av_E \quad \text{and} \quad iR_{in} = v_E$$

At point G, $i_1 = i_2 + i$, so that

$$\frac{v_i}{Z_1} - \frac{v_E}{Z_1} = \frac{v_E}{Z_2} - \frac{v_o}{Z_2} + \frac{v_E}{R_{in}}$$

From this

$$\frac{v_i}{Z_1} = -\frac{v_o}{Z_2} - \frac{v_o}{A}\left(\frac{1}{Z_1} + \frac{1}{Z_2} + \frac{1}{R_{in}}\right)$$

Hence,

$$v_o = -v_i \frac{Z_2}{Z_1}\left[\frac{1}{1 + \dfrac{1}{A}\left(\dfrac{Z_2}{Z_1} + 1 + \dfrac{Z_2}{R_{in}}\right)}\right] \qquad . \qquad . \qquad (6.33)$$

Typically $A = 10^7$, so that the error involved in neglecting the second term in the denominator is negligible, and to better than 1 part in 10^6,

$$v_o = -v_i \frac{Z_2}{Z_1} \qquad . \qquad . \qquad . \qquad . \qquad (6.34)$$

(*b*) *D.C. Amplifier Input Current.* From eqns. (6.32), the ratio of currents through Z_2 and R_{in} is given by

$$\frac{i_2}{i} = \frac{\dfrac{v_E}{Z_2} - \dfrac{v_o}{Z_2}}{\dfrac{v_E}{R_{in}}} = \frac{-\dfrac{v_o}{Z_2}\left(\dfrac{1}{A} + 1\right)}{-\dfrac{v_o}{AR_{in}}} = \frac{(A+1)R_{in}}{Z_2}$$

This shows that the current through the input impedance of the d.c. amplifier is only a small fraction, $Z_2/(A+1)R_{in}$, of the current through the feedback impedance, Z_2, so that we may assume that all of the input current, i_i, flows through Z_2.

The impedance looking into the circuit from the point G is

$$Z' = \frac{v_E}{i_2 + i} = \frac{v_E}{\dfrac{v_E}{Z_2} - \dfrac{v_o}{Z_2} + \dfrac{v_E}{R_{in}}}$$

$$= \frac{v_E}{\dfrac{v_E}{Z_2} + \dfrac{Av_E}{Z_2} + \dfrac{v_E}{R_{in}}}$$

$$\simeq \frac{Z_2}{A} \quad . \quad . \quad . \quad . \quad . \quad . \quad (6.35)$$

provided that $A \gg 1$.

(*c*) *Virtual Earth Point.* Since A is normally very large the voltage v_E will be only a small fraction of a volt, and the point G is therefore almost at earth potential. It is called a *virtual earth point*, and the operational amplifier is often referred to as a *virtual earth amplifier*. The actual impedance to earth at G is, of course, Z_2/A.

(*d*) *Input and Output Impedances.* Since G is a virtual earth, the input impedance of the operational amplifier is effectively Z_1, and the output impedance is Z_2 in parallel with the output impedance of the d.c. amplifier.

(*e*) *The Inverting Amplifier.* Fig. 6.14 (*b*) is a schematic of an *inverting amplifier*. The input and feedback impedances are two equal resistances R (typically 1 MΩ). From eqn. (6.34),

$$v_o = -v_i \quad . \quad . \quad . \quad . \quad (6.36)$$

so that the amplifier acts as a sign-reversing device.

(*f*) *The Summing Amplifier.* A *summing amplifier* is shown in Fig. 6.14 (*c*).

Assuming that G is a virtual earth point, and that negligible current flows into the d.c. amplifier, then

$$i = i_1 + i_2 + i_3$$

or

$$-\frac{v_o}{R} = \frac{v_1}{R_1} + \frac{v_2}{R_2} + \frac{v_3}{R_3}$$

which yields

$$v_o = -v_1\frac{R}{R_1} - v_2\frac{R}{R_2} - v_3\frac{R}{R_3} \quad . \quad (6.37)$$

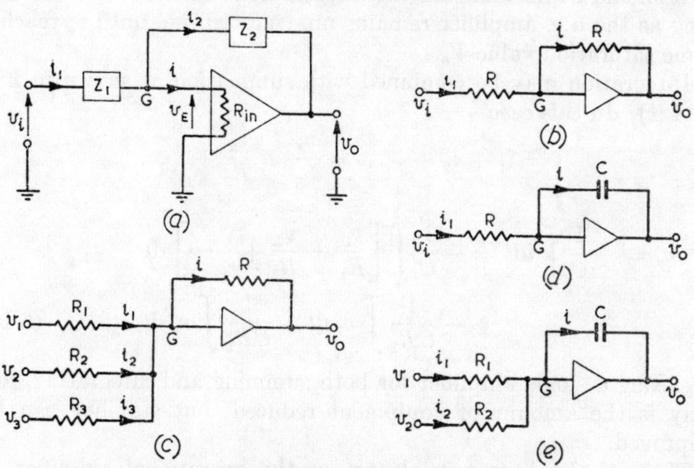

FIG. 6.14. OPERATIONAL AMPLIFIER CIRCUITS
(*a*) General
(*b*) Inverting
(*c*) Summing
(*d*) Summing and Integrating

Hence the output is the negative of the sum of the inputs taken in ratios which depend on the individual input resistances. Typically ratios of 1:1 or 10:1 are used. The advantage over the resistive summing junction of Section 6.3 is that gains of unity or more may be achieved.

(*g*) *The Integrating Amplifier*. In the *integrating amplifier*, Fig. 6.14 (*d*), the feedback impedance is provided by a capacitor. The current i is v_i/R, and the capacitor voltage is $v_C = (1/C)\int i \, dt$, so that

$$v_o = -v_C = -\frac{1}{C}\int \frac{v_i}{R} dt$$

$$= -\frac{1}{CR}\int v_i \, dt \quad . \quad . \quad . \quad . \quad (6.38)$$

The output voltage is thus the product of a time-constant (CR) and the time integral of the input voltage. Typically the CR constant is either 1 or 0·1. The result may be expressed in Laplace form direct from eqn. (6.34) by writing

$$\bar{v}_o = -\bar{v}_i \frac{\bar{Z}_2}{\bar{Z}_1} = -\frac{1}{sCR}\bar{v}_i \quad . \quad . \quad . \quad (6.39)$$

This is recognized as representing an output which is $-1/CR$ times the integral of the input (eqn. (6.38)).

It should be noticed that the integrating action continues only so long as the d.c. amplifier remains unsaturated, i.e. until v_o reaches some saturation value V_o.

Integration may be combined with summation as shown in Fig. 6.14 (*e*). In this case

$$i = i_1 + i_2 + \ldots = \frac{v_1}{R_1} + \frac{v_2}{R_2} + \ldots$$

or

$$v_o = -\frac{1}{C}\int i\mathrm{d}t = -\frac{1}{C}\int\left[\frac{v_1}{R_1} + \frac{v_2}{R_2} + \cdots\right]\mathrm{d}t$$

$$= -\frac{1}{CR_1}\int v_1\,\mathrm{d}t - \frac{1}{CR_2}\int v_2\,\mathrm{d}t - \ldots \quad (6.40)$$

By using a single amplifier for both summing and integrating, not only is the amount of equipment reduced, but stability can be improved.

Theoretically it is possible to use the operational amplifier to perform differentiation, by making Z_1 a capacitive reactance and Z_2 a resistance (development of the proof is left to the reader). In practice, differentiators are not used since they tend to be unstable and to be very sensitive to noise, and it is difficult to avoid ringing in the output circuit.

6.11. The Analogue Computer

Operational amplifiers combined with coefficient-setting voltage dividers can be used as *analogue computers*. In these, the problem to be solved is represented by an electrical analogue, the variables of the actual problem being represented by voltages. The response of the physical system to any given driving function or disturbance can then be observed as the electrical output of the computer when the input is an electrical signal corresponding to the driving function of the physical system. The computer can be used to solve problems whose complexity precludes a purely mathematical solution. It is much used in design studies, since the analogue parameters are easily varied by altering the setting of voltage dividers, and the effect of this alteration on the output is readily observed. Large analogue computers may employ hundreds of operational amplifiers. Two simple problems will be described to give some idea of the basic principles involved.

(*a*) *The Differential Analyser*. Consider the differential equation

$$a\frac{\mathrm{d}^2\theta}{\mathrm{d}t^2} + b\frac{\mathrm{d}\theta}{\mathrm{d}t} + c\theta = f(t)$$

This may be rewritten as

$$\frac{\mathrm{d}^2\theta}{\mathrm{d}t^2} = \frac{1}{a}f(t) - \frac{b}{a}\frac{\mathrm{d}\theta}{\mathrm{d}t} - \frac{c}{a}\theta \quad . \quad . \quad . \quad (6.41)$$

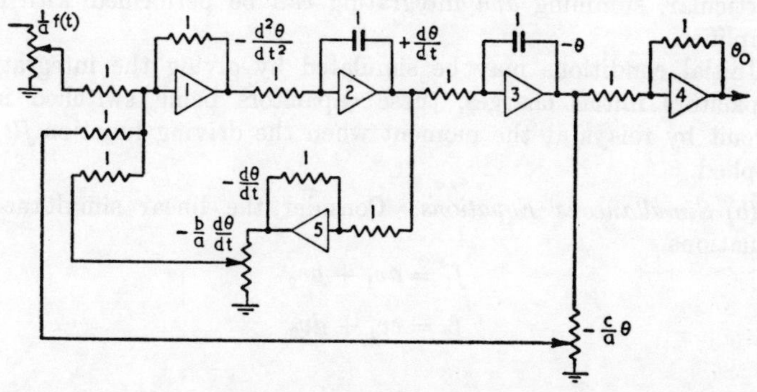

Fig. 6.15. Operational Amplifier to Solve a Second-order Differential Equation

$$\frac{\mathrm{d}^2\theta}{\mathrm{d}t^2} = \frac{1}{a}f(t) - \frac{b}{a}\frac{\mathrm{d}\theta}{\mathrm{d}t} - \frac{c}{a}\theta$$

An analogue of this is set up in Fig. 6.15. Unity coefficients are assumed for all operational elements. Amplifier 1 acts as a summing and inverting amplifier. The inputs are

$$\frac{1}{a}f(t) \qquad -\frac{b}{a}\frac{\mathrm{d}\theta}{\mathrm{d}t} \qquad \text{and} \qquad -\frac{c}{a}\theta$$

so that the output is

$$-\left(\frac{1}{a}f(t) - \frac{b}{a}\frac{\mathrm{d}\theta}{\mathrm{d}t} - \frac{c}{a}\theta\right)$$

which is equal to $-\mathrm{d}^2\theta/\mathrm{d}t^2$ (by eqn. (6.41)).

Amplifier 2 integrates this, and hence its output is $+\mathrm{d}\theta/\mathrm{d}t$. The output of this second amplifier is fed into an inverting amplifier (5) and a fraction, b/a, of its output is fed into amplifier 1.

Amplifier 3 integrates $d\theta/dt$ and so gives an output of $-\theta$. A fraction, c/a, of this is fed into the input of amplifier 1, while one further inverting amplifier (4) provides the desired output, θ, which is the required solution. Hence, for any input voltage, $f(t)$, the corresponding output voltage can be observed on an indicating instrument, a recorder or a cathode-ray oscillograph.

In practice it may be necessary to use amplifiers with voltage gains of 10 in order to give ratios b/a or c/a greater than unity, and care must be taken to check that the output voltage from each amplifier is neither too small nor exceeds the saturation value. In some cases it is necessary to change the time scale in order to obtain a solution. It should also be noted that the circuit shown is not the only one which will provide a solution to the equation. In particular, summing *and* integrating can be performed with one amplifier.

Initial conditions may be simulated by giving the integrating capacitors initial charges, these capacitors being switched into circuit by relays at the moment when the driving function $f(t)$ is applied.

(*b*) *Simultaneous Equations.* Consider the linear simultaneous equations

$$f_1 = av_1 + bv_2$$

$$f_2 = cv_1 + dv_2$$

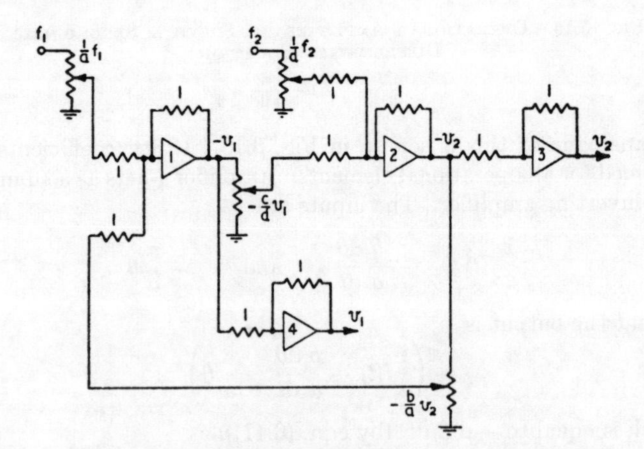

FIG. 6.16. ANALOGUE SOLUTION OF THE EQUATIONS

$$f_1 = av_1 + bv_2$$
$$f_2 = cv_1 + dv_2$$

where, for given values of f_1 and f_2, it is required to determine v_1 and v_2. The equations can be rewritten as

$$v_1 = \frac{1}{a}f_1 - \frac{b}{a}v_2$$

and

$$v_2 = \frac{1}{d}f_2 - \frac{c}{d}v_1$$

One possible analogue circuit is shown in Fig. 6.16. The input to amplifier 1 is $(1/a)f_1 - (b/a)v_2$, so that its output is $-v_1$. The input to amplifier 2 is $(1/d)f_2 - (c/a)v_1$, so that its output is $-v_2$. Inverting amplifiers 3 and 4 give outputs v_2 and v_1 for any given inputs f_1 and f_2.

Bibliography

Stockdale, *Servomechanisms* (Pitman).

Westwater and Waddell, *An Introduction to Servomechanisms* (English Universities Press).

Rogers and Connolly, *Analog Computation in Engineering Design* (McGraw-Hill).

Problems

6.1. A servo system for the positional control of a rotatable mass is stabilized by viscous damping which is less than that required for critical damping. Derive an expression for the output of the system if the input member is suddenly moved to a new position, the system being initially at rest.

Calculate the amount of the first overshoot if the undamped natural frequency is 5 c/s and the frictional forces are one-third of the forces required for critical damping. (*A.E.E.*, 1957)

(*Ans.* 0·31 × input step.)

6.2. Describe briefly the essentials of a simple remote-position-control servomechanism stabilized by direct velocity feedback, and show that its operation is characterized by a differential equation of the form

$$\frac{1}{\omega_n^2}\frac{d^2\theta_o}{dt^2} + T\frac{d\theta_o}{dt} + \theta_o = \theta_i$$

The position of a rotatable mass driven by an electric motor is controlled from a handwheel. The damping torques due to viscous friction and velocity feedback, respectively, are equal. The moment of inertia of the moving parts referred to the mass is 100 kg-m², and the undamped natural frequency is 2·5 c/s. If the motion is critically damped, calculate the feedback torque per unit angular velocity and the steady-state angular misalignment when the input angle is varied at the rate of 1 rad/sec. [*Hint.* Remember that there is error feedback as well as velocity feedback.] (*A.E.E.*, 1960)

(*Ans.* 1,570 N-m per rad/sec; 0·127 rad.)

6.3. The angular position of a rotatable mass is controlled from a hand-wheel by means of a closed-loop electrical servomechanism which is critically damped by viscous friction. Set up the differential equation for the system. Derive an expression for the angular position of the mass at any time if, with the system initially at rest, the control wheel is suddenly turned through an angle θ_1.

Given that the moment of inertia of the moving parts is 500 kg-m² and the motor torque is 2,000 N-m per radian of misalignment, and that $\theta_1 = 90°$, calculate the angle of misalignment 2 sec after moving the control wheel. (*A.E.E.*, 1958)

(*Ans.* 0·146 rad.)

6.4. An error-actuated electrical servomechanism is employed to control the angular position of a rotatable mass, subject to viscous damping, in response to the rotation of a control handle. Give a circuit diagram of a suitable scheme, and set up the equation of motion of the system.

In a particular case, the moment of inertia of the rotating parts referred to the mass under control is 2,000 kg-m², and the system is critically damped. The motor produces a torque of 1,800 N-m per minute of misalignment. Calculate the steady-state angular error if the control handle is continuously rotated at a speed of 2 rev/min.

(*Ans.* 0·00752 rad.) (*A.E.E.*, 1957)

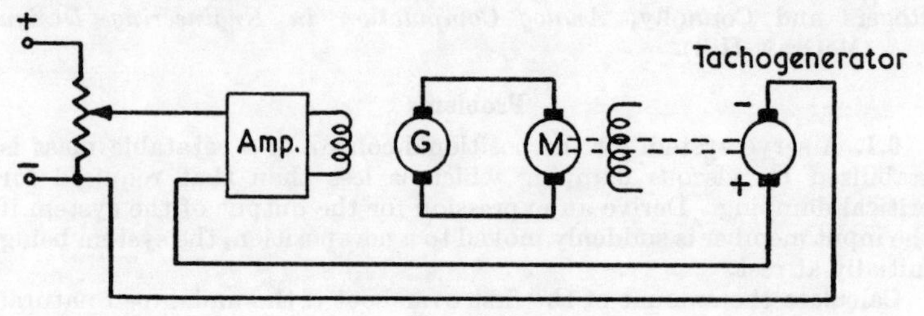

FIG. 6.17

6.5. In the speed-regulating system shown in Fig. 6.17 the generator field time-constant and that of the motor-generator armature are both negligible. The motor and load mechanical time-constant, τ_m, is 1 sec. The amplifier transconductance is 4 A/V, the generator e.m.f. is 50 V per field ampere, the steady speed is 0·25 V per rad/sec. Find the transfer function relating the output speed, $\bar{\omega}_o$, to the input voltage setting, $\bar{v}_i$. Find the time variation of the output speed for a sudden input of 5 V applied with the system initially at rest.

(*Ans.* $\bar{\omega}_o = \dfrac{50\bar{v}_i}{s+11}$; $\omega_0(t) = 22{\cdot}7(1 - e^{-11t})$.)

6.6. Fig 6.18 represents a proportional control system. An input signal θ_i passes through a summation device which subtracts from it the output signal θ_o to give an error signal θ. This error signal passes to an amplifier A, which has a gain K, and thence to the system under control represented by B, the output of which is θ_o. This system can be simulated by three independent RC exponential lags of 5, 2 and 1 minutes each.

Plot approximately to scale the magnitude ratio and phase difference of θ_o and θ using a logarithmic scale for the magnitude ratio and a linear scale for the phase difference, both being plotted to a logarithmic scale of frequency. Determine from these diagrams the magnitude of K which will cause the system to hunt. (*L.U. Part III Elect.*, 1957)

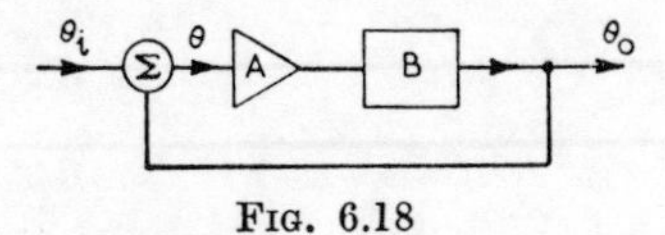

FIG. 6.18

[*Hint*. It is required to plot the Bode diagram for the *open loop* system (i.e. θ_o/θ), using the linear asymptotes. The condition for stability is that the magnitude ratio shall be less than unity when the phase angle is 180°.]

(*Ans.* 12·6.)

6.7. The signals from an error-detecting system using voltage dividers are fed to an amplifier by star adding through resistors of 100 kΩ. The star point is connected to earth through a 1-MΩ resistor. If each voltage divider is wound over 344° and has 30 V applied across it, find the input to the amplifier per radian error between the two voltage dividers.

(*Ans.* 2·38 V/rad.)

6.8. Find the overall, or closed-loop, transfer function, and hence derive the open-loop transfer functions of each of the servomechanisms described by the following equations.

(i) $\dfrac{d^2\theta_0}{dt^2} + 9\dfrac{d\theta_0}{dt} + 25\theta_o = 25\theta_i + 5\dfrac{d\theta_i}{dt}$ (*Ans.* $5(s+5)/s(s+4)$).

(ii) $\dfrac{d^2\theta_0}{dt^2} + 9{\cdot}6\dfrac{d\theta_0}{dt} + 16\theta_0 = 16\theta_i + 4\dfrac{d\theta_i}{dt}$ (*Ans*: $4(s+4)/s(s+5{\cdot}6)$)

(iii) $\dfrac{d^2\theta_0}{dt^2} + 3\dfrac{d\theta_0}{dt} + 10\theta_0 = 10\theta_i$ (*Ans*: $10/s(s+3)$).

(iv) $0{\cdot}7\dfrac{d^2\theta_0}{dt^2} + \dfrac{d\theta_0}{dt} + 5\theta_0 = 5\theta_i$ (*Ans*: $5/s(0{\cdot}7s+1)$).

6.9. The motor in an r.p.c. servomechanism has a moment of inertia of $3{\cdot}2 \times 10^{-6}$ kg-m² and is geared to the load through 100:1 gearing. The error-measuring device gives 1 V per degree error at the load, the motor torque constant is 10^{-4} N-m/mA, and the amplifier transconductance is 40 mA/V. Friction is negligible. Determine the natural frequency of oscillation of the system. [*Hint*. Inertia at load is $n^2 \times$ inertia at motor; torque at load is $n \times$ torque at motor.]

(*Ans.* 4·26 c/s.)

6.10. An r.p.c. servo uses synchros for error detection, giving an output of 1 V per degree error. The motor is coupled through a 100:1 reduction gear to the load. Given that the amplifier transconductance is 400 mA/V, the motor torque constant is 5×10^{-4} N-m/mA, the motor inertia is 30×10^{-6} kg-m² and the load inertia 0·5 kg-m², calculate (*a*) the viscous friction coefficient in newton-metre-seconds at the output shaft to give a damping ratio of 0·6, (*b*) the undamped natural frequency, and (*c*) the steady-state error in degrees when the input shaft is rotated at a constant speed of 12 rev/min.

(*Ans.* 36·3 N-m-sec; 6·03 c/s; 2·28°.)

6.11. Explain Nyquist's criterion to determine the stability of a servo system.

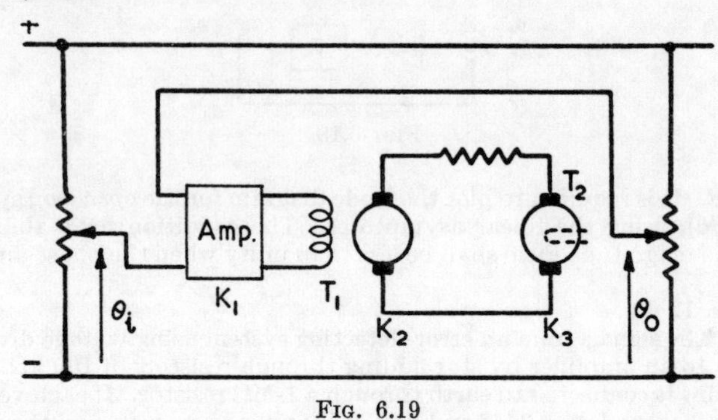

FIG. 6.19

The open-loop transfer function of the position-control system shown in Fig. 6.19 is given by

$$\frac{\theta_o}{\theta} = \frac{K_1K_2K_3}{s(1 + s\tau_1)(1 + s\tau_2)}$$

where τ_1 generator-field time-constant = 0·5 sec; τ_2 = mechanical time-constant of motor = 0·1 sec; K_1 = amplifier gain; K_2 = (generator armature voltage)/(generator field voltage) = 2; and K_3 = change of output voltage per second per volt applied to the motor armature = 0·01.

Find the maximum value of K_1 to give a stable system within the Nyquist criterion. (*I.E.E. Util.*, 1954)

(*Ans.* 240.)

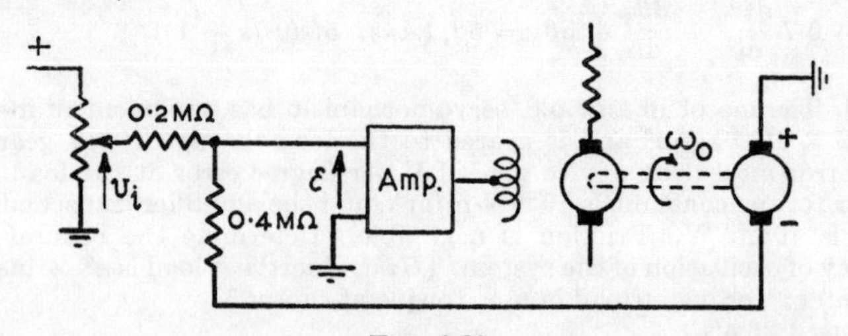

FIG. 6.20

6.12. The velodyne speed control shown in Fig. 6.20 has an amplifier with an infinite input impedance, and a transconductance of 500 mA/V. The motor develops a torque of 1·36 × 10⁻⁴ N-m per milliampere of field current, and the tachogenerator constant is 0·025 V per rev/min.

Find the steady speed at which the motor will run on no-load when the input voltage-divider is set at 18 V. If the load torque is 0·0678 N-m, find the percentage speed regulation when the input is 18 V. [*Hint.* develop an expression for the error voltage.] (*R.R.E.*)

(*Ans.* 1,440 rev/min; 1,320 rev/min; 8·3 per cent.)

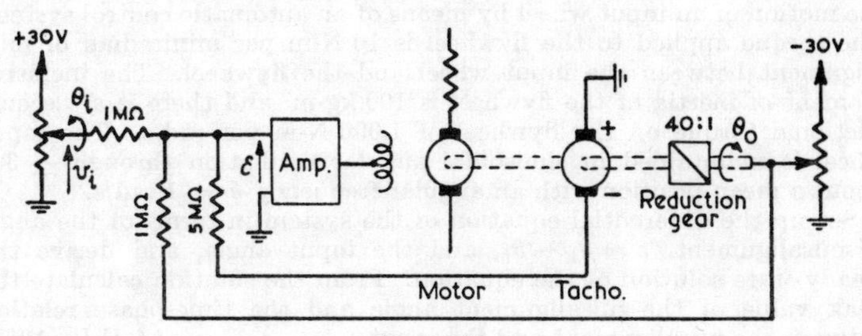

FIG. 6.21

6.13. The r.p.c. servo shown in Fig. 6.21 employs voltage dividers which subtend 300° and have 30 V across them. Find the tachometer constant in volts per 1,000 rev/min so that the system time-constant shall be 1/6 sec. [Note that if the system transfer function can be represented in the form $\bar{\theta}_o/\bar{\theta} = K/(1 + s\tau)$, the time-constant is τ.]

(*Ans.* 12·5 V per 1,000 rev/min.) (*R.R.E.*)

6.14. A torque-controlled servo uses a split-field motor. The damping ratio is to be 0·5, and the steady-state error is to be 0·5° for an input speed of 5 rev/min. Other details are: the transducer gives 18 V per radian error; the amplifier transconductance is 125 mA/V; the motor torque constant is 4×10^{-4} N-m/mA; the motor inertia is $1{\cdot}6 \times 10^{-5}$ kg-m² (at the motor shaft); the viscous friction coefficient is 6×10^{-5} N-m-sec at the motor shaft; and the gear ratio is 30:1.

Draw the block diagram showing the transfer function of each block, and design a phase-advance circuit to give the required performance. (*R.R.E.*)

[*Hint.* Refer all quantities to the output shaft. Remember that the inertia referred through the gearing becomes n^2J, the viscous friction coefficient is also multiplied by n^2, while the output torque is n times the torque at the motor shaft.]

(*Ans.* $\alpha = 0{\cdot}12$, $\tau = 0{\cdot}05$.)

6.15. The speed of a flywheel, driven by an electric motor, is to be controlled from the setting of an input voltage-divider using a closed-loop automatic speed-control system. Draw a labelled block diagram for such a system, and set up a differential equation for it assuming that there is viscous friction. Time lags in the motor and control equipment may be neglected.

The inclusive moment of inertia of the flywheel and motor is 100 kg-m² and a speed error of 1 rad/s produces a torque on the flywheel of 45 N-m. The frictional torque is 5 N-m when the flywheel velocity is 1 rad/s. With the system at rest the input voltage-divider setting is suddenly increased from zero to 100 rev min. Derive the relation between the subsequent flywheel velocity and time, and calculate the steady-state speed error of the flywheel ($\omega_i - \omega_0$). (*A.E.E.*, 1963)

(*Ans.* $90(1 - e^{-0\cdot 5t})$ rev/min; 10 rev/min.)

Note. The frictional torque gives rise to a speed error between input and output, which should not be confused with velocity lag.

6.16. A flywheel is driven by an electric motor and is made to follow the motion of an input wheel by means of an automatic control system. The torque applied to the flywheel is 10 N-m per milliradian of misalignment between the input wheel and the flywheel. The inclusive moment of inertia of the flywheel is 100 kg-m² and there is a viscous-frictional torque on the flywheel of 1,000 N-m per rad/s. The input wheel is maintained in sinusoidal angular oscillation through $\pm 30°$ about a mean position with an angular frequency $\omega = 1$ rad/s.

Set up the differential equation of the system in terms of the angle of misalignment, $\varepsilon = \theta_i - \theta_0$, and the input angle, and derive the steady-state solution of the equation. From the solution calculate the peak value of the misalignment angle and the time-phase relation between the misalignment and the input. (*A.E.E.*, 1963)

$$\left(Ans.\ J\frac{d^2\varepsilon}{dt^2} + F\frac{d\varepsilon}{dt} + K_A\varepsilon = J\frac{d^2\theta_i}{dt^2} + F\frac{d\theta_i}{dt};\right.$$

$$\left.\varepsilon_{max} = 3\cdot 03°;\ \varepsilon \text{ leads } \theta_i \text{ by } 1\cdot 57 \text{ sec.}\right)$$

CHAPTER 7

Transmission Lines

THE series resistance and inductance and the leakage conductance and capacitance to earth of a transmission line are distributed over the entire length of the line. The analysis of circuits in which the impedance parameters are distributed in this way is necessarily different from that of circuits in which the parameters are lumped. However, in lines whose physical length is short compared with the wavelength of the alternating currents and voltages which they handle, it is usually sufficient to represent the line in terms of equivalent lumped circuit parameters, as is done in Chapter 2. When the length of the line is of the same order as the wavelength this approach is no longer adequate, as will be seen in the following sections.

7.1. The Transmission Line Equations

In order to consider the relationship between the currents and voltages at any point on a transmission line and the currents and voltages which appear at the load, let us consider a sinusoidal generator feeding a load, Z_L, through a transmission line of overall length l, which has a distributed series impedance of $(R + j\omega L)$ ohms per unit length, and a distributed shunt admittance of $(G + j\omega C)$ mho per unit length (Fig. 7.1). In the small length δx we may consider that the parameters are lumped (provided that δx is chosen small enough) to give a series impedance over the section of $(R + j\omega L)\delta x$ and a shunt admittance at the end of the section of $(G + j\omega C)\delta x$, as shown in the diagram. If the current flowing out of this small section is I_x and the voltage at the output end is V_x, then the current flowing into the section is $I_x + \delta I_x$, and the voltage at the input end is $V_x + \delta V_x$, where δI_x represents the leakage current in the shunt admittance and δV_x is the voltage drop in the series impedance. Hence we can write

$$\delta V_x = (R + j\omega L)\,\delta x(I_x + \delta I_x) \simeq (R + j\omega L)I_x\,\delta x \quad . \quad (7.1)$$

assuming that $\delta I_x \ll I_x$.

Also

$$\delta I_x = (G + j\omega C)\,\delta x\, V_x \qquad (7.2)$$

In the limit as $\delta x \to 0$ these equations become

$$\frac{\partial V_x}{\partial x} = (R + j\omega L)I_x \qquad (7.3)$$

and

$$\frac{\partial I_x}{\partial x} = (G + j\omega C)V_x \qquad (7.4)$$

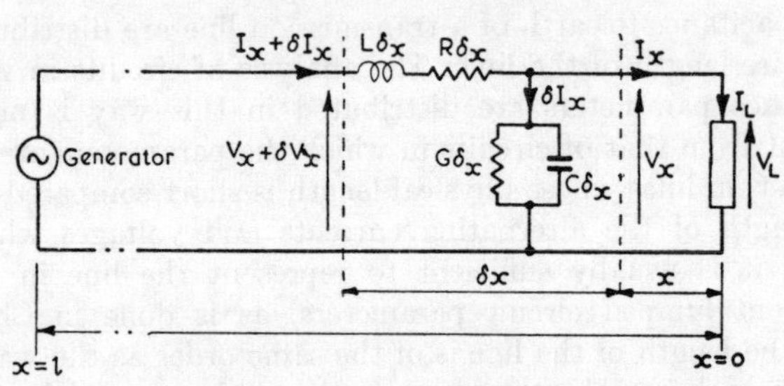

FIG. 7.1. THE DISTRIBUTED LINE

Partial derivatives are used here in order to indicate that only the variations with distance are being considered. The time variations are taken care of by the use of the complex notation. Note that we are measuring x from the load end of the line and not from the input end, as it turns out that by doing this the formulation is somewhat easier. Differentiating eqn. (7.4) with respect to x and substituting for $\partial V_x/\partial x$ from eqn. (7.3),

$$\frac{\partial^2 I_x}{\partial x^2} = (R + j\omega L)(G + j\omega C)I_x = \gamma^2 I_x \qquad (7.5)$$

Similarly,

$$\frac{\partial^2 V_x}{\partial x^2} = (R + j\omega L)(G + j\omega C)V_x = \gamma^2 V_x \qquad (7.6)$$

where $\gamma = \sqrt{[(R + j\omega L)(G + j\omega C)]}$. (7.7)

For a given line γ will be a constant.

Solving eqns. (7.5) and (7.6) gives

$$I_x = Ce^{\gamma x} + De^{-\gamma x} \qquad (7.8)$$

and

$$V_x = Ae^{\gamma x} + Be^{-\gamma x} \qquad (7.9)$$

where C, D, A and B are complex constants which must be determined from known boundary conditions. If the value of V_x given by eqn. (7.9) is substituted in eqn. (7.3), we obtain the corresponding expression for I_x as

$$I_x = \frac{\gamma}{(R + j\omega L)} A e^{\gamma x} - \frac{\gamma}{(R + j\omega L)} B e^{-\gamma x}$$

$$= \frac{A}{Z_0} e^{\gamma x} - \frac{B}{Z_0} e^{-\gamma x} \quad . \quad . \quad . \quad . \quad . \quad (7.10)$$

where
$$Z_0 = \sqrt{\frac{R + j\omega L}{G + j\omega C}} \quad . \quad . \quad . \quad (7.11)$$

A special case arises if the load impedance is equal to Z_0. In this case, at the load end ($x = 0$),

$$\frac{V_L}{I_L} = Z_0 = \frac{A + B}{A/Z_0 - B/Z_0}$$

from eqns. (7.9) and (7.10) with $x = 0$, and with V_L = load voltage and I_L = load current.

Cross-multiplying gives $(A - B) = (A + B)$, and it follows that B must be zero. Hence with this termination the input impedance, Z_{in}, at a point which is at a distance x from the load is

$$Z_{in} = \frac{V_x}{I_x} = \frac{A e^{\gamma x}}{(A/Z_0)e^{\gamma x}} = Z_0$$

By the definition on p. 49, Z_0 must be the characteristic impedance of the line, i.e. the impedance which when terminating the line gives rise to an input impedance equal to itself.

The propagation coefficient of a line is defined, in the same way as for a transmission network (p. 53), as the hyperbolic logarithm of the ratio of input to output voltage (or current) when the line is terminated in its characteristic impedance. The propagation coefficient per unit length of line may be similarly defined as the hyperbolic logarithm of the ratio of input to output voltage of a unit length of line when the line is terminated in Z_0.

At a distance x from the load (Z_0), eqn. (7.9) with $B = 0$ gives,

$$V_x = A e^{\gamma x}$$

At a distance $(x + 1)$ from the termination

$$V_{x+1} = A e^{\gamma(x+1)}$$

and hence

$$\log_e \frac{V_{x+1}}{V_x} = \log_e \frac{e^{\gamma(x+1)}}{e^{\gamma x}} = \log_e e^{\gamma} = \gamma$$

so that $\gamma = \sqrt{[(R + j\omega L)(G + j\omega C)]}$ is in fact the propagation coefficient per unit length of line.

For any other load condition (load impedance $= Z_L$) eqns. (7.9) and (7.10) give the load voltage and current, by putting $x = 0$, as

$$V_L = A + B \quad \text{and} \quad I_L = (A - B)/Z_0$$

so that $A = \frac{1}{2}(V_L + I_L Z_0)$ and $B = \frac{1}{2}(V_L - I_L Z_0)$. Substituting these values in eqns. (7.9) and (7.10),

$$V_x = V_L\left(\frac{e^{\gamma x} + e^{-\gamma x}}{2}\right) + I_L Z_0\left(\frac{e^{\gamma x} - e^{-\gamma x}}{2}\right) \quad . \quad (7.12)$$

and

$$I_x = \frac{V_L}{Z_0}\left(\frac{e^{\gamma x} - e^{-\gamma x}}{2}\right) + I_L\left(\frac{e^{\gamma x} + e^{-\gamma x}}{2}\right) . \quad . \quad (7.13)$$

where $V_L/I_L = Z_L$. These are the exponential solutions which represent the behaviour of the transmission line.

7.2. Hyperbolic Expressions for V_x and I_x

It follows directly from eqns. (7.12) and (7.13) that

$$V_x = V_L \cosh \gamma x + I_L Z_0 \sinh \gamma x \quad . \quad . \quad (7.14)$$

and

$$I_x = \frac{V_L}{Z_0} \sinh \gamma x + I_L \cosh \gamma x \; . \quad . \quad . \quad (7.15)$$

For an overall length of line of l the voltage, V_s, and current, I_s, at the input or sending end are

$$V_s = V_L \cosh \gamma l + I_L Z_0 \sinh \gamma l \quad . \quad . \quad (7.16)$$

and

$$I_s = \frac{V_L}{Z_0} \sinh \gamma l + I_L \cosh \gamma l \; . \quad . \quad . \quad (7.17)$$

so that the ABCD parameters of the line must be

$$\mathrm{A} = \mathrm{D} = \cosh \gamma l \quad . \quad . \quad . (7.18a)$$

$$\mathrm{B} = Z_0 \sinh \gamma l \quad . \quad . \quad . (7.18b)$$

and

$$\mathrm{C} = \frac{1}{Z_0} \cosh \gamma l \quad . \quad . \quad . (7.18c)$$

From this it follows that for a load Z_L the input impedance at any distance x from the load is given by

$$Z_{in} = \frac{V_x}{I_x} = \frac{\cosh \gamma x + (Z_0/Z_L) \sinh \gamma x}{(1/Z_0) \sinh \gamma x + (1/Z_L) \cosh \gamma x}$$

$$= Z_0 \frac{Z_L + Z_0 \tanh \gamma x}{Z_0 + Z_L \tanh \gamma x} \qquad (7.19)$$

The load voltage and current can be expressed in terms of the sending-end voltage and current by using the results of Section 2.4—

$$V_L = V_s \cosh \gamma l - I_s Z_0 \sinh \gamma l \qquad (7.20)$$

and

$$I_L = -\frac{V_s}{Z_0} \sinh \gamma l + I_s \cosh \gamma l \qquad (7.21)$$

The characteristic impedance and propagation coefficient of a line may be determined by measuring the input impedance with the line (*a*) open-circuited ($Z_L = \infty$), and (*b*) short-circuited ($Z_L = 0$). In general, one of these input impedances is inductive and the other is capacitive.

From eqn. (7.19), with $Z_L = \infty$,

$$Z_{in\ o.c.} = \frac{Z_0}{\tanh \gamma l} \qquad (7.22)$$

and with $Z_L = 0$,

$$Z_{in\ s.c.} = Z_0 \tanh \gamma l \qquad (7.23)$$

Multiplication of eqns. (7.22) and (7.23) leads to the following result, which has already been proved for a two-port network (and, of course, a transmission line is itself a form of two-port network)—

$$Z_0 = \sqrt{(Z_{in\ s.c.} Z_{in\ o.c.})} \qquad (7.24)$$

and hence, from eqn. (7.23),

$$\gamma l = \tanh^{-1} \sqrt{\frac{Z_{in\ s.c.}}{Z_{in\ o.c.}}} \qquad (7.25)$$

$$= \tfrac{1}{2} \log_e \frac{1 + \sqrt{\dfrac{Z_{in\ s.c.}}{Z_{in\ o.c.}}}}{1 - \sqrt{\dfrac{Z_{in\ s.c.}}{Z_{in\ o.c.}}}} \qquad (7.26)$$

$$= \tfrac{1}{2} \log_e \frac{\sqrt{Z_{in\ o.c.}} + \sqrt{Z_{in\ s.c.}}}{\sqrt{Z_{in\ o.c.}} - \sqrt{Z_{in\ s.c.}}} \qquad (7.27)$$

Example 7.1. The sending-end impedance of a 20-mile length of transmission line is $(294 + j170)$ ohms with the receiving end open-circuited and $(1{,}000 - j1{,}190)$ ohms with the receiving end short-circuited. Find the characteristic impedance and propagation coefficient per mile. (*A.E.E.*, 1958)

From eqn. (7.24), $Z_0 = \sqrt{(294 + j170)(1{,}000 - j1{,}190)}$

$$= \sqrt{(340\underline{/30^\circ} \times 1{,}555\underline{/-50^\circ})} = \underline{\underline{727\underline{/-10^\circ}\ \Omega}}$$

In this solution the second root represents a negative resistance and may be neglected.

Also,
$$\sqrt{Z_{in\ o.c.}} = 18{\cdot}4\underline{/15^\circ} = 17{\cdot}7 + j4{\cdot}76$$
$$\sqrt{Z_{in\ s.c.}} = 39{\cdot}4\underline{/-25^\circ} = 35{\cdot}7 - j16{\cdot}67$$

so that
$$\gamma l = \frac{1}{2}\log_e \frac{53{\cdot}4 - j11{\cdot}9}{-18{\cdot}0 + j21{\cdot}4} \qquad \text{(eqn. (7.27))}$$
$$= \tfrac{1}{2}\log_e 1{\cdot}96\underline{/217^\circ} = \tfrac{1}{2}\log_e 1{\cdot}96e^{j3{\cdot}78}$$
$$= \tfrac{1}{2}(\log_e 1{\cdot}96 + j3{\cdot}78) = 0{\cdot}335 + j1{\cdot}89$$

Hence
$$\gamma = \underline{\underline{0{\cdot}0168 + j0{\cdot}095 \text{ per mile}}}$$

A line that is very short compared with a wavelength can be represented by the *nominal* T- or π-circuit whose total series impedance is $(R + j\omega L)l$, and whose total shunt admittance is $(G + j\omega C)l$. This representation is normally sufficient in the case of power-frequency lines and cables. A more exact equivalent is obtained by equating the ABCD parameters of the actual line with those of the *equivalent* T- or π-network. Thus the nominal T-circuit equivalent to a line which has a series impedance $(R + j\omega L)l$ and a shunt admittance $(G + j\omega C)l$ has series arms $Z/2 = (R + j\omega L)l/2$ and a shunt arm $Y = (G + j\omega C)l$. On the other hand, the equivalent T-circuit is obtained by equating eqns. (2.8) and (7.18) to give

$$\text{A} = \text{D} = 1 + YZ/2 = \cosh \gamma l \qquad (7.28)$$
$$\text{B} = Z(1 + Y^2/4) = Z_0 \sinh \gamma l \qquad (7.29)$$

and
$$\text{C} = Y = \frac{1}{Z_0}\sinh \gamma l \qquad (7.30)$$

Substituting from eqn. (7.30) in eqn. (7.28),

$$1 + \frac{Z \sinh \gamma l}{2Z_0} = \cosh \gamma l$$

so that
$$Z = \frac{2Z_0(\cosh \gamma l - 1)}{\sinh \gamma l}$$

Thus both Y and Z for the equivalent T-circuit are found.

It is left as an exercise for the reader to deduce the components of the equivalent π-section.

7.3. Travelling Wave Interpretation of the Transmission Line Equations

The propagation coefficient will generally be a complex number, say

$$\gamma = \alpha + j\beta \quad . \quad . \quad . \quad . \quad (7.31)$$

For a sinusoidal input to the line, eqn. (7.9) can be written in instantaneous values as

$$v_x = A\mathrm{e}^{\gamma x} \sin \omega t + B\mathrm{e}^{-\gamma x} \sin (\omega t + \phi_B)$$

where ϕ_B is the phase angle of B relative to A. Hence

$$\begin{aligned} v_x &= A\mathrm{e}^{\alpha x}\mathrm{e}^{j\beta x} \sin \omega t + B\mathrm{e}^{-\alpha x}\mathrm{e}^{-j\beta x} \sin (\omega t + \phi_B) \\ &= A\mathrm{e}^{\alpha x} \sin (\omega t + \beta x) + B\mathrm{e}^{-\alpha x} \sin (\omega t + \phi_B - \beta x) \quad (7.32) \end{aligned}$$

since $\mathrm{e}^{j\beta x} = 1\underline{/\beta x}$ represents a phase change of βx.

The two terms on the right-hand side of eqn. (7.32) represent travelling waves, the first (the *incident wave*) moving towards the

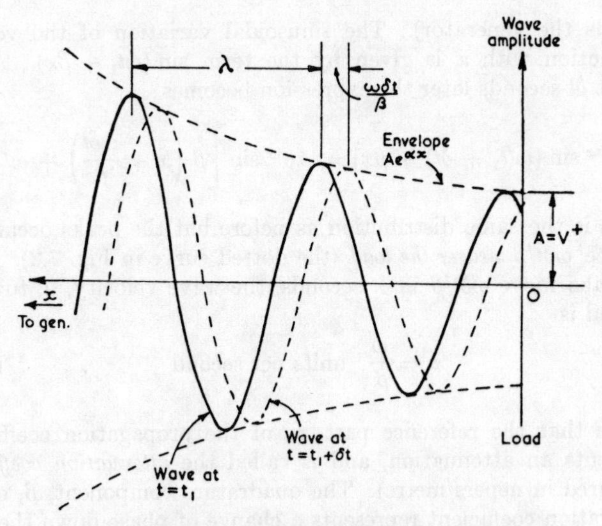

Fig. 7.2. Travelling-wave Interpretation of the Transmission-line Equations: Incident Wave

load and the second (the *reflected wave*) moving away from it. Consider first the term $Ae^{\alpha x} \sin (\omega t + \beta x)$. At the load ($x = 0$) this represents a sinusoidal variation of angular frequency ω and peak amplitude A. If we plot the voltage distribution along the line due to $Ae^{\alpha x} \sin (\omega t + \beta x)$ at a given instant, t_1, say, we obtain the sinusoidal form shown by the full curve in Fig. 7.2, where the envelope of the peaks is given by $Ae^{\alpha x}$ (i.e. increasing exponentially

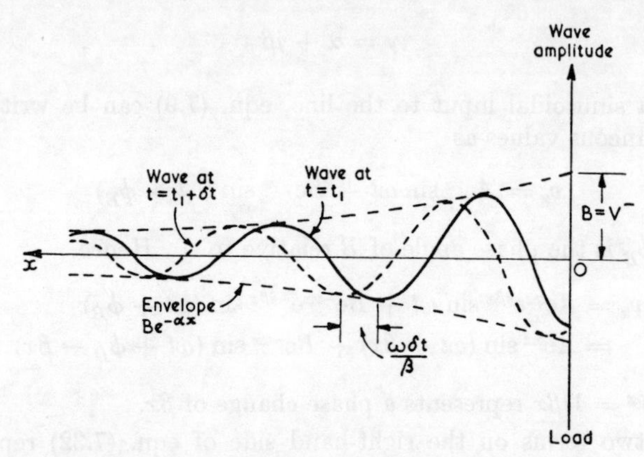

FIG. 7.3. TRAVELLING-WAVE INTERPRETATION OF THE TRANSMISSION-LINE EQUATIONS: REFLECTED WAVE

towards the generator). The sinusoidal variation of the voltage distribution with x is given by the term $\sin (\omega t_1 + \beta x)$. At an instant δt seconds later the expression becomes

$$Ae^{\alpha x} \sin [\omega(t_1 + \delta t) + \beta x] = Ae^{\alpha x} \sin \left[\beta \left(x + \frac{\omega \delta t}{\beta}\right) + \omega t_1\right]$$

This is the same distribution as before but the peaks occur at a distance $\omega\delta t/\beta$ *nearer the load* (the dotted curve in Fig. 7.2). Since the peaks move $\omega\delta t/\beta$ in δt seconds, the wave velocity, u, towards the load is

$$u = \frac{\omega}{\beta} \quad \text{units per second} \quad . \quad . \quad . \quad (7.33)$$

Note that the reference part, α, of the propagation coefficient represents an attenuation, and is called the *attenuation coefficient* (measured in nepers/metre). The quadrature component, β, of the propagation coefficient represents a change of phase down the line, and so β is called the *phase-change coefficient*. Also, when $\beta x = 2\pi$

the wave has moved through one wavelength, λ, so that, writing $x = \lambda$, (in metres),

$$\beta = \frac{2\pi}{\lambda} \text{ radians per metre} \quad . \quad . \quad . \quad (7.34)$$

The incident wave has a peak magnitude of $V^+ = A$ at the load.

If we now look at the second term in eqn. (7.32) we see that at a given instant, $t = t_1$, this second term represents a sinusoidal voltage distribution down the line, decreasing exponentially from the load towards the generator (owing to the term $e^{-\alpha x}$) and having a peak value $V^- = B$ at the load. At a time δt seconds later, the sine term becomes

$$\sin\left[\omega t_1 + \phi_B + \left(\frac{\omega \delta t}{\beta} - x\right)\beta\right]$$

which represents a sinusoidal distribution which has moved a distance $\omega\, \delta t/\beta$ *towards the generator* within the same exponential envelope (Fig. 7.3). Thus the second term in eqn. (7.32) represents a travelling wave, which is reflected from the load.

Similar relations will exist for the distribution of current down the line, so that we can rewrite eqns. (7.9) and (7.10) as

$$v_x = V^+ e^{\alpha x} \sin(\omega t + \beta x) + V^- e^{-\alpha x} \sin(\omega t + \phi_B - \beta x) \quad (7.35)$$

and $$i_x = \frac{V^+}{Z_0} e^{\alpha x} \sin(\omega t + \beta x) - \frac{V^-}{Z_0} e^{-\alpha x} \sin(\omega t + \phi_B - \beta x) \quad (7.36)$$

Expressions of the form $V \sin(\omega t - \beta x)$ or $V \sin(\omega t - \omega x/u)$ represent a voltage travelling wave which moves at a velocity $u = \omega/\beta$ in the direction of increasing x, while expressions of the form $V \sin(\omega t + \beta x)$ or $V \sin(\omega t + \omega x/u)$ represent a voltage travelling wave moving at a velocity of $u = \omega/\beta$ in the direction of decreasing x. These expressions are called *retarded functions.*

7.4. Distortion in Lines

If the input signal to a line contains sine waves of different frequencies (as it must do if information is to be transmitted), the output waveform will be a faithful reproduction of the input only if (*a*) the line attenuation is independent of frequency (or there will be frequency distortion), and (*b*) the line phase-change coefficient varies linearly with frequency (or the velocity, ω/β, will vary with frequency and there will be phase distortion which will cause the output waveshape to differ from that of the input). Phase distortion is of less importance in speech communication (since the human ear is not phase sensitive), but may seriously affect television transmissions.

The conditions for distortionless transmission can be deduced from eqns. (7.31) and (7.7). Thus

$$\gamma = \alpha + j\beta = \sqrt{[(R + j\omega L)(G + j\omega C)]}$$
$$= \sqrt{[RG - \omega^2 LC + j\omega(LG + RC)]}$$

Hence, if $LG = RC$,

$$\gamma = \sqrt{\left(\frac{R^2C}{L} + 2j\omega RC - \omega^2 LC\right)}$$
$$= \sqrt{\left[R\sqrt{\frac{C}{L}} + j\omega\sqrt{(LC)}\right]^2} = R\sqrt{\frac{C}{L}} + j\omega\sqrt{(LC)}$$
$$= \sqrt{(RG)} + j\omega\sqrt{(LC)} \quad . \quad (7.37)$$

so that $\alpha = \sqrt{(RG)}$ and $\beta = \omega\sqrt{(LC)}$ and the condition for distortionless transmission is fulfilled, i.e. the propagation velocity is independent of frequency. An important special case of a distortionless line occurs in the loss-free line in which $R = G = 0$.

7.5. The Loss-free Line

For high-frequency lines of short physical length the resistance and leakage conductance terms may usually be neglected. If $R = G = 0$, eqns. (7.31) and (7.7) reduce to

$$\gamma = j\omega\sqrt{(LC)}$$

so that $\alpha = 0$ and $\beta = \omega\sqrt{(LC)}$.

Also, from eqn. (7.11),

$$Z_0 = \sqrt{\frac{L}{C}} \quad . \quad . \quad . \quad . \quad (7.38)$$

For a twin line of conductor radius a, and spacing D, eqns. (11.42) and (10.30) give $L = (\mu/\pi) \log_e (D/a)$ henrys per metre, and $C = \pi\epsilon/\log_e (D/a)$ farads per metre. Substituting these values in the above expression for β gives $\beta = \omega\sqrt{(LC)} = \omega\sqrt{(\mu\epsilon)}$, and thus the velocity of propagation down the line is

$$u = \frac{\omega}{\beta} = \frac{1}{\sqrt{(LC)}} = \frac{1}{\sqrt{(\mu\epsilon)}}$$
$$= \frac{1}{\sqrt{(\mu_r\mu_0\epsilon_r\epsilon_0)}} = \frac{3 \times 10^8}{\sqrt{(\mu_r\epsilon_r)}} \text{ metres per second} \quad . \quad (7.39)$$

since $1/\sqrt{(\mu_0\epsilon_0)} = c = 3 \times 10^8$ m/sec.

The relative permeability of the dielectric is μ_r and its relative permittivity is ϵ_r. For air-spaced lines $\epsilon_r = 1$ and $\mu_r = 1$ and the velocity of propagation is the velocity of light, c.

Substituting the above values for L and C in eqn. (7.38) gives the characteristic impedance of the twin line as

$$Z_0 = \frac{1}{\pi}\sqrt{\frac{\mu}{\epsilon}} \log_e \frac{D}{a} \quad \text{ohms} \qquad (7.40)$$

For a concentric cable $L = (\mu/2\pi) \log_e (b/a)$ henrys per metre (eqn. (11.40)), and $C = 2\pi\epsilon/\log_e (b/a)$ farads per metre (eqn. (10.24)), where b and a are the sheath and core radii respectively. This again yields $u = 3 \times 10^8/\sqrt{(\mu_r \epsilon_r)}$ metres per second for the velocity of propagation, and gives a value for the characteristic impedance of

$$Z_0 = \frac{1}{2\pi}\sqrt{\frac{\mu}{\epsilon}} \log_e \frac{b}{a} \quad \text{ohms} \qquad (7.41)$$

The input impedance of the loaded loss-free line is found by substituting $j\beta$ for γ in eqn. (7.19), and making use of the relation $\tanh j\beta x = j \tan \beta x$. Thus

$$Z_{in} = Z_0 \frac{Z_L + jZ_0 \tan \beta x}{Z_0 + jZ_L \tan \beta x} \qquad (7.42)$$

It follows directly from eqn. (7.22) that the input impedance of an open-circuited loss-free line is

$$Z_{in\ o.c.} = \frac{Z_0}{\tanh j\beta x} = \frac{-jZ_0}{\tan \beta x}$$

$$= -jZ_0 \cot \beta x = -j\sqrt{\frac{L}{C}} \cot \omega\sqrt{(LC)}x \qquad (7.43)$$

$$= -jZ_0 \cot \frac{2\pi x}{\lambda} \qquad (7.44)$$

and that the input impedance for a short-circuited termination is obtained from eqn. (7.23) as

$$Z_{in\ s.c.} = Z_0 \tanh j\beta x = jZ_0 \tan \beta x \qquad (7.45)$$

$$= jZ_0 \tan \omega\sqrt{(LC)}x \qquad (7.46)$$

$$= jZ_0 \tan \frac{2\pi x}{\lambda} \qquad (7.46a)$$

From eqns. (7.43) and (7.45) it can be seen that the input impedances of open- or short-circuited loss-free lines are always reactive, and vary from $+j\infty$ to $-j\infty$ as the phase βx changes with either length or frequency. The variation in reactance with length for a fixed signal frequency is shown in Fig. 7.4. From this it should be particularly noted that (*a*) the input impedance of a quarter-wave ($x = \lambda/4$) open-circuited line is zero (i.e. a short-circuit), (*b*) the input impedance of a quarter-wave short-circuited line is infinite (i.e. an open-circuit), and (*c*) increasing the length of the line by exactly half a wavelength does not alter the input impedance.

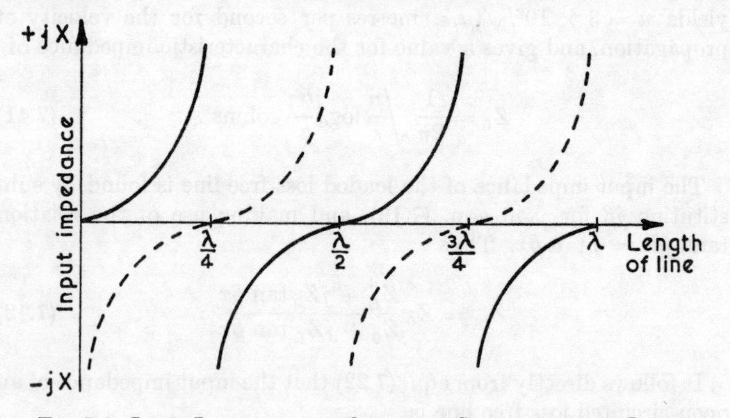

FIG. 7.4. INPUT IMPEDANCES OF OPEN- AND SHORT-CIRCUITED LINES

Similar variations of input reactance occur if the frequency of the input signal fed into a fixed length of line is varied. If the frequency of the input is low, $\beta x\ [= \omega\sqrt{(LC)}x]$ will be very small (provided that x is not too large), and we can write $\tan \beta x \simeq \beta x$. Hence

$$Z_{in\ o.c.} = -j\sqrt{\frac{L}{C}}\Big/\omega\sqrt{(LC)}x = \frac{1}{j\omega Cx}$$

and

$$Z_{in\ s.c.} = j\sqrt{\frac{L}{C}}\,\omega\sqrt{(LC)}x = j\omega Lx$$

so that from measurements of low-frequency input impedance the distributed capacitance and inductance of the line per unit length can be determined.

A loss-free line one-quarter wavelength long terminated in an impedance Z_L has (by eqn. (7.42)) an input impedance of

$$Z_{in} = Z_0 \frac{Z_L + jZ_0 \tan(\beta\lambda/4)}{Z_0 + jZ_L \tan(\beta\lambda/4)} \simeq Z_0 \frac{jZ_0 \tan(\pi/2)}{jZ_L \tan(\pi/2)}$$

$$= \frac{Z_0^2}{Z_L} \quad . \quad . \quad . \quad (7.47)$$

since $\beta = 2\pi/\lambda$, $Z_0 \tan(\pi/2) \gg Z_L$, and $Z_L \tan(\pi/2) \gg Z_0$.

This shows that a line $\lambda/4$ long (or $\lambda/4 + n\lambda/2$) may be used as an impedance-matching device, and is often referred to as an *impedance transformer*.

7.6. The Low-loss Line

Suppose that the shunt conductance, G, of a line is negligible and that $\omega L \gg R$. Then

$$\gamma = \sqrt{[(R + j\omega L)j\omega C]} = j\omega \sqrt{\left[LC\left(1 - \frac{jR}{\omega L}\right)\right]}$$

$$\simeq j\omega\left(1 - j\frac{R}{2\omega L}\right)\sqrt{(LC)}$$

and so

$$\alpha = \frac{R}{2}\sqrt{\frac{C}{L}} \quad \text{and} \quad \beta = \omega\sqrt{(LC)} \quad . \quad . \quad (7.48)$$

The phase properties are seen to be the same as those of the loss-free line, but in this case there will be some attenuation. Also,

$$Z_0 = \sqrt{\frac{R + j\omega L}{j\omega C}} = \sqrt{\left[\frac{L}{C}\left(1 - j\frac{R}{\omega L}\right)\right]}$$

$$\simeq \left(1 - j\frac{R}{2\omega L}\right)\sqrt{\frac{L}{C}} \quad . \quad . \quad . \quad . \quad (7.49)$$

If the resistance of the line is finite, and if a wide frequency band is covered, there will be some frequency distortion owing to the frequency dependence of the characteristic impedance.

Example 7.2. The primary constants per loop mile of a loaded telephone cable are: resistance 30 Ω, inductance 20 mH, capacitance 0·06 μF, leakage negligible. For an angular frequency of 5,000 rad/sec, calculate (*a*) Z_0, (*b*) the attenuation coefficient in decibels per mile, and (*c*) the velocity of propagation. (*A.E.E.*)

From the given data $\omega L = 100\ \Omega$ and $\omega C = 3 \times 10^{-4}$ mho.

(*a*) From eqn. (7.11),

$$Z_0 = \sqrt{\frac{30 + j100}{j3 \times 10^{-4}}} = \underline{\underline{591\underline{/-8{\cdot}4^\circ}\ \Omega}}$$

It should be noted that, since R is not negligible with respect to ωL, eqn. (7.49) would yield inaccurate results for the characteristic impedance.

(*b*) $\gamma = \sqrt{[(30 + j100)j3 \times 10^{-4}]} = 0{\cdot}177\underline{/81{\cdot}7^\circ}$

$$= 0{\cdot}0253 + j0{\cdot}175 \text{ per mile}$$

Hence $\alpha = 0{\cdot}0253$ Np/mile

$$= \frac{20}{2{\cdot}303} \times 0{\cdot}0253 = \underline{\underline{0{\cdot}22 \text{ dB/mile}}}$$

since decibels $= 20 \log_{10}(V_1/V_2)$ and nepers $= \log_e(V_1/V_2)$.

(*c*) From eqn. (7.33), the velocity of propagation is

$$u = \omega/\beta = \frac{5{,}000}{0{\cdot}175} = \underline{\underline{28{,}600 \text{ miles/sec}}}$$

Example 7.3. A telephone line 100 miles long has a propagation coefficient of $\gamma = 0{\cdot}006 + j0{\cdot}039$ per mile, and is open-circuited at the receiving end. Find the magnitude and phase angle of the voltage at a distance of 40 miles from the sending end when a sinusoidal voltage of 1 V (r.m.s.) is applied to the sending end. (*I.E.E. Line*)

For an overall length of line of l miles and a voltage V_L at the load, eqn. (7.12) gives the input voltage of the line, V_S, as

$$V_S = \tfrac{1}{2}V_L(e^{\gamma l} + e^{-\gamma l}) = \tfrac{1}{2}V_L(e^{0{\cdot}6+j3{\cdot}9} + e^{-0{\cdot}6-j3{\cdot}9})$$

$$= 0{\cdot}97V_L\underline{/207^\circ}$$

At a distance x from the load the voltage on the line is

$$V_x = \tfrac{1}{2}V_L(e^{\gamma x} + e^{-\gamma x})$$

For a distance of 40 miles from the sending end of a 100-mile line, $x = 100 - 40 = 60$. Hence

$$V_{60} = \tfrac{1}{2}V_L(e^{0{\cdot}36}\underline{/2{\cdot}34} + e^{-0{\cdot}36}\underline{/-2{\cdot}34})$$

$$= \frac{V_S\, 1{\cdot}58\underline{/160^\circ}}{2 \times 0{\cdot}97\underline{/207^\circ}} = \underline{0{\cdot}815\underline{/-47^\circ}}$$

The phase angle relative to the input voltage is lagging by $207^\circ + 47^\circ = 254^\circ$, or leading by 106°.

7.7. Telephone Lines and Cables

In some telephone lines and cables the leakage conductance and inductive reactance may be negligible, especially at low frequencies. This will give rise to a certain amount of frequency and phase distortion, and it may be expedient to increase the value of L artificially by means of inductive loading coils. On overhead lines the added inductance can only be in the form of lumped coils, which give the line the characteristics of a low-pass filter. In this case there will be a limit to the maximum frequency which can be transmitted down the line.

If $G \to 0$ and $L \to 0$, then

$$\gamma = \sqrt{(Rj\omega C)} = \sqrt{(\omega CR\underline{/90^\circ})}$$

$$= \sqrt{\frac{\omega CR}{2}} + j\sqrt{\frac{\omega CR}{2}} \quad . \quad . \quad . \quad (7.50)$$

Hence $\alpha = \beta = \sqrt{(\omega CR/2)}$, and the velocity of propagation is $u = \omega/\beta = \sqrt{(2\omega/CR)}$, which varies with frequency, so that phase distortion results. With added inductive loading which gives $CR = LG$ a distortionless line is obtained.

7.8. The Loaded Line—Reflection Coefficient

The complexor equations for the operation of the loaded transmission line are eqns. (7.9) and (7.10), and if the complex constants A and B are replaced by the complexors representing the incident and reflected voltages at the load ($x = 0$) we obtain

$$V_x = V^+e^{\gamma x} + V^-e^{-\gamma x} \quad . \quad . \quad . \quad (7.51)$$

and

$$I_x = \frac{V^+}{Z_0}e^{\gamma x} - \frac{V^-}{Z_0}e^{-\gamma x} \quad . \quad . \quad . \quad (7.52)$$

where γ is the propagation coefficient per unit length of line, and Z_0 is the characteristic impedance.

Suppose that we have such a line of length l terminated in a

FIG. 7.5. THE TERMINATED LINE

load Z_L, as shown in Fig. 7.5. At the load $x = 0$, and hence we obtain

$$V_L = V^+ + V^- \quad . \quad . \quad . \quad (7.53)$$

and

$$I_L = \frac{V^+}{Z_0} - \frac{V^-}{Z_0} \quad \text{or} \quad I_L Z_0 = V^+ - V^- \quad . \quad (7.54)$$

Adding these two equations,

$$2V^+ = V_L + I_L Z_0$$

and subtracting,

$$2V^- = V_L - I_L Z_0$$

Hence

$$\frac{V^+}{V^-} = \frac{V_L + I_L Z_0}{V_L - I_L Z_0} = \frac{(V_L/I_L) + Z_0}{(V_L/I_L) - Z_0} = \frac{Z_L + Z_0}{Z_L - Z_0}$$

and we may write

$$V^- = \frac{Z_L - Z_0}{Z_L + Z_0} V^+ = \rho V^+ \quad . \quad . \quad . \quad (7.55)$$

where

$$\rho = \frac{Z_L - Z_0}{Z_L + Z_0} \quad . \quad . \quad . \quad (7.56)$$

ρ is called the *reflection coefficient*, and gives the ratio of reflected to incident wave *at the load*. It will normally be a complex number. Substituting eqn. (7.55) in eqns. (7.51) and (7.52),

$$V_x = V^+ e^{\gamma x} + \rho V^+ e^{-\gamma x} \quad . \quad . \quad . \quad (7.57)$$

and

$$I_x = \frac{V^+}{Z_0} e^{\gamma x} - \frac{\rho V^+}{Z_0} e^{-\gamma x} \quad . \quad . \quad . \quad (7.58)$$

If the incident current at the load is I^+,

$$I^+ = \frac{V^+}{Z_0} \quad . \quad . \quad . \quad (7.59)$$

and if I^- is the reflected current at the load,

$$I^- = \frac{-\rho V^+}{Z_0} = -\rho I^+ \quad . \quad . \quad . \quad (7.60)$$

We can therefore write

$$I_x = I^+ e^{\gamma x} - \rho I^+ e^{-\gamma x} \quad . \quad . \quad . \quad (7.61)$$

Note also that

$$I_L = I^+ + I^- \quad . \quad . \quad . \quad (7.62)$$

Relations similar to those just derived will exist at any junction between lines of different characteristic impedances. The transmitted currents and voltages will correspond to the load values in the above derivation.

Interesting particular cases occur for open- and short-circuited, matched and reactively loaded lines.

(*a*) *Matched line* ($Z_L = Z_0$). In this case $\rho = 0$. There is no reflected voltage or current, so that $V_L = V^+$ and $I_L = I^+$. Hence the sending-end voltage and current are

$$V_S = V_L e^{\gamma l} \quad . \quad . \quad . \quad . \quad (7.63)$$

and

$$I_S = I_L e^{\gamma l} \quad . \quad . \quad . \quad . \quad (7.64)$$

(*b*) *Open-circuited line* ($Z_L = \infty$). Substituting in eqn. (7.56) gives $\rho = 1$. This means that there is total reflection of both voltage and current, the reflected voltage being in phase with the incident voltage at the load. The reflected current (from eqn. (7.62)) is, however, in antiphase with the incident current, so that the resultant current at the load is zero, as we should expect. Hence

$$V_L = V^+ + V^- = 2V^+ \qquad \text{and} \qquad I_L = I^+ + I^- = 0$$

(*c*) *Short-circuited line* ($Z_L = 0$). In this case $\rho = -1$, and there is again total reflection, this time with voltage cancellation and current doubling at the load, i.e.

$$V_L = V^+ + V^- = 0 \qquad \text{and} \qquad I_L = 2I^+$$

(*d*) *Purely reactive load* ($Z_L = \pm jX$). For the case where Z_L is a pure reactance and Z_0 is a pure resistance (loss-free line), we have

$$\rho = \frac{\pm jX - Z_0}{\pm jX + Z_0} = \frac{\sqrt{(Z_0^2 + X^2)} \Big/ \tan^{-1} \dfrac{\pm X}{-Z_0}}{\sqrt{(Z_0^2 + X^2)} \Big/ \tan^{-1} \dfrac{\pm X}{Z_0}} = 1\underline{/\phi}$$

where $\phi = \tan^{-1}(\pm X/-Z_0) - \tan^{-1}(\pm X/Z_0) = -2\tan^{-1}(\pm X/Z_0)$

(*e*) *General case*. For any load impedance other than those considered above there will be a reflection coefficient which is less than unity, showing that there will be partial reflection of the incident voltage and current waves. If $0 < \rho < 1$ the incident and reflected voltages will be additive while the currents will subtract to give the load conditions. If $-1 < \rho < 0$, the currents are additive, and the voltages subtract at the load.

Example 7.4. A 10-mile transmission line has the following constants per loop mile: $R = 50\ \Omega$, $L = 1$ mH, $C = 0{\cdot}06\ \mu$F. Find Z_0. If the line is matched and a p.d. of 5 V at a frequency of $5{,}000/2\pi$ c/s is applied at the sending end, calculate (i) the magnitude of the received current, (ii) the wavelength, and (iii) the velocity of propagation. (*A.E.E.*, 1958)

The characteristic impedance is found from eqn. (7.11) as

$$Z_0 = \sqrt{\frac{R + j\omega L}{G + j\omega C}} = \sqrt{\frac{50 + j5}{j3 \times 10^{-4}}} = 410\underline{/-42 \cdot 2^\circ}\ \Omega$$

In order to solve the rest of the problem the propagation coefficient is first determined. Since the line constants are given per mile, γ can readily be found per mile as

$$\gamma = \sqrt{[(R + j\omega L)(G + j\omega C)]} = \sqrt{(50 \cdot 3\underline{/5 \cdot 7^\circ} \times 3 \times 10^4\underline{/90^\circ})}$$
$$= 0 \cdot 123\underline{/47 \cdot 9^\circ}$$
$$= 0 \cdot 0826 + j0 \cdot 091 \text{ per mile}$$

For the matched line $Z_L = Z_0$, and eqn. (7.64) yields

$$I_S = I_L e^{\gamma l} \text{ or } I_L = I_S e^{-\gamma l} = I_S e^{-0 \cdot 826}\underline{/-0 \cdot 91 \text{ rad}}$$

since the line is 10 miles long.

Also, since the line is matched, the input impedance will be equal to Z_0 and so

$$I_S = \frac{V_S}{Z_0} = \frac{5\underline{/0}}{410\underline{/-42 \cdot 2^\circ}}$$

Hence

(i) $$I_L = \frac{5\underline{/0^\circ}}{410\underline{/-42 \cdot 2^\circ}} e^{-0 \cdot 826}\underline{/52 \cdot 7^\circ} = 5 \cdot 37\underline{/94 \cdot 9^\circ} \text{ mA}$$

(ii) From eqn. (7.34), $\lambda = 2\pi/\beta = 2\pi/0 \cdot 091 = 69$ miles

(iii) From eqn. (7.33), $u = \omega/\beta = 5{,}000/0 \cdot 091 = 55{,}000$ miles/sec

Example 7.5. An open-wire transmission line has $Z_0 = 600\ \Omega$ and negligible losses. It is 50 m long and is open-circuited at one end. At the other end it is connected to a generator which generates 100 V at 15 Mc/s, and has an internal impedance of $(200 + j200)\ \Omega$. Non-inductive loads of 600 Ω each are connected across the line at distances of 30 and 35 m from the source. Calculate the currents in the two loads. (*A.E.E.*, 1959)

For an open-wire line $\mu_r = 1$ and $\epsilon_r = 1$, so that the velocity of propagation is the velocity of light, and the wavelength is related to the frequency by $\lambda = c/f = (3 \times 10^8)/(15 \times 10^6) = 20$ m. Hence $\lambda/4 = 5$ m.

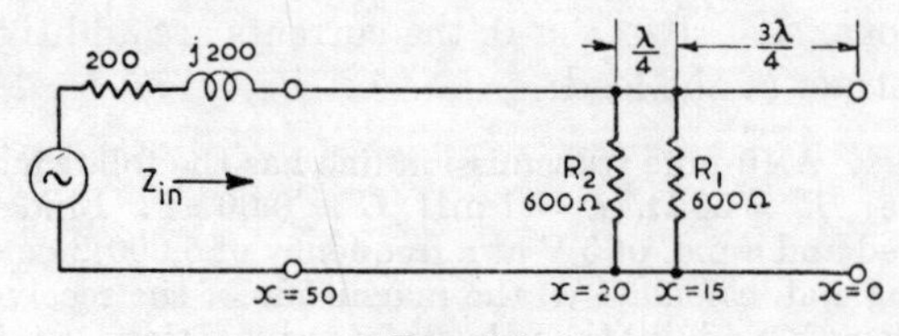

FIG. 7.6

The line is shown in Fig. 7.6. The open-circuit at the end of the line is 15 m ($=\lambda/4 + \lambda/2$) from the load, R_1. The results of Section 7.5 show that the open-circuit appears as a short-circuit at the position of R_1, and hence there will be no current in R_1.

The apparent short-circuit at R_1 is 5 m ($=\lambda/4$) from the second load, R_2, and therefore appears as an open-circuit at R_2. From the generator the line appears to be 30 m long and terminated in its characteristic impedance. Hence the input impedance is Z_0 ($=600\ \Omega$), and the input current is

$$I_S = \frac{100}{200 + j200 + 600} = 0{\cdot}121\underline{/-14^\circ}\text{ A}$$

From eqn. (7.64) the load current is

$$I_L = I_S e^{-\gamma l} = I_S e^{-\alpha l}\underline{/-\beta l}$$

Since the attenuation coefficient is zero, the magnitude of the load current will be the same as that of the input current, so that the current in R_2 must be 0·121 A.

7.9. Voltage Standing Waves on a Loss-free Line

For the loss-free line $\gamma = j\beta$ since α is zero, and eqn. (7.57) becomes

$$V_x = V^+ e^{j\beta x} + \rho V^+ e^{-j\beta x}$$

If ρ is represented in polar form as $\rho'\underline{/\theta}$ this equation can be written

$$V_x = V^+\underline{/\beta x} + \rho' V^+\underline{/\theta - \beta x}$$

where the reference complexor is V^+ at $x = 0$. If the incident voltage at the position x is taken as the reference complexor, we obtain

$$\begin{aligned} V_x &= V^+\underline{/0} + \rho' V^+\underline{/\theta - 2\beta x} \\ &= V^+\underline{/0} + \rho' V^+\underline{/\theta - 4\pi x/\lambda} \qquad (7.65) \end{aligned}$$

This equation can be interpreted as meaning that V_x is composed of a fixed complexor $V^+\underline{/0}$ to which is added a complexor of constant magnitude, $\rho' V^+$, whose phase becomes linearly more negative as x increases. In Fig. 7.7, OC represents V^+, while CP represents the reflected wave. P_0 represents the conditions at the load ($x = 0$), i.e. CP_0 makes an angle θ with V^+, and $OP_0 = V_L$. As x increases towards the generator the phase angle of the reflected wave increases in the negative sense (clockwise) by $(2\pi/\lambda)2x$, so that point P moves round the circle of centre C and radius $\rho' V^+$. Each time x increases by $\lambda/2$ the phase angle changes by 2π and we are back at P_0, so that a linear scale can be constructed round the outside circle representing wavelengths moved, once round being equivalent to half a wavelength. Moving towards the generator, P moves clockwise; moving towards the load (x negative), P moves anti-clockwise round the diagram. It is usual to start this scale of "wavelengths moved" at the origin, O. Fig. 7.7 shows distances moved towards the generator from O.

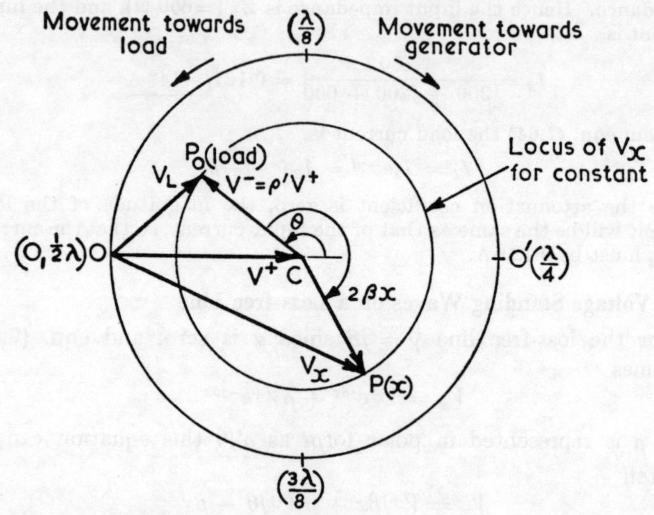

Fig. 7.7. Complexor Relations for Incident and Reflected Waves

The peak magnitude of the alternating voltage, V_x, at any point, x, on the line is seen to vary with x. The line is said to support a standing wave. The maximum value of the standing wave on the line will be $V^+ + \rho' V^+$ and occurs when the reflected wave is in phase with the incident wave. The minimum value of the standing wave will be $V^+ - \rho' V^+$ and will occur at those points where the reflected wave is in antiphase to the incident wave. The ratio of this maximum value to the minimum value is called the *voltage standing wave ratio* (V.S.W.R.), and is given the symbol S. Thus for a loss-free line,

$$S = \frac{1 + \rho'}{1 - \rho'} \quad . \quad . \quad . \quad . \quad (7.66)$$

Sometimes the reciprocal is used, and is represented by the small letter s:

$$s = \frac{1 - \rho'}{1 + \rho'} \quad . \quad . \quad . \quad . \quad (7.67)$$

The magnitude of the reflection coefficient can be expressed in terms of the V.S.W.R. as

$$\rho' = \frac{S-1}{S+1} \qquad . \quad . \quad . \quad . \ (7.68)$$

In the case of a matched line, $\rho' = 0$, and $S = s = 1$, there is no reflected wave and hence no standing wave pattern (which is an interference pattern between incident and reflected waves) and the line supports only the incident travelling wave. For a loss-free line on open- or short-circuit, or with a purely reactive load, $\rho' = 1$, and the locus of V_x is the outside circle shown in Fig. 7.7. In this case zeros occur at intervals of $\lambda/2$ along the line, while voltage doubling occurs between the zeros. Within this outer circle any load condition can be represented. Note also that circles with centre C will represent the loci of V_x for constant values of the V.S.W.R.

For a given V.S.W.R. the magnitude of V_x is given by the cosine rule from triangle OCP in Fig. 7.7 as

$$\begin{aligned} V_x^2 &= (V^+)^2 + \rho'^2(V^+)^2 + 2\rho'(V^+)^2 \cos(\theta - 2\beta x) \\ &= (V^+)^2[1 + \rho'^2 + 2\rho' \cos(\theta - 2\beta x)] \qquad . \quad . \ (7.69) \end{aligned}$$

Voltage standing wave patterns for an open-circuited line, a short-circuited line and a loaded line are shown in Fig. 7.8. These are obtained from eqn. (7.69) by using the appropriate values of ρ'.

Example 7.6. A line of characteristic impedance $600\underline{/0°}$ Ω is terminated in a load Z_L. The V.S.W.R. measured on the line is 1·5 and the first maximum occurs 20 cm from the load. The line is open wire and is fed at 300 Mc/s. Find the value of the load impedance. (*R.R.E.*)

Since the line is open wire we may write

$$\lambda = c/f = (3 \times 10^8)/(3 \times 10^8) = 1 \text{ m}$$

Therefore 20 cm = 0·2 wavelength.

The angle between V^+ and V^- at the load is θ, the phase angle of ρ. Hence, since once round the diagram (Fig. 7.7) represents 0·5 λ, 0·2 λ must represent a phase angle of (0·2/0·5)360° = 144°. It follows that $\theta = 144°$. The fact that this angle is positive means that the reflected wave leads the incident wave at the load by 144°. Had a minimum occurred first, θ would have been negative.

From eqn. (7.68),

$$\rho' = \frac{S-1}{S+1} = \frac{0{\cdot}5}{2{\cdot}5} = 0{\cdot}2$$

so that the complex expression for ρ is

$$\rho = 0{\cdot}2\underline{/144°} = -0{\cdot}162 + j0{\cdot}117 = \frac{Z_L - Z_0}{Z_L + Z_0} \qquad \text{(from eqn. (7.56))}$$

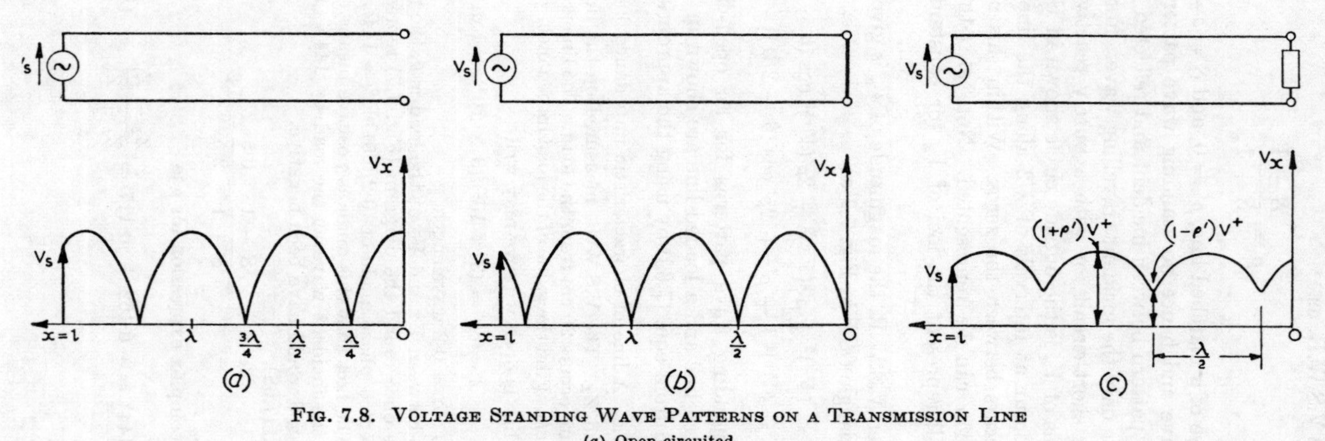

FIG. 7.8. VOLTAGE STANDING WAVE PATTERNS ON A TRANSMISSION LINE

(a) Open-circuited
(b) Short-circuited
(c) Loaded

Therefore $$Z_L(1{\cdot}162 - j0{\cdot}117) = Z_0(0{\cdot}838 + j0{\cdot}117)$$

or $$Z_L = \frac{600(0{\cdot}0838 + j0{\cdot}117)}{1{\cdot}162 - j0{\cdot}117}$$

$$= 436\underline{/13{\cdot}7^\circ} = \underline{\underline{(424 + j103)\ \Omega}}$$

After reading Section 7.12 the reader may verify that the above problem can be solved more simply by using the Smith chart.

7.10. Standing Waves of Current in a Loss-free Line

We can develop expressions for the standing waves of current in a loss-free line in the same way as in the previous section. Thus, from eqn. (7.61),

$$I_x = I^+e^{\gamma x} - \rho I^+e^{-\gamma x}$$

$$= I^+\underline{/\beta x} - \rho' I^+\underline{/\theta - \beta x}$$

since $\gamma = j\beta$.

The reference complexor is I^+ at the load. If the complexor $-I^+$ at x is taken as reference we can show that the same circle diagram can be used to give both V_x and I_x. In this case

$$-I_x = -I^+\underline{/0} + \rho' I^+\underline{/\theta - 2\beta x} \quad . \quad . \quad (7.70)$$

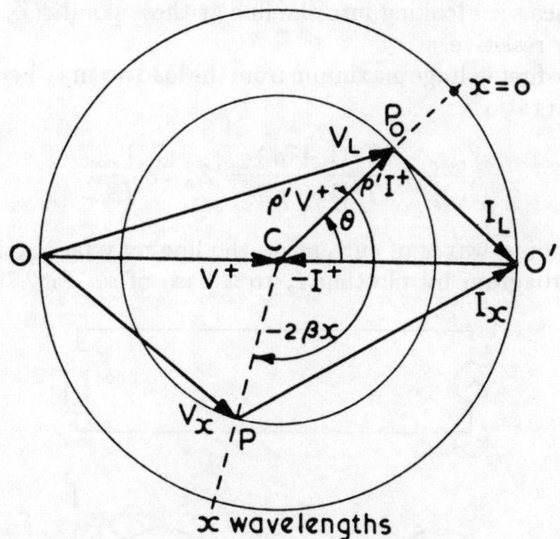

Fig. 7.9. Circle Diagram with both Voltages and Currents

In Fig. 7.9 let $O'C$ represent $-I^+$ to a suitable scale; then to the same scale $O'P_0$ will represent $-I^+ + \rho' I^+\underline{/\theta}$ (i.e. the condition at the load, where $x = 0$), and P_0O' will therefore represent I_L (i.e.

I_x at $x = 0$) in magnitude and phase. At any distance x from the load, the end point of I_x has moved clockwise along the circumference of the circle of centre C and radius $\rho' I^+$, through an angle $2\beta x$ from P_0. Thus at the point P, $OP = V_x$ and $PO' = I_x$ to suitable scales, so that the same circle diagram gives both the current and voltage complexors, voltage being measured from O, and current being measured to O'.

The following points should be noted—

1. Motion clockwise represents movement towards the generator, once round the circle being equivalent to a half-wavelength moved.
2. Motion anti-clockwise must therefore represent movement towards the load.
3. At a voltage maximum (where the constant-V.S.W.R. circle cuts CO') there is a current minimum, and at a voltage minimum there is a current maximum.
4. Adjacent voltage minima are one half-wavelength apart, as are adjacent current minima.
5. Minima are sharper than maxima.
6. At maxima and minima V_x and I_x are in phase, so that the impedance seen looking into the line at these points ($Z_x = V_x/I_x$) is purely resistive.
7. The first voltage maximum from the load occurs when $2\beta x = \theta$.
8. The ratio

$$\frac{V_{max}}{I_{max}} = \frac{V^+(1 + \rho')}{I^+(1 + \rho')} = Z_0 = \frac{V_{min}}{I_{min}}$$

The standing waves of current on the line may be obtained from the circle diagram by plotting I_x to a base of x. Fig. 7.10 shows

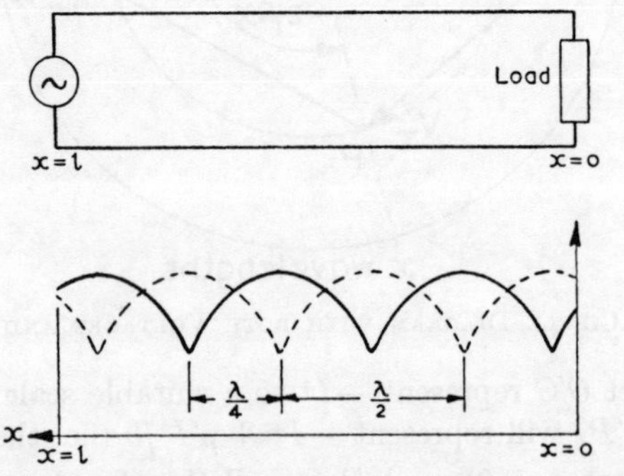

FIG. 7.10. CURRENT AND VOLTAGE STANDING WAVES

typical current and voltage standing-wave patterns. It is important to remember that these patterns represent the peak or (to a suitable scale) r.m.s. values of the alternating current or voltage which would be found at the given points on the line.

The maximum input impedance which a loaded line will present occurs when the length of the line is such that a voltage maximum occurs at its input end. In this case,

$$Z_{in\ max} = \frac{V_{max}}{I_{min}} = \frac{V^+(1+\rho')}{I^+(1-\rho')} = Z_0 S \quad . \quad . \quad (7.71)$$

Similarly the minimum input impedance will be

$$Z_{in\ min} = \frac{V_{min}}{I_{max}} = \frac{Z_0}{S} \quad . \quad . \quad . \quad (7.72)$$

7.11. The Impedance Diagram

Since every point within the circle diagrams of the previous two sections represents a definite complex relation between voltage and current on a transmission line, they must also represent definite values of impedance. In Fig. 7.11 the circle diagram is redrawn

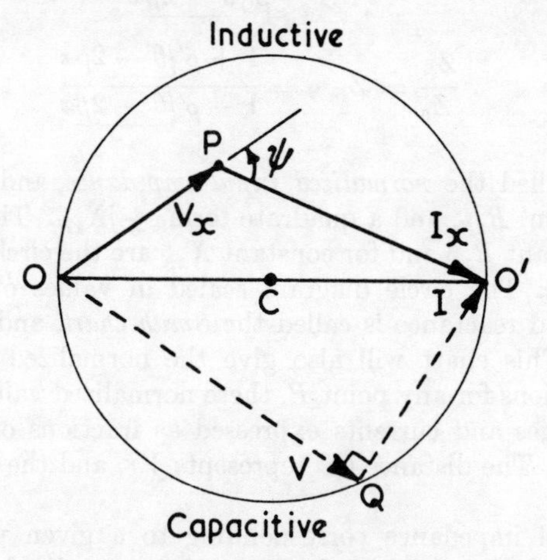

FIG. 7.11. DERIVING THE IMPEDANCE CHART

with point P representing conditions at a distance x from the load on a line. The impedance seen when looking into the line at P is

$$Z_P = \frac{V_x}{I_x} \underline{/\psi} = R_x + jX_x$$

so that each point within the circle represents some unique value of impedance. Around the circumference V_x and I_x must (by geometry) be perpendicular to one another, and hence the impedances on the circumference must be purely reactive, with a zero at O $(Z = 0/I_x)$ and ∞ at O' $(Z = V_x/0)$. If I_x lags behind V_x the reactance is positive (inductive), and so the upper half of the diagram represents positive reactance. In the same way the lower half of the diagram represents negative (capacitive) reactance.

If V_x and I_x are in phase, P must lie somewhere on the diameter OO', which can therefore be scaled in resistance values.

For loss-free lines eqns. (7.57) and (7.58) can be written

$$V_x = V^+ e^{j\beta x} + \rho' V^+ e^{-j\beta x} \underline{/\theta}$$

and

$$I_x = I^+ e^{j\beta x} - \rho' I^+ e^{-j\beta x} \underline{/\theta}$$

so that

$$Z_{in} = \frac{V_x}{I_x} = \frac{V^+ \, 1\underline{/\beta x} + \rho' \underline{/\theta - \beta x}}{I^+ \, 1\underline{/\beta x} - \rho' \underline{/\theta - \beta x}}$$

$$= Z_0 \frac{1 + \rho' \underline{/\theta - 2\beta x}}{1 - \rho' \underline{/\theta - 2\beta x}}$$

and therefore

$$\frac{Z_{in}}{Z_0} = Z_{in\,N} = \frac{1 + \rho' \underline{/\theta - 2\beta x}}{1 - \rho' \underline{/\theta - 2\beta x}} \qquad . \quad . \quad (7.73)$$

$Z_{in\,N}$ is called the *normalized input impedance*, and will have a reference term, R_{xN}, and a quadrate term, $\pm jX_{xN}$. The complexor loci for constant R_{xN} and for constant X_{xN} are the circles derived in Section 3.12. The circle diagram scaled in values of normalized resistance and reactance is called the *Smith chart*, and is shown in Fig. 7.12. This chart will also give the normalized voltage and current relations for any point P, these normalized values being the actual voltages and currents expressed as fractions of V^+ and I^+ respectively. The distance OC represents V^+, and the distance CO' represents I^+.

The actual impedance corresponding to a given point on the Smith chart is found by multiplying the normalized value taken from the chart by the characteristic impedance.

Circles with centre C represent loci of constant V.S.W.R., and the relation between the scales for normalized resistance and for V.S.W.R. can be found as follows. Consider a point on the line at which the impedance seen looking into the line is purely resistive, say R_P. For the loss-free line the characteristic impedance is itself purely resistive, and between P and the generator the line will behave as if it were terminated at P by the resistance R_P. This gives the reflection coefficient ρ as $(R_P - Z_0)/(R_P + Z_0)$, with zero phase angle if $R_P > Z_0$.

Hence
$$\rho = \rho' = \frac{(R_P/Z_0) - 1}{(R_P/Z_0) + 1} = \frac{R_{PN} - 1}{R_{PN} + 1} \quad . \quad . \quad (7.74)$$

where $R_{PN} = R_P/Z_0$ is the normalized resistance. But from eqn. (7.68),

$$\rho' = \frac{S - 1}{S + 1}$$

and, comparing this with eqn. (7.74), we see that $S = R_{PN}$, i.e. the scale for V.S.W.R. is the same as the normalized resistance scale along the line CO'. If the alternative definition for V.S.W.R. is used ($s = 1/S$), the normalized resistance scale from O to C will also be the scale of V.S.W.R.

7.12. Use of the Impedance Chart

Some important applications of the impedance chart will now be considered.

(*a*) *Length of a short-circuited line for a given input reactance.* In this case the load is the short-circuit (zero impedance) and the V.S.W.R. is infinite. The load is represented by the point O in Fig. 7.12. The constant V.S.W.R. circle is the outer circle of the Smith chart, and moving from the load towards the generator we progress round this outside circle clockwise. For any distance (in wavelengths) moved, the corresponding normalized input reactance can be immediately read off. The input impedance can thus never have any resistive component.

Example 7.7. A transmission line is air-spaced and has a characteristic impedance of $150\underline{/0^\circ}$ Ω. Find the length of short-circuited line operating at 300 Mc/s which will give an input impedance of $-j300$ Ω. What input impedance is presented if the line is 0·6 λ long? (*R.E.E.*)

The normalized input impedance is $Z_{in\,N} = -j300/150 = -j2$, and the load point is O (Fig. 7.13). Moving from O clockwise round the outer circle to the point $-j2$ gives a distance of $l_1 = 0{\cdot}324\lambda$. For an air-spaced line at 300 Mc/s one wavelength is $\lambda = c/f = 3 \times 10^8/3 \times 10^8 = 1$ m, so that the required length of line is 0·324 m, or 0·324 plus any whole number of half-wavelengths.

$0{\cdot}6\lambda$ is $(0{\cdot}5 + 0{\cdot}1)\lambda$, and therefore if we move $0{\cdot}6\lambda$ from the load at O we move once completely round the outer circle plus a further

$0{\cdot}1\lambda$. The reactance at this point is read off as $+j0{\cdot}725$ (normalized). The actual input impedance is $j0{\cdot}725 \times 150 = j108{\cdot}8$.

Fig. 7.13 shows the method of construction in outline, and this construction should be applied to the chart of Fig. 7.12 to obtain the actual numerical values.

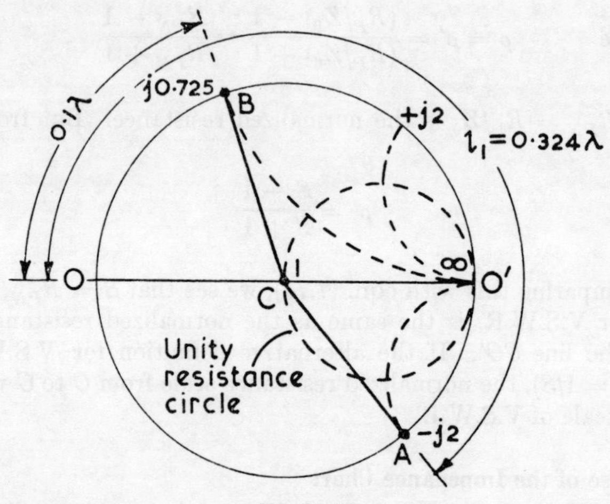

Fig. 7.13

(*b*) *Length of open-circuited line for a given input impedance.* This follows the same method as for the short-circuited line, but here the load impedance is infinite so that we must start from the point O' in Fig. 7.12. It should be noted that, owing to radiation from the end of an open-circuited high-frequency line, it is almost impossible to obtain a true open-circuit.

(*c*) *To find the V.S.W.R. for a given load.* The load impedance is first normalized, and this normalized impedance is then located on the Smith chart. The circle with centre C, and with radius the distance from C to the load point, is the constant V.S.W.R. locus. The point where this circle cuts the line CO' gives the value of the V.S.W.R. (Fig. 7.12). The impedance looking into the line at any distance from the load is found by moving the required number of wavelengths clockwise from the load point on the constant V.S.W.R. circle.

Example 7.8. A 70-Ω line is loaded at its end by an impedance of $(70 - j140)$ Ω. Find (i) the input impedance if the line is $0{\cdot}8\ \lambda$ long, (ii) the V.S.W.R.

The normalized impedance is $Z_{LN} = (70 - j140)/70 = 1 - j2$. This point is shown as A in Fig. 7.14. The circle with centre C and radius

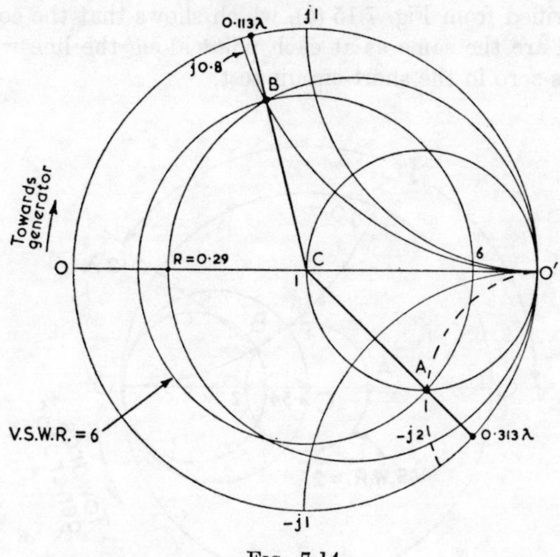

Fig. 7.14

CA is the required constant V.S.W.R. circle. This circle cuts the line CO' at resistance scale point 6, so that the required V.S.W.R. is 6.

The line from C through A cuts the wavelength scale at 0·313. Adding 0·8 to this gives 1·113, which leaves 0·113 when the integral number of half-wavelengths is subtracted. The required input impedance is read off from the intersection of the line from C to 0·113λ with the constant V.S.W.R. circle (point B in Fig. 7.14). From the chart,

$$Z_{in\,N} = 0{\cdot}29 + j0{\cdot}8$$

and hence

$$Z_{in} = \underline{\underline{20{\cdot}3 + j56\ \Omega}}$$

(*d*) *Impedance measurement at very high frequencies.* A transmission line may be used to measure impedance at very high frequencies. The line is first short-circuited, and the standing wave pattern is investigated using a probe which is coupled through a crystal rectifier to a d.c. microammeter. The positions of the zeros are noted, and the short-circuit is then replaced by the unknown impedance. The new value of V.S.W.R. is measured and the distance, y, which the minima shift *towards the load* is noted. The V.S.W.R. circle for the load can now be drawn on the Smith chart (Fig. 7.15 (*a*)). The chart is entered on this circle at a voltage minimum (point A), and the value of the load impedance is then found at the point y wavelengths *towards the generator* from the minimum on the V.S.W.R. circle. That this is in fact the load impedance may be verified from Fig. 7.15 (b), which shows that the conditions at the load are the same as at each point along the line where the voltage was zero in the short-circuit test.

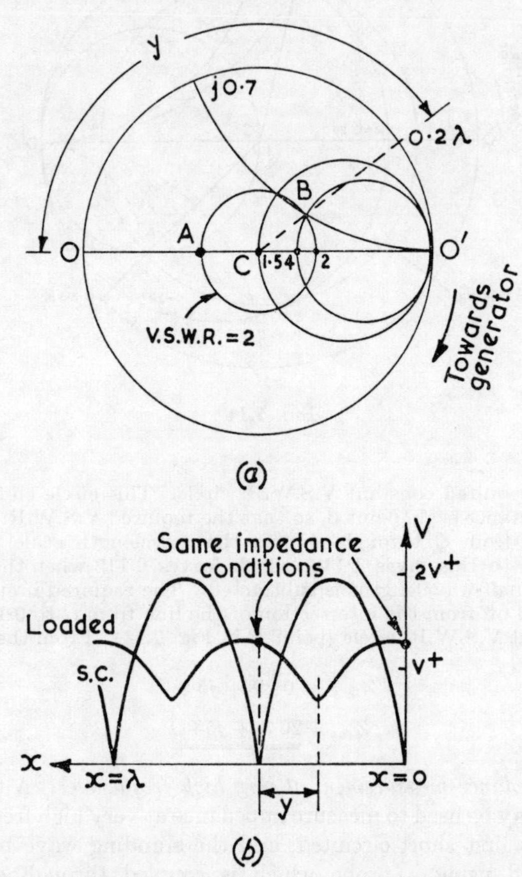

Fig. 7.15. Measurement of Impedance at Very High Frequencies

Example 7.9. When a certain concentric 250-Ω line is terminated in an unknown load impedance a V.S.W.R. of 2 is measured, and the voltage minima are found to occur at a distance of $0{\cdot}2\lambda$ nearer the load than was the case when the line was short-circuited. Determine the load impedance.

In Fig. 7.15(a) the V.S.W.R. = 2 circle is drawn. Entering the chart on this circle at a voltage minimum (point A) and moving $0{\cdot}2\lambda$ towards the generator, we arrive at B, where the normalized impedance co-ordinates are $(1{\cdot}54 + j0{\cdot}7)$. Hence the load impedance is

$$Z_L = 250(1{\cdot}54 + j0{\cdot}7) = \underline{\underline{(385 + j175)\ \Omega}}$$

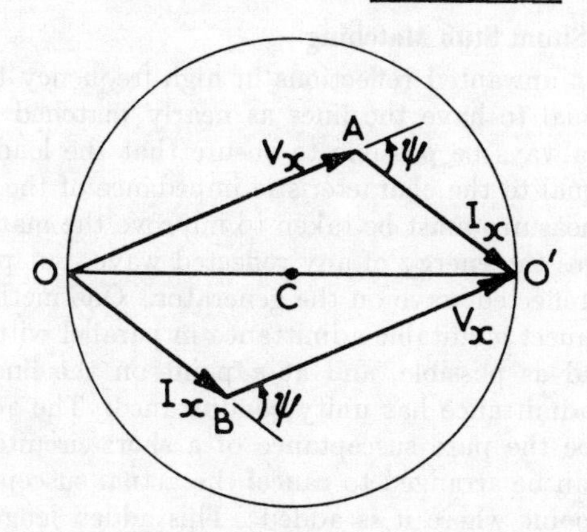

FIG. 7.16. THE ADMITTANCE DIAGRAM

7.13. The Admittance Chart

The Smith chart has been shown to give the impedance at any point on a loss-free transmission line, all values of impedance from zero to infinity being represented within the outer circle. The same diagram with the same scales can also represent admittance. Thus in Fig. 7.16 the point A represents a normalized impedance $Z_N\underline{/\psi}$. Numerically,

$$Z_N = \frac{OA}{AO'}$$

Consider now the point B which is diametrically opposite A, so that $OAO'B$ is a parallelogram. Then the point B represents a function which has a magnitude of

$$\frac{OB}{BO'} = \frac{AO'}{OA} = \frac{1}{Z_N} = Y_N$$

and a phase angle of $-\psi$ (the angle between OB and BO').

This means that the numerical values at B give the magnitude and phase of the admittance represented by the point A. The Smith chart can therefore be taken as an admittance diagram, the resistance scale representing parallel conductance, and the reactance scale representing parallel susceptance, provided that we consider voltages to be measured from O' and currents from O. Thus O' represents infinite admittance (zero voltage), while O represents zero admittance (zero current).

7.14. Single Shunt Stub Matching

To prevent unwanted reflections in high-frequency transmission lines it is usual to have the lines as nearly matched as possible. It may not always be possible to ensure that the load impedance is exactly equal to the characteristic impedance of the line, and in such cases measures must be taken to improve the matching. This not only saves the energy of any reflected wave, but prevents any action by a reflected wave on the generator. One method of doing this is to connect a suitable admittance in parallel with the line as near the load as possible, and at a point on the line where the normalized admittance has unity conductance. The added admittance may be the pure susceptance of a short-circuited length of line which can be arranged to cancel the actual susceptance of the line at the point where it is added. This added length of line is called a *shunt stub*.

Consider the circle diagram of Fig. 7.17 (a) used as an admittance diagram. For an incorrectly matched loaded line there will be a constant V.S.W.R. circle to represent the conditions at any point on the line (circle $AFED$). This circle crosses the unity conductance circle at the two points F and D where the normalized line admittances are $(1 + jB)$ and $(1 - jB)$ respectively. If we add a parallel pure normalized susceptance of $-jB$ at F (which is l_1 wavelengths towards the load from a voltage minimum at E), the resultant normalized admittance at F becomes unity—in other words, the line is matched between point F and the generator. Similarly, if a normalized admittance of $+jB$ is added at the point on the line represented by D on the circle diagram (i.e. a distance l_1 wavelengths nearer the generator than a voltage minimum), the resulting admittance at D becomes unity and the line is matched. Usually matching would be done at F rather than D, since this would involve a shorter length of stub. The physical arrangement is shown in Fig. 7.17 (b).

Example 7.10. Find the length and position (relative to a voltage minimum) of the shunt stub required to match a load of $(162{\cdot}5 + j200)\ \Omega$ to a loss-free twin line of characteristic impedance 250 Ω. The generator frequency is 200 Mc/s, and the line is air-spaced. (*R.R.E.*)

In this (as in other problems involving the circle diagram) the figure number quoted gives the construction required for the given problem. The reader should refer the construction to Fig. 7.12 to obtain actual numerical results.

From the data, the normalized load impedance is $(0{\cdot}65 + j0{\cdot}8)$

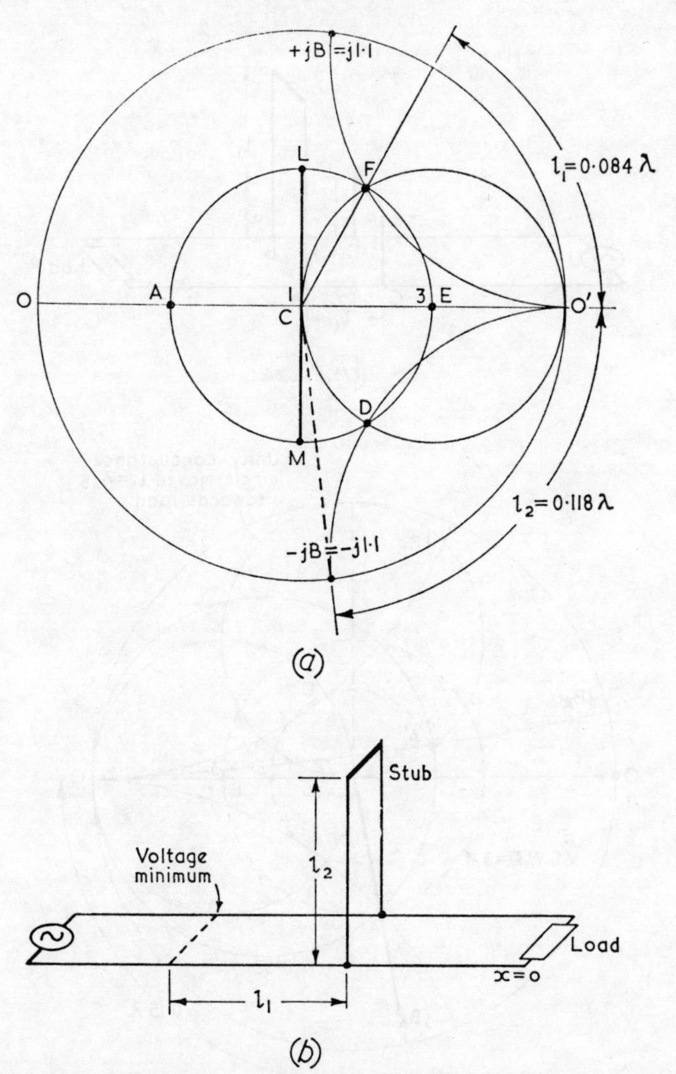

FIG. 7.17. SINGLE SHUNT STUB MATCHING

(point L in Fig. 7.17(a)), and hence the normalized load admittance is $(0{\cdot}62 - j0{\cdot}75)$ (point M). The circle, of centre C and radius CL $(=CM)$ is the required constant V.S.W.R. circle, and cuts the line CO' at the point 3, so that the V.S.W.R. at the load is 3. The constant V.S.W.R. circle cuts the unity-conductance circle at F, and the measured susceptance at this point is $+jB = j1{\cdot}1$. The distance moved from a voltage minimum is $l_1 = 0{\cdot}084\lambda$ towards the load.

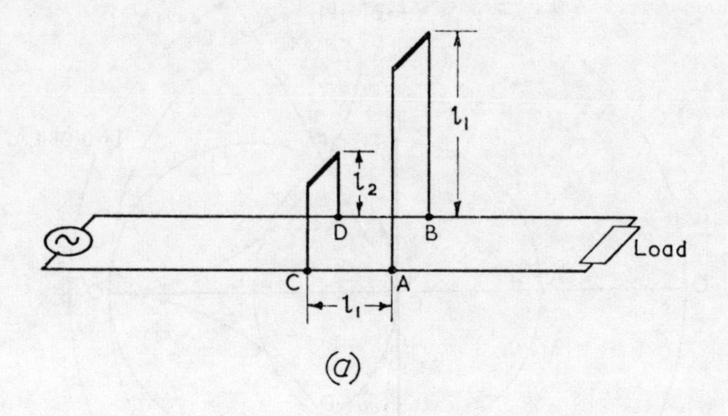

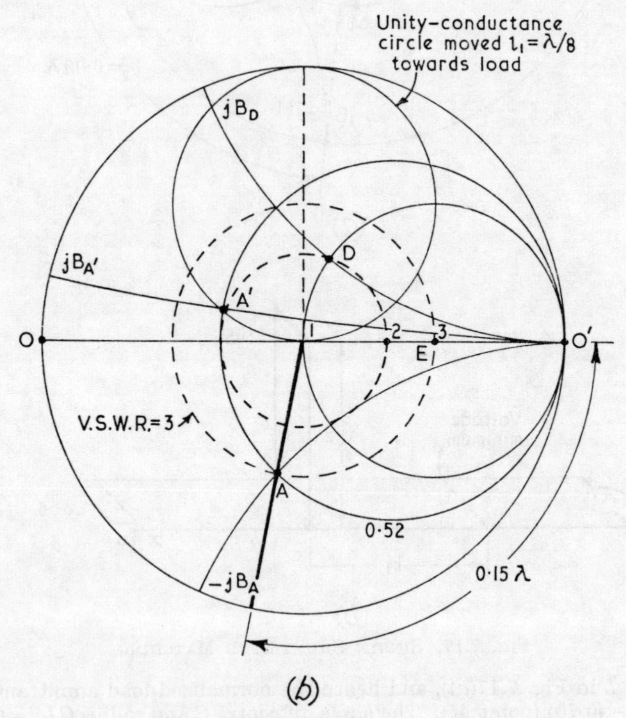

FIG. 7.18. DOUBLE STUB MATCHING

The short-circuited stub to be added at F must therefore have a susceptance of $-j1{\cdot}1$, and thus must have a length of $l_2 = 0{\cdot}118\lambda$ (i.e. measured from the short-circuit at O' towards the generator).

Since $f = 200$ Mc/s, $\lambda = c/f = 1{\cdot}5$ m

and hence $l_1 = 1{\cdot}5 \times 0{\cdot}084 = \underline{\underline{0{\cdot}126 \text{ m}}}$

and $l_2 = 1{\cdot}5 \times 0{\cdot}118 = \underline{\underline{0{\cdot}177 \text{ m}}}$

A stub of length 0·177 m placed at a point 0·126 m nearer the load than a voltage minimum will therefore match the line between that point and the generator.

In practice, a stub may only "flatten" the line—i.e. it may not reduce the V.S.W.R. to exactly unity. In such cases the stub is placed in its calculated position and its length is varied until a voltage maximum (or minimum) occurs exactly $\lambda/4$ from the stub (the reactive components then being exactly cancelled). If there is a maximum at the $\lambda/4$ point, the impedance is too low at the matching point and the stub should be moved towards the load. If a minimum occurs, the impedance at the matching point is too high and the stub should be moved towards the generator until a more satisfactory match is obtained.

7.15. Double Stub Matching

It is evident that single-stub matching involves altering both the length and position of the stub if either the load or the frequency changes and the matching is upset. This is obviously not convenient to do physically in concentric lines or in waveguides (to which the foregoing ideas of matching also apply). In these cases it is better to have two fixed stubs which are normally located $\lambda/8$ or $\lambda/8 + n\lambda/2$ apart. The exact spacing is not critical, except that it must not be $\lambda/2$, since this is equivalent to a single stub. Correct matching is not always possible with double-stub matching as will be seen.

The set-up is shown in Fig. 7.18 (*a*). The length of stub 1 at AB is adjusted until the reference part of the admittance seen at CD is equal to $Y_0 = 1/Z_0$. The length of the second stub is then adjusted to cancel the reactive part of the total admittance at CD, and the line will then be matched from that point to the generator. With double-stub matching it is also possible to use series stubs, in which case the Smith chart is used as an impedance diagram. The following example illustrates the method for double shunt stubs.

Example 7.11. Two short-circuited stubs of characteristic impedance Z_0 are fixed $\lambda/8$ apart in shunt with a concentric line of the same characteristic impedance. The load gives a V.S.W.R. of 3 on the line when the length of the stubs is such that they give zero admittance at the line. The first stub is located at $0{\cdot}15\lambda$ towards the generator from a voltage minimum. Determine the stub lengths required to match the line. (*R.R.E.*)

The constant V.S.W.R. circle for $S = 3$ is drawn as shown in Fig. 7.18(*b*), using the chart as an admittance diagram. From the voltage minimum at E move $0{\cdot}15\lambda$ towards the generator to point A. The normalized admittance at this point is $Y_{AN} = G_{AN} - jB_{AN} = 0{\cdot}52 - j0{\cdot}6$.

We now redraw the unity-conductance circle moved round by $\lambda/8$ towards the load (corresponding to the spacing of the stubs). From A move along the constant-conductance circle to the point A', where it intersects the redrawn unity-conductance circle. At A' the admittance is $Y_{A'N} = G_{AN} + jB_{A'N} = 0{\cdot}52 + j0{\cdot}12$. The difference between this and the value obtained for Y_{AN} represents the normalized susceptance of the first stub, i.e.

$$Y_1 = Y_{A'N} - Y_{AN} = 0{\cdot}52 + j0{\cdot}12 - 0{\cdot}52 + j0{\cdot}6 = j0{\cdot}72$$

From the chart the length of the short-circuited stub is $0{\cdot}348\lambda$.

At A' the admittance is that of the load as seen from AB plus that of the first stub. A' is on the V.S.W.R. circle for $S = 2$, and this is then the V.S.W.R. in the length of line between the stubs. Now we move from A' along the V.S.W.R. $= 2$ circle to D on the true unity-conductance circle, this representing the movement $\lambda/8$ from stub 1 to stub 2.

At point D the normalized admittance is $Y_{DN} = 1 + jB_{DN} = 1 + j0{\cdot}7$, so that, if the second stub has a susceptance of $-j0{\cdot}7$, the line will be matched. From the Smith chart the length of short-circuited stub required is $0{\cdot}17\lambda$.

The reader will notice that, if the constant-conductance circle from A does not cut the moved unity-conductance circle, a match will not be possible. This would have been the case, for example, if a voltage minimum had occurred at the position of the first stub. The larger the mismatch (i.e. the higher the V.S.W.R. on the mismatched line), the more likely is it that double stub matching will not be effective, but the condition may be overcome by using three stubs.

7.16. The Circle Diagram Applied to Lines with Small Losses

The voltage and current at any distance, x, from the load in a lossy line are given by eqns. (7.57) and (7.61) as

$$V_x = V^{+}e^{\gamma x} + \rho V^{+}e^{-\gamma x}$$

and

$$I_x = I^{+}e^{\gamma x} - \rho I^{+}e^{-\gamma x}$$

Hence

$$Z_{in} = \frac{V_x}{I_x} = \frac{V^{+}}{I^{+}}\,\frac{e^{\alpha x}\underline{/\beta x} + \rho' e^{-\alpha x}\underline{/\theta - \beta x}}{e^{\alpha x}\underline{/\beta x} - \rho' e^{-\alpha x}\underline{/\theta - \beta x}}$$

since $\gamma = \alpha + j\beta$, and $\rho = \rho'\underline{/\theta}$.

Thus
$$Z_{in} = Z_0 \frac{e^{\alpha x} + \rho' e^{-\alpha x} \underline{/\theta - 2\beta x}}{e^{\alpha x} - \rho' e^{-\alpha x} \underline{/\theta - 2\beta x}}$$

$$= Z_0 \frac{1 + \rho' e^{-2\alpha x} \underline{/\theta - 2\beta x}}{1 - \rho' e^{-2\alpha x} \underline{/\theta - 2\beta x}} \quad . \quad . \quad (7.75)$$

Comparing this with eqn. (7.73) for the loss-free line, we see that it is similar but that ρ' may be considered to be modified by a dissipation factor $e^{-2\alpha x}$. As x increases, $\rho' e^{-2\alpha x}$ decreases and the V.S.W.R. therefore becomes nearer unity. Indeed, short lengths of lossy line are sometimes used to match loads to loss-free lines over wide frequency ranges.

If we have a lossy line with a purely reactive termination (so that the magnitude of the reflection coefficient is unity), then at the load the V.S.W.R. is infinite. This corresponds to the outside circle on the Smith chart. At a distance x from the load the reflection coefficient becomes $\rho' e^{-2\alpha x} = e^{-2\alpha x}$, so that

$$S_x = \frac{1 + e^{-2\alpha x}}{1 - e^{-2\alpha x}} \qquad \text{(from eqn. (7.58))}$$

Suppose now that the attenuation between two points a and b which are a distance x apart is P decibels. Then $P = 20 \log_{10} (V_a/V_b)$.

Hence
$$\log_e \frac{V_a}{V_b} = \log_e 10 \times \log_{10} \frac{V_a}{V_b} = 0{\cdot}115P = \alpha x$$

by definition. From this,

$$S_x = \frac{1 + e^{-0{\cdot}23P}}{1 - e^{-0{\cdot}23P}} \quad . \quad . \quad . \quad (7.76)$$

Eqn. (7.76) enables the following table relating the V.S.W.R. to the attenuation P to be drawn up.

V.S.W.R.	S	∞	8·65	4·39	3·0	2·32	1·93	1·67	1·49	1·22	1·06
	s	0	0·115	0·228	0·332	0·431	0·52	0·6	0·67	0·82	0·94
P (dB)		0	1	2	3	4	5	6	7	10	15

If we know the attenuation for a given length of line, then with a reactive load the conditions at any point on the line can be determined from the Smith chart by moving in a spiral inwards from the load in a clockwise direction, the V.S.W.R. changing with the length of line according to the above relation. For any other load condition and for known attenuation, it is evident that we can arrange a spiral from the outside circle to pass through the known load point. The continuation of this spiral towards the centre of the chart then gives the conditions on the line from the load to the generator. If the line losses are high, V^+ and I^+ are no longer in phase, and the Smith chart method becomes inaccurate.

A graph of the relation between V.S.W.R. and attenuation is shown in Fig. 7.19.

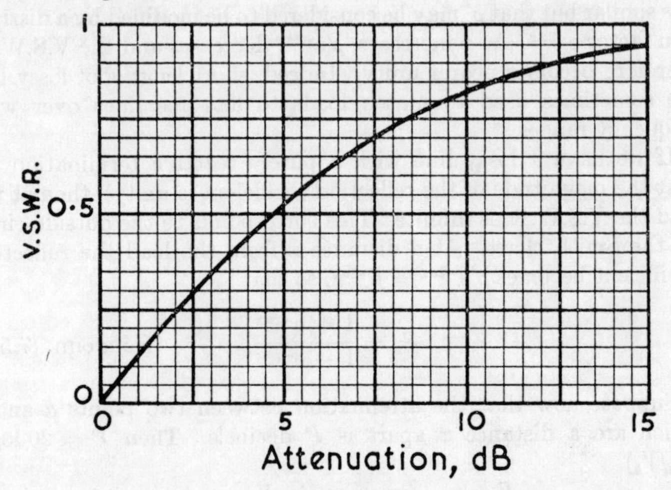

FIG. 7.19. RELATION BETWEEN V.S.W.R. AND ATTENUATION ON A LOW-LOSS LINE

Example 7.12. A transmission line has a characteristic impedance of 70 Ω and the V.S.W.R. measured at about 20λ from the load is 2·0. A voltage minimum occurs 20·19λ from the load. If the attenuation between the point at which the V.S.W.R. is measured and the load is 1·4 dB, find the load impedance. (*R.R.E.*)

From the graph of Fig. 7.19, the attenuation corresponding to a V.S.W.R. of 2·0 is 4·8 dB. Hence the attenuation at the load will be 4·8 − 1·4 = 3·4 dB, which corresponds to a V.S.W.R. of 2·7.

Now, moving 20·19λ towards the load from x is equivalent on the circle diagram to moving 0·19λ, provided that we move from the 2·0 V.S.W.R. circle to the 2·7 V.S.W.R. circle at the same time. Properly, we should have moved from x round the diagram 40 times plus 0·19 λ in a gradually increasing spiral, which comes to the same thing. The result is shown in Fig. 7.20, where it can be seen that at the load point the normalized impedance is $Z_{LN} = 1{\cdot}43 - j1{\cdot}2$. This gives the actual load impedance as

$$Z_L = 70(1{\cdot}43 - j1{\cdot}2) = \underline{\underline{(100{\cdot}1 - j84)\ \Omega}}$$

A further example shows the time-saving which can be achieved by the use of the attenuation scale.

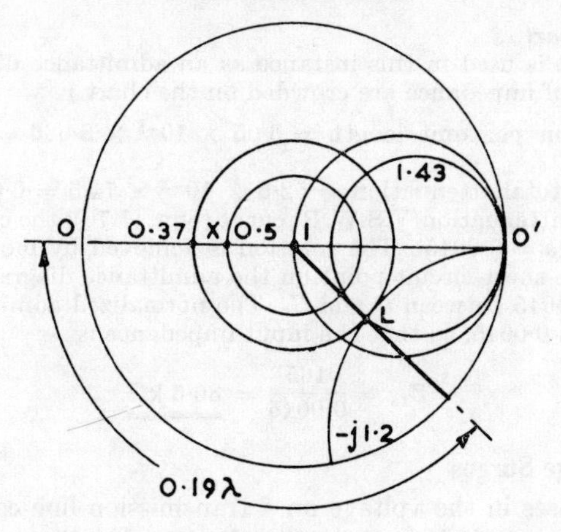

FIG. 7.20

Example 7.13. A uniform transmission line of length $\lambda/4$ is short-circuited at the receiving end. The constants of the line per centimetre length are: $R = 0{\cdot}02\ \Omega$; $L = 5{\cdot}5 \times 10^{-9}$ H; $C = 0{\cdot}202 \times 10^{-12}$ F; $G = 0$. Derive an expression for, and determine the value of, the impedance at the input end at a frequency of 100 Mc/s.

(*I.E.E. Line*, 1957)

For a low-loss line,

$$Z_0 \simeq \sqrt{(L/C)} = \sqrt{(5{\cdot}5 \times 10^{-9})/(0{\cdot}202 \times 10^{-12})} = 165\ \Omega$$

From eqn. (7.48),

$$\alpha \simeq R/2Z_0 = 0{\cdot}02/(2 \times 165) = 6{\cdot}05 \times 10^{-5}\ \text{Np/cm}$$

Since $\beta l = 2\pi l/\lambda$, it follows that

$$\beta l = \pi/2 \text{ for } l = \lambda/4$$

Also,

$$\lambda = 1/f\sqrt{(LC)} = 1/10^8\sqrt{(5{\cdot}5 \times 0{\cdot}202 \times 10^{-21})} = 2{\cdot}98\ \text{m}$$

Thus $$l = \lambda/4 = 74{\cdot}5\ \text{cm}$$

(*a*) *By calculation*

$$Z_{in} = Z_0 \tanh \gamma l \qquad \text{(from eqn. (7.23))}$$

$$= Z_0 \frac{\sinh(\alpha + j\beta)l}{\cosh(\alpha + j\beta)l}$$

$$= Z_0 \frac{\sinh \alpha l \cos \beta l + j \cosh \alpha l \sin \beta l}{\cosh \alpha l \cos \beta l + j \sinh \alpha l \sin \beta l}$$

$$= Z_0 \frac{\cosh \alpha l}{\sinh \alpha l}$$

since $\beta l = \pi/2$.

Hence $$Z_{in} = 165 \times \frac{1}{45 \times 10^{-4}} = \underline{\underline{36{\cdot}6\ \text{k}\Omega}}$$

(*b*) *By the chart*

(The chart is used in this instance as an admittance diagram, since high values of impedance are crowded on the chart.)

Attenuation per unit length $= 6{\cdot}05 \times 10^{-5} \times 8{\cdot}686 = 52{\cdot}5 \times 10^{-5}$ dB/cm.

Therefore total attenuation $= 52{\cdot}5 \times 10^{-5} \times 74{\cdot}5 = 0{\cdot}0391$ dB.

From the attenuation/V.S.W.R. curve (eqn. (7.76)) the corresponding V.S.W.R. is $s = 0{\cdot}0045$. The solution is achieved by moving from O' (which is the short-circuit point on the admittance diagram) by $\lambda/4$ to the point 0·0045 between O and C. The normalized admittance at the input is thus 0·0045, so that the input impedance is

$$Z_{in} = \frac{165}{0{\cdot}0045} = \underline{\underline{36{\cdot}6\ \text{k}\Omega}}$$

7.17. Voltage Surges

Sudden rises in the voltage on a transmission line can be due to many causes, including lightning discharges (direct or indirect) and switching transients. Such surges are characterized by a sudden very steep rise in voltage (the *surge front*) followed by a gradual fall (the *surge tail*), and in the following section we shall consider that in fact the surge consists of a step-function rise in voltage. The Fourier expansion of such a step function shows it to consist of a continuous frequency spectrum, and the results of Section 7.4 indicate that, if the line has resistance and leakage conductance, each frequency component will travel down the line at a different velocity. This will mean that the surge waveform will alter as the surge moves down the line, and in general the sharp-rising surge front tends to be smoothed as the surge progresses. In the following analysis it will be assumed that the line is loss-free so that no distortion occurs.

As in Section 7.5, the surge velocity will be given by

$$u = \frac{1}{\sqrt{(LC)}} = \frac{3 \times 10^8}{\sqrt{(\mu_r \epsilon_r)}} \text{ metres per second} \quad . \quad (7.77)$$

where μ_r and ϵ_r are the relative permeability and permittivity of the dielectric, while the characteristic impedance (called the *surge impedance* in this case) will be

$$Z_0 = \sqrt{\frac{L}{C}}$$

If the magnitude of the surge voltage step is V^+, the current associated with this voltage will be

$$I^+ = \frac{V^+}{Z_0} \quad . \quad . \quad . \quad . \quad (7.78)$$

As the surge travels down the line the surge energy is stored progressively in the electric and magnetic fields of the line. If L and C refer to unit length of line, and the surge velocity is u metres per second, then in one second a length u metres stores $W_e = \frac{1}{2}Cu(V^+)^2$ joules as dielectric energy and $W_m = \frac{1}{2}Lu(I^+)^2$ joules as magnetic energy.

Hence, from eqns. (7.77) and (7.78),

$$W_e = \tfrac{1}{2}C\frac{1}{\sqrt{(LC)}}(V^+)^2 = \frac{(V^+)^2}{2Z_0}$$

and
$$W_m = \tfrac{1}{2}L\frac{1}{\sqrt{(LC)}}(I^+)^2 = \tfrac{1}{2}Z_0\left(\frac{V^+}{Z_0}\right)^2 = \frac{(V^+)^2}{2Z_0}$$

i e. the surge energy is stored equally in the electric and magnetic fields.

Reflection of each component frequency and hence of the total surge occurs at any discontinuity in the line. The results of Section 7.8 may be used, therefore, to determine the values of the reflected voltages and currents. An alternative method using power relations, however, enables us to treat the reflection of surges more generally, as will be seen from the following paragraphs.

(*a*) *Reflection of surges from a resistive termination.* Consider a line with surge impedance Z_0 terminated in a pure resistance, R. If the incident surge voltage is V^+ then the surge current is $I^+ = V^+/Z_0$. On arrival at the termination the incident energy per second is $V^+I^+ = (V^+)^2/Z_0$. If the resultant voltage at the termination is V_L, the energy absorbed in the load per second is V_L^2/R, and the reflected energy per second is $(V^-)^2/Z_0$ (where V^- is the reflected voltage). Hence

$$\frac{(V^+)^2}{Z_0} = \frac{V_L^2}{R} + \frac{(V^-)^2}{Z_0}$$

or
$$(V^+)^2 - (V^-)^2 = (V^+ + V^-)(V^+ - V^-) = \frac{V_L^2Z_0}{R} \quad . \quad (7.79)$$

But
$$V^+ + V^- = V_L \quad . \quad . \quad . \quad (7.80)$$

so that, from eqn. (7.79),

$$V^+ - V^- = \frac{V_LZ_0}{R} \quad . \quad . \quad . \quad (7.81)$$

Adding these two equations,

$$2V^+ = V_L \frac{R + Z_0}{R}$$

or

$$V_L = \frac{2RV^+}{R + Z_0} \quad . \quad . \quad . \quad (7.82)$$

In the same way the load current, I_L, is given by

$$I_L = \frac{V_L}{R} = \frac{2V^+}{R + Z_0} \quad . \quad . \quad . \quad (7.83)$$

From the above relations the reflected surge voltage is

$$V^- = V^+ - \frac{V_L Z_0}{R} = \frac{R - Z_0}{R + Z_0} V^+ = \rho V^+ \quad . \quad (7.84)$$

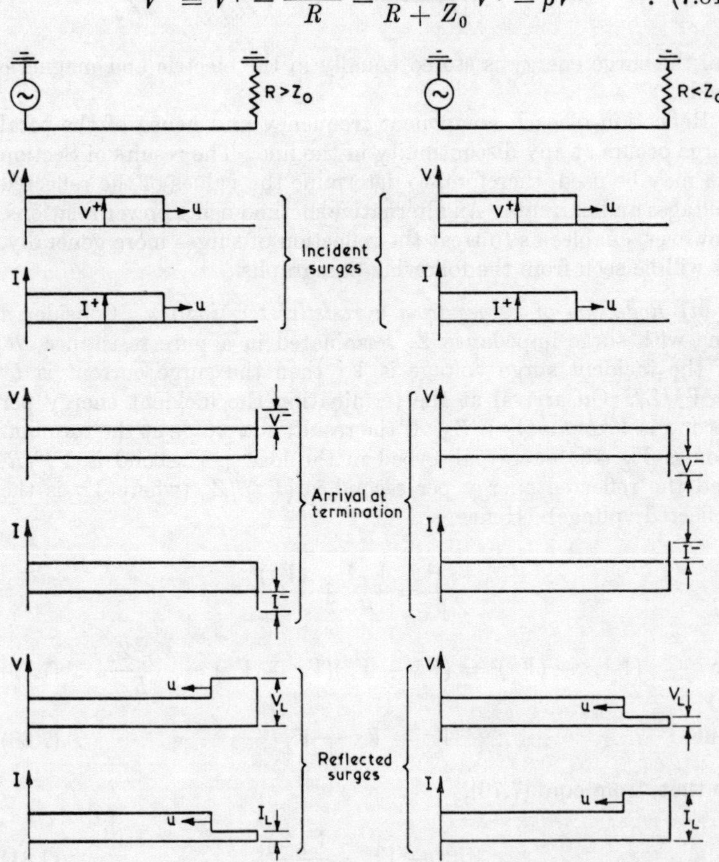

Fig. 7.21. Reflection of Surges at a Resistive Termination

and the reflected surge current is

$$I^- = I - I^+ = \frac{2V^+}{R + Z_0} - \frac{V^+}{Z_0} = -\frac{R - Z_0}{R + Z_0}\frac{V^+}{Z_0}$$

or
$$I^- = -\frac{R - Z_0}{R + Z_0} I^+ = -\rho I^+ \qquad . \quad . \quad . \quad . \quad (7.85)$$

where ρ is the reflection coefficient.

Eqns. (7.84) and (7.85) show that, if $R > Z_0$, the reflected voltage surge is of the same polarity as the incident surge, while the reflected current surge partly cancels the incident current surge. If $R < Z_0$, the reflected current surge adds to the incident current surge, while the reflected voltage surge partly cancels the incident voltage surge. In the matched condition ($R = Z_0$) there is no reflection, while open- and short-circuit terminations give total reflection. The conditions for $R < Z_0$ and $R > Z_0$ are illustrated in Fig. 7.21.

(*b*) *Reflections at a parallel reactive termination.* If the termination of a line is reactive, a transient will occur when a surge arrives there.

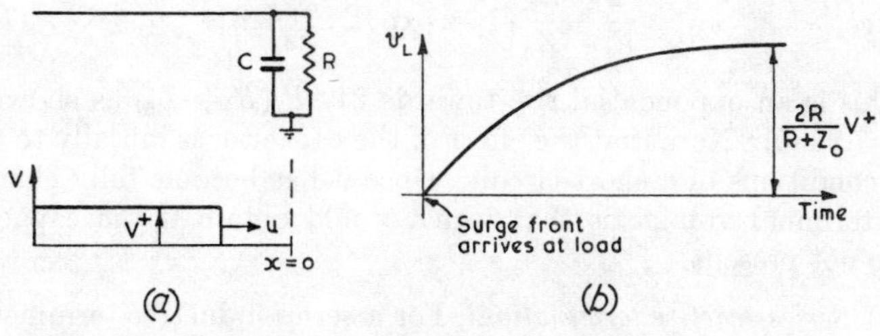

FIG. 7.22. EFFECT OF A REACTIVE PARALLEL TERMINATION

Consider the parallel CR network shown in Fig. 7.22 (*a*). The incident energy per second is, as before, $(V^+)^2/Z_0$. If the instantaneous load voltage is v_L, the rate at which the load absorbs energy is $v_L^2/R + v_L C(\mathrm{d}v_L/\mathrm{d}t)$ joules per second. For an instantaneous reflected voltage v^- we therefore have

$$\text{Input power} = \begin{Bmatrix}\text{Energy dissipation and}\\ \text{storage rate in load}\end{Bmatrix} + \text{Reflected power}$$

Hence
$$\frac{(V^+)^2}{Z_0} = \frac{v_L^2}{R} + v_L C\frac{\mathrm{d}v_L}{\mathrm{d}t} + \frac{(v^-)^2}{Z_0}$$

This yields
$$(V^+)^2 - (v^-)^2 = (V^+ + v^-)(V^+ - v^-)$$
$$= v_L^2\frac{Z_0}{R} + v_L C Z_0\frac{\mathrm{d}v_L}{\mathrm{d}t}$$

But
$$v_L = V^+ + v^- \qquad (7.86)$$

so that
$$V^+ - v^- = v_L \frac{Z_0}{R} + CZ_0 \frac{\mathrm{d}v_L}{\mathrm{d}t} \qquad (7.87)$$

Adding these two equations,

$$2V^+ = v_L\left(\frac{R+Z_0}{R}\right) + CZ_0 \frac{\mathrm{d}v_L}{\mathrm{d}t}$$

Taking Laplace transforms,

$$\frac{2V^+}{s} = \left(\frac{R+Z_0}{R} + sCZ_0\right)\bar{v}_L$$

(since the input is a step function its Laplace transform will be V^+/s). Rearranging, and splitting into partial fractions,

$$\bar{v}_L = \frac{2V^+}{CZ_0}\left[\frac{1}{s} - \frac{1}{s + \dfrac{R+Z_0}{RZ_0C}}\right]\frac{RZ_0C}{R+Z_0}$$

Hence,
$$v_L = \frac{2V^+R}{R+Z_0}\left[1 - \exp\left\{-\frac{(R+Z_0)t}{RZ_0C}\right\}\right] \qquad (7.88)$$

This is an exponential rise towards $2V^+R/(R+Z_0)$ as shown in Fig. 7.22 (*b*). Note that the effect of the capacitor is initially to give the conditions of a short-circuit. Once it has become fully charged the terminal voltage is that which would obtain if the capacitor were not present.

(*c*) *Series reactive termination.* For a series inductive termination it is simpler to work in terms of incident surge current. Thus, for the series RL termination shown in Fig. 7.23 (*a*), we can write

Incident power = Absorbed power + Reflected power

$$(I^+)^2Z_0 = i_L{}^2R + i_L L\frac{\mathrm{d}i_L}{\mathrm{d}t} + (i^-)^2Z_0$$

where i_L is the load current, given by

$$i_L = I^+ + i^- \qquad (7.89)$$

and i^- is the instantaneous reflected current. Hence

$$(I^+)^2 - (i^-)^2 = (I^+ + i^-)(I^+ - i^-) = i_L{}^2\frac{R}{Z_0} + i_L\frac{L}{Z_0}\frac{\mathrm{d}i_L}{\mathrm{d}t}$$

so that
$$I^+ - i^- = i_L\frac{R}{Z_0} + \frac{L}{Z_0}\frac{\mathrm{d}i_L}{\mathrm{d}t} \qquad (7.90)$$

Adding eqns. (7.89) and (7.90),

$$2I^+ = i_L\left(\frac{R + Z_0}{Z_0}\right) + \frac{L}{Z_0}\frac{\mathrm{d}i_L}{\mathrm{d}t}$$

and taking Laplace transforms (I^+ being a step function),

$$\bar{i}_L = \frac{2I^+}{s\left(\frac{R + Z_0}{Z_0} + \frac{sL}{Z_0}\right)}$$

Solving, by splitting into partial fractions and taking inverse transforms,

$$i_L = \frac{2I^+Z_0}{R + Z_0}[1 - \mathrm{e}^{-(R+Z_0)t/L}] \qquad . \qquad . \quad (7.91)$$

The time variation of i_L is shown in Fig. 7.23 (*b*).

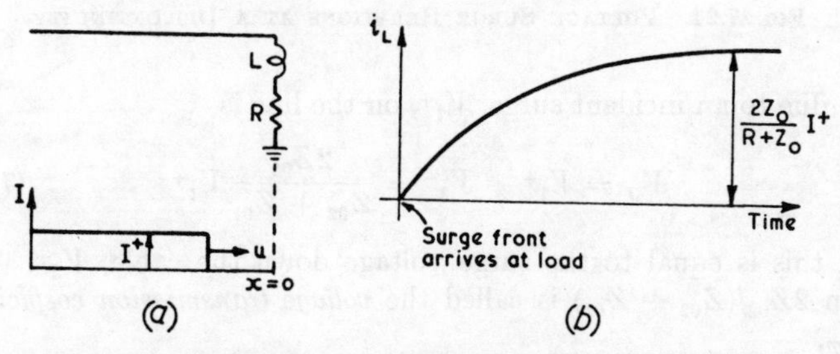

FIG. 7.23. EFFECT OF AN INDUCTIVE SERIES TERMINATION

(*d*) *Reflection of practical surge waveform.* In practice, a surge waveform may consist of a steeply rising surge front followed by a more slowly falling tail. This can be expressed as the difference between two exponential functions as

$$v^+ = V(\mathrm{e}^{-at} - \mathrm{e}^{-bt})$$

where $b > a$. Eqns. (7.82) and (7.83) will still give the instantaneous voltage and current at a resistive termination. Thus

$$v_L = \frac{2RV}{R + Z_0}(\mathrm{e}^{-at} - \mathrm{e}^{-bt})$$

and

$$i_L = \frac{2V}{R + Z_0}(\mathrm{e}^{-at} - \mathrm{e}^{-bt})$$

The energy absorbed in the termination R in time t is

$$W = \int_0^t i_L{}^2R\,\mathrm{d}t \text{ joules}$$

7.18. Transmission Coefficient

The junction between two lines (or between a line and a cable) with different characteristic impedances is equivalent to an unmatched termination, and reflection of energy will take place. The surge which is transmitted down the second line will correspond to the resultant load voltage in the previous section. Thus, consider a line of characteristic impedance Z_{01} joined to a cable of characteristic impedance Z_{02} as shown in Fig. 7.24. The voltage at the junction,

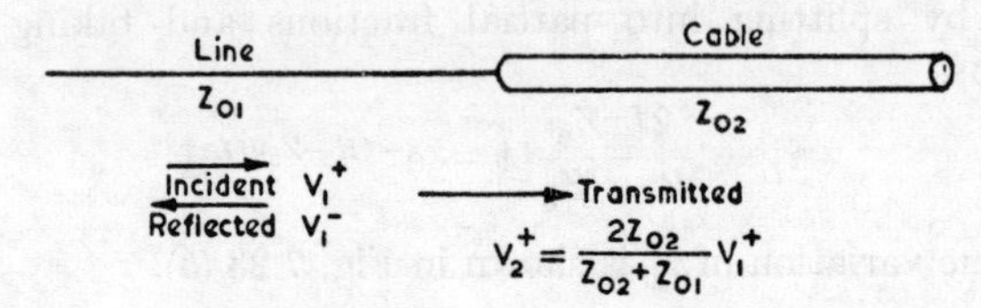

FIG. 7.24. VOLTAGE SURGE RELATIONS AT A DISCONTINUITY

V_J, due to an incident surge, $V_1{}^+$, on the line is

$$V_J = V_1{}^+ + V_1{}^- = \frac{2Z_{02}}{Z_{02} + Z_{01}} V_1{}^+ \quad . \quad . \quad (7.92)$$

and this is equal to the surge voltage down the cable, $V_2{}^+$. The term $2Z_{02}/(Z_{02} + Z_{01})$ is called the *voltage transmission coefficient*, K_{VT}.

The current surge in the cable will be

$$I_2{}^+ = \frac{V_2{}^+}{Z_{02}} = \frac{2V_1{}^+}{Z_{02} + Z_{01}} = \frac{2Z_{01}I_1{}^+}{Z_{02} + Z_{01}} \quad . \quad . \quad (7.93)$$

where the term $2Z_{01}/(Z_{02} + Z_{01})$ may be called the *current transmission coefficient*, K_{IT}.

Since the characteristic impedance of a cable is generally much lower than that of a line, only a small fraction of any voltage surge which travels down the line will be transmitted down the cable. For this reason apparatus at the end of an overhead power line is sometimes connected to the line by a short length of surge-minimizing cable.

Example 7.14. A long-tailed unit-function 500-kV surge voltage on an overhead line of surge impedance 400 Ω arrives at a point where the line continues into a cable, 1 km long and having a total inductance of 265 μH and total capacitance of 0·165 μF. At the end of the cable connexion is made to a transformer of surge impedance 1,000 Ω. Find the surge voltage distribution 12 μsec after the surge arrives at the line-cable junction.

The surge impedance of the cable is $Z_{02} = \sqrt{(L/C)} = 40 \cdot 1\ \Omega$, and the velocity of propagation down the cable is given by eqn. (7.77) as

$$u = \frac{1}{\sqrt{(LC)}} = \frac{1}{\sqrt{(0 \cdot 265 \times 0 \cdot 165 \times 10^{-15})}} = 151 \times 10^6 \text{ m/sec}$$

In 12 μsec the surge can therefore travel a distance of $12 \times 0 \cdot 151 = 1 \cdot 82$ km in the cable. It will therefore have reached the end of the cable and have been reflected back a distance of 0·82 m towards the line–cable junction. From eqn. (7.92) the incident voltage surge, V_2^+, in the cable is

$$V_2^+ = \frac{2Z_{02}}{Z_{02} + Z_{01}} V_1^+ = \frac{2 \times 40 \cdot 1}{440 \cdot 1} \times 500 = 91 \cdot 2 \text{ kV}$$

At the transformer end of the cable reflection takes place, the reflected voltage being given by

$$V_2^- = \rho V_2^+ = \frac{1{,}000 - 40 \cdot 1}{1040 \cdot 1} \times 91 \cdot 2 = 84 \cdot 2 \text{ kV}$$

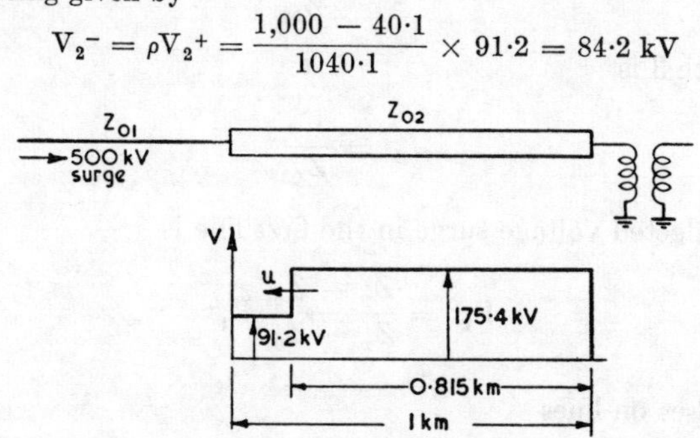

Fig. 7.25. Voltage Distribution in a Surge-limiting Cable

Hence the voltage on the cable between the transformer and the point 0·82 km towards the line will be the sum of the incident and reflected voltages, i.e. $(91 \cdot 2 + 84 \cdot 2) = 175 \cdot 4$ kV. From this point up to the line–cable junction, only the incident voltage surge will be present, i.e. the cable voltage will be 91·2 kV. The conditions are illustrated in Fig. 7.25.

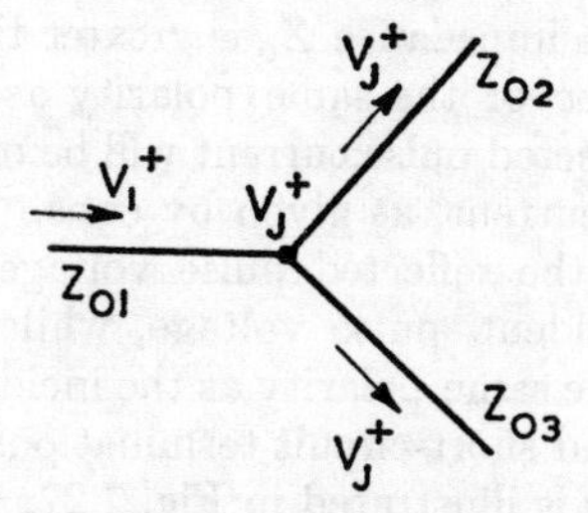

Fig. 7.26. Division of Surge Energy

At the junction of three lines, as shown in Fig. 7.26, a surge travelling down any one line (say line 1) will be reflected, the reflection and transmission depending on the parallel impedance of the two other lines, Z_p. The resultant surge voltage at the junction will be transmitted down the two other lines (2 and 3). Thus if Z_{01}, Z_{02} and Z_{03} are the surge impedances of the three lines, then for a surge V_1^+ travelling down line 1 towards the junction, the voltage at the junction will be

$$V_J = \frac{2Z_p}{Z_p + Z_{01}} V_1^+ = K_{VT} V_1^+ \quad . \quad . \quad (7.94)$$

where $Z_p = Z_{02}Z_{03}/(Z_{02} + Z_{03})$. The current surge in line 2 is

$$I_2^+ = \frac{V_J}{Z_{02}} \quad . \quad . \quad . \quad . \quad (7.95)$$

and in line 3 is

$$I_3^+ = \frac{V_J}{Z_{03}} \quad . \quad . \quad . \quad . \quad (7.96)$$

The reflected voltage surge in the first line is

$$V_1^- = \frac{Z_p - Z_{01}}{Z_p + Z_{01}} V_1^+ \quad . \quad . \quad . \quad (7.97)$$

7.19. Pulses on lines

Since a pulse may be considered as the sum of a positive-going and an equal but time-displaced negative-going step function, the analysis of the previous sections will also apply to the case of pulses on transmission lines. If the pulse waveform is to be preserved, the lines used must be as loss-free as possible. Pulses are reflected from terminations and discontinuities by the same rules as those which we have determined for step functions. Thus, if the resistance, R, terminating a line of surge impedance Z_0 is greater than Z_0, the reflected pulse voltage will be of the same polarity as that of the incident pulse, while the reflected pulse current will be of opposite polarity to the incident pulse current, as given by eqns. (7.84) and (7.85). If, however, $R < Z_0$, the reflected pulse voltage will be of opposite polarity to the incident pulse voltage, while the reflected pulse current will be of the same polarity as the incident pulse current. In the case of open- and short-circuit terminations, total reflection will take place, and this is illustrated in Fig. 7.27.

If the width of a pulse which is applied to a short-circuited line is longer than twice the time taken for the pulse to travel from the generator to the end of the line, the reflected voltage pulse when it arrives back at the generator will cancel the incident voltage. If the

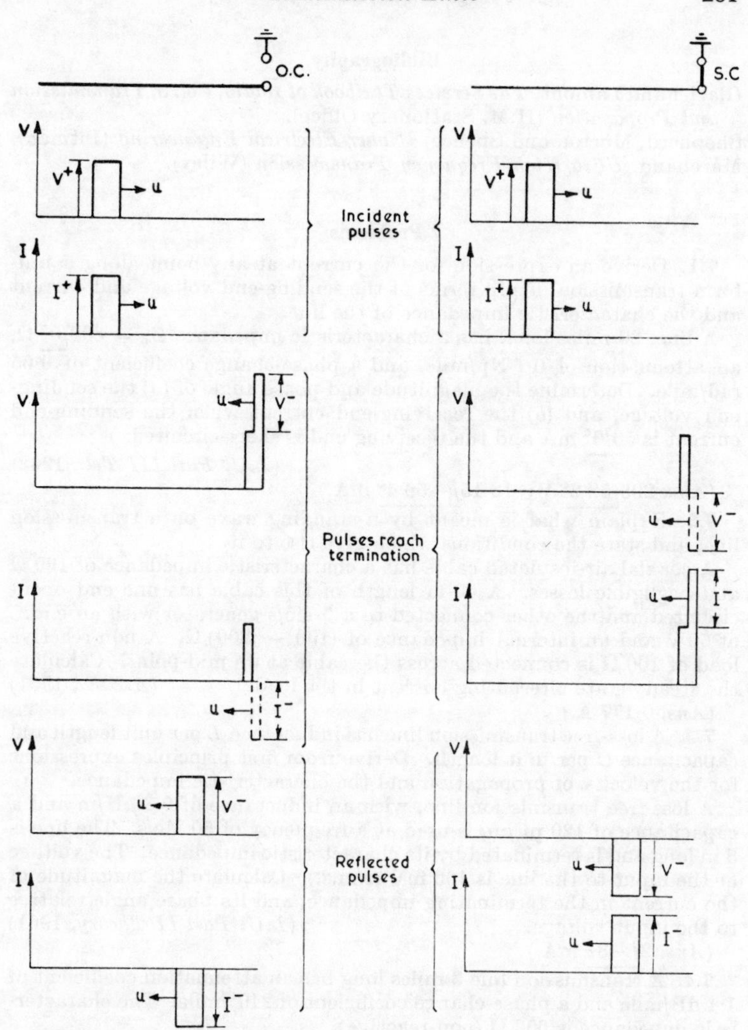

FIG. 7.27. REFLECTION OF PULSES FROM OPEN- AND SHORT-CIRCUITS

generator impedance is matched to the line the reflected wave will be absorbed completely by this impedance. The resultant voltage waveform at the generator terminals will thus consist of a pulse of duration equal to twice the delay time of the line. This is one method of producing pulses of very short and accurately known duration.

Bibliography

Glazier and Lamont, *The Services Textbook of Radio, Vol. 5, Transmission and Propagation* (H.M. Stationery Office).
Shepherd, Morton and Spence, *Higher Electrical Engineering* (Pitman).
Marchand, *Ultra High Frequency Transmission* (Wiley).

Problems

7.1. Derive an expression for the current at any point along a uniform transmission line in terms of the sending-end voltage and current and the characteristic impedance of the line.

A line, 20 miles long, has a characteristic impedance $Z_0 = 600\underline{/0°}\ \Omega$, an attenuation of 0·1 Np/mile, and a phase-change coefficient of 0·05 rad/mile. Determine the magnitude and phase angle of (*a*) the sending-end voltage, and (*b*) the receiving-end current when the sending-end current is $50\underline{/0°}$ mA and the receiving end is short-circuited.

(*L.U. Part III Tel.*, 1962)

(*Ans.* $295\underline{/58·2°}$ V; $13·15\underline{/-56·4°}$ mA.)

7.2. Explain what is meant by a standing wave on a transmission line, and state the conditions which give rise to it.

A coaxial air-insulated cable has a characteristic impedance of 100 Ω and negligible losses. A 60-m length of this cable has one end open-circuited and the other connected to a 5-Mc/s generator with an e.m.f. of 50 V and an internal impedance of $(100 + j200)\ \Omega$. A non-reactive load of 100 Ω is connected across the cable at its mid-point. Calculate the steady-state alternating current in the load. (*A.E.E.*, 1961)

(*Ans.* 0·177 A.)

7.3. A loss-free transmission line has inductance L per unit length and capacitance C per unit length. Derive from first principles expressions for the velocity of propagation and the characteristic impedance.

A loss-free transmission line, with an inductance of 0·3 μH/m and a capacitance of 120 pF/m, is used at a frequency of 50 Mc/s. The line is 5 m long and is terminated by its characteristic impedance. The voltage at the input to the line is 100 mV (r.m.s.). Calculate the magnitude of the current in the terminating impedance, and its phase angle relative to the input voltage. (*L.U. Part II Theory*, 1961)

(*Ans.* $2\underline{/-3\pi}$ mA.)

7.4. A transmission line 3 miles long has an attenuation coefficient of 1·4 dB/mile and a phase-change coefficient of 240°/mile. The characteristic impedance is 600 Ω (non-reactive).

The sending end of the line is fed from a matched a.c. generator having an e.m.f. of 1 V, and the receiving end is joined to a similar generator having an e.m.f. of 0·5 V. The frequencies of the two generators are the same, and there is no phase difference between them.

Find the current in the receiving-end generator. (*A.E.E.*, 1954)

(*Ans.* 98 μA.)

7.5. Determine from first principles in terms of the characteristic impedance and the propagation coefficient the values of the arms of a T-section equivalent to a transmission line. Hence find the values of the arms of a T-section of resistors to be equivalent at a very low frequency

to 8 miles of a $600\underline{/0^\circ}$-Ω line with an attenuation of 2·5 dB/mile when terminated in its characteristic impedance. (*A.E.E.*, 1957)
(*Ans.* Series arms, 490 Ω; shunt arm, 121 Ω.)

7.6. A cable 8 miles long has the following constants per loop-mile: resistance, 45 Ω; inductance, 0·35 mH; capacitance, 0·13 μF. The shunt conductance may be neglected.

Calculate the characteristic impedance of the line at 5,000 c/s. If the load impedance at the receiving end has this value, and a voltage of 200 V 5,000 c/s is applied at the sending end, calculate (*a*) the magnitude of the received current, (*b*) the wavelength, and (*c*) the velocity of propagation. (*A.E.E.*, 1960)
(*Ans.* $(84 - j66)$ Ω; 0·22 A; 18·3 miles; 91,500 miles/sec.)

7.7. Explain what is meant by the terms nominal-π and equivalent-π when applied to a uniform transmission line.

Determine from first principles the values of the elements of an equivalent-π section for a transmission line having a characteristic impedance Z_0 and a propagation coefficient γ. (*A.E.E.*, 1961)

7.8. Explain why the velocity of propagation of electromagnetic waves along a transmission line is normally less than that in free space.

A line has $R = 15$ Ω; $G = 0$; $C = 0{\cdot}02$ μF and $L = 5$ mH per loop-mile. Calculate $\alpha + j\beta$ per mile (do not approximate since $R \simeq \omega L$) and the velocity of propagation if the signal frequency is 796 c/s. (*A.E.E.*, 1962)
(*Ans.* $0{\cdot}0143 + j0{\cdot}0518$; 96,500 miles/sec.)

7.9. A uniform transmission line is sometimes represented by lumped circuit-elements. Discuss the limitations of such a representation.

For the purpose of calculating the phase shift between input and output voltages, a length of loss-free transmission line is represented by a T-section which has a total series inductance of L henrys and a total shunt capacitance of C farads. Calculate the error in the result at a frequency of $1/\pi\sqrt{(LC)}$ cycles per second. Both line and section are terminated in their respective characteristic impedances. The propagation coefficient γ of the section is given by

$$\cosh \gamma = 1 - (1/2)\omega^2 LC$$

(*Ans.* 1·14 rad.) (*A.E.E.*, 1962)

7.10. The series loop inductance of a certain cable is small compared with the loop resistance at low frequencies, while the shunt leakance is small compared with the shunt capacitance. Show from first principles that at such frequencies the attenuation and wavelength coefficients approximate to values proportional to the square root of the operating frequency.

A cable has a loop resistance of 88 Ω per mile, and a shunt capacitance of 0·66 μF per mile. Calculate for a steady-state sine transmission at 50 c/s (*a*) the attenuation and phase-change coefficients, (*b*) the time of transmission over a 50-mile length of the cable. (*A.E.E.*, 1963)
(*Ans.* 0·0279 Np/mile; 0·0279 rad/mile, 4·44 msec.)

7.11. A uniform transmission line has the following primary constants per unit length: resistance, R ohms; inductance, L henrys; conductance, G mhos; and capacitance C farads. Derive from first principles an expression for the characteristic impedance of the line.

A telephone line has the following constants per loop-mile: $R = 40$ Ω;

$L = 50$ mH; $G = 10$ micromhos; $C = 0{\cdot}06\ \mu$F. Determine the frequency at which the magnitude of the characteristic impedance is 950 Ω. (*L.U. Part III Tel.*, 1960)

(*Ans.* 305 c/s.)

7.12. A coaxial cable 100 m long has an inner conductor of 0·35 cm diameter and an outer conductor of 2 cm inner diameter, the annular space being filled with a dielectric of relative permittivity 2·5.

Derive an expression for the capacitance per metre length of the cable. Find the velocity of wave propagation along it, explaining your solution. Assume that the losses in the cable can be neglected.

A sinusoidal p.d. having an r.m.s. value of 100 V at a frequency of 1 Mc/s is applied at one end, the other end being correctly terminated to prevent reflection. Find the r.m.s. value of the current in the termination, and also the value of this current at the instant when the sending-end p.d. is at its positive maximum. (*L.U. Part III Tel.*, 1961)

(*Ans.* $1{\cdot}9 \times 10^8$ m/sec; 0·93 A; −0·59 A.)

7.13. Explain the terms travelling wave and standing wave as applied to transmission lines.

Two long transmission lines each having a surge impedance of 400 Ω are connected by a cable having a surge impedance of 50 Ω. If a short pulse having a magnitude of 10 kV travels along the first line towards the junction, determine from first principles the magnitude of the first and second pulses entering the second line. State any assumptions made. (*A.E.E.*, 1962)

(*Ans.* 3·95 kV; 2·25 kV.)

7.14. Explain the term characteristic (surge) impedance when applied to a transmission line. Indicate how this quantity depends upon the spacing and size of the conductors in the case of an open-wire line.

A cable with surge impedance of 100 Ω is terminated in two parallel-connected open-wire lines having surge impedances of 600 Ω and 1,000 Ω respectively. If a steep-fronted voltage wave of 1,000 V travels along the cable, find from first principles the voltage and current in the cable and open-wire lines immediately after the travelling wave has reached the transition point. The voltage wave may be assumed to be of infinite length. (*A.E.E.*, 1957)

(*Ans.* In cable, 1,578 V, 4·22 A; line 1, 1,578 V, 2·63 A; line 2, 1,578 V, 1·578 A.)

7.15. Explain what is meant by (*a*) a travelling wave, and (*b*) a standing wave on a transmission line.

Two long transmission lines with different characteristic impedances are connected together. Derive expressions for the reflection and transmission coefficients for voltage and current.

A cable with a characteristic impedance of 80 Ω is joined in series with an open-wire line having a characteristic impedance of 700 Ω. If, as a consequence of connecting a direct voltage to a cable, a steep-fronted voltage wave of 1·2 kV travels along it, determine the voltage and current in the cable and open-wire line immediately after the travelling wave has reached the junction. Assume the lines to be loss-free. (*A.E.E.*, 1959)

(*Ans.* 2·155 kV, 3·08 A.)

7.16. Show from first principles that the ratio of voltage to current in a surge on a uniform loss-free transmission line of inductance L henrys and capacitance C farads per unit length is given by $v/i = Z_0 = \sqrt{(L/C)}$.

An overhead transmission line with a surge impedance of 500 Ω has a load comprising a 10-kΩ resistor in parallel with a 0·005 μF capacitor connected across the far end. A surge voltage of 10-kV magnitude and unit-function form travels along the line. Derive an expression for the time variation of the voltage across the load, and calculate this voltage 5 μsec after the arrival of the wavefront of the surge. State any assumptions made. (*A.E.E.*, 1961)

(*Ans.* $19{\cdot}05(1 - e^{-421{,}000t})$; 16·7 kV.)

7.17. A transmission line of surge impedance 400 Ω is terminated in a resistance of 1·2 kΩ. If a voltage surge of magnitude $250\ (e^{-0{\cdot}05t} - e^{-t})$ kV, where t is in microseconds, originates in the line and travels towards the terminating impedance, determine the voltage at the termination at a time $t = 2$ μsec after the arrival of the wave, and the energy dissipated in the terminating resistance during this interval. (*I.E.E. Supply*, 1959)

(*Ans.* 289 kV, 73 J.)

7.18. A long single-conductor overhead line, having inductance and capacitance of 2·4 mH/km and 0·01 μF/km, respectively, is connected to terminal apparatus through a length of single-phase cable having constants of 1·1 mH/km and 0·5 μF/km, respectively. A travelling wave originating in the line travels towards the junction with the cable. Discuss the effects produced during a short interval of time after the arrival at the junction of the travelling wave.

Calculate the energy transmitted into the cable during a period of 3 μsec after the arrival, at the junction, of a travelling wave of vertical wavefront and infinite wavetail of 50 kV magnitude; determine also the initial magnitude of the voltage wave reflected into the line if a 100-Ω resistor is connected from the junction of the line and cable to earth. (*I.E.E. Supply*, 1958)

(*Ans.* 4·88 J; −44 kV.)

7.19. Describe the impulse voltage waveshape commonly used for the testing of electrical power plant, and the way in which the variation of a surge voltage with time may conveniently be expressed as the difference of the two voltages, each of which varies exponentially with time.

A transmission line AB of characteristic impedance 400 Ω is connected to a length of cable, BC, of characteristic impedance 45 Ω. A fault of zero impedance occurs at C, and the resulting fault current is subsequently cleared by a circuit-breaker near C which "chops" the current abruptly from 700 A to zero. Determine the surge voltages and currents which are initiated in the line by the interruption of the fault current. (*I.E.E. Supply*, 1957)

(*Ans.* −56·6 kV, −141·4A.)

7.20. A 600-Ω transmission line supplies a terminating impedance at a frequency of 100 Mc/s. Measurement discloses a standing wave, of ratio 3, with a voltage minimum at 60 cm from the termination. Calculate the effective values of the resistive and reactive components of the terminating impedance. Assume that the wavelength on the line is equal to the free-space value. (*I.E.E. Radio*, 1959)

(*Ans.* $(1{,}020 - j864)$ Ω.)

7.21. A 600-Ω air-spaced balanced feeder connects a 100-Mc/s transmitter to a dipole aerial. If the aerial impedance is $(75 + j0)$ Ω and the power delivered to the aerial is 1 kW at a frequency of 100 Mc/s, calculate

(*a*) The standing wave ratio on the feeder.

(*b*) The maximum and minimum values of voltage and current along the feeder.

(*c*) The point of attachment and the length of a short-circuited 600-Ω stub line which will eliminate standing waves from that portion of the feeder lying between the transmitter and the stub line.

Neglect attenuation in the feeder and stub line. (*I.E.E. Radio*, 1960)

(*Ans.* (*a*) 8; (*b*) 2,190 V, 274 V, 3·65 A, 0·456 A; (*c*) 21·3 cm stub 1·32 m from load.)

7.22. Derive a formula in terms of Z_0 and Z_2 for the impedance Z_1 presented by a quarter-wavelength section of loss-free line of characteristic impedance Z_0 when it is terminated in an impedance Z_2.

The characteristic impedance of a balanced 2-wire loss-free line is given by $Z_0 = 275 \log_{10} (2s/d)$ ohms, where s is the spacing between the wires and d is the wire diameter.

A 2-wire transmission line of 600-Ω impedance and spacing 6 in. is to be matched to a dipole aerial of 250-Ω impedance by means of a quarter-wavelength matching section using wire of 0·10 in. diameter. Determine: (*a*) the diameter of the wire for the 600-Ω transmission line, (*b*) the impedance of the quarter-wavelength matching section, and (*c*) the spacing of the wires for the matching section.

Describe a simple method for determining whether the 2-wire transmission line is accurately matched to the aerial. (*I.E.E.*, *Radio*, 1962)

(*Ans.* (*a*) 0·047 in.; (*b*) 386·5 Ω; (*c*) 3·04 in.)

7.23. A loss-free transmission line is terminated in a normalized admittance $y = 2 + j0{\cdot}5$. Use a Smith circle diagram to design a matching system using short-circuited stubs of the same characteristic admittance as the feeder in the following cases: (*a*) a single stub adjustable in length and position, and (*b*) two stubs fixed in position at $\lambda/4$ and $5\lambda/8$, respectively, from the termination.

Compare the merits of the two methods, and comment on the choice of fixed positions in case (*b*).

(*Note:* In this context normalized admittance is the ratio of an admittance to the characteristic admittance of the transmission line.)

(*I.E.E. Radio*, 1962)

(*Ans.* (*a*) stub length $0{\cdot}143\lambda$ located $0{\cdot}405\lambda$ nearer generator than a voltage minimum (or $0{\cdot}428\lambda$ from the termination); (*b*) load stub $0{\cdot}29\lambda$, generator stub $0{\cdot}143\lambda$.)

CHAPTER 8

Communication Theory

WHEN information has to be transmitted between two points by means of an electrical signal, a range of frequencies is required, depending on the type of signal and also on the speed of signalling. This bandwidth may vary from a few cycles per second for telegraphy to over 6 Mc/s for 625-line television. Speech frequencies vary from some 75 c/s up to perhaps 10 kc/s, and music often contains frequencies higher than 15 kc/s. It is difficult to transmit such signals directly as electromagnetic waves owing to the physical size and power rating of the aerial systems which would be involved (*see* Chapter 12), but they may readily be superimposed on a high-frequency *carrier* in such a way that the original signal may be recovered at the receiver. The carrier is said to be *modulated* by the signal, and the process of recovery is called *demodulation*, or detection. This same process is used to transmit many speech channels through a single telephone circuit (carrier telephony).

In carrier communication systems a restriction of the bandwidth is often imposed on economic grounds. Thus medium-wave radio is normally restricted to a signal bandwidth of 4·5 kc/s, while an internationally agreed standard for telephony is from 300 c/s to 3,400 c/s (commercial speech). These bandwidths represent compromises between the ideal and the minimum required for intelligibility. Obviously the narrower the bandwidth which is needed the more communication channels can be accommodated in a given carrier frequency range. High-quality speech and music broadcast transmissions are made in the v.h.f. band (carrier frequencies ranging from 80 to 100 Mc/s), allowing an audio bandwidth of 15 kc/s to be economically achieved.

There are three main groups of modulation process: (*a*) *amplitude modulation*, where the amplitude of the carrier is varied by the signal, (*b*) *frequency and phase modulation*, where the instantaneous frequency or phase of the carrier is varied by the signal, and (*c*) *pulse code modulation*, where the information is transmitted in the form of carrier pulses. The discussion in this chapter will be limited to some of the available modulation processes.

8.1. The Amplitude-modulated Wave

Consider a carrier wave represented by $v_c = V_{cm} \sin \omega t$, where V_{cm} is the amplitude of the unmodulated carrier voltage and ω is its angular frequency. In order that the modulated carrier shall convey information about both the frequency and the magnitude of the signal, and in addition be easily detected, it is necessary to vary the carrier amplitude about its mean value, so that, assuming a sinusoidal signal of frequency $\omega_s/2\pi$, the modulated carrier is represented by

$$v = V_{cm}(1 + m \sin \omega_s t) \sin \omega t \quad . \quad . \quad (8.1)$$

The waveform represented by eqn. (8.1) is shown in Fig. 8.1 (*a*). The factor m varies with the magnitude of the signal, and is variously

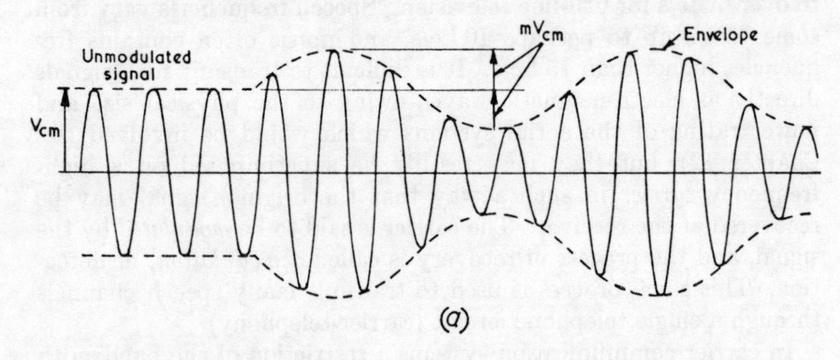

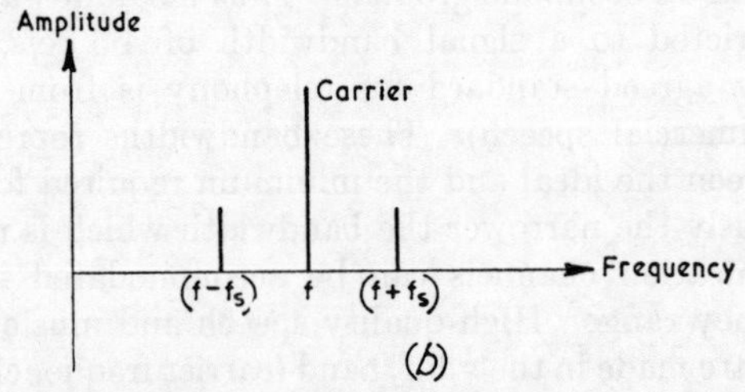

FIG. 8.1. AMPLITUDE-MODULATED WAVE, WITH ITS FREQUENCY SPECTRUM

called the depth of modulation, or *modulation factor*. It is to be noted that the envelope of the wave varies at the signal frequency. Eqn. (8.1) may readily be expanded to

$$v_c = V_{cm} \sin \omega t + V_{cm} m \sin \omega t \sin \omega_s t$$

$$= V_{cm} \sin \omega t + \frac{V_{cm} m}{2} \cos (\omega - \omega_s)t - \frac{V_{cm} m}{2} \cos (\omega + \omega_s)t \quad (8.2)$$

The first term in this equation represents the unmodulated carrier while the second and third terms are called the *lower* and *upper sidebands* respectively. The frequency spectrum of the modulated wave (Fig. 8.1 (*b*)) shows the amplitude of the carrier and the sidebands plotted to a base of frequency. The bandwidth required (i.e. the difference between the highest and lowest transmitted frequency) is seen to be twice the highest signal frequency.

Since the power transmitted at each frequency is proportional to the square of the voltage at that frequency, we can define a quantity called the *transmission efficiency* as

$$\frac{\text{Sideband power}}{\text{Total power}} = \frac{2(V_{cm}{}^2 m^2/4)}{V_{cm}{}^2 + 2(V_{cm}{}^2 m^2/4)} = \frac{m^2}{2 + m^2} \quad . \tag{8.3}$$

For 100 per cent modulation ($m = 1$) this ratio is one-third. It can be seen from eqn. (8.1) that each sideband contains both the signal amplitude and frequency, and hence if only one sideband is transmitted with the carrier and the second sideband suppressed (thus improving the transmission efficiency), it should be possible to obtain the signal information at the receiving end. This system is called the *single-sideband* system and is used in carrier telephony. Another possibility is to transmit one sideband, part of the carrier, and only a small fraction of the second sideband. This means that part of the original carrier is available at the receiver and makes demodulation easier. The system is then called a *vestigial-sideband* system, and is widely used in television broadcasting. In both these systems the overall bandwidth requirement is approximately halved, so that for the same spacing between adjacent carriers each channel can handle twice the signal frequency of the corresponding double-sideband system, or alternatively for the same signal frequency twice the number of separate channels can be provided in a given overall band of frequencies. The double-sideband system, in which the complete modulated wave is transmitted, is used in normal sound broadcasting.

Example 8.1. A thermal ammeter is connected in series with the capacitor in a tuned circuit of an amplifier stage. When the input to the stage consists of a carrier amplitude modulated to a depth of 80 per cent by a single frequency, the ammeter reads 0·85 A. When the same carrier is subjected to two-tone modulation the ammeter reads 0·91 A. If the depth of modulation due to one tone is 50 per cent, what is the depth of modulation due to the other tone? (*I.E.E. App. El.*, 1963)

The thermal ammeter will read r.m.s. values of the current passing through it. With single-frequency modulation there are three components in the current wave (eqn. (8.2)), and hence, if I_{0m} is the amplitude of the carrier current the ammeter reading is

$$0{\cdot}85 = (I_{0m}/\sqrt{2})\sqrt{\left(1^2 + \frac{m^2}{4} + \frac{m^2}{4}\right)} = \frac{I_{0m}}{\sqrt{2}}\sqrt{\left(1 + \frac{0{\cdot}8^2}{2}\right)}$$

This gives $I_{0m}/\sqrt{2} = 0{\cdot}85\sqrt{1{\cdot}32}$.

With two-tone modulation there are five components in the current wave and the ammeter reading will be

$$0{\cdot}91 = (I_{0m}/\sqrt{2})\sqrt{\left(1^2 + \frac{m_1^{\,2}}{4} + \frac{m_1^{\,2}}{4} + \frac{m_2^{\,2}}{4} + \frac{m_2^{\,2}}{4}\right)}$$

$$= (0{\cdot}85/\sqrt{1{\cdot}32})\sqrt{\left(1^2 + \frac{0{\cdot}5^2}{2} + \frac{m_2^{\,2}}{4}\right)}$$

where m_1 is the depth of modulation of the first tone and m_2 is that of the second. Solving for m_2 gives $m_2 = 0{\cdot}881$.

The depth of modulation of the second tone is therefore 88·1 per cent.

8.2. Phase and Group Velocities

The velocity of propagation of a sinusoidal wave down a transmission line, which was derived in Chapter 7, is the velocity with which one particular point on the sine wave (e.g. the peak) is transmitted—in other words, it is the velocity at which the phase of the wave is transmitted. This velocity is called the *phase velocity*, u_{ph}. From eqn. (7.33), we have

$$u_{ph} = \lambda f = \frac{\omega}{\beta}$$

where β is the phase-change coefficient of the line.

Suppose now that we transmit a modulated wave down the line and that the signal frequency, $\omega_s/2\pi$, is only a small fraction of the carrier frequency, $\omega/2\pi$. Let $\omega_s = \delta\omega$; then eqn. (8.2) for the input voltage to the line becomes

$$v_{in} = V_{cm}\sin\omega t + \frac{V_{cm}}{2}m\cos(\omega - \delta\omega)t - \frac{V_{cm}m}{2}\cos(\omega + \delta\omega)t$$

If β is the phase-change coefficient at the carrier frequency, then for a length of line l metres long the carrier phase delay is βl, while that of the lower sideband is $(\beta - (\partial\beta/\partial\omega)\delta\omega)l$ and that of the upper sideband is $(\beta + (\partial\beta/\partial\omega)\delta\omega)l$. The voltage at distance l from the input is thus

$$v_l = V_{cm} \sin(\omega t - \beta l) + \frac{V_{cm}m}{2} \cos\left[(\omega - \delta\omega)t - \left(\beta - \frac{\partial\beta}{\partial\omega}\delta\omega\right)l\right]$$

$$- \frac{V_{cm}m}{2} \cos\left[(\omega + \partial\omega)t - \left(\beta + \frac{\partial\beta}{\partial\omega}\delta\omega\right)l\right]$$

$$= V_{cm} \sin(\omega t - \beta l) + \frac{V_{cm}m}{2} \cos\left[(\omega t - \beta l) - \delta\omega\left(t - \frac{\partial\beta}{\partial\omega}l\right)\right]$$

$$- \frac{V_{cm}m}{2} \cos\left[(\omega t - \beta l) + \delta\omega\left(t - \frac{\partial\beta}{\partial\omega}l\right)\right]$$

$$= V_{cm} \sin(\omega t - \beta l) + V_{cm}m \sin(\omega t - \beta l) \sin \delta\omega\left(t - \frac{l}{\partial\omega/\partial\beta}\right)$$

$$= V_{cm}\left[1 + m \sin \omega_s\left(t - \frac{l}{\partial\omega/\partial\beta}\right)\right] \sin(\omega t - \beta l) \quad . \quad . \quad (8.4)$$

Thus the envelope, $V_{cm}[1 + m \sin \omega_s (t - l\partial\beta/\partial\omega)]$, is a retarded function with a velocity $\partial\omega/\partial\beta$. This is called the *group velocity*, u_g, and represents the speed of transmission of the information by the modulated wave down the line. Only if β is independent of frequency (for instance in the lossless line) is the group velocity equal to the phase velocity. In general the group velocity, $\partial\omega/\partial\beta$ is less than the phase velocity ω/β.

8.3. Production of an Amplitude-modulated Wave

An amplitude-modulated wave will be produced if carrier and signal frequency voltages are applied in series to a device which has a non-linear characteristic. Let us suppose the characteristic of this device is represented by the equation

$$v_{out} = c + av_{in} + bv_{in}^2 \quad . \quad . \quad . \quad (8.5)$$

If

$$v_{in} = V_c \sin \omega t + V_s \sin \omega_s t$$

then the output voltage is

$$v_{out} = c + aV_c \sin \omega t + aV_s \sin \omega_s t$$

$$+ b(V_c^2 \sin^2 \omega t + 2V_cV_s \sin \omega t \sin \omega_s t + V_s^2 \sin^2 \omega_s t)$$

$$= c + aV_c \sin \omega t + aV_s \sin \omega_s t + \frac{bV_c^2}{2}(1 - \cos 2\omega t)$$

$$+ bV_cV_s[\cos(\omega - \omega_s)t - \cos(\omega + w_s)t] + \frac{bV_s^2}{2}(1 - \cos 2\omega_s t)$$

In order to produce the amplitude-modulated wave this output voltage is applied to a selective circuit (usually a tuned circuit, resonant to the carrier frequency) which passes the carrier and the two sidebands only, so that the resultant output is

$$v_0 = aV_c \sin \omega t + bV_cV_s \cos (\omega - \omega_s)t + bV_cV_s \cos (\omega + \omega_s)t$$

which is the equation of the modulated wave (eqn. (8.2)).

8.4. Frequency Changing

In some cases it is necessary to convert a modulated wave from one carrier frequency to another, as, for example, in the superheterodyne radio receiver. This is called *frequency changing*, or mixing, and is achieved by feeding a non-linear device with the original modulated signal and also with a sine wave from a separate oscillator (the local oscillator). The process is similar to that described for the production of the original amplitude-modulated wave. If we assume again that the equation of the non-linear device is eqn. (8.5) where the input this time is

$$v_{in} = V_{cm}(1 + m \sin \omega_s t) \sin \omega t + V_{lm} \sin \omega_l t$$

ω_l being the angular frequency, and V_{lm} the amplitude of the local-oscillator voltage, then the output is

$$v_{out} = c + av_{in} + b\left[V_{cm} \sin \omega t + \frac{V_{cm}m}{2} \cos (\omega - \omega_s)t - \frac{V_{cm}m}{2} \cos (\omega + \omega_s)t + V_{lm} \sin \omega_l t\right]^2 \quad (8.6)$$

The linear term, av_{in}, gives rise to components at each input frequency (carrier, two sidebands and local oscillator). The squared term in brackets yields (*a*) self-products, which by the previous section gives rise to d.c. and double-frequency terms, (*b*) cross-products, which give rise to sum and difference terms. Thus components with the following angular frequencies will appear in the output—

ω_s, ω_l, ω, $\omega + \omega_s$, $\omega - \omega_s$, $2\omega_l$, 2ω, $2(\omega + \omega_s)$, $2(\omega - \omega_s)$, $2\omega - \omega_s$, $2\omega + \omega_s$, $\omega + \omega_l$, $\omega - \omega_l$, $\omega - \omega_l - \omega_s$, $\omega - \omega_l + \omega_s$ and $\omega + \omega_l + \omega_s$.

Since normally the signal frequency is a small fraction of any carrier, $\omega \gg \omega_s$, $\omega_l \gg \omega_s$, $(\omega - \omega_l \gg \omega_s)$ and a resonant circuit tuned to $\omega - \omega_l$ will also accept components of angular frequency $\omega - \omega_l + \omega_s$ and $\omega - \omega_l - \omega_s$, but will reject all the other frequencies in the output of the non-linear device, which in this case is called the *mixer*. Neglecting the selective effects of the tuned circuit on the three accepted components, the output will consist of the terms

$$\frac{bV_{cm}V_{lm}}{2}\cos(\omega - \omega_l)t - \frac{bV_{cm}V_{lm}m}{4}\sin(\omega - \omega_l - \omega_s)t$$

$$+ \frac{bV_{cm}V_{lm}m}{4}\sin(\omega - \omega_l + \omega_s)t$$

$$= \frac{bV_{cm}V_{lm}}{2}(1 + m\sin\omega_s t)\cos(\omega - \omega_l)t \quad . \quad (8.7)$$

This represents a modulated cosine wave. The frequency of the carrier is the difference between that of the original carrier and that of the local oscillator. The modulation factor and modulation frequency are the same as those of the original wave. The same would be true for the sum-frequency components, $\omega + \omega_l$.

In practice, the selectivity of the tuned circuit makes the side-bands relatively smaller in relation to the new carrier than they were in relation to the original carrier, but this change in the depth of modulation, though frequency dependent, can usually be made small enough not to introduce undue distortion of the signal. It is obvious from the analysis that the same results are obtained whether the local-oscillator frequency is higher or lower than that of the original carrier.

Practical mixers include diodes, transistors, and hexodes.

8.5. Demodulation of an Amplitude-modulated Wave

The previous sections have shown that, when sine waves of different frequencies are applied to a non-linear device, the output includes, not only terms of the original frequencies and second-harmonics of these, but also terms representing sum and difference frequencies. Hence, if the modulated carrier

$$v = V_{cm}\sin\omega t + \frac{mV_{cm}}{2}\cos(\omega - \omega_s)t - \frac{mV_{cm}}{2}\cos(\omega + \omega_s)t$$

is applied to a non-linear device (with characteristic given by eqn. (8.5)), difference terms of signal frequency will arise from the cross-products—

$$bmV_{cm}^2\sin\omega t\cos(\omega - \omega_s)t - bmV_{cm}^2\sin\omega t\cos(\omega + \omega_s)t$$

$$= \frac{bmV_{cm}^2}{2}[\sin(2\omega - \omega_s)t + \sin\omega_s t$$

$$- \sin(2\omega + \omega_s)t + \sin\omega_s t] \quad . \quad . \quad . \quad (8.8)$$

If the high-frequency terms in the total output are filtered off (by a simple CR low-pass circuit, for example) there remains

$$bmV_{cm}^2 \sin \omega_s t = \text{Constant} \times \text{Original signal}$$

Note also, however, that the cross-product of the sidebands gives

$$\frac{bm^2V_{cm}^2}{2}[\cos(\omega - \omega_s)t \cos(\omega + \omega_s)t]$$

$$= \frac{bm^2V_{cm}^2}{4}(\cos 2\omega t + \cos 2\omega_s t) \quad . \quad (8.9)$$

Hence there will be considerable second-harmonic distortion which would normally be unacceptable. For this reason "square-law detection" is not now much used. Better results are achieved by applying the modulated wave to a diode rectifier followed by a low-pass CR filter. In this case the diode is the non-linear device. The results of Example 4.5 show that the half-wave rectified output of an *unmodulated carrier* of peak magnitude V_m is

$$v_o = \frac{V_m}{\pi} + \tfrac{1}{2}V_m \cos \omega t + \cdots$$

If the signal frequency is much lower than that of the carrier, we may consider that, since the amplitude of the modulated carrier varies according to the expression $V_m(1 + m \sin \omega_s t)$, the d.c. term in the expression for v_o can be written approximately as $(V_m/\pi)(1 + m \sin \omega_s t)$, so that the signal can be recovered by a suitable low-pass filter without the introduction of second-harmonic distortion.

8.6. Bessel Coefficients

In order to determine the frequency spectrum of a phase or frequency modulated wave it is convenient to be able to use Bessel functions. These are tabulated functions which may be found in some advanced books of mathematical tables. Consider the function

$$f = \exp \tfrac{1}{2}z(x - 1/x)$$

for all integral values of x, but excluding $x = 0$. Expanding gives

$$f = \exp \tfrac{1}{2}zx \exp -\tfrac{1}{2}z/x$$

$$= \left[1 + \frac{zx}{2} + \frac{1}{2!}\left(\frac{zx}{2}\right)^2 + \frac{1}{3!}\left(\frac{zx}{2}\right)^3 + \cdots\right]$$

$$\left[1 - \frac{z}{2x} + \frac{1}{2!}\left(\frac{z}{2x}\right)^2 - \cdots\right] \quad . \quad (8.10)$$

The coefficient of x^0 in this expansion is

$$\left(1 - \frac{z^2}{2^2} + \frac{z^4}{2^2 4^2} - \frac{z^6}{2^2 4^2 6^2} + \ldots\right)$$

This is known as Bessel's function of the first kind and of zero order, and is given the symbol $J_0(z)$. If z is known then $J_0(z)$ may be found from tables of Bessel functions (or from the graph of Fig. 8.2).

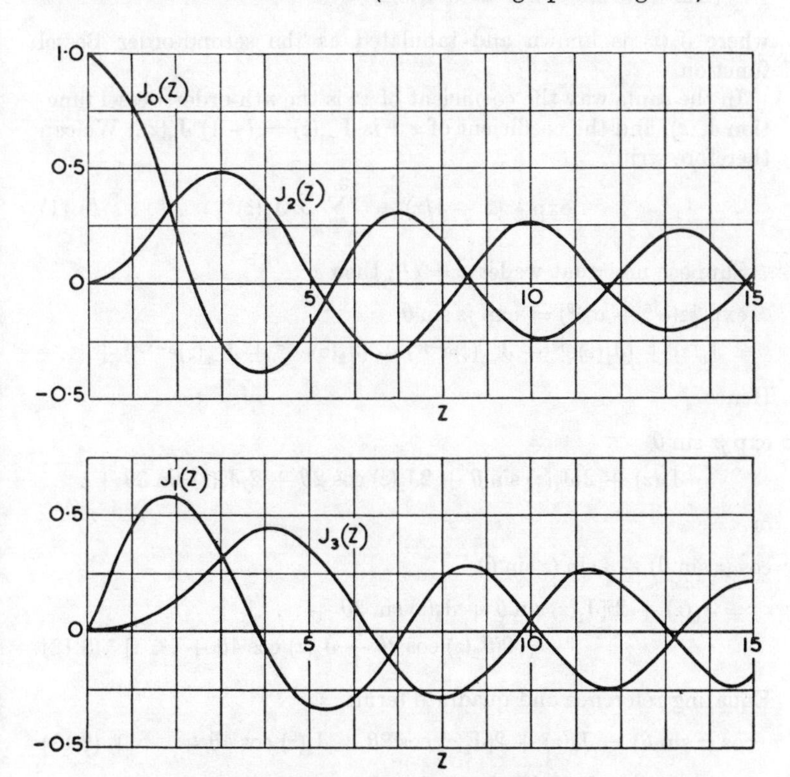

FIG. 8.2. BESSEL FUNCTIONS OF THE FIRST KIND AND OF ZERO, FIRST, SECOND AND THIRD ORDER

The coefficient of x^1 in eqn. (8.10) is

$$\left(+\frac{z}{2} - \frac{z^3}{1\,!2\,!2^3} + \frac{z^5}{2\,!3\,!2^5} - \frac{z^7}{3\,!4\,!2^7} + \ldots\right)$$

This is known and tabulated as the first-order Bessel function $J_1(z)$.

The coefficient of x^{-1} in eqn. (8.10) is

$$\left(-\frac{z}{2}+\frac{z^3}{1\,!2\,!2^3}-\frac{z^5}{2\,!3\,!2^5}+\frac{z^7}{3\,!4\,!2^7}-\ldots\right)=\mathrm{J}_{-1}(z)=-\mathrm{J}_1(z)$$

The coefficients of x^2 and of x^{-2} are both

$$\left(\frac{z^2}{2\,!2^2}-\frac{z^4}{3\,!2^4}+\frac{z^6}{2\,!4\,!2^6}-\frac{z^8}{3\,!5\,!2^8}-\ldots\right)=\mathrm{J}_2(z)=\mathrm{J}_{-2}(z)$$

where $\mathrm{J}_2(z)$ is known and tabulated as the second-order Bessel function.

In the same way the coefficient of x^n is the nth-order Bessel function $\mathrm{J}_n(z)$, and the coefficient of x^{-n} is $\mathrm{J}_{-n}(z)=(-1)^n\mathrm{J}_n(z)$. We can therefore write

$$\exp \tfrac{1}{2}z(x-1/x)=\sum_{n=-\infty}^{\infty} x^n\mathrm{J}_n(z) \quad . \quad . \quad (8.11)$$

Suppose now that we let $x=\mathrm{e}^{j\theta}$, then

$$\exp \tfrac{1}{2}z(\mathrm{e}^{j\theta}-\mathrm{e}^{-j\theta})=\exp jz\sin\theta$$
$$=\mathrm{J}_0(z)+(\mathrm{J}_1(z)\mathrm{e}^{j\theta}+\mathrm{J}_{-1}(z)\mathrm{e}^{-j\theta})+(\mathrm{J}_2(z)\mathrm{e}^{j2\theta}+\mathrm{J}_{-2}(z)\mathrm{e}^{-j2\theta})+\ldots$$

Hence

$$\exp jz\sin\theta$$
$$=\mathrm{J}_0(z)+2j\mathrm{J}_1(z)\sin\theta+2\mathrm{J}_2(z)\cos 2\theta+2j\mathrm{J}_3(z)\sin 3\theta+\ldots$$

or

$$\cos(z\sin\theta)+j\sin(z\sin\theta)$$
$$=\mathrm{J}_0(z)+2j[\mathrm{J}_1(z)\sin\theta+\mathrm{J}_3(z)\sin 3\theta+\ldots]$$
$$+2[\mathrm{J}_2(z)\cos 2\theta+\mathrm{J}_4(z)\cos 4\theta+\ldots] \quad (8.12)$$

Equating reference and quadrate terms,

$$\cos(z\sin\theta)=\mathrm{J}_0(z)+2[\mathrm{J}_2(z)\cos 2\theta+\mathrm{J}_4(z)\cos 4\theta+\ldots] \quad (8.13)$$

$$\sin(z\sin\theta)=2[\mathrm{J}_1(z)\sin\theta+\mathrm{J}_3(z)\sin 3\theta+\ldots] \quad (8.14)$$

It is worth emphasizing that in the above expansions the Bessel functions for given values of z are simply numbers which may be found from tables.

8.7. Phase Modulation

Let us consider a modulated wave represented by

$$v=V_m\cos(\omega_c t+\theta)$$

where θ is the phase of the wave relative to that of an unmodulated carrier, $V_m \cos \omega_c t$, and where ω_c is the carrier frequency. In phase modulation θ depends on the signal, and for sinusoidal modulation we can write

$$\theta = m_p \sin \omega_s t$$

In this expression m_p is the *phase-modulation index*, and corresponds to the phase change produced in the carrier by the positive or negative peaks of the largest signal. Hence the phase-modulated signal may be written

$$v = V_m \cos (\omega_c t + m_p \sin \omega_s t) \quad . \quad . \quad . \quad . \quad . \quad . \quad (8.15)$$

$$= V_m[\cos \omega_c t \cos (m_p \sin \omega_s t) - \sin \omega_c t \sin (m_p \sin \omega_s t)]$$

From eqns (8.13) and (8.14),

$$v = V_m[\cos \omega_c t(\mathrm{J}_0(m_p) + 2J_2(m_p) \cos 2\omega_s t + . \; . \; .)$$

$$- 2 \sin \omega_c t(\mathrm{J}_1(m_p) \sin \omega_s t + \mathrm{J}_3(m_p) \sin 3\omega_s t + . \; . \; .)]$$

$$= V_m[\mathrm{J}_0(m_p) \cos \omega_c t - \mathrm{J}_1(m_p)(\cos (\omega_c - \omega_s)t - \cos (\omega_c + \omega_s)t)$$

$$+ \mathrm{J}_2(m_p)(\cos) \; \omega_c - 2\omega_s)t + \cos (\omega_c + 2\omega_s)t)$$

$$- J_3(m_p)(\cos (\omega_c - 3\omega_s)t - \cos (\omega_c + 3\omega_s)t) + . \; . \; .] \quad . \quad (8.16)$$

The wave is thus seen to be made up of a carrier with an infinite number of sidebands which are spaced at signal-frequency intervals above and below the carrier. If the phase-modulation index, m_p, is less than unity, however, the Bessel functions $\mathrm{J}_2(m_p)$, $\mathrm{J}_3(m_p)$, etc., are all very small, and the wave consists essentially of a carrier and two sidebands as in the case of the amplitude-modulated wave. It is interesting to note that if $m_p = 2{\cdot}405$, then $\mathrm{J}_0(m_p) = 0$ and there is no carrier.

8.8. Frequency Modulation

In a frequency-modulated wave the instantaneous frequency of the carrier varies according to the amplitude of the signal, and the periodic time of this variation is the period of the signal (assuming a sinusoidal signal). The mathematical expression for the frequency-modulated (f.m.) wave is thus

$$v = V_m \cos 2\pi(f_c + \delta f \cos \omega_s t)t \quad . \quad . \quad (8.17)$$

where f_c is the carrier frequency and δf is the frequency deviation corresponding to the frequency shift of the carrier for the peak amplitude of the signal ($\cos \omega_s t = \pm 1$). Thus for a carrier of 10·7 Mc/s and a frequency deviation of 75 kc/s the instantaneous carrier frequency is 10·775 Mc/s at the positive peak of the signal and 10·625 Mc/s at the negative peak. The instantaneous frequency of the f.m. signal is $f_{inst} = f_c + \delta f \cos \omega_s t$.

Let us look more closely at the idea of instantaneous frequency. With respect to a wave given by $v = V_m \cos(\omega_c t + \theta)$ whose phase, θ, is changing, we can define instantaneous frequency as the time rate of change of phase angle, i.e.

$$f_{inst} = \frac{1}{2\pi}\frac{\mathrm{d}}{\mathrm{d}t}(\omega_c t + \theta) = f_c + \frac{1}{2\pi}\frac{\mathrm{d}\theta}{\mathrm{d}t} \quad . \quad . \quad (8.18)$$

Comparing this expression with the instantaneous frequency of the f.m. wave above, we can write

$$f_{inst} = f_c + \delta f \cos \omega_s t = f_c + \frac{1}{2\pi}\frac{\mathrm{d}\theta}{\mathrm{d}t}$$

Hence

$$\delta f \cos \omega_s t = \frac{1}{2\pi}\frac{\mathrm{d}\theta}{\mathrm{d}t}$$

Integrating this yields

$$\theta = \int 2\pi\, \delta f \cos \omega_s t \,\mathrm{d}t = \frac{\delta f}{f_s} \sin \omega_s t$$

so that the f.m. wave may be expressed mathematically in the form

$$v = V_m \cos\left(\omega_c t + \frac{\delta f}{f_s} \sin \omega_s t\right) = V_m \cos(\omega_c t + m_f \sin \omega_s t) \quad . \quad (8.19)$$

By doing this we have been able to express the f.m. wave in the same form as the expression in the previous section for the phase-modulated wave. Hence eqn. (8.19) may also be expanded in terms of Bessel functions. The difference between the two is that, for the frequency-modulated wave, the modulation index is $m_f = \delta f/f_s$ which is inversely proportional to the signal frequency, whereas in the phase-modulated wave the modulation index is independent of the signal frequency.

A frequency-modulated wave may be generated by connecting across a circuit, which is tuned to the carrier, a device whose reactance can be varied by altering the voltage applied to it. If this applied voltage is the signal, the total reactance and hence the frequency of the tuned circuit will depend on the amplitude of the signal, and the number of alternations of this frequency per second will depend on the signal frequency. Practical devices which can simulate a reactance which varies linearly with an input voltage include the semiconductor varactor and the thermionic reactance valve.

As with a phase-modulated wave, if the modulation index of the f.m. wave ($\delta f/f_s$) is small, only the first-order sidebands at $\pm f_s$ from the carrier are important and the bandwidth required is $2f_s$. If, however, m_f is high (as is normally the case, since this gives much better noise rejection in a communication system), the bandwidth required is approximately $2\delta f$. For $\delta f = 75$ kc/s this means that an overall bandwidth of at least 150 kc/s is required for a satisfactory f.m. communication channel. It is interesting to note that frequency modulation was originally investigated as a means of obtaining bandwidth compression (for which purpose it was proved to be unsuitable).

If we have a constant-amplitude variable-frequency signal, the modulation index produced by phase modulation is a constant, but that produced by frequency modulation is inversely proportional to the frequency of the signal. This means that "white noise," which has a continuous frequency spectrum, will cause more interference with the high signal frequencies in an f.m. system than with those in the corresponding phase-modulated system. For this reason pre-emphasis, in which the high-frequency signal components are increased in amplitude by an amount depending on the signal frequency, is used in f.m. transmitters. This may be considered as turning the upper part of the signal frequency range into phase modulation, and so increasing the effective depth of modulation of these components. At the receiver the signal is passed through a de-emphasis circuit to recover the original signal. Practical pre-emphasis and de-emphasis circuits may consist of simple CR high-pass and low-pass filters respectively. An improvement of between 4 and 5 dB in the signal/noise ratio can be achieved by this means.

8.9. Sampled Signals—the Sampling Theorem

Any electrical signal may be defined by taking a number of samples of the signal waveform, and expressing the instantaneous values which are sampled in a pulse code. Three main types of pulse coding exist: (*a*) amplitude coding, where the amplitude of the instantaneous signal is represented by the height of a pulse of constant width, (*b*) pulse-width coding, where the amplitude of the instantaneous signal is represented by the width of a pulse whose height is constant, and (*c*) pulse-position coding, where the position in time with respect to a reference position of a pulse of constant width and amplitude depends on the instantaneous value of the sampled signal.

The theory of sampling is used to determine the relationship which must exist between the maximum frequency of the sampled signal, f_{sm}, and the pulse-repetition frequency of the sampling pulses. The *sampling theorem* states that, if a function of time, $f(t)$, contains no frequencies higher than f_{sm}, it may be completely specified by giving its ordinates at a series of points spaced $1/2f_{sm}$ seconds apart.

Thus, consider a waveform of period T which does not contain any harmonics having frequencies higher than f_{sm} cycles per second. The waveform can be represented by a Fourier series whose fundamental frequency is $1/T$ cycles per second, and whose maximum harmonic frequency is f_{sm}. The harmonic number of the highest harmonic is $f_{sm}/(1/T) = Tf_{sm}$.

Now, a Fourier series is defined if the coefficients of both its sine and cosine terms are known (the coefficients a_n and b_n of the expansion of eqn. (4.12)). Thus two parameters are required to specify any harmonic, and it therefore follows that for Tf_{sm} harmonics a total of $2Tf_{sm}$ parameters are necessary per cycle of the fundamental. Hence the number of parameters required per second is $2Tf_{sm}/T = 2f_{sm}$, so that if the signal is sampled every $1/2f_{sm}$th of a second, the required number of parameters is obtained. It follows directly that the minimum pulse-repetition frequency of the sampling pulses is $2f_{sm}$ per second.

The theory of sampling has applications both to communication systems and to sampling oscilloscopes.

8.10. Frequency Spectrum for Pulse-amplitude Modulation (P.A.M.)

A pulse-amplitude-modulated wave may be produced by the arrangement shown in Fig. 8.3 (*a*). The signal is superimposed on a direct voltage and fed into point A. The gating pulses which are fed into the electronic switch alternately connect point B to earth or isolate it. Hence the voltage at B is zero when the electronic switch is closed, and is the same as the voltage at A when the switch is open. The form of the voltage at A is shown in Fig. 8.3 (*c*), and the output voltage at B is shown at (*b*).

Suppose that the interval between the gating pulses is τ_R, and that the output pulse width is τ. Then, from eqn. (4.16), the Fourier series for the unmodulated output pulse train is

$$e_1(t) = \frac{V\tau}{\tau_R} + \frac{2V\tau}{\tau_R}\sum_{n=1}^{\infty}\frac{\sin(n2\pi\tau/2\tau_R)}{n2\pi\tau/2\tau_R}\cos\frac{2\pi n}{\tau_R}t$$

Since τ_R is much smaller than the signal period, T, the Fourier

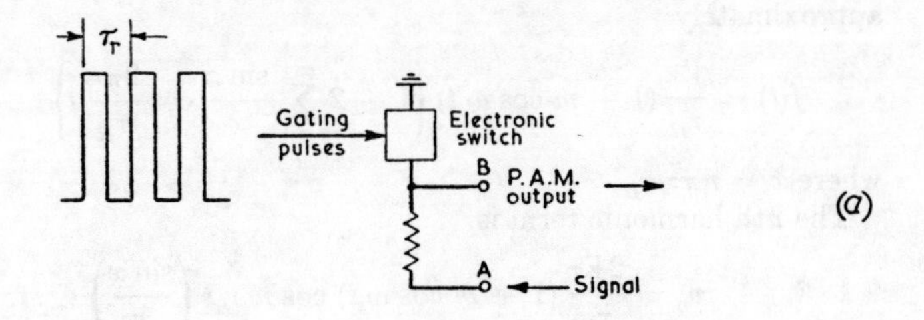

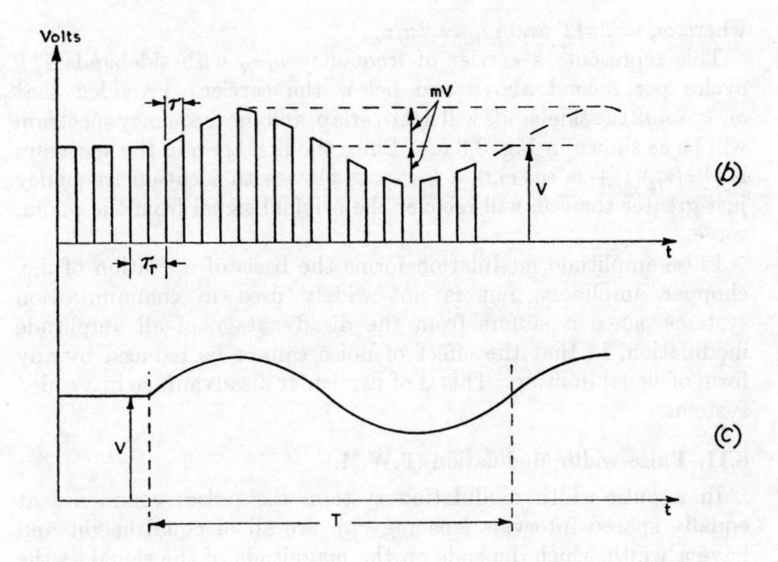

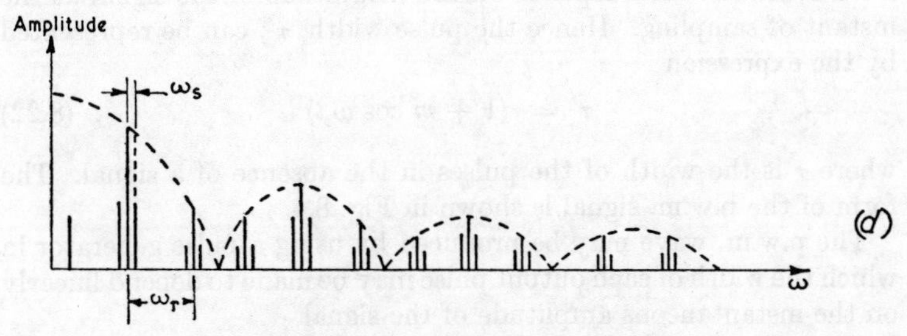

FIG. 8.3. PULSE-AMPLITUDE MODULATION

series for the wave modulated to a depth of modulation m is approximately

$$f(t) = \frac{V\tau}{\tau_R}(1 + m \cos \omega_s t)\left\{1 + 2\sum_{n=1}^{\infty} \frac{\sin x}{x} \cos \frac{2\pi n}{\tau_R} t\right\} \quad . \quad (8.20)$$

where $x = n\pi\tau/\tau_R$.

The nth harmonic term is

$$v_n = \frac{2V\tau}{\tau_R}(1 + m \cos \omega_s t) \cos n\omega_R t \left(\frac{\sin x}{x}\right) \quad . \quad (8.21)$$

where $\omega_s = 2\pi/T$ and $\omega_R = 2\pi/\tau_R$.

This represents a carrier of frequency n/τ_R with sidebands $1/T$ cycles per second above and below the carrier. Provided that $\omega_s < \omega_R/2$ the sidebands will not overlap, and the frequency spectrum will be as shown in Fig. 8.3 (d). Since the first term in the spectrum is $(V\tau/\tau_R)(1 + m \cos \omega_s t)$, a low-pass filter with a cut-off frequency just greater than ω_s will recover the original signal from the p.a.m. wave.

Pulse-amplitude modulation forms the basis of operation of d.c. chopper amplifiers, but is not widely used in communication systems, since it suffers from the disadvantage of all amplitude modulation, in that the effect of noise cannot be reduced by any form of signal limiting. This is of particular disadvantage in a coded system.

8.11. Pulse-width Modulation (P.W.M.)

In a pulse-width modulation system, the pulses commence at equally spaced intervals (spacing τ_R), are all of equal height and have a width which depends on the magnitude of the signal at the instant of sampling. Hence the pulse width, τ', can be represented by the expression

$$\tau' = \tau(1 + m \cos \omega_s t) \quad . \quad . \quad . \quad (8.22)$$

where τ is the width of the pulses in the absence of a signal. The form of the p.w.m. signal is shown in Fig. 8.4.

The p.w.m. wave may be produced by using a pulse generator in which the width of each output pulse may be made to depend linearly on the instantaneous amplitude of the signal.

Eqn. (4.16) gives the Fourier series for an unmodulated pulse train, in which the reference zero is taken at the mid-point of a pulse. For the p.w.m. wave it is the leading edge of each pulse which has a fixed relation to all other leading edges, and hence this point must be used as the reference zero. This means that the zero

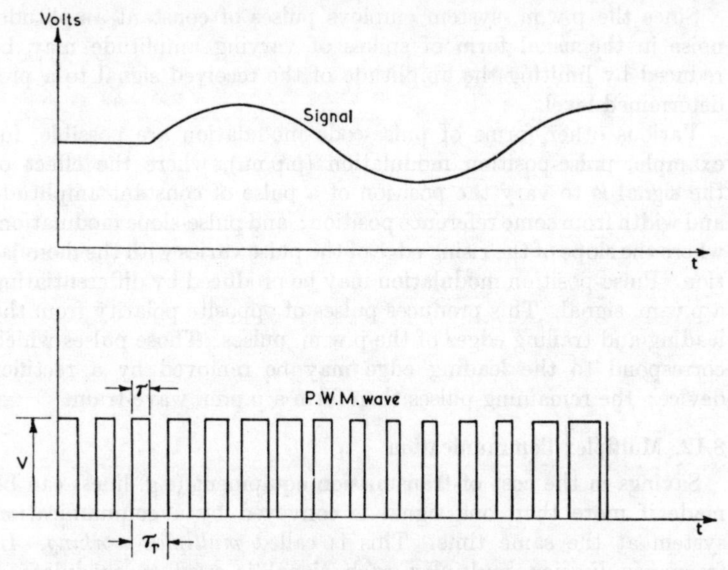

FIG. 8.4. PULSE-WIDTH MODULATION

must be shifted by an amount $\tau/2$ in time, and gives the modified Fourier series

$$f(t) = \frac{V\tau}{\tau_R} + V\frac{2\tau}{\tau_R}\sum_{n=1}^{\infty}\frac{\sin(n2\pi\tau/(2\tau_R))}{n2\pi\tau/(2\tau_R)}\cos\frac{n2\pi}{\tau_R}\left(t - \frac{\tau}{2}\right). \quad (8.23)$$

Assuming that τ_R and τ are both very small compared with the signal period, $1/f_s$, we may, to a reasonable approximation, write the Fourier series of the p.w.m. wave by replacing τ in eqn. (8.23) by τ' from eqn. (8.22) and so obtain

$$f(t) = \frac{V\tau}{\tau_R}(1 + m\cos\omega_s t)$$

$$\times\left[1 + 2\sum_{n=1}^{\infty}\frac{\sin x}{x}\cos\frac{n2\pi}{\tau_R}\left(t - \frac{\tau}{2}(1 + m\cos\omega_s t)\right)\right]. \quad (8.24)$$

where $x = n\pi\tau(1 + m\cos\omega_s t)/\tau_R$. The first term in eqn. (8.24) contains the signal, which may therefore be recovered by a low-pass filter as in pulse-amplitude modulation. Notice that each harmonic term in the expansion is both amplitude and phase modulated, and the frequency spectrum will therefore consist of the harmonic plus a large number of sidebands spaced at frequency intervals of f_s.

Since the p.w.m. system employs pulses of constant amplitude, noise in the usual form of spikes of varying amplitude may be reduced by limiting the amplitude of the received signal to a predetermined level.

Various other forms of pulse-code modulation are possible, for example, pulse-position modulation (p.p.m.), where the effect of the signal is to vary the position of a pulse of constant amplitude and width from some reference position; and pulse-slope modulation, where the slope of the rising edge of the pulse varies with the modulation. Pulse-position modulation may be produced by differentiating a p.w.m. signal. This produces pulses of opposite polarity from the leading and trailing edges of the p.w.m. pulses. Those pulses which correspond to the leading edge may be removed by a rectifier device; the remaining pulses then form a p.p.m. wave-train.

8.12. Multiplex Communication

Savings in the cost of transmission equipment (e.g. lines) can be made if more than one signal is conveyed by a communication system at the same time. This is called *multiplex working*. In frequency-division multiplex each signal is used to modulate a different frequency carrier, so that a large number of carriers (with their respective signals) may be passed along a line at the same time. The number of channels possible is limited by the bandwidth of the line. Each carrier is separated from the others at the receiving end, and then demodulated. Normally amplitude modulation is used in this type of system, which is widely employed in line telephony and telegraphy.

In a time-division multiplex system, pulse-code modulation is used, with pulse-widths which are narrow compared with their period. Hence several other pulse-coded signals may be inserted in the space between successive pulses representing the same sampled signal. At the receiving end the pulses are routed by suitable gating circuits to the required number of output channels. This type of multiplex is common in data-transmission systems, and is sometimes used in multi-channel speech communication.

Bibliography

Starr, *Telecommunications* (Pitman).
Goldman, *Frequency Analysis, Modulation, and Noise* (McGraw-Hill).

Problems

8.1. Explain what is meant by amplitude modulation of a sine-wave carrier. A signal voltage $v_1 = 2 \sin 157t$ and a carrier voltage $v_2 = 5 \sin 6{,}280t$ are applied in series between the grid and cathode of a valve for which the anode-current/grid-voltage characteristic under the conditions of operation is

$$i_a = 6 + 2v_g + 0{\cdot}1\, v_g^2$$

where i_a is in milliamperes and v_g is in volts.

Calculate (*a*) the frequency and amplitude of the various components of the anode current, and (*b*) the depth of modulation produced. Roughly sketch the anode-current waveform. (*A.E.E.*, 1961)

(*Ans.* D.C., 7·45 mA; 25 c/s, 4 mA; 50 c/s, 0·2 mA; 975 c/s, 1mA; 1,000 c/s, 10 mA; 1,025 c/s, 1 mA; 2,000 c/s, 1·25 mA; (*b*) 0·2.)

8.2. Show that the processes of modulation, demodulation and frequency changing are essentially the same.

The dynamic characteristic of a triode with a 10-kΩ resistive load is represented by

$$i_a = 2{\cdot}5\,(V_{gk} + 5)^2 + 0{\cdot}2\,(V_{gk} + 5)^2 \qquad \text{milliamperes}$$

The valve is operated with a fixed bias of −3V, and sinusoidal signals of amplitudes 1V and 0·5V at frequencies of 2 kc/s and 5 kc/s, respectively. Determine the amplitudes and frequencies of the various components of voltage across the load. (*I.E.E., Eln.*, 1954)

(*Ans.* D.C., 59·5 V; 2 kc/s, 33 V; 4 kc/s, 1 V; 5 kc/s, 16·5 V; 10 kc/s, 0·5 V; 3 kc/s, 1 V; 7 kc/s, 1 V.)

8.3. A non-linear device has the current/voltage relation

$$i = a + bv + cv^2 + dv^3$$

Show that the third-order term leads to cross-modulation between two amplitude-modulated sinusoidal signal voltages.

A carbon microphone, when subjected to a sinusoidal sound wave of frequency ω, has a resistance given by $100\,(1 + 0{\cdot}2 \sin \omega t)$. Find the percentage second-harmonic distortion current in its output.

(*A.E.E.*, 1956)

(*Ans.* 10 per cent.)

[*Note.* Cross-modulation refers to the modulation of carrier 1 by signal 2, and of carrier 2 by signal 1, when two modulated carriers are present.]

8.4. Explain what is meant by amplitude modulation.

The output current i and input voltage v of a device are related by the expression $i = av + bv^2$. If sinusoidal signal and carrier voltages are applied in series to such a device, show that amplitude modulation is produced. Show also that the depth of modulation produced is, in general, not equal to the signal-carrier voltage ratio.

If the values of a and b are, respectively, 1·5 and 0·02, calculate the signal voltage to produce 10 per cent modulation of a carrier of 5 V peak. What frequencies will be present in the output if the carrier frequency is 1 kc/s and the signal frequency is 50 c/s? (*A.E.E.*, 1963)

(*Ans.* 3·75 V; 50, 100, 950, 1,000, 1,050, 2,000 and 0 c/s.)

8.5. An amplifier is driven simultaneously with sinusoidal signals of 1·0 V at 1,000 c/s, and 0·1 V at 850 c/s. The output is found to contain components of several frequencies, including one of 100 V at 1,000 c/s, and one of 1·0 V at 150 c/s. Derive an expression showing how these and other frequencies arise, and calculate the percentage second-harmonic distortion of the 1,000-c/s signal (third- and higher-order components can be ignored). (The values given are the maximum values of the sinusoidal voltages.) (*I.E.E. Eln.*, 1957)

(*Ans.* 5 per cent.)

8.6. A waveform, sinusoidally modulated at 5 kc/s to 60 per cent, has a peak value of 80 mV and a carrier frequency of 908 kc/s. This voltage is applied to the signal grid of a hexode, and an oscillator voltage of peak value 5·0 V and frequency 1,373 kc/s is applied to the oscillator grid. The anode-current characteristic is given by

$$i_a = (2 + 0{\cdot}4\ V_{gs})(4 + 0{\cdot}625\ V_{go}) \qquad \text{milliamperes}$$

where V_{gs} and V_{go} are, respectively, the total voltage between signal grid and cathode, and oscillator grid and cathode. The anode circuit contains a single tuned circuit of resonant frequency 465 kc/s, and resonant impedance 20 kΩ. The Q-factor of the circuit is 46·5. Calculate the i.f. output carrier voltage and its percentage modulation. (*I.E.E. Radio*, 1958)

(*Ans.* 1 V; 42·4 per cent.)

8.7. Explain the single-sideband suppressed-carrier system of amplitude modulation and discuss its merits.

Why is a double-sideband suppressed-carrier system not practicable?

The output stage of a certain radio transmitter can deliver a maximum of 10 kW of radio-frequency power into the aerial. The transmitter is modulated to a depth of 40 per cent by a sinusoidal signal. Compare the possible power in the sidebands when the carrier and both sidebands are radiated, with that when the single-sideband suppressed-carrier system is employed. (*L.U. Part III Tel.*, 1961)

(*Ans.* 0·741 kW; 10 kW.)

8.8. A sinusoidal carrier voltage of peak amplitude 10 V and frequency 1 Mc/s is amplitude modulated at a frequency of 1 kc/s. Show that sidebands arise and calculate their frequencies. If the voltage is applied to a 1,000-Ω resistance, find from first principles the power dissipated at the carrier frequency and at each sideband frequency when the percentage modulation is (i) 10, (ii) 100. Comment on the significance of these results as regards the efficiency with which a speech signal may be transmitted. (*L.U. Part II, Elect.*, 1961)

(*Ans.* 999, 1,001 kc/s; (i) 5 W, 0·0125 W; (ii) 5 W, 1·25 W.)

CHAPTER 9

Symmetrical Components

SYMMETRICAL component analysis may be applied to the solution of unbalanced 3-phase networks. The analysis is performed by replacing an unbalanced system of 3-phase voltages or currents by three balanced systems, called the *positive*, *negative* and *zero phase-sequence components* of the original unbalanced system.

Unbalanced 3-phase network problems may then be solved by considering one phase only of each of three balanced systems. The required steady-state current or voltage response is obtained by adding together the corresponding symmetrical components of each phase.

The method of symmetrical components depends on the principle of superposition and is applicable only to those networks where the assumption of linearity is justified.

9.1. The 120° Operator (a)

In balanced 3-phase systems the complexors representing voltages or currents are displaced from one another by 120°, so that it is convenient to have an operator which rotates a complexor through this angle. This operator is given the symbol a or h (we shall use a). Any complexor when operated on by a remains unchanged in modulus and has 120° added to its phase angle; that is, it is rotated by 120° in an anti-clockwise direction. Thus

$$a = 1\underline{/120^\circ} \quad \quad (9.1)$$

or

$$a = -\frac{1}{2} + j\frac{\sqrt{3}}{2} \quad \quad (9.2)$$

Two successive operations on a complexor by a are represented by a^2. Thus $a^2 = a \times a = 1\underline{/120^\circ} \times 1\underline{/120^\circ}$, or

$$a^2 = 1\underline{/240^\circ}$$

or

$$a^2 = 1\underline{/-120^\circ} \quad \quad (9.3)$$

or

$$a^2 = -\frac{1}{2} - j\,\frac{\sqrt{3}}{2} \quad . \quad . \quad . \quad (9.4)$$

In the same way, three successive operations by a are represented by a^3. Thus

$$a^3 = a \times a \times a = 1\underline{/360^\circ} = 1\underline{/0^\circ} \quad . \quad . \quad (9.5)$$

also

$$a^4 = 1\underline{/480^\circ} = 1\underline{/120^\circ} = a \quad . \quad . \quad . \quad (9.6)$$

and

$$a^5 = 1\underline{/600^\circ} = 1\underline{/240^\circ} = a^2 \quad . \quad . \quad . \quad (9.7)$$

When a complexor is operated on by $-a$ it is rotated by 60° in a *clockwise* direction. This may be verified as follows.

$$-a = a \times (-1) = 1\underline{/120^\circ} \times 1\underline{/180^\circ} = 1\underline{/300^\circ} = 1\underline{/-60^\circ} \quad (9.8)$$

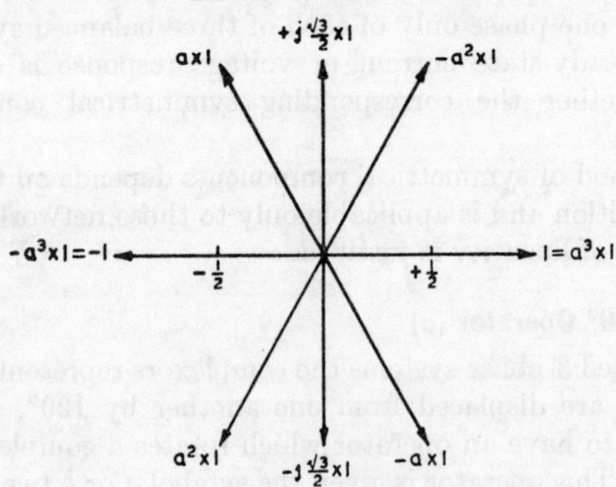

FIG. 9.1. OPERATING ON A UNIT COMPLEXOR WITH POWERS OF THE OPERATOR a

Fig. 9.1 shows the results of various operations on the complexor $(1 + j0)$ (i.e. a unit complexor) by the operator a.

The following results may readily be verified using eqns. (9.2) and (9.4).

$$1 + a + a^2 = 0 = 1 + a^4 + a^2 \quad . \quad . \quad (9.9)$$

$$a - a^2 = j\sqrt{3} \quad . \quad . \quad . \quad (9.10)$$

$$a^2 - a = -j\sqrt{3} \quad . \quad . \quad . \quad (9.11)$$

$$1 - a = \frac{3}{2} - j\,\frac{\sqrt{3}}{2} \quad . \quad . \quad . \quad (9.12)$$

$$1 - a^2 = \frac{3}{2} + j\,\frac{\sqrt{3}}{2} \quad . \quad . \quad . \quad (9.13)$$

$$2 - a - a^2 = 3 + j0 \quad . \quad . \quad . \quad . \quad (9.14)$$

The operator a can be used in the representation of balanced 3-phase currents or voltages. Thus for a balanced positive phase sequence V_R, V_Y, Y_B (i.e. maxima occur in the order R, Y, B), we can write

$$V_R = V_1 \qquad V_Y = a^2V_1 = V_1\underline{/-120°} \qquad V_B = aV_1 = V_1\underline{/120°}$$

For a negative phase sequence (i.e. for maxima occurring in the order R, B, Y) we may write

$$V_R = V_2 \qquad V_Y = aV_2 = V_2\underline{/120°} \qquad V_B = a^2V_2 = V_2\underline{/-120°}$$

9.2. The Symmetrical Components of an Unbalanced Three-phase System

Any unbalanced system of 3-phase currents may be represented by the sum of—

1. A balanced system of 3-phase currents having the same phase sequence as the original unbalanced system and called the *positive-sequence* system.
2. A balanced system of 3-phase currents having a phase sequence opposite to that of the original unbalanced system and called the *negative-sequence* system.
3. A system of three currents all equal in magnitude and in phase and called the *zero-sequence* system.

Thus, consider a positive sequence of currents I_1, a^2I_1 and aI_1 flowing in the red, yellow and blue lines respectively of a 3-phase system, a negative sequence I_2, aI_2 and a^2I_2 again flowing in the red, yellow and blue lines respectively, and a zero sequence, I_0, flowing in all three lines. Then, by superposition, and assuming linear elements only, the actual currents in the red, yellow and blue lines are

$$I_R = I_1 + I_2 + I_0 \quad . \quad . \quad . \quad (9.15a)$$

$$I_Y = a^2I_1 + aI_2 + I_0 \quad . \quad . \quad . \quad (9.15b)$$

and

$$I_B = aI_1 + a^2I_2 + I_0 \quad . \quad . \quad . \quad (9.15c)$$

The positive-sequence component, I_1, in the red line can be expressed in terms of the line currents by multiplying eqn. (9.15*b*) by a, multiplying eqn. (9.15*c*) by a^2 and adding these to eqn. (9.15*a*). This gives

$$I_R + aI_Y + a^2I_B$$
$$= I_1(1 + a^3 + a^3) + I_2(1 + a^2 + a^4) + I_0(1 + a + a^2)$$

But, from eqn. (9.9), $1 + a^2 + a^4 = 1 + a + a^2 = 0$, while from eqn. (9.5), $1 + a^3 + a^3 = 3$, and hence

$$I_1 = \tfrac{1}{3}(I_R + aI_Y + a^2 I_B) \qquad . \quad . \; (9.16)$$

In the same way, by multiplying eqn. (9.15*b*) by a^2 and eqn. (9.15*c*) by a and adding to eqn. (9.15*a*), we obtain the negative-sequence component, I_2, in the red line as

$$I_2 = \tfrac{1}{3}(I_R + a^2 I_Y + aI_B) \qquad . \quad . \; (9.17)$$

Adding eqns. (9.15*a*), (9.15*b*) and (9.15*c*) gives the zero-sequence component, I_0, as

$$I_0 = \tfrac{1}{3}(I_R + I_Y + I_B) \qquad . \quad . \quad . \; (9.18)$$

These results verify the statements made at the beginning of the section, since for any conceivable values of I_R, I_Y and I_B, it must always be possible to determine values for I_1, I_2 and I_0.

Note that the zero-sequence current is one-third of the current in the neutral of a 3-phase 4-wire system. It follows that in a 3-wire system, where the complex sum of the line currents is zero, there can

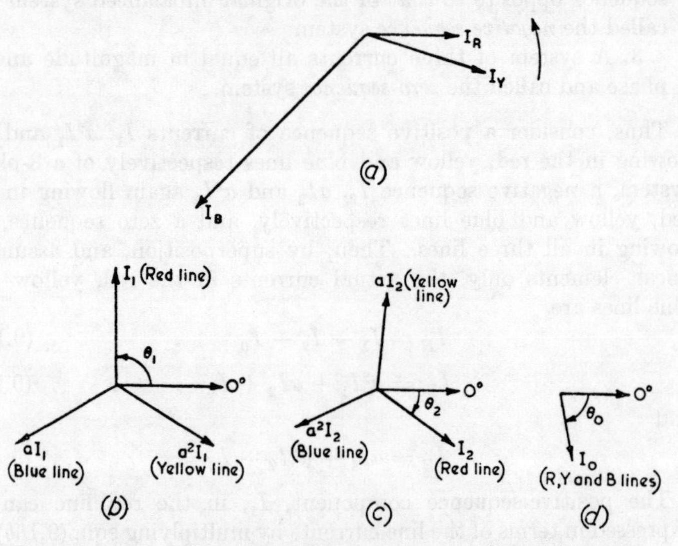

FIG. 9.2. ANY UNBALANCED THREE-PHASE SYSTEM OF CURRENTS CAN BE REPRESENTED BY THE SUM OF BALANCED POSITIVE-, NEGATIVE- AND ZERO-SEQUENCE COMPONENTS

(*a*) Unbalanced 3-phase currents
(*b*) Balanced positive phase sequence
(*c*) Balanced negative phase sequence
(*d*) Balanced zero phase sequence

be no zero-sequence component. In the same way, since the complex sum of the line voltages in any 3-phase system is zero, there cannot be any zero-sequence component of line voltage.

The symmetrical components of line or phase voltages in a 3-phase system can obviously be found in a manner similar to that used for currents.

The symmetrical components of an unbalanced system of currents are shown in Fig. 9.2 (*b*), (*c*) and (*d*), together with the unbalanced current system which they represent.

Usually the symmetrical components of red-phase quantities are found, since these are sufficient to synthesize the current or voltage values of all three phases. However, since each component forms part of a balanced system, the corresponding components for the blue and yellow phases are readily found.

Example 9.1. Find the symmetrical components of a system of unbalanced 3-phase currents given by $I_R = 200\underline{/0°}$ A, $I_Y = 100\underline{/-90°}$ A, $I_B = 50\underline{/45°}$ A.

The complexors I_R, I_Y, I_B, aI_Y, a^2I_Y, aI_B and a^2I_B are first found in rectangular form. Thus

$I_R = 200\underline{/0°} = 200 + j0$; $I_Y = 100\underline{/-90°} = 0 - j100$; $I_B = 50\underline{/45°} = 35{\cdot}4 + j35{\cdot}4$.

It follows that

$$aI_Y = 100\underline{/30°} = 36{\cdot}6 + j50.$$

$$a^2I_Y = 100\underline{/-210°} = -86{\cdot}6 + j50.$$

$$aI_B = 50\underline{/165°} = -48{\cdot}3 + j12{\cdot}9.$$

$$a^2I_B = 50\underline{/-75°} = 12{\cdot}9 - 48{\cdot}3.$$

Hence from eqn. (9.18), the zero-sequence component is

$$I_0 = \tfrac{1}{3}(I_R + I_Y + I_B) = \tfrac{1}{3}(235{\cdot}4 - j64{\cdot}6) = \underline{\underline{81{\cdot}3\underline{/-16°}\text{ A}}}$$

The positive-sequence component is (eqn. (9.16))

$$I_1 = \tfrac{1}{3}(I_R + aI_Y + a^2I_B) = \tfrac{1}{3}(299{\cdot}5 + j1{\cdot}7) = \underline{\underline{99{\cdot}8\underline{/0{\cdot}3°}\text{ A}}}$$

The negative-sequence component is (eqn. (9·17))

$$I_2 = \tfrac{1}{3}(I_R + a^2I_Y + aI_B) = \tfrac{1}{3}(65{\cdot}1 + 62{\cdot}9) = \underline{\underline{30{\cdot}2\underline{/44°}\text{ A}}}$$

These values are the sequence components in the red line.

A graphical solution is shown in Fig. 9.3. At (*a*) the original complexors I_R, I_Y and I_B are drawn. Complexors aI_Y, a^2I_Y, aI_B and a^2I_B are then obtained by rotating the complexors I_Y and I_B by 120° in the appropriate directions. The complexor additions which give $3I_0$, $3I_1$ and $3I_2$ from which the symmetrical components may readily be found are shown in Figs. 9.3(*b*), (*c*) and (*d*).

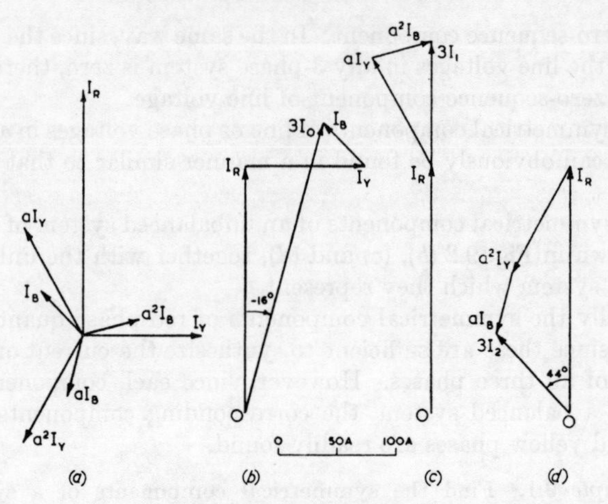

FIG. 9.3. GRAPHICAL SOLUTION OF EXAMPLE 9.1

$$3I_0 = I_R + I_Y + I_B$$
$$3I_1 = I_R + aI_Y + a^2I_B$$
$$3I_2 = I_R + a^2I_Y + aI_B$$

Example 9.2. The positive, negative and zero sequence of currents in an unbalanced 3-phase system are 100 A, 50 A and 20 A respectively. They are all in phase in the yellow line. Determine the actual line currents.

It is convenient in this case to take the current in the yellow line as the reference complexor. Then, in the red line,

$$I_1 = a100;\ I_2 = a^2 50;\ I_0 = 20.$$

Hence from eqn. (9.15*a*), and the above values,

$$I_R = I_1 + I_2 + I_0 = 100\left(-\tfrac{1}{2} + j\sqrt{\frac{3}{2}}\right) + 50\left(-\tfrac{1}{2} - j\sqrt{\frac{3}{2}}\right) + 20$$

$$= -55 + j43{\cdot}25 = \underline{\underline{70/142^\circ\ \text{A}}}$$

From eqn. (9.15*b*),

$$I_Y = a^2I_1 + aI_2 + I_0 = 100 + 50 + 20 = \underline{\underline{170/0^\circ\ \text{A}}}$$

and, from eqn. (9.15*c*),

$$I_B = aI_1 + a^2I_2 + I_0 = a^2 100 + a^4 50 + 20$$

$$= 100\left(-\tfrac{1}{2} - j\sqrt{\frac{3}{2}}\right) + 50\left(-\tfrac{1}{2} + j\sqrt{\frac{3}{2}}\right) + 20$$

$$= -55 - j43{\cdot}25 = \underline{\underline{70/218^\circ\ \text{A}}}$$

It is left as an exercise for the reader to check this result graphically.

9.3. Zero Phase-sequence Components of Current and Voltage

Any network which permits the flow of positive phase-sequence currents will also permit the flow of negative phase-sequence currents, since these are similar. It has already been seen, however, that a fourth wire is required if zero-sequence components are to flow in the lines of a 3-phase system. Let us look at some typical 3-phase connexions with reference to zero-sequence components.

(*a*) *Four-wire Star Connexion.* Since there is a fourth wire, zero phase-sequence currents may flow. The neutral carries only zero-sequence components, these being the sum of the zero-sequence components in the three lines. Since the line voltages sum to zero, there can be no zero-sequence component of line voltage.

(*b*) *Three-wire Star Connexion.* No zero-sequence components of current can flow, since there is no fourth wire for their return. We may consider that the impedance to zero phase-sequence currents is infinite, and that this impedance is situated between the star point of the generator and the star point of the load. If these two star points were joined by a neutral, only zero phase-sequence currents would flow in it, so that only a zero-sequence voltage can exist between the two star points. It also follows that no zero phase-sequence component of voltage appears across the phase loads.

(*c*) *Three-wire Mesh Connexion.* Since there is no fourth wire, zero phase-sequence components of current cannot be fed into the mesh-connected load. However, although the line currents must sum to zero, the phase currents need not do so, and hence it is possible to have a zero-sequence component of current circulating in the delta-connected load. Such a situation may arise, for example, in a delta-star connected transformer, in which the star point is earthed. This permits the flow of zero phase-sequence currents in the star-connected winding, and a zero phase-sequence current will then circulate round the closed delta, so maintaining m.m.f. balance between corresponding primary and secondary phase windings.

9.4. The Unbalanced Star Load

Let us suppose that a star of impedances Z_R, Z_Y and Z_B with no mutual couplings is fed from an unsymmetrical 3-phase 3-wire supply, the line voltages being V_{RY}, V_{YB} and V_{BR}. The problem of finding the phase currents may be solved using Millman's theorem (Section 1.22), by taking one line as the reference point and determining the star-point voltage with respect to this reference. It may also be solved using the method of symmetrical components. Thus, if the star point is S, the phase voltages V_{RS}, V_{YS} and V_{BS} will have symmetrical components V_{p1}, V_{p2} and V_{p0} given by

$$V_{p0} = \tfrac{1}{3}(V_{RS} + V_{YS} + V_{BS}) \text{ (from eqn. (9.18) applied to voltages)}$$
$$= \tfrac{1}{3}(I_R Z_R + I_Y Z_Y + I_B Z_B)$$

where I_R, I_Y and I_B are the line currents. Since we have a 3-wire supply there can be no zero phase-sequence component of line current, and if I_1 and I_2 are the positive and negative phase-sequence components in the red line, eqn. (9.15) gives $I_R = I_1 + I_2$, $I_Y = a^2 I_1 + aI_2$ and $I_B = aI_1 + a^2 I_2$. Hence

$$V_{p0} = \tfrac{1}{3}[I_1(Z_R + a^2 Z_Y + aZ_B) + I_2(Z_R + aZ_Y + a^2 Z_B)] \quad (9.19)$$

Similarly,

$$V_{p1} = \tfrac{1}{3}(V_{RS} + aV_{YS} + a^2 V_{BS})$$
$$= \tfrac{1}{3}(I_R Z_R + aI_Y Z_Y + a^2 I_B Z_B)$$
$$= \tfrac{1}{3}[I_1(Z_R + Z_Y + Z_B) + I_2(Z_R + a^2 Z_Y + aZ_B)] \quad . \quad (9.20)$$

and

$$V_{p2} = \tfrac{1}{3}[I_2(Z_R + Z_Y + Z_B) + I_1(Z_R + aZ_Y + a^2 Z_B)] \quad . \quad (9.21)$$

The symmetrical components of phase voltage can be expressed in terms of the line voltages by using the relations

$$V_{RY} = V_{RN} - V_{YN}$$
$$= V_{p1} + V_{p2} + V_{p0} - a^2 V_{p1} - aV_{p2} - V_{p0} \text{ (from eqns. (9.15)}$$
$$= (1 - a^2)V_{p1} + (1 - a)V_{p2}$$

and

$$V_{YB} = V_{YN} - V_{BN}$$
$$= a^2 V_{p1} + aV_{p2} + V_{p0} - aV_{p1} - a^2 V_{p2} - V_{p0}$$
$$= (a^2 - a)V_{p1} + (a - a^2)V_{p2}$$

From these,

$$(a^2 - a)V_{RY} - (1 - a^2)V_{YB}$$
$$= (1 - a)(a^2 - a)V_{p2} - (a - a^2)(1 - a^2)V_{p2}$$
$$= 3(a^2 - a)V_{p2}$$

so that

$$V_{p2} = \frac{1}{3}\left(V_{RY} - \frac{1-a^2}{a^2-a}V_{YB}\right) = \frac{1}{3}\left(V_{RY} + \left(\frac{1}{2} - j\sqrt{\frac{3}{2}}\right)V_{YB}\right) \quad . \quad . \quad . \quad (9.22)$$

In the same way,

$$V_{p1} = \frac{1}{3}\left(V_{RY} + \left(\frac{1}{2} + j\sqrt{\frac{3}{2}}\right)V_{YB}\right) \quad . \quad . \quad (9.23)$$

Substitution of these expressions in eqns. (9.20) and (9.21) enables the values of I_1 and I_2, and hence of the line currents, to be found. Simplifications arise if the load is balanced, or if the line voltages are symmetrical.

Example 9.3. Three equal 100-Ω resistors are connected in star to an unsymmetrical 3-phase 3-wire system whose line voltages are $V_{RY} = 300$ V, $V_{YB} = 400$ V, and $V_{BR} = 500$ V. Find the current in the red line.

Since the line voltages form a closed triangle their phase relations may be determined either graphically or by trigonometry. In this case the line voltages form a right-angled triangle, with an angle of $\tan^{-1}(4/3) = 37°$. Hence if $V_{RY} = 300\underline{/0°}$, then $V_{YB} = 400\underline{/-90°} = -j400$, and $V_{BR} = 500\underline{/127°}$.

Method (*a*). With the nomenclature used above, eqn. (9.23) gives

$$V_{p1} = \tfrac{1}{3}\left[300 + \left(\frac{1}{2} + j\sqrt{\frac{3}{2}}\right)(-j400)\right] = 215{\cdot}3 - j66{\cdot}7$$

and eqn. (9.22) gives

$$V_{p2} = \tfrac{1}{3}\left[300 + \left(\frac{1}{2} - j\sqrt{\frac{3}{2}}\right)(-j400)\right] = -15{\cdot}3 - j66{\cdot}7$$

There is, of course, no zero-sequence component.

Also $\quad Z_R + Z_Y + Z_B = 300$

$\quad Z_R + a^2Z_Y + aZ_B = 0$

(since $Z_R = Z_Y = Z_B$, and $1 + a + a^2 = 0$)

and $\quad Z_R + aZ_Y + a^2Z_B = 0$

so that eqns. (9.20) and (9.21) become

$$215{\cdot}3 - j66{\cdot}7 = 100I_1$$

and

$$-15{\cdot}3 - j66{\cdot}7 = 100I_2$$

Using eqn. (9·15*a*) (with $I_0 = 0$),

$$I_R = I_1 + I_2 = 2{\cdot}153 - j066{\cdot}7 - 0{\cdot}153 - j0{\cdot}667 = 2 - j1{\cdot}334$$

$$= \underline{\underline{2{\cdot}4\underline{/-34°}\text{ A}}}$$

Method (*b*). Using Millman's theorem and letting the red line terminal, R, be the reference point, the voltage between R and the star point, S, is

$$V_{SR} = \frac{V_{YR}Y_Y + V_{BR}Y_B}{Y_R + Y_Y + Y_B}$$

$$= \frac{300\underline{/180^\circ}\,\frac{1}{100} + 500\underline{/127^\circ}\,\frac{1}{100}}{0{\cdot}03}$$

$$= \frac{-6 + j4}{0{\cdot}03} = 240\underline{/146^\circ}$$

so that $V_{RS} = 240\underline{/-34^\circ}$, and $I_R = 2{\cdot}4\underline{/-34^\circ}$ A

9.5. Impedance to Symmetrical Components of Current

The main application of the theory of symmetrical components is to the analysis of 3-phase networks which are subjected to asymmetrical faults. In such networks the impedances of each phase *up to the fault* are equal, so that writing $Z_R = Z_Y = Z_B = Z$, and remembering that $1 + a + a^2 = 0$, eqns. (9.20) and (9.21) become

$$V_{p1} = I_1 Z \quad \text{and} \quad V_{p2} = I_2 Z$$

In other words, positive-sequence components of current give rise to positive-sequence voltage drops only, while negative-sequence components of current give rise to negative-sequence voltage drops only. We shall see shortly that the same applies to zero-sequence components. This leads to considerable simplification in the calculation of asymmetrical fault currents, since each sequence can be considered separately.

The ratio of phase-sequence voltage to phase-sequence current can be defined as the phase-sequence impedance (Z_1 for positive, Z_2 for negative, and Z_0 for zero phase sequences). Note that, although the impedances *of each phase* to one particular sequence must all be the same, they are not necessarily equal to each other. In static plant (transformers, lines, etc.) the positive and negative phase-sequence impedances are equal, but the zero phase-sequence impedance depends on the type of neutral connexion. We shall now consider some specific cases.

(*a*) *Transmission Lines with Mutual Coupling*. Suppose we have a 3-phase 3-wire line, the self-impedance of the lines being Z_s, and the mutual impedances being Z_m (assume the same for each pair of lines). If the line currents are I_R, I_Y, I_B, the *voltage drops* in the lines are

$$V_A = I_R Z_s + I_Y Z_m + I_B Z_m$$
$$V_B = I_Y Z_s + I_R Z_m + I_B Z_m$$
$$V_C = I_B Z_s + I_R Z_m + I_Y Z_m$$

Since these are phase voltage drops, they will have a zero phase-sequence voltage component, referred to line A, of

$$V_0 = \tfrac{1}{3}(V_A + V_B + V_C) = \tfrac{1}{3}(Z_s + 2Z_m)(I_R + I_Y + I_B)$$

But from eqn. (9.18), $\tfrac{1}{3}(I_R + I_Y + I_B) = I_0$, so that

$$V_0 = I_0(Z_s + 2Z_m)$$

i.e. V_0 depends on the zero-sequence component of current only. The zero sequence impedance is

$$Z_0 = (Z_s + 2Z_m) \quad . \quad . \quad . \quad (9.24)$$

The positive-sequence voltage component, referred to line A is

$$\begin{aligned} V_1 &= \tfrac{1}{3}(V_A + aV_B + a^2V_C) \\ &= \tfrac{1}{3}Z_s(I_R + aI_Y + a^2I_B) + \tfrac{1}{3}Z_m[(a + a^2)I_R + (1 + a^2)I_Y \\ &\qquad + (1 + a)I_B] \\ &= \tfrac{1}{3}Z_sI_1 + \tfrac{1}{3}Z_m(-I_R - aI_Y - a^2I_B) \text{ (from eqns. (9.16) and (9.9))} \\ &= I_1(Z_s - Z_m) \end{aligned}$$

Hence the positive-sequence impedance is

$$Z_1 = Z_s - Z_m \quad . \quad . \quad . \quad . \quad (9.25)$$

In the same way we can show that

$$V_2 = I_2(Z_s - Z_m)$$

so that the negative-sequence impedance is

$$Z_2 = Z_s - Z_m \quad . \quad . \quad . \quad . \quad (9.26)$$

Note that $Z_1 = Z_2$.

(*b*) *Transformers.* Since the transformer looks the same to positive and negative phase sequences, the positive and negative phase-sequence impedances are equal. The zero phase-sequence impedance depends on the connexion of a neutral, and is infinite if there is no neutral connexion.

(*c*) *Rotating Machines.* The voltage drop in rotating machines is made up of the voltage drop due to the winding resistance and leakage reactance plus the voltage drop due to armature reaction. For normal operation (positive phase sequence) the armature reaction effect is large and may be represented by the voltage drop due to an inductance (the synchronous inductance). The positive phase-sequence impedance, Z_1, is therefore approximately the synchronous impedance of the machine. However, the armature reaction effect of currents of negative phase sequence is negligible, and hence the negative phase-sequence impedance, Z_2, will normally be considerably less than Z_1. This is more particularly the case in synchronous than in asynchronous machines (where the slip is greater than zero, and hence where negative-sequence currents can have greater armature reaction effect).

The zero-sequence impedance, Z_0, is approximately equal to the leakage reactance since all three phases of the armature carry equal currents which are in phase and do not produce a rotating field. Generally Z_2 and Z_0 are about one-third of Z_1 in synchronous machines.

The phase-sequence impedances may be measured fairly readily.

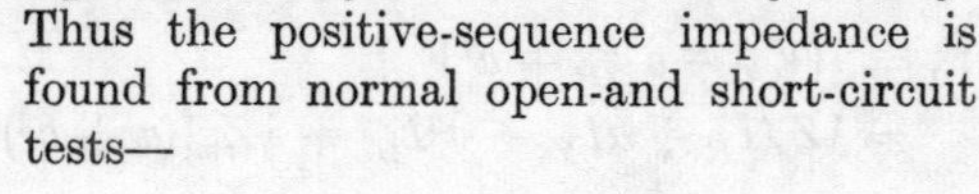

Thus the positive-sequence impedance is found from normal open-and short-circuit tests—

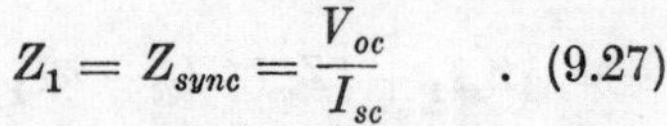

$$Z_1 = Z_{sync} = \frac{V_{oc}}{I_{sc}} \quad . \quad (9.27)$$

FIG. 9.4. MEASUREMENT OF NEGATIVE-SEQUENCE IMPEDANCE OF AN ALTERNATOR

One method of determining the negative-sequence impedance is to run the machine as a generator with reduced excitation (Fig. 9.4) with two phases short-circuited through an ammeter (current, I) and with a voltage V between the third phase and one of the others. Then $I_R = 0$, $I_Y = -I_B$, and since B and Y are short-circuited, $V_B = V_Y$, while $V = V_R - V_Y$ (these being phase voltages). Hence

$$I_2 = \tfrac{1}{3}(I_R + a^2 I_Y + a I_B) = \tfrac{1}{3}(-a^2 + a)I_B = \frac{jI_B}{\sqrt{3}} = -\frac{jI}{\sqrt{3}}$$

and

$$V_2 = \tfrac{1}{3}(V_R + a^2 V_Y + a V_B) = \tfrac{1}{3}(V_R - V_Y) = \tfrac{1}{3}V$$

It therefore follows that

$$Z_2 = \frac{V_2}{I_2} = \frac{\sqrt{3}V}{-j3I} = \frac{jV}{\sqrt{3}I} \quad . \quad . \quad (9.28)$$

The zero phase-sequence impedance is found from a static test of the voltage V' and current I' when a normal-frequency single-phase voltage is applied to all three phases in parallel. Then

$$Z_0 = \frac{V'}{I'/3} = \frac{3V'}{I'} \quad . \quad . \quad (9.29)$$

9.6. Asymmetrical Fault Analysis

In the following analysis it will be assumed that (*a*) the impedance of the fault is negligible, (*b*) load currents in the system are negligible in comparison to fault currents, (*c*) the generator e.m.f. is of positive sequence only, and (*d*) the network impedances up to the fault are balanced, so that the phase-sequence components are independent of one another.

The generated phase e.m.f.s are E_{RN}, E_{YN} ($= a^2E_{RN}$) and E_{BN} ($= aE_{RN}$), so that the phase sequence e.m.f.s in the red line are

$$E_0 = \tfrac{1}{3}(E_{RN} + E_{YN} + E_{BN}) = \tfrac{1}{3}(1 + a^2 + a)E_{RN} = 0 \quad . \quad (9.30)$$

$$E_1 = \tfrac{1}{3}(E_{RN} + aE_{YN} + a^2E_{BN}) = \tfrac{1}{3}(1 + 1 + 1)E_{RN} = E_{RN} \quad (9.31)$$

and

$$E_2 = \tfrac{1}{3}(E_{RN} + a^2E_{YN} + aE_{BN}) = \tfrac{1}{3}(1 + a^4 + a^2)E_{RN} = 0 . \quad (9.32)$$

These equations show that the open-circuit phase voltages are of positive phase sequence only, as we should expect.

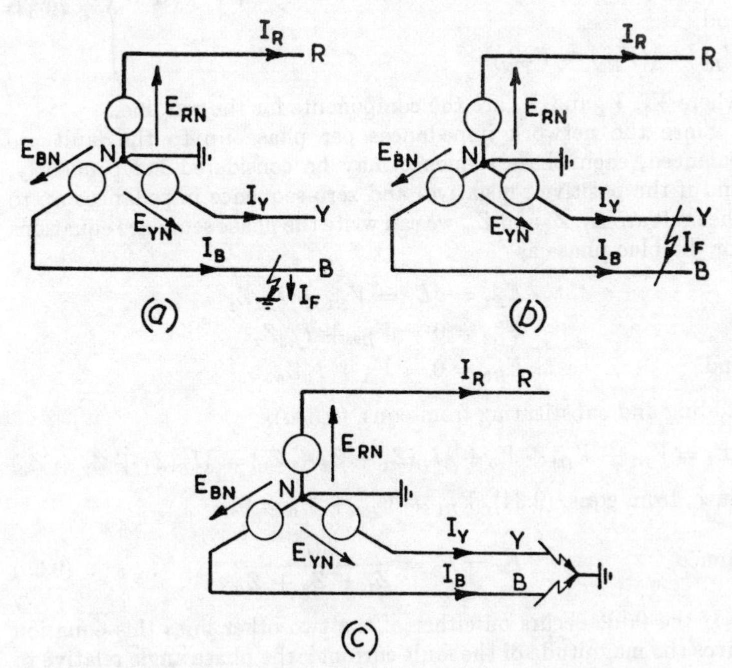

Fig. 9.5. Asymmetrical Faults on a Three-phase System

9.6.1. Line-to-earth Fault

The circuit of a 3-phase system with a fault to earth on one line (the blue line) is shown in Fig. 9.5 (*a*). If the generator star point is solidly earthed, the conditions at the fault are $V_{BN} = 0$, $I_R = I_Y = 0$, and the fault current, I_F, is equal to I_B. The phase-sequence components of current in the red line are (from eqns. (9.16) to (9.18))

$$\left.\begin{aligned} I_1 &= \tfrac{1}{3}(I_R + aI_Y + a^2I_B) = \tfrac{1}{3}a^2I_B \\ I_2 &= \tfrac{1}{3}(I_R + a^2I_Y + aI_B) = \tfrac{1}{3}aI_B \\ \text{and} \quad I_0 &= \tfrac{1}{3}(I_R + I_Y + I_B) = \tfrac{1}{3}I_B \end{aligned}\right\} \quad . \quad . \quad (9.33)$$

In the blue (faulty) line these components are

$$I_{B1} = aI_1 = \tfrac{1}{3}I_B \quad I_{B2} = a^2I_2 = \tfrac{1}{3}I_B \quad I_{B0} = I_0 = \tfrac{1}{3}I_B \quad .(9.33a)$$

The symmetrical components of the phase voltage of the blue line at the fault are

$$\left.\begin{aligned} V_{B1} &= aV_1 = \tfrac{1}{3}a(V_{RN} + aV_{YN}) = \tfrac{1}{3}(aV_{RN} + a^2V_{YN}) \\ V_{B2} &= a^2V_2 = \tfrac{1}{3}a^2(V_{RN} + a^2V_{YN}) = \tfrac{1}{3}(a^2V_{RN} + aV_{YN}) \\ &\text{and} \\ V_{B0} &= \tfrac{1}{3}(V_{RN} + V_{YN}) \end{aligned}\right\} \quad . \quad (9.34)$$

where V_1, V_2 and V_0 are the components for the red line.

Since the network impedances per phase up to the fault are balanced, each phase sequence may be considered independently, and if the positive-, negative- and zero-sequence impedances up to the fault are Z_1, Z_2 and Z_0, we can write the phase sequence equations for the blue phase as

$$\begin{aligned} E_{B1} &= aE_1 = V_{B1} + I_{B1}Z_1 \\ E_{B2} &= 0 = V_{B2} + I_{B2}Z_2 \\ \text{and} \quad E_{B0} &= 0 = V_0 + I_0Z_0 \end{aligned}$$

Adding and substituting from eqns. (9.33*a*),

$$aE_1 = V_{B1} + V_{B2} + V_0 + \tfrac{1}{3}I_B(Z_1 + Z_2 + Z_0) = \tfrac{1}{3}I_B(Z_1 + Z_2 + Z_0)$$

since, from eqns. (9.34), $V_{B1} + V_{B2} + V_0 = 0$.

Hence
$$I_F = I_B = \frac{3aE_1}{Z_1 + Z_2 + Z_0} \quad . \quad . \quad . \quad (9.35)$$

If the fault occurs on either of the two other lines this equation gives the magnitude of the fault current; the phase angle relative to E_{RN} will, of course, be different.

9.6.2. Short-circuit between Two Lines

Fig. 9.5 (*b*) shows a fault between the yellow and blue lines of a 3-phase system. The conditions at the fault are $I_R = 0$, $V_{BN} = V_{YN}$ and $I_Y = -I_B$. The phase-sequence components of current in the red line are

$$\left.\begin{aligned} I_1 &= \tfrac{1}{3}(I_R + aI_Y + a^2 I_B) = \tfrac{1}{3}(a - a^2)I_Y \\ I_2 &= \tfrac{1}{3}(I_R + a^2 I_Y + aI_B) = \tfrac{1}{3}(a^2 - a)I_Y = -I_1 \\ \text{and} \quad I_0 &= \tfrac{1}{3}(I_R + I_Y + I_B) = 0 \end{aligned}\right\} \quad (9.36)$$

As would be expected, since there is no earth return path there is no zero-sequence component of current.

The symmetrical components of the red phase voltage are

$$\left.\begin{aligned} V_1 &= \tfrac{1}{3}(V_{RN} + aV_{YN} + a^2 V_{BN}) = \tfrac{1}{3}(V_{RN} + (a + a^2)V_{YN}) \\ V_2 &= \tfrac{1}{3}(V_{RN} + a^2 V_{YN} + aV_{BN}) \\ &= \tfrac{1}{3}(V_{RN} + (a + a^2)V_{YN}) = V_1 \\ V_0 &= \tfrac{1}{3}(V_{RN} + V_{YN} + V_{BN}) = \tfrac{1}{3}(V_{RN} + 2V_{YN}) \end{aligned}\right\} \quad (9.37)$$

The phase-sequence voltage equations for the red phase are

$$E_1 = V_1 + I_1 Z_1 \quad . \quad . \quad . \quad (9.38)$$

$$E_2 = 0 = V_2 + I_2 Z_2 \quad . \quad . \quad . \quad (9.39)$$

$$E_0 = 0 = V_0 + I_0 Z_0$$

Subtracting eqn. (9.39) from eqn. (9.38),

$$\begin{aligned} E_1 &= V_1 - V_2 + I_1 Z_1 - I_2 Z_2 \\ &= I_1 Z_1 - I_2 Z_2 \qquad \text{(since } V_1 = V_2\text{)} \\ &= I_1(Z_1 + Z_2) \qquad \text{(since } I_2 = -I_1\text{)} \end{aligned}$$

Hence

$$I_1 = \frac{E_1}{Z_1 + Z_2} = -I_2$$

and the fault current, I_F, is

$$\begin{aligned} I_F = I_Y &= a^2 I_1 + aI_2 \\ &= (a^2 - a)\frac{E_1}{Z_1 + Z_2} \\ &= \frac{-j\sqrt{3}\,E_{RN}}{Z_1 + Z_2} \quad . \quad . \quad . \quad (9.40) \end{aligned}$$

9.6.3. Two Lines Short-circuited to Earth

Fig. 9.5 (*c*) shows the yellow and blue lines of a 3-phase system short-circuited to earth. The conditions at the fault are $I_R = 0$, $V_{YN} = V_{BN} = 0$, and fault current $I_F = I_Y + I_B$. The voltage V_{RN} of the red line to earth at the fault depends on this fault current. The symmetrical components of voltage for the red line *at the fault* are

$$V_0 = \tfrac{1}{3}V_{RN} = V_1 = V_2$$

Also, since the current in the red line is zero,

$$I_0 + I_1 + I_2 = 0$$

The sequence voltage equations for the red phase are

$$E_1 = E_{RN} = V_1 + I_1 Z_1$$
$$E_2 = 0 = V_2 + I_2 Z_2$$
$$E_0 = 0 = V_0 + I_0 Z_0$$

Hence

$$I_1 = \frac{E_{RN}}{Z_1} - \frac{V_1}{Z_1} \qquad I_2 = -\frac{V_2}{Z_2} = -\frac{V_1}{Z_2} \qquad I_0 = -\frac{V_0}{Z_0} = -\frac{V_1}{Z_0}$$

so that

$$I_1 + I_2 + I_0 = 0 = \frac{E_{RN}}{Z_1} - V_1 \frac{Z_1 Z_2 + Z_1 Z_0 + Z_2 Z_0}{Z_1 Z_2 Z_0}$$

and

$$V_1 = \frac{E_{RN} Z_2 Z_0}{Z_1 Z_2 + Z_1 Z_0 + Z_2 Z_0} = \frac{E_{RN}}{1 + (Z_1/Z_0) + (Z_1/Z_2)} \quad . \quad (9.41)$$

I_1 can now be expressed in terms of E_{RN} from

$$I_1 = \frac{E_{RN}}{Z_1} - \frac{V_1}{Z_1} = \frac{E_{RN}}{Z_1}\left[1 - \frac{Z_2 Z_0}{Z_1 Z_2 + Z_1 Z_0 + Z_2 Z_0}\right]$$

$$= \frac{E_{RN}(Z_2 + Z_0)}{Z_1 Z_2 + Z_1 Z_0 + Z_2 Z_0} = \frac{E_{RN}}{Z_1 + Z_2 Z_0/(Z_2 + Z_0)} \quad . \quad . \quad . \quad (9.42)$$

Similarly,

$$I_2 = -\frac{V_1}{Z_2} = \frac{-E_{RN} Z_0}{Z_1 Z_2 + Z_1 Z_0 + Z_2 Z_0} = \frac{-I_1 Z_0}{Z_2 + Z_0} \quad . \quad (9.43)$$

and

$$I_0 = -\frac{V_1}{Z_0} = \frac{-E_{RN} Z_2}{Z_1 Z_2 + Z_1 Z_0 + Z_2 Z_0} = \frac{-I_1 Z_2}{Z_2 + Z_0} \quad . \quad (9.44)$$

The fault current and the voltage between the sound line and earth can readily be found from these equations.

Example 9.4. A 3-phase 10-MVA 11-kV star-connected alternator which has its star point solidly earthed supplies a feeder. The relevant per-unit symmetrical components of impedance are

	Generator (Z_G)	*Feeder* (Z_F)
Positive sequence impedance (p.u.) .	$j0{\cdot}16$	$j0{\cdot}1$
Negative sequence impedance (p.u.)	$j0{\cdot}08$	$j0{\cdot}1$
Zero sequence impedance (p.u.) .	$j0{\cdot}06$	$j0{\cdot}3$

Determine the fault current and line-to-neutral voltages at the generator terminals for two lines short-circuited to earth at the distant end of the feeder.

$$\text{Rated phase voltage} = \frac{11 \times 10^3}{\sqrt{3}} = 6{\cdot}35 \times 10^3 \text{ V}$$

$$\text{Rated phase current} = \frac{10 \times 10^6}{\sqrt{3} \times 11 \times 10^3} = 5{\cdot}24 \times 10^2 \text{ A}$$

The symmetrical components of impedance up to the fault are

$$Z_1 = Z_{G1} + Z_{F1} = j0{\cdot}26 \qquad Z_2 = j0{\cdot}18 \qquad Z_0 = j0{\cdot}36$$

Let the generator red phase voltage be taken as the reference: $E_{RN} = E_1 = 1\underline{/0^\circ}$ p.u., and suppose that the fault occurs on the yellow and blue lines. Then, from eqn. (9.42), the positive-sequence component of current in the red line is

$$I_1 = \frac{1\underline{/0^\circ}}{j0{\cdot}26 + \dfrac{j0{\cdot}18 \times j0{\cdot}36}{j0{\cdot}54}} = -j2{\cdot}63 \text{ p.u.}$$

Similarly, eqn. (9.43) gives the negative-sequence component:

$$I_2 = \frac{j2{\cdot}63 \times j0{\cdot}36}{j0{\cdot}54} = j1{\cdot}75 \text{ p.u.}$$

and eqn. (9.44) gives the zero sequence component:

$$I_0 = \frac{j2{\cdot}63 \times j0{\cdot}18}{j0{\cdot}54} = j0{\cdot}88 \text{ p.u.}$$

Note that, as a check, $I_1 + I_2 + I_0 = 0$.

The fault current in the yellow line is (from eqn. (9.15*b*))

$$\begin{aligned} I_Y &= a^2 I_1 + a I_2 + I_0 \\ &= (2{\cdot}63\underline{/-90^\circ} \times 1\underline{/240^\circ}) + (1{\cdot}75\underline{/90^\circ} \times 1\underline{/120^\circ}) + 0{\cdot}88\underline{/90^\circ} \\ &= -3{\cdot}8 + j1{\cdot}32 = 4{\cdot}02\underline{/160^\circ} \text{ p.u.} \end{aligned}$$

and that in the blue line is

$$\begin{aligned} I_B &= a I_1 + a^2 I_2 + I_0 \qquad \text{(eqn. (9.15}c\text{))} \\ &= (2{\cdot}63\underline{/-90^\circ} \times 1\underline{/120^\circ}) + (1{\cdot}75\underline{/90^\circ} \times 1\underline{/240^\circ}) + j0{\cdot}88 \\ &= 3{\cdot}8 + j1{\cdot}32 = 4{\cdot}02\underline{/20^\circ} \text{ p.u.} \end{aligned}$$

The actual fault currents are found by multiplying the per-unit values by the rated current to give

$$I_Y = 2{,}110\underline{/160^\circ}\ \text{A}$$

and

$$I_B = 2{,}110\underline{/20^\circ}\ \text{A}$$

The fault current to earth is

$$I_F = I_Y + I_B = 2{,}110\underline{/160^\circ} + 2{,}110\underline{/20^\circ} = 660\underline{/90^\circ}\ \text{A}$$

At the generator terminals the symmetrical components of voltage for the red line are

$$V_{G1} = E_1 - I_1 Z_{G1} = 1 + (j2{\cdot}63 \times j0{\cdot}16) = 0{\cdot}58\underline{/0^\circ}\ \text{p.u.}$$
$$V_{G2} = 0 - I_2 Z_{G2} = -j1{\cdot}75 \times j0{\cdot}08 = 0{\cdot}14\underline{/0^\circ}\ \text{p.u.}$$
$$V_{G0} = 0 - I_0 V_{G0} = -j0{\cdot}88 \times j0{\cdot}06 = 0{\cdot}053\underline{/0^\circ}\ \text{p.u.}$$

so that the red phase terminal voltage is

$$V_{GR} = V_{G1} + V_{G2} + V_{G0} = 0{\cdot}773\ \text{p.u.} = 4{\cdot}9\underline{/0^\circ}\ \text{kV}$$

the yellow phase voltage at the generator terminals is

$$V_{GY} = a^2 V_{G1} + a V_{G2} + V_0 = 0{\cdot}58\underline{/240^\circ} + 0{\cdot}14\underline{/120^\circ} + 0{\cdot}053$$
$$= -0{\cdot}306 - j0{\cdot}382\ \text{p.u.} = 3{\cdot}11\underline{/232^\circ}\ \text{kV}$$

and the blue phase voltage is

$$V_{GB} = a V_{G1} + a^2 V_{G2} + V_0 = 0{\cdot}58\underline{/120^\circ} + 0{\cdot}14\underline{/240^\circ} + 0{\cdot}053$$
$$= 0{\cdot}306 + j0{\cdot}382\ \text{p.u.} = 3{\cdot}11\underline{/128^\circ}\ \text{kV}$$

Note that at the *fault*, the symmetrical components for the red line are

$$V_{R1} = 1 - (2{\cdot}63 \times 0{\cdot}26) = 0{\cdot}316 \qquad V_{R2} = -I_2 Z_2 = 0{\cdot}316$$
$$V_{R0} = I_0 Z_0 = 0{\cdot}316$$

so that $V_{RN} = 0{\cdot}948$, $V_{YN} = 0$ and $V_{BN} = 0$.

The additional voltage in the red line is induced by the mutual coupling between the lines.

9.6.4. Effect of Fault Impedance

If the fault on a system does not have zero impedance this can be taken into account in the formulation of the equations of the system. Thus, in the case of a fault between one line and earth shown in Fig. 9.5 (*a*), suppose that the impedance of the fault is Z_f. Then, at the fault,

$$V_{BN} = I_B Z_f \qquad I_R = I_Y = 0$$

This yields the phase-sequence components of current in the red line as

$$I_{R1} = \tfrac{1}{3} a^2 I_B \qquad I_{R2} = \tfrac{1}{3} a I_B \qquad I_{R0} = \tfrac{1}{3} I_B$$

as before (eqn. (9.33)).

The symmetrical components of blue phase voltage at the fault are

$$V_{B1} = aV_{R1} = \tfrac{1}{3}a(V_{RN} + aV_{YN} + a^2V_{BN})$$
$$= \tfrac{1}{3}(aV_{RN} + a^2V_{YN} + I_BZ_f)$$
$$V_{B2} = a^2V_{R2} = \tfrac{1}{3}(a^2V_{RN} + aV_{YN} + I_BZ_f)$$

and

$$V_{B0} = \tfrac{1}{3}(V_{RN} + V_{YN} + I_BZ_f)$$

Writing the voltage-drop equations for each phase sequence (referred to the blue line),

$$E_{B1} = aE_{ph} = V_{B1} + I_{B1}Z_1 = \tfrac{1}{3}(aV_{RN} + a^2V_{YN} + I_BZ_f) + \tfrac{1}{3}I_BZ_1$$
$$E_{B2} = 0 = V_{B2} + I_{B2}Z_2 = \tfrac{1}{3}(a^2V_{RN} + aV_{YN} + I_BZ_f) + \tfrac{1}{3}I_BZ_2$$

and

$$E_{B0} = 0 = V_{B0} + I_{B0}Z_0 = \tfrac{1}{3}(V_{RN} + V_{YN} + I_BZ_f) + \tfrac{1}{3}I_BZ_0$$

where Z_1, Z_2 and Z_0 are the phase-sequence impedances up to the fault, and $I_{B1} = I_{B2} = I_{B0} = \tfrac{1}{3}I_B$. Adding the above three equations,

$$aE_{ph} = I_BZ_f + I_B\tfrac{1}{3}(Z_1 + Z_2 + Z_0)$$

or

$$I_B = \frac{aE_{ph}}{Z_f + \tfrac{1}{3}(Z_1 + Z_2 + Z_0)}$$

Expressions for other types of fault can be obtained in a similar fashion.

9.7. Effect of Transformer Connexions on Zero-sequence Impedance

It was pointed out in Section 9.5 that the zero-sequence impedance of a transformer may well be different from its positive- and negative-sequence impedances. This phenomenon will now be considered in greater detail.

The equivalent circuit of a single-phase double-wound transformer may be considered to consist of a series or leakage impedance as measured by a short-circuit test and a shunt or exciting admittance (or impedance) as measured by an open-circuit test. In fault calculations the latter may be frequently, though not always, neglected.

To simplify the discussion, 3-phase transformers will first be considered as if they consisted of a bank of three separate single-phase transformers, each phase thus having an independent magnetic circuit. The effect of the type of construction of 3-phase transformers on the zero-sequence impedance will be considered separately.

The zero-sequence impedance of such a bank of three single-phase transformers may be either (*a*) the same as the positive- and negative-sequence impedance and thus equal to the series or leakage impedance, or (*b*) equal to the magnetizing or exciting impedance. It has the former value if a zero phase-sequence component of current may flow in the primary when a fault occurs on the secondary side of the transformer, and the latter if such a component of current may not flow in the primary.

It should be noted that zero-sequence current will flow only if the side of the transformer feeding into the fault is star-connected and has its star point earthed so that a neutral or earth-return path exists for the zero phase-sequence component of current (*see* Section 9.3).

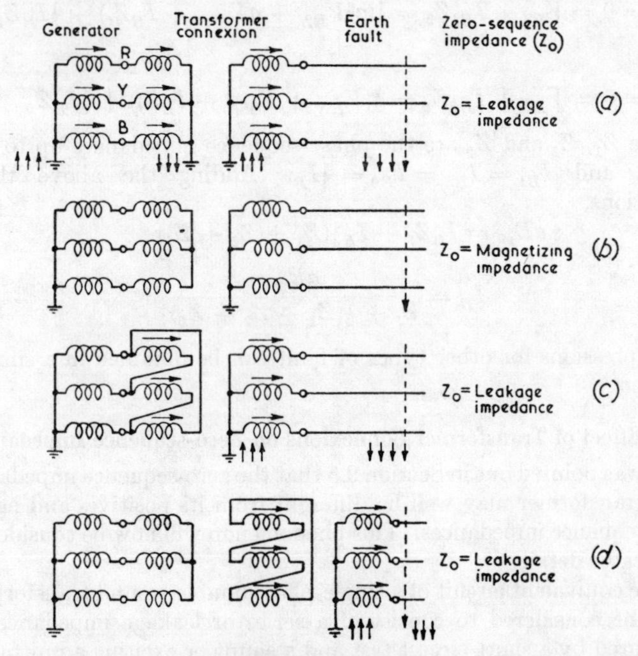

FIG. 9.6. EFFECT OF TRANSFORMER CONNEXION ON ZERO-SEQUENCE IMPEDANCE

Fig. 9.6 shows some common 3-phase transformer connexions. In each diagram a star-connected generator having its star point earthed is shown supplying the transformer. The fault is supposed to occur, in each case, on the side of the transformer remote from the generator. In each case the distribution of the zero phase-sequence component of current is indicated.

In Fig. 9.6 (*a*) a path for zero-sequence current exists on the secondary side of the transformer if an earth fault occurs. Further, a path for zero-sequence current exists between the generator and the transformer primary, so that under earth-fault conditions the generator is able to supply zero-sequence current to the primary, thus maintaining m.m.f. balance (indicated on the diagram by the dot notation) between the primary and secondary, neglecting magnetizing current. The zero-sequence impedance of the transformer is thus equal to the series or leakage impedance for this connexion. In this case the zero-sequence impedance of the generator would also require to be represented in the equivalent circuit, since the generator passes zero-sequence current.

In Fig. 9.6 (*b*) the generator cannot supply zero-sequence current to the transformer primary, and as a result the secondary can supply only a small zero-sequence current under an earth-fault condition as compared with the circuit connexion of Fig. 9.6 (*a*). The zero-sequence current which can flow is a magnetizing current, and the zero-sequence impedance is equal to the exciting impedance. In fault calculations this current is often negligible, the zero-sequence impedance then being assumed infinite. Since in this case the generator does not supply zero-sequence current, the zero-sequence impedance of the generator would not appear in the equivalent circuit.

In Fig. 9.6 (*c*) the secondary is able to supply a relatively large zero-sequence current to an earth fault, since a closed path on the secondary side exists and m.m.f. balance between the transformer windings may be maintained by a zero-sequence current circulating in the closed delta-connected primary, so that the zero-sequence impedance is equal to the series or leakage impedance. In this case the zero-sequence impedance of the generator should not appear in the equivalent circuit since it supplies no zero-sequence current.

The transformer connexion of Fig. 9.6 (*d*) includes a delta-connected tertiary winding. The conditions are similar to those of Fig. 9.6 (*c*) and the zero-sequence impedance is again equal to the series or leakage impedance. The value of this impedance is, however, modified by the presence of the tertiary winding.

If any star point through which zero-sequence current flows is earthed through an impedance Z_E, an element $3Z_E$ must appear in the zero-sequence impedance, since the zero-sequence current of each phase flows through it.

If a 3-phase transformer replaces the three single-phase transformers assumed in the foregoing discussion, the above remarks apply equally to Figs. 9.6 (*a*), (*c*) and (*d*), but must be modified in respect of Fig. 9.6 (*b*). Although in this case the zero-sequence impedance is still equal to the exciting impedance, the exciting impedance for zero sequence may differ from that for the positive and negative sequences.

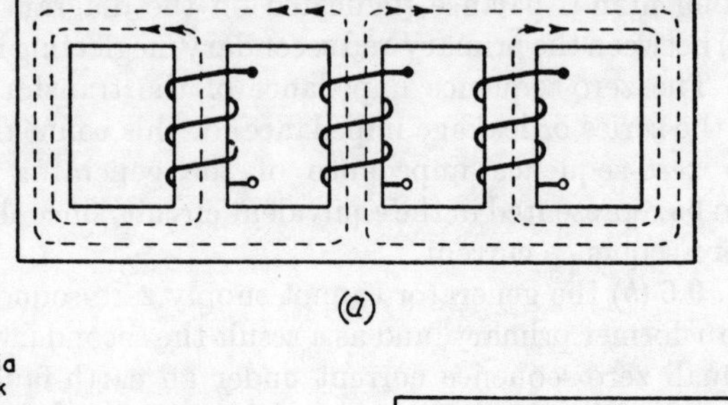

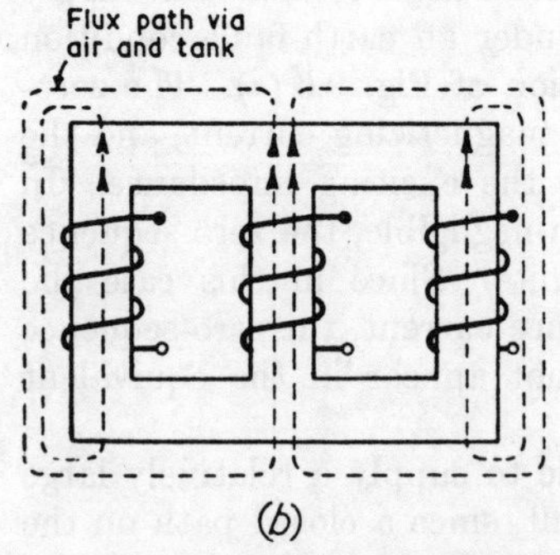

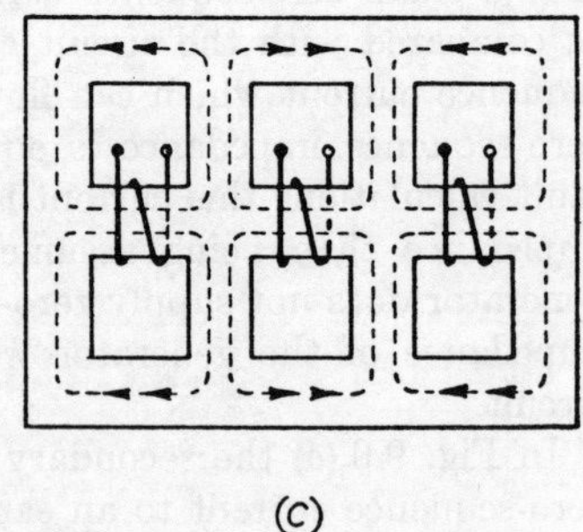

Fig. 9.7. Flux Distribution in Three-phase Transformers under Conditions of Zero-sequence Excitation

(*a*) Five-limb core type
(*b*) Three-limb core type
(*c*) Shell type

Three-phase transformers may be of the 5-limb core type (Fig. 9.7 (*a*)), the 3-limb core-type (Fig. 9.7 (*b*)) or the shell type of construction (Fig. 9.7 (*c*)). The magnetic circuit for each phase of a 5-limb core-type transformer is complete. Under single-phase or zero phase-sequence excitation the outer unwound limbs provide return paths for the fluxes in the three wound limbs, and as a result the magnetizing impedance of such a transformer is the same for zero phase sequence as for positive and negative phase sequences.

With 3-limb core-type transformers no such return path exists within the core for the fluxes in the three limbs under zero sequence excitation. Such flux as exists in any core limb must cross the air-gaps between the core and the steel tank of the transformer. As a result the reluctance of the flux paths under zero-sequence excitation is much higher and the exciting impedance is much lower than under positive- or negative-sequence excitation.

Example 9.5. Three 3-phase 10-MVA 11-kV star-connected alternators are joined to 3-phase busbars to which is connected an 11/66-kV 10-MVA delta-star h.v. transformer. One generator has its star point solidly earthed, the star points of the two other generators being isolated. The star point of the h.v. transformer winding is earthed through an effective resistance of 20 Ω.

The impedances of each generator and the transformer are as follows—

	Generator	*Transformer (referred to 11 kV)*
	Z_G	Z_T
Positive-sequence impedance (ohms)	$j2$	$j2$
Negative-sequence impedance (ohms)	$j1{\cdot}8$	$j2$
Zero-sequence impedance (ohms)	$j0{\cdot}8$	$j2$

Determine the current fed into an earth fault (*a*) on a busbar, and (*b*) at an h.v. terminal of the transformer. The transformer may be assumed to be open-circuited at the h.v. terminals prior to the imposition of the fault.

(*a*) *Busbar fault.* A busbar fault is not affected by the transformer. The positive-sequence impedance up to this fault therefore includes the positive-sequence impedances of the three generators in parallel, and the negative-sequence impedance includes the three negative-sequence impedances of the generators in parallel. The zero-sequence impedance, however, consists of only the zero-sequence impedance of the generator which has its star-point earthed. The two other generators with isolated star points cannot supply zero phase-sequence current to the fault.

If the generator phase e.m.f.s are E_{RN}, E_{YN} and E_{BN}, then with E_{RN} as reference $(= 11 \times 10^3/\sqrt{3}) = 6{\cdot}35 \times 10^3\underline{/0^\circ}$ V), eqns. (9.30)–(9.32) give $E_0 = 0$, $E_1 = 6{\cdot}35 \times 10^3\underline{/0^\circ}$, $E_2 = 0$. Also eqn. (9.35) gives the fault current for a fault on the blue line as

$$I_F = \frac{3aE_1}{Z_1 + Z_2 + Z_0}$$

where $\qquad Z_1 = j\frac{2}{3} \qquad Z_2 = j\frac{1{\cdot}8}{3} \qquad Z_0 = j0{\cdot}8$

Hence
$$I_F = \frac{3\underline{/120^\circ} \times 6{\cdot}35 \times 10^3\underline{/0^\circ}}{j2{\cdot}07} = \underline{\underline{9{,}200\underline{/30^\circ}\text{ A}}}$$

(*b*) *Fault at h.v. terminal of transformer.* The positive and negative phase-sequence impedances of the transformer are added in series with the generator impedances to give the total positive and negative phase-sequence impedances up to the fault. The generators are unable to supply zero-sequence current to the mesh-connected primary of the transformer, so that the total zero-sequence impedance consists of the zero-sequence impedance of the transformer in series with three times the earth resistance. The earth resistance must be multiplied by three because the zero phase-sequence current of each phase passes through it, i.e. each phase component may be considered to flow in a separate earth path of three times the actual earth-path impedance. In this instance the calculation is most easily performed by referring all quantities to the h.v. side of the transformer, the turns ratio being

$$n = \frac{66/\sqrt{3}}{11} = 2\sqrt{3}$$

Because of the delta-star connexion, the generator e.m.f. referred to the h.v. side of the transformer is $66/\sqrt{3}$ kV/phase, so that, since the actual generator e.m.f. is $11/\sqrt{3}$ kV/phase, the effective transformer step-up ratio is 6:1. Hence the referred generator phase sequence impedances are

$$Z_{G1} = j\frac{2}{3} \times 6^2 = j24 \qquad Z_{G2} = j\frac{1{\cdot}8}{3} \times 6^2 = j21{\cdot}9$$

The transformer phase-sequence impedances referred to the high-voltage side are

$$Z_{T1} = Z_{T2} = Z_{T0} = j2(2\sqrt{3})^2 = j24$$

so that the total phase-sequence impedances up to the fault are

$$Z_1 = j48 \qquad Z_2 = j45{\cdot}9 \qquad Z_0 = 60 + j24$$

From eqn. (9.35), and since the positive-sequence component of transformer secondary phase voltage is $E_1 = 66/\sqrt{3}$ kV,

$$I_F = \frac{3\underline{/120^\circ} \times 66{,}000/\sqrt{3}}{j48 + j45{\cdot}9 + 60 + j24} = \underline{\underline{877\underline{/57^\circ}\text{ A}}}$$

9.8. Measurement of Symmetrical Components of Current

Bridge circuits with suitably chosen components, and with current transformers, can be used to measure the positive-, negative- and zero-sequence symmetrical components of current. In the circuit

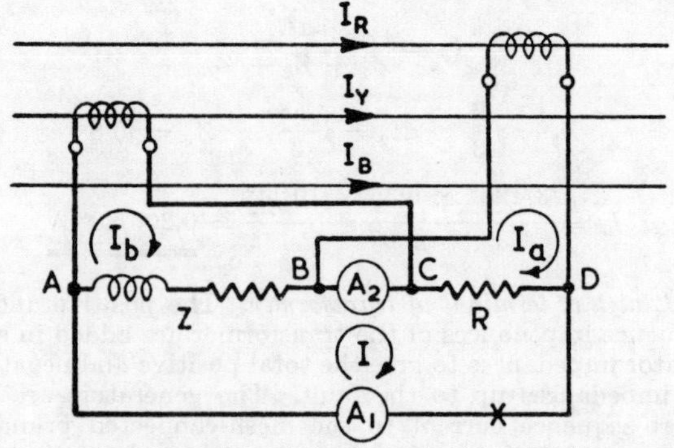

FIG. 9.8. BRIDGE FOR MEASUREMENT OF SYMMETRICAL-COMPONENT CURRENTS IN A THREE-WIRE SYSTEM

shown in Fig. 9.8 it will be shown that ammeter A_1 reads positive-sequence current only, while ammeter A_2 reads negative-sequence current only. Thus, for a current transformer ratio of $n:1$,

$$I_a = \frac{1}{n} I_R = \frac{1}{n}(I_1 + I_2) \quad \text{(from eqn. (9.15 (a))}$$

since the zero-sequence component is zero. I_1 and I_2 are the positive- and negative-sequence components in the red line. Similarly,

$$I_b = \frac{1}{n} I_Y = \frac{1}{n}(a^2 I_1 + a I_2)$$

If the impedance between B and D is Z_{BD} (including the ammeter impedance $R_A + jX_A$), then

$$Z_{BD} = R + R_A + jX_A$$

If now the impedance between A and C is chosen so that $Z_{AC} = Z + R_A + jX_A = (R + R_A + jX_A)\underline{/60^\circ}$, then the Thévenin voltage across a break at X is

$$V_T = I_b(R + R_A + jX_A)\underline{/60^\circ} + I_a(R + R_A + jX_A)$$
$$= (I_b\underline{/60^\circ} + I_a)Z_{BD}$$

If the two ammeters are identical the impedance in series with this voltage is

$$Z_T = R_A + jX_A + Z + Z_{BD}$$
$$= Z_{BD}\underline{/60^\circ} + Z_{BD} = \left(\frac{3}{2} + j\frac{\sqrt{3}}{2}\right) Z_{BD}$$

The current, I, normally flowing through the break is

$$I = \frac{V_T}{Z_T} = \frac{I_b\underline{/60^\circ} + I_a}{(3/2) + j(\sqrt{3}/2)}$$
$$= \frac{1}{n} \cdot \frac{(a^2 I_1 + a I_2)\underline{/60^\circ} + I_1 + I_2}{\sqrt{3}\underline{/30^\circ}}$$
$$= \frac{1}{n} \cdot \frac{I_1(1\underline{/300^\circ} + 1) + I_2(1\underline{/180^\circ} + 1)}{\sqrt{3}\underline{/30^\circ}}$$
$$= \frac{I_1}{n}\underline{/-60^\circ} \quad . \quad . \quad . \quad . \quad . \quad (9.45)$$

i.e. ammeter A_1 reads positive-sequence current only.

Ammeter A_2 reads

$$I_{A2} = I_a + I_b - I$$

$$= \frac{1}{n}(I_1 + I_2 + I_1\underline{/240^\circ} + I_2\underline{/120^\circ} - I_1\underline{/-60^\circ})$$

$$= \frac{I_1}{n}\left(1 - \frac{1}{2} - j\frac{\sqrt{3}}{2} - \frac{1}{2} + j\frac{\sqrt{3}}{2}\right) + \frac{I_2}{n}(1 + 1\underline{/120^\circ})$$

$$= \frac{I_2}{n}\underline{/60^\circ} \quad . \quad . \quad . \quad . \quad . \quad . \quad . \quad . \quad (9.46)$$

i.e. ammeter A_2 reads negative-sequence components only.

Cross-connected current transformers may be used in 4-wire systems to exclude zero-sequence currents; these cause circulating currents in the windings which do not appear in the measuring network. The current supplied to the network is $\sqrt{3}$ times the current supplied in the 3-wire case (Fig. 9.9). The zero-sequence component of current is simply one-third of the current in the neutral.

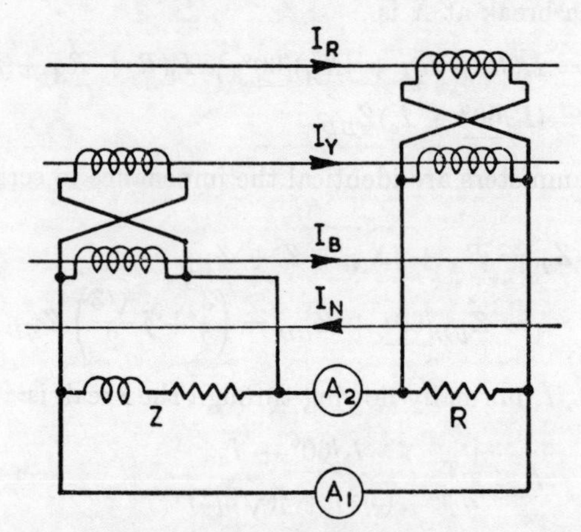

Fig. 9.9. Measurement of Symmetrical Components of Current in a Four-wire System

9.9. Single Sequence Current Component Bridge

A bridge network which will measure positive- or negative-sequence components of line current in a 3-phase system is shown in Fig. 9.10. The arrangement will carry negative-sequence components only in the detector branch YP. If any two inputs to the bridge are interchanged, positive-sequence current only will flow in YP. A protective relay in the branch YP can be set to indicate a fault involving negative-sequence currents.

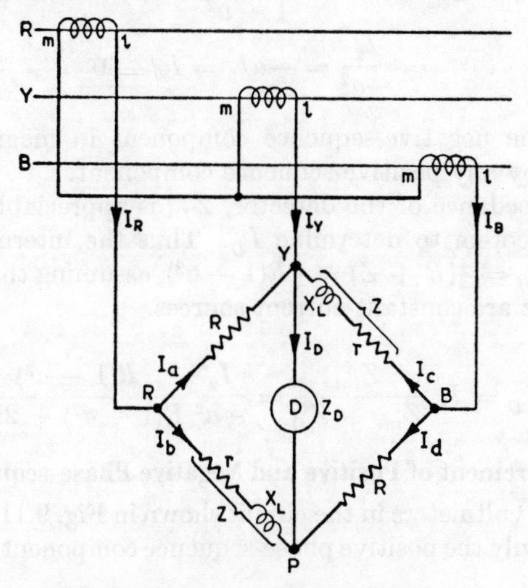

FIG. 9.10. PHASE-SEQUENCE BRIDGE

For the circuit as shown, let YP be a short-circuit. Then, since the current transformers can be considered as constant-current sources, the short-circuit current is

$$I_{sc} = I_b + I_d$$

Also, since Y and P are at the same potential,

$$I_b = I_R \frac{R}{R+Z} \qquad \text{and} \qquad I_d = I_B \frac{Z}{R+Z}$$

so that

$$I_{sc} = I_R \frac{R}{R+Z} + I_B \frac{Z}{R+Z}$$

If we choose values for the bridge elements so that $Z = R\underline{/60^\circ} = -a^2R$, then

$$I_{sc} = \frac{I_R - a^2 I_B}{1 - a^2}$$

But, for a 3-wire system, eqn. (9.15) gives

$$I_R = I_1 + I_2 \qquad \text{and} \qquad I_B = aI_1 + a^2I_2$$

Hence

$$I_{sc} = \frac{I_1 + I_2 - a^3I_1 - a^4I_2}{1 - a^2}$$

$$= \frac{I_2}{-a^2} = -aI_2 = I_2\underline{/-60^\circ} \quad . \quad . \quad . \quad (9.47)$$

i.e. I_{sc} is the negative-sequence component in magnitude and is unaffected by any positive-sequence component.

If the impedance of the detector, Z_D, is appreciable, we can use Norton's theorem to determine I_D. Thus the internal impedance at YP is $Z_{int} = \frac{1}{2}(R + Z) = \frac{1}{2}R(1 - a^2)$, assuming that the current transformers are constant-current sources.

Hence

$$I_D = I_{sc}\frac{Z_{int}}{Z_{int} + Z_D} = \frac{I_2}{-a^2}\,\frac{R(1 - a^2)}{R(1 - a^2) + 2Z_D} \quad . \quad (9.48)$$

9.10. Measurement of Positive and Negative Phase-sequence Voltages

The three voltmeters in the circuit shown in Fig. 9.11 can be shown to indicate only the positive phase-sequence component of the voltage

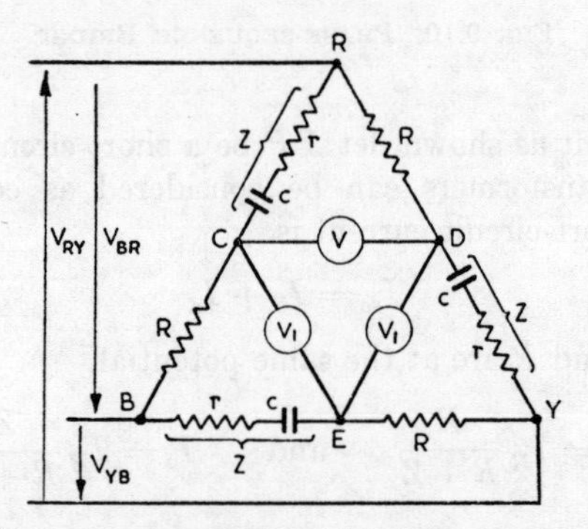

Fig. 9.11. Bridge Arranged to Measure Positive-sequence Components of Voltage

of the 3-phase system. If the line-to-neutral voltages of the three lines are V_{RN}, V_{YN} and V_{BN}, then by analogy from eqn. (9.15), these may be expressed as symmetrical components referred to the red phase as

$$V_{RN} = V_1 + V_2 + V_0$$

$$V_{YN} = a^2V_1 + aV_2 + V_0$$

and

$$V_{BN} = aV_1 + a^2V_2 + V_0$$

Hence

$$V_{RY} = V_{RN} - V_{YN} = (1 - a^2)V_1 + (1 - a)V_2$$

and

$$V_{YB} = V_{YN} - V_{BN} = (a^2 - a)V_1 + (a - a^2)V_2$$

Also

$$V_{DY} = \frac{V_{RY}(r + 1/j\omega C)}{(R + r + 1/j\omega C)}$$

and

$$V_{YE} = \frac{V_{YB}R}{R + r + 1/j\omega C}$$

so that

$$V_{DE} = \frac{(r + 1/j\omega C)}{(R + r + 1/j\omega C)}\left(V_{RY} + \frac{R}{(r + 1/j\omega C)}V_{YB}\right)$$

The bridge elements are chosen so that

$$\frac{R}{r + 1/j\omega C} = -a^2 = \frac{1}{2} + j\frac{\sqrt{3}}{2}$$

i.e.

$$R = \frac{r}{2} + \frac{\sqrt{3}}{2\omega C} + \frac{1}{j2\omega C} + j\frac{\sqrt{3}r}{2}$$

Equating quadrate terms,

$$\frac{1}{\omega C} = \sqrt{3}r \quad . \quad . \quad . \quad . \quad (9.49)$$

Equating reference terms,

$$R = \frac{r}{2} + \frac{\sqrt{3}}{2\omega C} = \frac{r}{2} + \frac{3r}{2} = 2r \quad . \quad . \quad (9.50)$$

It follows that

$$\frac{r + 1/j\omega C}{R + r + 1/j\omega C} = \frac{r(1 - j\sqrt{3})}{r(3 - j\sqrt{3})} = \frac{1}{\sqrt{3}}\underline{/-30^\circ}$$

and

$$V_{DE} = \frac{1\underline{/-30^\circ}}{\sqrt{3}}[(1 - a^2)V_1 + (1 - a)V_2 - a^2(a^2 - a)V_1 - a^2(a - a^2)V_2]$$

$$= \frac{1\underline{/-30^\circ}}{\sqrt{3}}(1 - a^2 - a + a^3)V_1 \qquad (\text{since } 1 - a = a^2(a - a^2))$$

$$= \sqrt{3}V_1\underline{/-30^\circ} \quad . \quad . \quad . \quad . \quad (9.51)$$

Hence, if the voltmeter between D and E has infinite impedance it measures $\sqrt{3}$ times the positive phase-sequence component of phase voltage. Similarly all the voltmeters will read positive-sequence components only.

If the elements are reversed, but the same relationship is maintained between them, the high-impedance voltmeter connected as

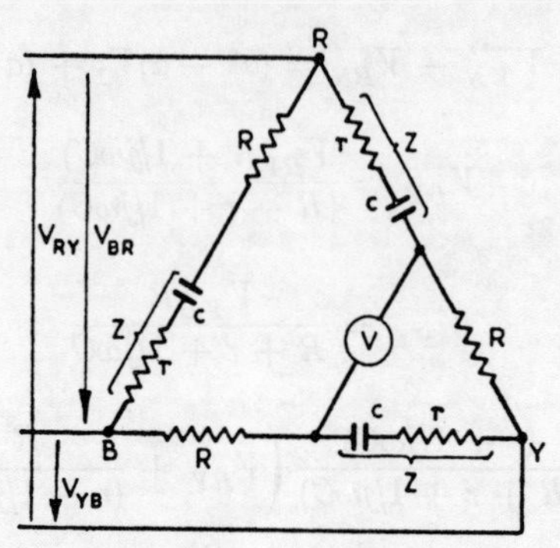

FIG. 9.12. BRIDGE ARRANGED TO MEASURE NEGATIVE-SEQUENCE COMPONENTS OF VOLTAGE

shown in Fig. 9.12 will measure $\sqrt{3}$ times the negative-sequence component of phase voltage.

9.11. Measurement of Zero Phase-sequence Voltage

By analogy with eqn. (9.18), the zero phase-sequence voltage is

$$V_0 = \tfrac{1}{3}(V_{RS} + V_{YS} + V_{BS})$$

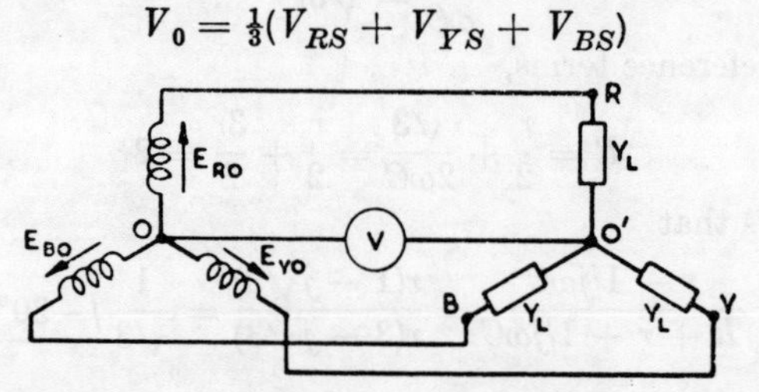

FIG. 9.13. MEASUREMENT OF ZERO-SEQUENCE VOLTAGE

Fig. 9.13 indicates one method of obtaining a measurement of the zero phase-sequence component of voltage. Referring to this diagram,

$$V_{0'0} = \frac{V_{R0}Y_L + V_{Y0}Y_L + V_{B0}Y_L}{3Y_L} = \tfrac{1}{3}(V_{R0} + V_{Y0} + V_{B0}) = V_0$$

The voltmeter therefore reads the zero phase-sequence component of voltage.

Bibliography

Grover, *Principles of Symmetrical Components* (Classifax).
Rissik, *The Calculation of Unsymmetrical Short Circuits* (Pitman).

Problems

9.1. The positive, negative, and zero phase sequence of phase voltage at a given point in a 3-phase transmission system are $V_1 = 200\angle 30°$ $V_2 = 100\angle -30°$ and $V_0 = 60\angle 0°$, respectively, for the red phase. Determine the actual phase voltages.
(*Ans.* $V_{RN} = 323{\cdot}5\angle 9°$; $V_{YN} = 116{\cdot}7\angle -59°$; $V_{BN} = 206\angle 166°$).

9.2. A 3-phase 3-wire transmission line of positive phase-sequence RYB has positive and negative symmetrical components of current in the yellow line of $920 - j1{,}600$ A and $350 + j600$ A respectively. Determine the three line currents in magnitude and phase.
(*Ans.* $I_R = 1{,}615\angle 38°$ A; $I_Y = 1{,}615\angle -38°$ A; $I_B = 2{,}540\angle 180°$ A.)

9.3. Show that an unsymmetrical system of 3-phase voltages or currents may be represented by three symmetrical systems, and find expressions for the symmetrical components.

Under fault conditions in a 3-phase system, the following currents were recorded in the R, Y and B lines: $I_R = 3{,}000\angle 0°$ A, $I_Y = 2{,}000\angle 270°$ A, $I_B = 1{,}000\angle 120°$ A. Calculate the positive, negative and zero phase-sequence components, referred to the red line current.
(*Ans.* $1{,}940\angle 9{\cdot}8°$ A; $261\angle 9{\cdot}7°$ A; $914\angle -24{\cdot}4°$ A.) (*A.E.E.*, 1957)

9.4. Show that an unbalanced system of 3-phase voltages or currents may be represented by three symmetrical systems, and find expressions for the symmetrical components.

The following currents were recorded in the R, Y and B lines of a 3-phase system under abnormal conditions: $I_R = 300\angle 300°$ A, $I_Y = 500\angle 240°$ A, $I_B = 1{,}000\angle 60°$ A. Calculate the positive, negative and zero phase-sequence components. (*A.E.E.*, 1959)
(*Ans.* 535 A; 371 A; 145 A.)

9.5. Derive expressions for the symmetrical components corresponding to an unbalanced system of 3-phase voltages V_a, V_b and V_c, where V_a, V_b and V_c are given in complex notation and h is the sequence operator.

In a 3-phase 4-wire system the currents in the R, Y and B lines under abnormal conditions of loading were as follows: $I_R = 100\angle 30°$ A, $I_Y = 50\angle 300°$ A, $I_B = 30\angle 180°$ A. Calculate the positive, negative and zero phase-sequence currents in the R line, and the return current in the neutral conductor. (*A.E.E.*, 1960)
(*Ans.* $I_{1R} = 42{\cdot}2 + j39{\cdot}8$ amperes; $I_{2R} = 17{\cdot}2 + j8$ amperes; $I_0 = 27{\cdot}2 + j2{\cdot}2$ amperes; return current $= 81{\cdot}7$ A.)

9.6. A 3-phase 10-MVA 11-kV generator with a solidly-earthed neutral point supplies a feeder. The relevant impedances of the generator and feeder are as follows—

	Generator	*Feeder*
To positive-sequence currents, Z_1 (ohms) .	$j1{\cdot}2$	$j1{\cdot}0$
To negative-sequence currents, Z_2 (ohms) .	$j0{\cdot}9$	$j1{\cdot}0$
To zero-sequence currents, Z_0 (ohms) .	$j0{\cdot}4$	$j3{\cdot}0$

A fault from one phase to earth occurs at the far end of the feeder. Determine the voltage to neutral of the faulty phase at the terminals of the generator. The fault current is given by the expression

$$I = 3E/(Z_1 + Z_2 + Z_0)$$

(*A.E.E.*, 1956)

(*Ans.* 4·22 kV.)

9.7. An unbalanced star-connected 3-phase supply, having negligible internal impedance, is connected by three conductors to three identical voltmeters also in star connexion. Each voltmeter has a resistance of 10,000 Ω and negligible reactance. The supply voltages of the three phases from line to neutral are, respectively, $E_R = 100\underline{/0°}$ V, $E_Y = 200\underline{/270°}$ V and $E_B = 100\underline{/120°}$ V. Using the method of symmetrical components, determine EITHER graphically OR algebraically the reading of the voltmeter connected to the yellow line. (*A.E.E.*, 1957)

(*Ans.* 163 V.)

9.8. The phase voltage between the terminals *R*, *Y* and *B* of a 3-phase 3-wire supply and the inaccessible neutral point of the system consists of a positive-sequence component of 200 V and a negative-sequence component of 50 V; these are in phase for the *Y*-terminal. Determine, neglecting source impedances, the current which will flow in each of three 100-Ω non-reactive resistors connected in delta across the terminals.

Determine also the power which would be consumed if these resistors were reconnected in star to the same terminals.

(*L.U. Part III Power*, 1957)

(*Ans.* 3·97 A; 3·97 A; 2·6 A; 1,270 W.)

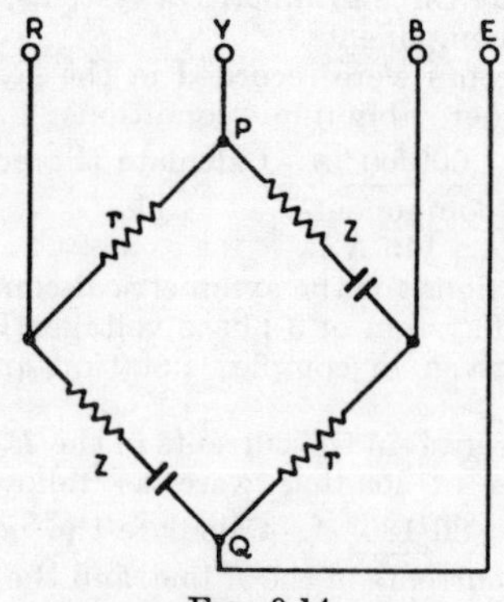

FIG 9.14

9.9. The bridge circuit shown in Fig. 9.14 is made up with two equal resistors *r* and two equal impedances *Z* as shown, such that $r = 2\Omega$ and $Z = (1 - j\sqrt{3})\ \Omega$. The bridge is supplied with unbalanced currents I_R, I_Y and I_B from a 3-phase 4-wire symmetrical supply and the phase sequence is *RYB*. In the *R*-line the positive, negative and zero sequence components of current are respectively $(0 + j\sqrt{50})$, $5(3 + j\sqrt{3})$ and $(5 + j0)$ A. Calculate the potential difference between the points *P* and *Q* of the bridge. (*A.E.E.*, 1962)

(*Ans.* 28·6 V.)

9.10. Explain briefly what is meant by the symmetrical components of a 3-phase system of currents or voltages.

Each phase of a 2-phase 3-wire system supplies 120 A at 200 V, unity power factor. Draw to scale complexors of the currents in, and the p.d.s between, the three line conductors. Hence determine the positive and negative sequence components of current and voltage, regarding the system as unbalanced 3-phase.

Verify that the total power is the same, considered both from 2-phase and 3-phase aspects. (*L.U. Part III Power*, 1958)

(*Ans.* $I_{1R} = 134\underline{/15^\circ}$ A, $I_{2R} = 36\underline{/-75^\circ}$ A; $V_{1R} = 223\underline{/45^\circ}$ V, $V_{2R} = 59{\cdot}6\underline{/-45^\circ}$ A.)

9.11. Explain with the help of complexor diagrams how an unsymmetrical system of 3-phase voltages or currents can be replaced by an equivalent system of symmetrical components.

Phase voltages E_a, E_b and E_c, across three identical star-connected impedances, are unsymmetrical. The neutral point of the load is isolated. Using the principle of symmetrical components, find an expression for the current in phase *a* and check the result by some other method. (*L.U. Part III Theory*, 1958)

9.12. Explain the terms positive-, negative- and zero-sequence components of a 3-phase system.

A star-connected load consists of three equal resistors each of 1 Ω resistance. When the load is connected to an unsymmetrical 3-phase supply, the line voltages are 200 V, 346 V and 400 V. Find the magnitude of the current in any one phase by the method of symmetrical components. (*L.U. Part III Theory*, 1961)

(*Ans.* 177 A.)

9.13. The star point of a 3-kV 3-MVA 3-phase alternator is solidly earthed. Its positive-, negative- and zero-sequence reactances are 2·4, 0·45 and 0·3 Ω respectively. The alternator operating unloaded sustains a resistive fault between the *R* phase and earth: this fault has a resistance of 1·2 Ω. Using the method of symmetrical components, calculate the fault current and the voltage to earth of the red phase, justifying any expression used. The internal resistance of the alternator may be neglected. (*L.U. Part III Power*, 1961)

(*Ans.* 1,090 A; 1,310 V.)

9.14. A star-connected alternator, with earthed neutral, supplies an unloaded 3-phase network through a transformer. The transformer has a star-connected secondary having the neutral point earthed, and a delta-connected primary. The phase sequence is *RYB*. An earth fault of negligible impedance occurs near the *R*-phase terminal of the transformer secondary.

(*a*) Explain why it is possible to calculate the fault current by using symmetrical-sequence networks although the actual network is unbalanced by the fault.

(*b*) Sketch and explain the zero-sequence network which would be used in the calculation.

(*c*) Will the transformer impedance used in the calculation be the same for all the sequence networks? Justify your answer.

(*L.U. Part III Power*, 1962)

CHAPTER 10

Electric Fields

THE regions of space surrounding static electric charges may be mapped by drawing lines of electric flux and orthogonal equipotential lines. These lines delineate the electric field surrounding the charges. It is proposed in this chapter to develop the theory of electric fields from the experimental observation of the force between charged bodies.

10.1. Force between Point Charges

It has been confirmed by direct experiment that the force, F, between two charged bodies varies directly as the product of the charges and inversely as the square of the distance between them, i.e.

$$F \propto \frac{Q_1 Q_2}{r^2}$$

If the charges, Q, are in coulombs and the distance, r, is in metres, the force is

$$F = \frac{Q_1 Q_2}{4\pi\epsilon r^2} \text{ newtons} \qquad . \quad . \quad . \; (10.1)$$

where ϵ is a constant, called the *permittivity*, which depends on the medium surrounding the charges. For a vacuum the permittivity is given the symbol ϵ_0, and has a value of $10^7/4\pi c^2$, where $c = 2{\cdot}998 \times 10^8$ m/sec is the velocity of light. When c is taken as 3×10^8 we obtain $\epsilon_0 = 1/(36\pi \times 10^9)$. For any other medium, $\epsilon = \epsilon_r\epsilon_0$, where ϵ_r is the *relative permittivity* of the medium. For air $\epsilon_r = 1{\cdot}000576$ at standard temperature and pressure, and is usually taken to be unity. The value of ϵ_r for most solid dielectrics does not normally exceed 10, but for the ferro-electric materials such as the titanates of barium and strontium, values of ϵ_r of several thousand can readily be obtained. Note that, by the theory of relativity, eqn. (10.1) can be derived from Ampère's law for the force between two current elements and does not rest on experimental proof alone.

The force is one of attraction for unlike charges, and repulsion for like charges.

10.2. Electric Field Strength, E

The *electric field strength*, or electric force, at any point in an electric field is defined as the force on a unit charge placed at that point. Electric field strength is a vector quantity whose direction is that of the force on the unit charge. At a point distant r from a point charge, Q, the electric field strength in a radial direction from the point charge will be

$$E = \frac{Q}{4\pi\epsilon r^2} \quad \text{newtons/coulomb} \qquad . \quad . \ (10.2)$$

The electric field strength in an insulating material is called *electric stress.*

10.3. Electric Potential, V

The *electric potential*, V, at a point in an electric field is defined as the work done in joules in bringing a unit positive charge from infinity (or a point where there is no force on the charge) up to the point. Potential is measured in joules per coulomb, or volts, and since it is a measure of work it is a scalar quantity. If a unit charge is moved a small distance δa in an electric field where the component of electric field strength E in the direction of a is E_a, the work done is $E_a\,\delta a$ joules, and hence the potential at a point r is

$$V_r = -\int_{\infty}^{r} E_a \,\mathrm{d}a \qquad . \quad . \ (10.3)$$

The negative sign indicates that the potential increases positively as the unit charge is moved *against* the direction of the electric field strength.

Now, the change in potential when a unit charge is moved a small distance δa against a field strength E_a is $\delta V = -E_a\,\delta a$. Hence

$$E_a = -\left.\frac{\delta V}{\delta a}\right|_{\delta a \to 0} = -\frac{\mathrm{d}V}{\mathrm{d}a} \qquad . \quad . \ (10.4)$$

This means that the component of field strength in a given direction is the rate of change of potential in that direction, and hence the actual field strength at a point is

$$E = -\frac{\mathrm{d}V}{\mathrm{d}s} \qquad . \quad . \quad . \ (10.5)$$

where the direction of E is the direction, ds, of the *greatest* rate of change of potential ta the point. The electric field strength is simply the potential gradient of the electric field. Eqn. (10.5) shows that another unit of E is the volt per metre, and this is more commonly used than the newton per coulomb.

The potential difference, V_{pq}, between two points p and q in an electric field is defined as the work done in moving a unit positive charge from q to p. Mathematically,

$$V_{pq} = -\int_q^p E_a \, \mathrm{d}a \quad . \quad . \quad . \quad (10.6)$$

$$= -\int_q^p \left(-\frac{\mathrm{d}V}{\mathrm{d}a}\right) \mathrm{d}a \quad \text{(from eqn. (10.4))}$$

$$= \int_q^p \mathrm{d}V = V_p - V_q \quad . \quad . \quad . \quad (10.7)$$

This shows that the work done in moving a unit charge between two points in an electric field depends on the potentials at the two end points only and is independent of the actual path taken between them.

Note that, if the electric field is uniform (i.e. E is constant), the potential difference between two points separated by a distance x on the line of E is simply

$$V = -Ex \quad \text{volts}$$

10.4. Electric Flux, Ψ, and Electric Flux Density, D

Electric charge is associated with a space flux which is independent of the medium in which the charge is situated. The *electric flux*, Ψ, emanating from a charge, Q, is equal in magnitude to the charge and is also measured in coulombs. The *electric flux density*, D, is the flux per unit area. This is a directed or vector quantity whose direction at any point is the direction of the electric field strength, E, at the point. It is measured in coulombs per square metre.

For an isolated point charge, Q, we may assume that the electric flux emanates uniformly in all directions. Hence the flux density on the surface of a sphere of radius r surrounding the charge is

$$D_r = \frac{\Psi}{4\pi r^2} = \frac{Q}{4\pi r^2} \quad \text{coulombs/metre}^2 \quad . \quad . \quad (10.8)$$

From eqn. (10.2), the electric field strength at this radius is

$$E_r = \frac{Q}{4\pi\epsilon r^2} \quad \text{newtons/coulomb or volts/metre}$$

and hence

$$\frac{D_r}{E_r} = \epsilon \quad \frac{\text{coulombs/metre}^2}{\text{volts/metre}}, \text{ or coulombs/volt-metre} \quad . \quad (10.9)$$

By the principle of superposition this will give the relation between D and E for any field configuration.

The electric field may be mapped by drawing lines to represent the boundaries of tubes of electric flux, each tube representing the same amount of flux. The lines are usually called *lines of electric flux.*

10.5. Equipotential Lines and Surfaces

An equipotential line or surface is one joining all points in an electric field which have the same potential. From the definition of potential, no work will be done in moving an electric charge along an equipotential line or surface. Hence any force on the charge must be directed at right angles to the direction of the equipotential. This means that the direction of electric field strength and flux density must be at right angles to the direction of the equipotentials, and the electric field can be represented by drawing orthogonal flux and equipotential lines.

Since there can be no potential gradient along the surface of a perfect conductor, the electric flux must emerge at right angles to such a surface.

10.6. Gauss' Theorem

An important theorem due to Gauss states that the total outward normal flux from any closed surface is numerically equal to the sum of the enclosed charges. This theorem is almost self-evident but may be proved as follows.

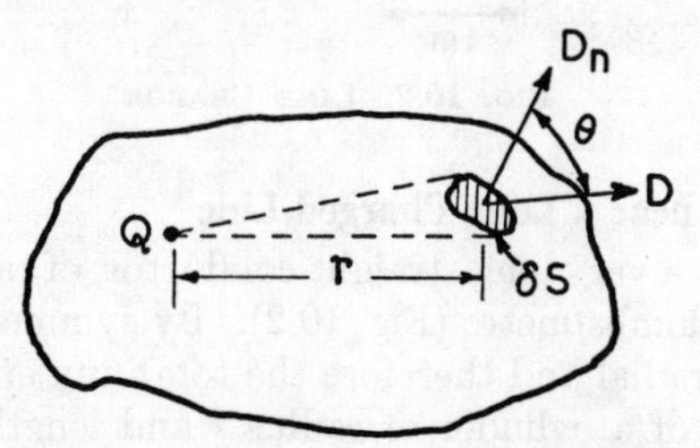

FIG. 10.1. ILLUSTRATING GAUSS' THEOREM

Consider a point charge, Q, at a point surrounded by a surface S (Fig. 10.1). For the element of area δS at a distance r from Q the flux density is

$$D = \frac{Q}{4\pi r^2}$$

The flux density normal to δS is

$$D_n = D \cos \theta$$

where θ is the angle between D and the normal to δS. The outward normal flux from δS is thus

$$\delta\Psi = D_n\,\delta S$$

and the total outward normal flux is

$$\Psi = \int_{surf} D_n\,\mathrm{d}S = \int_{surf} \frac{Q}{4\pi r^2}\cos\theta\,\mathrm{d}S = \frac{Q}{4\pi}\int_{surf}\frac{\mathrm{d}S\cos\theta}{r^2}$$

Now, $\mathrm{d}S\cos\theta/r^2$ is the solid angle subtended by $\mathrm{d}S$ at Q, and the integral of this over the surface is therefore 4π, so that the theorem is proved for the charge Q.

By the principle of superposition the result will apply for all charges enclosed by the surface S.

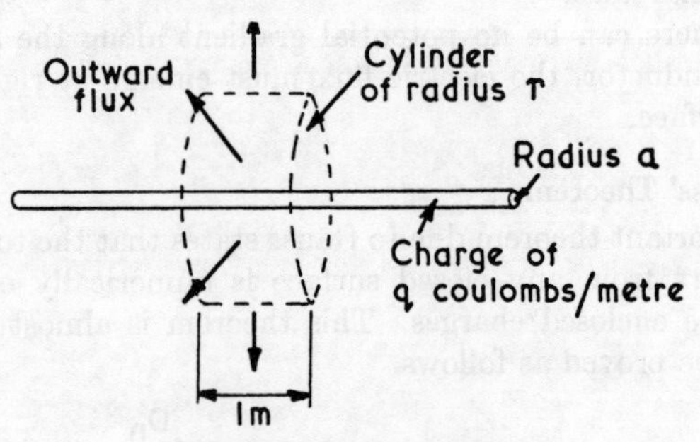

FIG. 10.2. LINE CHARGE

10.7. Potential near a Long Charged Line

Suppose that a very long straight conductor, of radius a, has a line charge of q coulombs/metre (Fig. 10.2). By symmetry, the outward electric flux is radial and therefore the total flux emerging from the curved surface of a cylinder of radius r and length 1 metre which is coaxial with the line is q coulombs. The flux density at this curved surface is hence

$$D_r = \frac{q}{2\pi r \times 1} \quad \text{coulombs/metre}^2$$

and the electric field strength is

$$E_r = \frac{q}{2\pi\epsilon r} \quad \text{volts/metre}$$

The potential at the surface is found by calculating the work done in moving a unit charge from a point of zero potential, at a distance z, up to r:

$$V_r = -\int_z^r \frac{q}{2\pi\epsilon r}\,dr = -\frac{q}{2\pi\epsilon}(\log_e r - \log_e z) \quad . \quad (10.10)$$

The term $(q/2\pi\epsilon)\log_e z$ is a constant for all values of r, and the term $(-q/2\pi\epsilon)\log_e r$ is often called the *effective potential* at r.

Example 10.1. Two long parallel wires, of radius small compared with their separation of 50 cm, carry line charges of $\pm 1\mu$C/m. Calculate the electric field strength and potential at a point P which is 50 cm from one line in a direction perpendicular to the plane containing the two lines (Fig. 10.3).

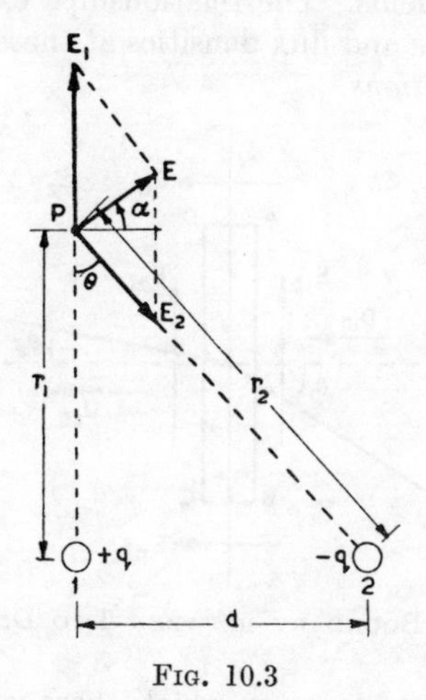

FIG. 10.3

Since potential is a scalar quantity, the potential at P will be the algebraic sum of the potentials due to lines 1 and 2. Thus from eqn. (10.10),

$$V_P = -\frac{q}{2\pi\epsilon}(\log_e r_1 - \log_e z) + \frac{q}{2\pi\epsilon}(\log_e r_2 - \log_e z)$$

$$= \frac{q}{2\pi\epsilon}\log_e \frac{r_2}{r_1}$$

$$= 1 \times 10^{-6} \times 18 \times 10^9 \log_e \sqrt{2} = \underline{\underline{6{,}240 \text{ V}}}$$

The electric field strength, E_1, due to the charge on line 1 is directed away from line 1. The field strength, E_2, due to line 2 is likewise $q/2\pi\epsilon r_2$ directed towards line 2; and the resultant stress is the vector sum of E_1 and E_2. Inserting numerical values,

$$E_1 = 36 \text{ kV/m} \quad \text{and} \quad E_2 = 25{\cdot}4 \text{ kV/m}$$

Resolving vertically,

$$E_v = E_1 - E_2 \cos\theta = 36 - 25{\cdot}4/\sqrt{2} = 18 \text{ kV/m (since } \theta = 45^\circ)$$

Resolving horizontally,

$$E_h = E_2 \sin\theta = 25{\cdot}4/\sqrt{2} = 18 \text{ kV/m}$$

Hence

$$E = \sqrt{(E_v^2 + E_h^2)} = \underline{\underline{25{\cdot}4 \text{ kV/m}}}$$

directed at an angle $\alpha = \tan^{-1}(E_v/E_h)$ $(= 45^\circ)$ to the horizontal.

10.8. Boundary Conditions

Boundaries between conductors and dielectrics, and between dielectrics of different permittivities, are the locations of discontinuities in electric fields. The relationships existing between the electric field strengths and flux densities at these discontinuities are called *boundary conditions.*

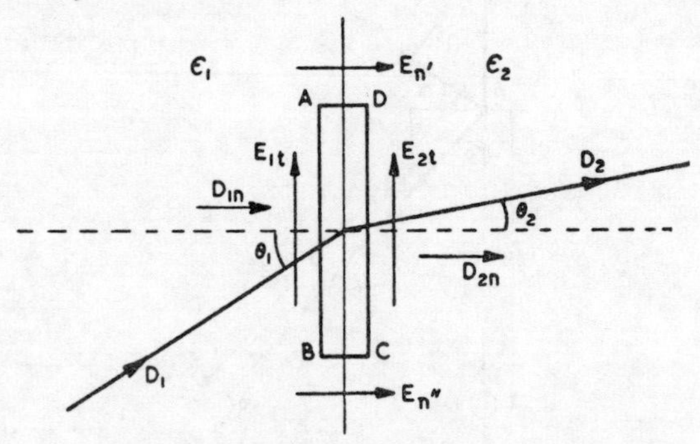

Fig. 10.4. Boundary between Two Dielectrics

Consider a plane interface (on which there is no surface charge) between two dielectrics of permittivities ϵ_1 and ϵ_2 as shown in Fig. 10.4. Suppose that electric flux approaches the boundary at an angle θ_1 to the normal, and leaves at an angle θ_2. Now imagine a coin-shaped element of volume enclosing the boundary. If the thickness of this element tends to zero, there can be no flux emerging from the curved surface, and hence, if the normal components of flux density are D_{1n} and D_{2n} and the area of the element is δS, then by Gauss' theorem,

$$D_{1n}\,\delta S = D_{2n}\,\delta S$$

or

$$D_{1n} = D_{2n} \quad . \quad . \quad . \quad . \quad (10\,11)$$

Hence one boundary condition is that the normal component of flux density is continuous across a surface.

Now let us move a unit charge along the path $ABCD$. Along BA the electric field strength is E_{1t}, along AD it is $E_{n'}$, along DC it is E_{2t}, and along BC it is $E_{n''}$ as shown. Hence

$$\text{Work done} = E_{1t} \,.\, AB + E_{n'} \,.\, BC - E_{2t} \,.\, CD - E_{n''} \,.\, DA$$
$$= E_{1t} \,.\, AB - E_{2t} \,.\, CD$$

if $BC = DA \rightarrow 0$.

This work done is the potential difference between the points at which we start and finish, which in this case are the same point A, so that the net work done must be zero. Hence

$$E_{1t} \,.\, AB = E_{2t} \,.\, DC$$

or

$$E_{1t} = E_{2t} \quad (\text{since } AB = DC) \qquad . \qquad . \quad (10.12)$$

This is a second boundary condition, i.e. that the tangential electric field strength is continuous across a boundary.

The relation between the angles θ_1 and θ_2 can be found as follows. The normal component of flux density in medium 1 is $D_{1n} = D_1 \cos \theta_1$ and in medium 2 is $D_{2n} = D_2 \cos \theta_2$. The tangential field strengths are $E_{1t} = D_1 \sin \theta_1/\epsilon_1$ and $E_{2t} = D_2 \sin \theta_2/\epsilon_2$. Hence

$$\frac{D_{1n}}{E_{1t}} = \frac{\epsilon_1}{\tan \theta_1} \quad \text{and} \quad \frac{D_{2n}}{E_{2t}} = \frac{\epsilon_2}{\tan \theta_2}$$

Since $D_{1n} = D_{2n}$ and $E_{1t} = E_{2t}$,

$$\frac{\tan \theta_1}{\epsilon_1} = \frac{\tan \theta_2}{\epsilon_2}$$

or

$$\frac{\tan \theta_1}{\tan \theta_2} = \frac{\epsilon_1}{\epsilon_2} \qquad . \qquad . \qquad . \quad (10.13)$$

This is the law of electric flux refraction at a boundary. Obviously if $\epsilon_1 > \epsilon_2$, then $\theta_1 > \theta_2$.

Note that at the boundary between a dielectric (medium 1) and a perfect conductor in which there can be no electric field strength (medium 2), Fig. 10.5, we have in the conductor $E_{2t} = 0$, and hence in the dielectric $E_{1t} = 0$ and $D_{1t} = 0$. The flux in the dielectric is therefore normal to the surface. The conductor must have a surface charge density, σ coulombs per square metre,

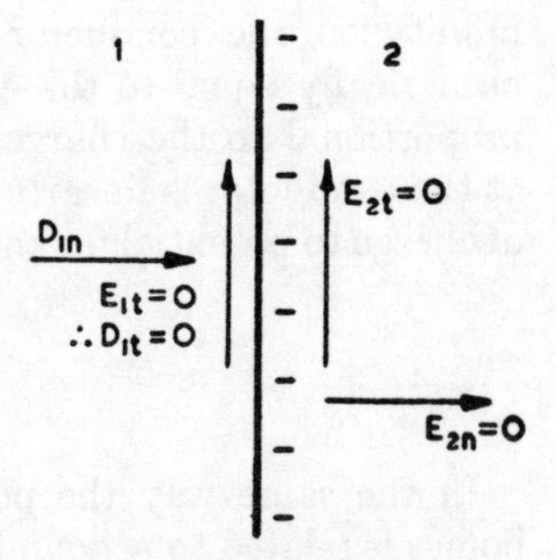

Fig. 10.5.
Dielectric–Conductor Boundary Conditions

which is equal to the normal flux density in the dielectric, since there is no electric field strength and hence no flux density in the conductor.

10.9. Electric Field due to an Electron Beam

A uniform beam of electrons is equivalent to a charged conductor, and we may calculate the electric field strength due to this beam. If the beam density is such that at any instant there is a charge of q coulombs per metre, and the electrons are moving at a velocity u metres per second, the equivalent current is

$$I = qu \quad \text{amperes} \qquad (10.14)$$

Outside the electron beam the electric field strength will be the same as that due to a line charge $q = I/u$. At a radius r greater than the beam radius, a,

$$E_{ext} = \frac{I}{2\pi\epsilon r u} \qquad (10.15)$$

directed radially.

Inside the beam there will also be an electric field strength given by

$$E_{int} = \frac{I(r^2/a^2)}{2\pi\epsilon r u} \qquad \text{where } r < a$$

$$= \frac{Ir}{2\pi\epsilon a^2 u} \quad \text{volts/metre} \qquad (10.16)$$

10.10. Capacitance

The potential at the surface of a charged conductor has been defined as the work done in bringing a unit positive charge from infinity to the conductor surface. The force on this charge is numerically equal to the electric field strength and this is directly proportional to the charge on the conductor. Hence the potential at the conductor is linearly related to its charge. The constant ratio of charge to potential for an isolated body is called *self-capacitance*, C:

$$C = \frac{Q}{V} \qquad (10.17)$$

In the same way the potential difference between two charged bodies is related to a bound charge by eqn. (10.17), where C is now the capacitance between the bodies. With Q in coulombs and V in volts, C is measured in farads.

Rewriting eqn. (10.17) as $Q = CV$ and differentiating with respect to time,

$$\frac{dq}{dt} = i = C\frac{dv}{dt} \quad . \quad . \quad . \quad (10.18)$$

For two parallel plates, of electrode area A square metres and separation t metres, separated by a dielectric of permittivity ϵ, let the bound charge be Q coulombs. This arrangement is usually called a *capacitor*. Neglecting end effects, the electric flux between the plates is $\Psi = Q$ coulombs, and the flux density is $D = Q/A$ coulombs per square metre. Hence the electric stress is $E = Q/\epsilon A$ volts per metre. Assuming that A is large compared with t, the electric stress will be uniform across the dielectric, and the p.d. between the plates will be simply Et,

i.e.
$$V = Et = \frac{Qt}{\epsilon A}$$

Hence
$$C = \frac{Q}{V} = \frac{\epsilon A}{t} \text{ farads} \quad . \quad . \quad . \quad (10.19)$$

Note that the unit of C is the coulomb per volt, or *farad*, and hence from eqn. (10.9) the unit of ϵ is the *farad per metre.*

Example 10.2. A capacitor has an electrode area of 1·5 m² and there is a composite dielectric consisting of 1 mm thickness of material of relative permittivity 10, and 0·5 mm of material of relative permittivity 1. Find the capacitance and electric stress in each part of the dielectric when the applied voltage is 200 V (Fig. 10.6).

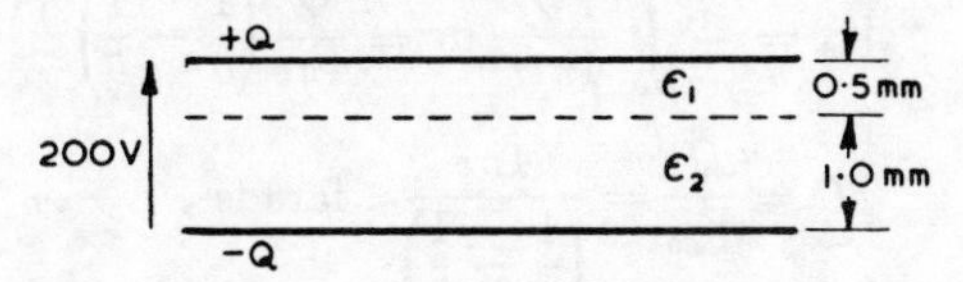

FIG. 10.6. PARALLEL-PLATE CAPACITOR WITH COMPOSITE DIELECTRIC

The flux density between the plates is $D = Q/A$.

Hence the electric stress in the 0·5 mm dielectric is $E_1 = Q/\epsilon_1 A$, and in the 1-mm dielectric, $E_2 = Q/\epsilon_2 A$.

The potential difference between the electrodes is

$$V = E_1 t_1 + E_2 t_2 = \frac{Q}{A}\left(\frac{t_1}{\epsilon_1} + \frac{t_2}{\epsilon_2}\right)$$

and the capacitance is

$$C = \frac{Q}{V} = \frac{A}{\dfrac{t_1}{\epsilon_1} + \dfrac{t_2}{\epsilon_2}} = \frac{\varepsilon_0 A}{\dfrac{t_1}{\epsilon_{r1}} + \dfrac{t_2}{\epsilon_{r2}}}$$

Numerically,

$$C = \frac{1{\cdot}5}{36\pi \times 10^9 [(0{\cdot}5 \times 10^{-3}) + (1 \times 10^{-3}/10)]} = \underline{\underline{0{\cdot}0221\ \mu\text{F}}}$$

For $V = 200$, $Q = CV = 0{\cdot}0221 \times 10^{-6} \times 200$, and hence

$$E_1 = \frac{4{\cdot}41 \times 10^{-6}}{\epsilon_0 \times 1{\cdot}5} = \underline{\underline{333\ \text{kV/m}}}$$

and $$E_2 = \frac{4{\cdot}41 \times 10^{-6}}{\epsilon_{r2}\epsilon_0 \times 1{\cdot}5} = \underline{\underline{33{\cdot}3\ \text{kV/m}}}$$

The example illustrates the fact that the greater electric stress occurs in the medium of lower permittivity.

10.11. Capacitance of Concentric Spheres

Although of minimal practical importance, the calculation of the capacitance between two concentric spheres of radii a and b (with $b > a$) illustrates a general method of approach. Suppose the bound charge is Q. There will be no electric field within the inner sphere, or outside the outer sphere. The electric flux between the spheres will be $\Psi = Q$, and hence at radius r $(a < r < b)$

$$D_r = \frac{Q}{4\pi r^2} \quad \text{and} \quad E_r = \frac{Q}{4\pi\epsilon r^2}$$

directed radially, where ϵ is the permittivity of the dielectric.

By eqn. (10.6),

$$V_{ab} = -\int_b^a \frac{Q}{4\pi\epsilon r^2}\,\text{d}r = \frac{Q}{4\pi\epsilon}\left(\frac{1}{a} - \frac{1}{b}\right) \quad . \quad (10.20)$$

Therefore, $$C = \frac{Q}{V_{ab}} = \frac{4\pi\epsilon}{\left(\frac{1}{a} - \frac{1}{b}\right)} \text{ farads} \quad . \quad . \quad . \quad (10.21)$$

The electric stress can be expressed in terms of the voltage between the spheres, since, from eqn. (10.20),

$$Q = \frac{4\pi\epsilon V_{ab}}{\left(\frac{1}{a} - \frac{1}{b}\right)}$$

and thus $$E_r = \frac{V_{ab}}{r^2\left(\frac{1}{a} - \frac{1}{b}\right)} \text{ volts/metre}$$

At the inner sphere, $r = a$ and the electric stress has a maximum value of

$$E_a = \frac{V_{ab}}{a^2\left(\frac{1}{a} - \frac{1}{b}\right)} \text{ volts/metre} \quad . \quad . \quad (10.22)$$

10.12. Capacitance of an Isolated Sphere

This may be deduced from the previous case by letting the radius of the outer sphere tend to infinity. Then, from eqn. (10.20),

$$V_a = \frac{Q}{4\pi\epsilon a} \quad \text{and} \quad C = 4\pi\epsilon a \quad . \quad . \quad (10.23)$$

10.13. Capacitance of a Concentric Line

Concentric lines are much used particularly at radio frequencies and above, since the outer conductor forms a screen which prevents radiation from the line, and screens the line from external fields.

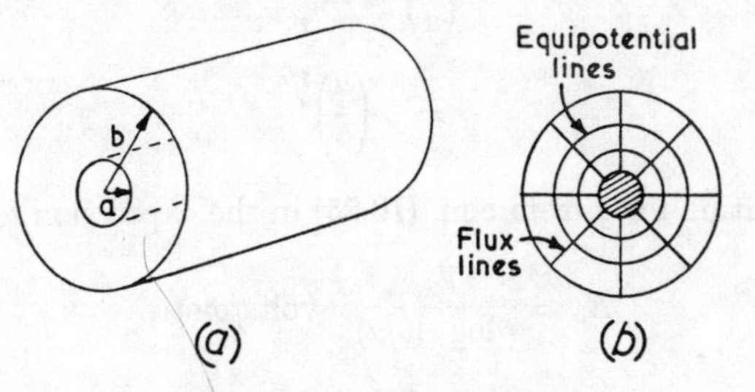

Fig. 10.7. Concentric Line

Consider the line shown in Fig. 10.7 (*a*) with core radius a and external radius b. For a line charge of q coulombs per metre the electric flux between the cylinders is q coulombs per metre run of line. Hence the flux density on the surface of a cylinder radius of r $(a < r < b)$, and 1 metre long is $D_r = q/(2\pi r \times 1)$ directed radially. Therefore

$$E_r = \frac{q}{2\pi\epsilon r} \quad \text{and} \quad V_{ab} = -\int_b^a \frac{q}{2\pi\epsilon r}\,dr = \frac{q}{2\pi\epsilon}\log_e\frac{b}{a} \quad (10.24)$$

This gives

$$C = \frac{q}{V_{ab}} = \frac{2\pi\epsilon}{\log_e\frac{b}{a}} \text{ farads/metre run} \quad . \quad (10.25)$$

The electric field pattern in the dielectric is shown in Fig. 10.7 (*b*). The surfaces of the inner conductor and of the sheath are equipotentials, and the electric flux lines are radial. Further equipotentials may be drawn at intermediate radii, since at radius r,

$$V_{rb} = -\int_b^r \frac{q}{2\pi\epsilon r}\,\mathrm{d}r$$

$$= -\int_b^r \frac{V_{ab}}{r \log_e (b/a)}\,\mathrm{d}r \qquad \text{(from eqn. (10.25))}$$

$$= V_{ab}\frac{\log_e (b/r)}{\log_e (b/a)} \quad . \quad . \quad . \quad (10.26)$$

Thus for $V_{rb} = V_{ab}/n$,

$$\frac{1}{n}\log_e \frac{b}{a} = \log_e \frac{b}{r}$$

Therefore
$$\left(\frac{b}{a}\right)^{1/n} = \frac{b}{r}$$

or
$$r = b\left(\frac{a}{b}\right)^{1/n} \quad . \quad . \quad . \quad (10.27)$$

Substituting for q from eqn. (10.25) in the expression for E_r,

$$E_r = \frac{V_{ab}}{r \log_e (b/a)} \quad \text{volts/metre}$$

This has a maximum value of $V_{ab}/a \log_e (b/a)$ at the surface of the core (where $r = a$) and a minimum value at the sheath. For a given voltage between core and sheath and a fixed sheath radius, the stress at the core will have a minimum value as a varies when $a \log_e (b/a)$ is a maximum,

i.e.
$$\frac{\mathrm{d}}{\mathrm{d}a}\, a \log_e \frac{b}{a} = 0 = \log_e \frac{b}{a} - a\frac{a}{b}\frac{b}{a^2}$$

Hence
$$\log_e \frac{b}{a} = 1 \quad \text{or} \quad \frac{b}{a} = \mathrm{e} \quad . \quad . \quad . \quad (10.28)$$

10.14. Capacitance between Parallel Wires

The capacitance between parallel round wires may be readily evaluated, provided that the wire radius, a, is small compared with the spacing, d, between them. The potential at the surface of each wire due to its own charge and to the charge on the other conductor is found. The relation between the line charge and the p.d. between the lines then gives the capacitance. Thus, for the system shown in Fig. 10.8 (*a*), let the line charges be $+q$ and $-q$ coulombs per metre.

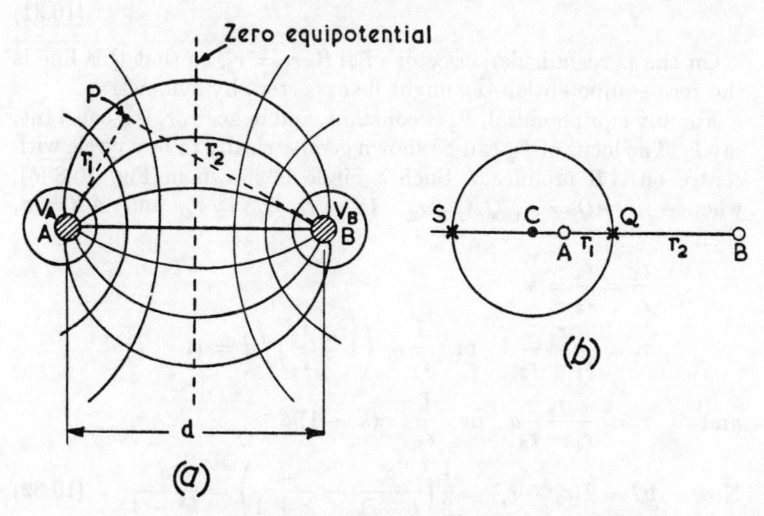

Fig. 10.8. Electric Field Between Parallel Wires

$$\text{Potential at surface of } A \text{ due to its own charge} = -\int_z^a \frac{q}{2\pi\epsilon r}\,dr$$

$$\text{Potential at surface of } A \text{ due to charge on } B = -\int_z^d \frac{-q}{2\pi\epsilon r}\,dr$$

where z is the distance to a zero-potential point and $a \ll d$.

Hence, $$V_A = \frac{q}{2\pi\epsilon}\log_e\frac{z}{a} + \frac{q}{2\pi\epsilon}\log_e\frac{d}{z} = \frac{q}{2\pi\epsilon}\log_e\frac{d}{a}$$

Similarly, $$V_B = \frac{-q}{2\pi\epsilon}\log_e\frac{d}{a}$$

so that $$V_{AB} = V_A - V_B = \frac{q}{\pi\epsilon}\log_e\frac{d}{a} \quad . \quad . \quad (10.29)$$

and $$C = \frac{q}{V_{AB}} = \frac{\pi\epsilon}{\log_e\dfrac{d}{a}} \quad \text{farads/metre} \quad . \quad . \quad (10.30)$$

At any point P which is distant r_1 from A and r_2 from B, the potential is

$$V_p = -\int_z^{r_1} \frac{q}{2\pi\epsilon r}\,\mathrm{d}r - \int_z^{r_2} \frac{-q}{2\pi\epsilon r}\,\mathrm{d}r = \frac{q}{2\pi\epsilon}\log_e \frac{r_2}{r_1} \quad . \quad . \quad . \quad (10.31)$$

On the perpendicular bisector of AB, $r_1 = r_2$, so that this line is the zero equipotential—as might be expected, by symmetry.

For any equipotential, V_p is constant, and hence r_2/r_1 is a constant, say k. The locus of V_p can be shown geometrically to be a circle with centre on AB produced. Such a circle is shown in Fig. 10.8 (*b*), whence, if $AQ = r_1$, $QB = r_2$, $AS = r_3$, $BS = r_4$, and $AB = d$, then

$$\frac{r_2}{r_1} = \frac{r_4}{r_3} = k$$

$$r_1 = \frac{r_1}{r_1 + r_2} d \quad \text{or} \quad \frac{1}{r_1} = \left(1 + \frac{r_2}{r_1}\right)\bigg/ d = (1 + k)/d$$

and $$r_3 = \frac{r_3}{r_4 - r_3} d \quad \text{or} \quad \frac{1}{r_3} = (k - 1)/d$$

Now $$AC = \tfrac{1}{2}(r_3 - r_1) = \frac{1}{2}\left(\frac{d}{k-1} - \frac{d}{k+1}\right) = \frac{d}{k^2 - 1} \quad . \quad (10.32)$$

This locates the centres of the equipotential circles.

Since the flux lines are orthogonal to the equipotentials, these will be arcs of circles with centres on the perpendicular bisector of AB.

On the line AB, the electric field strength at any point Q is

$$E_Q = \frac{q}{2\pi\epsilon r_1} + \frac{q}{2\pi\epsilon r_2} = \frac{q}{2\pi\epsilon}\left(\frac{1}{r_1} + \frac{1}{r_2}\right)$$

$$= \frac{q}{2\pi\epsilon}\frac{d}{r_1 r_2} \quad \text{directed towards } B \quad . \quad . \quad (10.33)$$

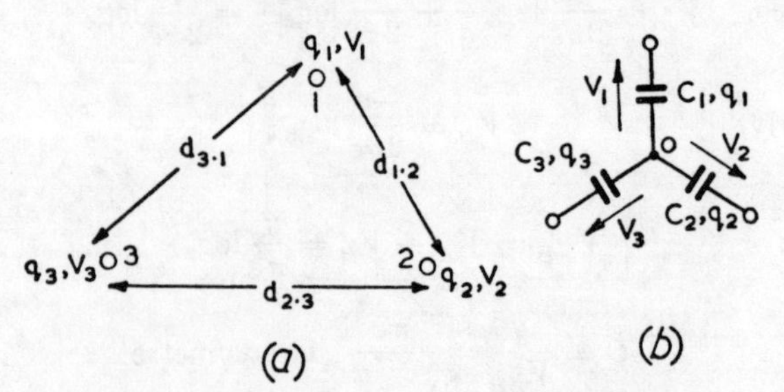

FIG. 10.9. CAPACITANCES OF A THREE-PHASE LINE

At each conductor surface this has a maximum value of

$$qd/2\pi\epsilon a(d - a) = V_{AB}d/[2a(d - a) \log_e (d/a)] \quad \text{volts per metre}$$

10.15. Capacitances of an Isolated Three-phase Line

In a 3-wire 3-phase system the instantaneous sum of the line currents and line charges is zero. For the system shown in Fig. 10.9 (*a*),

$$q_1 + q_2 + q_3 = 0$$

The potential at the surface of line 1 due to its own charge and to the charges on the other lines is

$$V_1 = \frac{q_1}{2\pi\epsilon} \log_e \frac{z}{a} + \frac{q_2}{2\pi\epsilon} \log_e \frac{z}{d_{12}} + \frac{q_3}{2\pi\epsilon} \log_e \frac{z}{d_{13}}$$

$$= \frac{q_1}{2\pi\epsilon} \log_e \frac{z}{a} + \frac{q_2}{2\pi\epsilon} \log_e \frac{z}{d_{12}} - \frac{q_1 + q_2}{2\pi\epsilon} \log_e \frac{z}{d_{13}}$$

$$= \frac{q_1}{2\pi\epsilon} \log_e \frac{d_{13}}{a} + \frac{q_2}{2\pi\epsilon} \log_e \frac{d_{13}}{d_{12}}$$

In the same way,

$$V_2 = \frac{q_2}{2\pi\epsilon} \log_e \frac{d_{23}}{a} + \frac{q_1}{2\pi\epsilon} \log_e \frac{d_{23}}{d_{12}}$$

Hence

$$V_1 - V_2 = \frac{q_1}{2\pi\epsilon} \log_e \frac{d_{13}d_{12}}{ad_{23}} - \frac{q_2}{2\pi\epsilon} \log_e \frac{d_{23}d_{12}}{ad_{13}} \quad . \quad (10.34)$$

Consider now the equivalent star of capacitors shown in Fig. 10.9 (*b*).

$$V_1 - V_2 = \frac{q_1}{C_1} - \frac{q_2}{C_2}$$

Comparing this equation with eqn. (10.34) we may write

$$C_1 = \frac{2\pi\epsilon}{\log_e \dfrac{d_{13}d_{12}}{ad_{23}}} \quad \text{farads/metre} \quad . \quad . \quad . \quad (10.35)$$

$$C_2 = \frac{2\pi\epsilon}{\log_e \dfrac{d_{23}d_{12}}{ad_{13}}} \quad \text{farads/metre} \quad . \quad . \quad . \quad (10.36)$$

And in a similar way,

$$C_3 = \frac{2\pi\epsilon}{\log_e \dfrac{d_{23}d_{13}}{ad_{12}}} \text{ farads/metre} \quad . \quad . \quad . \quad (10.37)$$

Example 10.3. Calculate the charging current in line 1 of a 3-phase 3-wire 33-kV 50-c/s system with conductor spacing $d_{12} = 8$ ft, $d_{23} = 6$ ft and $d_{31} = 9$ ft. The conductor radius is $\frac{1}{2}$ in. and the line is 30 miles long.

From eqn. (10.35),

$$C_1 = \frac{2\pi}{36\pi \times 10^9 \log_e \dfrac{8 \times 9 \times 12}{6 \times 0{\cdot}5}} = 9{\cdot}81 \text{ pF/m}$$

It should be noted that quite large variations in spacing or wire radius affect this result only slightly. Making the assumption that point O in Fig. 10.9(b) is approximately at zero potential,

$$I_1 = j\omega C_1 V_{1ph} l$$

where l is the line length in metres. Therefore

$$|I_1| = 2\pi \times 50 \times 9{\cdot}81 \times 10^{-12} \times (33 \times 10^3/\sqrt{3}) \times 30 \times 5{,}280 \times 0{\cdot}304$$

$$= \underline{\underline{2{\cdot}84 \text{ A}}}$$

10.16. Images in a Perfect Conductor

Since, by definition, there is no electric field strength along an equipotential surface, a conductor may be inserted along such a surface without affecting the external field. Hence, if a conducting plate is inserted along the zero equipotential of the twin-line system of Fig. 10.10 (a), the fields on either side will be unaffected, but there

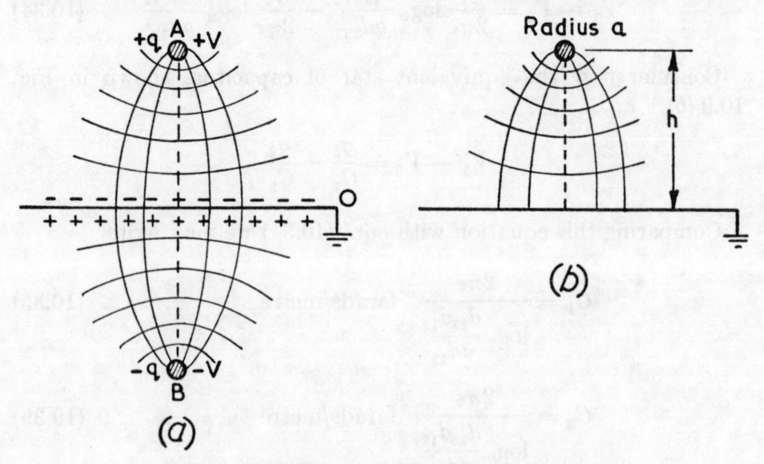

Fig. 10.10. Images in a Perfect Conductor

will be bound charges of $-q$ and $+q$ per metre on the conductor surfaces. The conductor screens the two halves of the field from one another, and if conductor B is removed, the field between conductor A and the earthed plate remains unaltered. Hence

$$V_{AO} = \tfrac{1}{2}V_{AB} = \frac{1}{2}\frac{q}{\pi\epsilon}\log_e\frac{2h}{a} \text{ (from eqn. (10.29))}$$

and

$$C = \frac{2\pi\epsilon}{\log_e \dfrac{2h}{a}} \text{ farads/metre.} \qquad (10.38)$$

where h is the height of the conductor above the earthed plate.

The capacitance of any system of conductors above an earthed plane can thus be found by determining the potentials due to the conductors and their negative images in the plane.

Example 10.4. A twin line has conductors of radius a at heights h_1 and h_2 above a plane conducting earth. The wire spacing is x. Calculate the capacitance of the system if $x \gg a$.

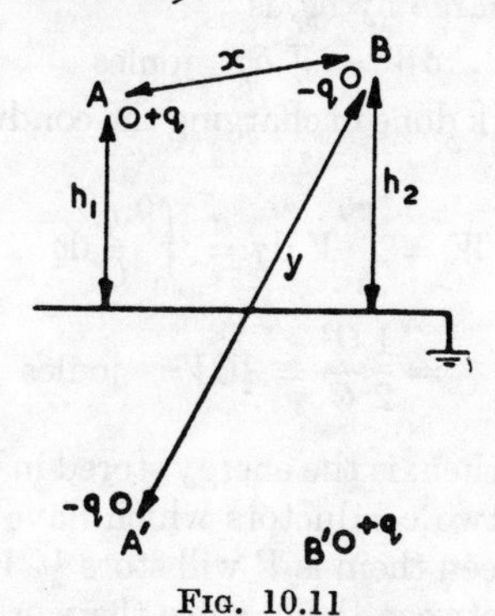

Fig. 10.11

The system with its image conductors is shown in Fig. 10.11. Suppose the line charge is q coulombs per metre, and $A'B = AB' = y$.

The potential at the surface of A due to its own charge and the charges on B, A' and B' is

$$V_A = \frac{q}{2\pi\epsilon}\log_e\frac{z}{a} - \frac{q}{2\pi\epsilon}\log_e\frac{z}{2h_1} - \frac{q}{2\pi\epsilon}\log_e\frac{z}{x} + \frac{q}{2\pi\epsilon}\log_e\frac{z}{y}$$

$$= \frac{q}{2\pi\epsilon}\log_e\frac{2h_1x}{ay}$$

where z is the distance to the zero-potential point.

Similarly,

$$V_B = \frac{-q}{2\pi\epsilon}\log_e\frac{2h_2x}{ay}$$

Hence

$$V_{AB} = V_A - V_B = \frac{q}{2\pi\epsilon}\log_e \frac{4h_1h_2x^2}{a^2y^2} = \frac{q}{\pi\epsilon}\log_e \frac{2x\sqrt{(h_1h_2)}}{ay}$$

and
$$C = \frac{q}{V_{AB}} = \frac{\pi\epsilon}{\log_e \dfrac{2x\sqrt{(h_1h_2)}}{ay}} \quad \text{farads/metre}$$

Note that the presence of the earth plane increases the capacitance slightly above the value for an isolated system.

From the geometry of the diagram,

$$y = \sqrt{[(h_2 + h_1)^2 + x^2 - (h_2 - h_1)^2]} = \sqrt{(x^2 + 4h_1h_2)}$$

10.17. Energy Stored in Electric Fields

Work must be done to increase the charge on a conductor. This work may be considered to be stored as energy in the field.

Consider an isolated conductor with an instantaneous charge of q coulombs and a surface potential of V volts. The self-capacitance of the conductor is $C = Q/V$ farads. The work done in bringing an increment of charge δq from a point of zero potential to the body, and so of increasing its charge by δq, is

$$\delta W = V\,\delta q \quad \text{joules}$$

Hence the total work done in charging the conductor to Q coulombs is

$$W = \int_0^Q V\,\mathrm{d}q = \int_0^Q \frac{q}{C}\,\mathrm{d}q \qquad (10.39)$$

$$= \frac{1}{2}\frac{Q^2}{C} = \tfrac{1}{2}CV^2 \quad \text{joules} \qquad (10.40)$$

since $Q = CV$. This, then, is the energy stored in the electric field.

In the same way, two conductors which have a bound charge of Q when the p.d. between them is V will store $\frac{1}{2}CV^2$ joules in the field between them, since the work done in moving a charge δq from one to the other is $V\,\delta q$, and hence the total work done is

$$W = \int_0^Q V\,\mathrm{d}q$$

as above.

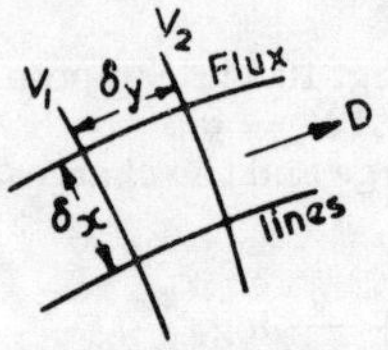

Fig. 10.12. Relating to Stored Energy Density

Now consider the small section of electric field shown in Fig. 10.12 between the equipotentials V_1 and V_2. For unit depth of field, the capacitance of the small section bounded by the two flux lines and the equipotentials is

$$C = \frac{\epsilon\,\delta x\,1}{\delta y}$$

and the stored energy is

$$W = \frac{1}{2}\frac{\epsilon\,\delta x}{\delta y}(V_1 - V_2)^2$$

$$= \frac{1}{2}\frac{\epsilon\,\delta x}{\delta y}(E\,\delta y)^2 = \tfrac{1}{2}\epsilon E^2\,\delta x\,\delta y \text{ joules}$$

where E is the electric field strength between the equipotentials.

The stored energy density is thus given by

$$w = \tfrac{1}{2}\epsilon E^2 \text{ joules/metre}^3 \quad . \quad . \quad . \quad (10.41)$$

$$= \frac{1}{2}\frac{D^2}{\epsilon} \quad . \quad . \quad . \quad . \quad . \quad (10.41a)$$

10.18. Potential and Capacitance Coefficients

Since potential is a scalar quantity directly proportional to charge, we may apply the principle of superposition to determine the potential at the surface of a conductor due to its own charge and to any number of related charges. Thus we may write that the potential, V_1, at body 1 due to its own charge Q_1 and to charges Q_2, Q_3, . . . on related conductors is

$$V_1 = p_{11}Q_1 + p_{12}Q_2 + p_{13}Q_3 + \ldots \quad . \quad (10.42)$$

where the p's are constants called *potential coefficients*. In general, p_{nm} is the potential at n per unit charge on m.

Similarly,

$$V_2 = p_{21}Q_1 + p_{22}Q_2 + p_{23}Q_3 + \ldots \quad . \quad (10.43)$$

and so on. Potential coefficients have the reciprocal property $p_{12} = p_{21}$ as will now be shown.

Suppose that body 1 is charged to Q_1 with all other conductors in the system uncharged. The energy required is given by eqns. (10.39) and (10.42) as

$$W_1 = \int_0^{Q_1} p_{11}q_1\,\mathrm{d}q = \tfrac{1}{2}p_{11}Q_1^{\,2}$$

Now let body 2 be charged to Q_2 with charge Q_1 on body 1 present. The energy required is

$$W_2 = \int_0^{Q_2} v_2\,\mathrm{d}q = \int_0^{Q_2}(p_{21}Q_1 + p_{22}q_2)\,\mathrm{d}q$$

$$= p_{21}Q_1Q_2 + \tfrac{1}{2}p_{22}Q_2^{\,2}$$

Hence

$$W_1 + W_2 = \tfrac{1}{2}p_{11}Q_1^2 + p_{21}Q_1Q_2 + \tfrac{1}{2}p_{22}Q_2^2 \quad . \quad (10.44)$$

If the procedure is reversed and body 2 is first charged to Q_2, then body 1 to Q_1, the energy required is

$$W_2' + W_1' = \tfrac{1}{2}p_{22}Q_2^2 + p_{12}Q_1Q_2 + \tfrac{1}{2}p_{11}Q_1^2 \quad . \quad (10.45)$$

The energies represented by eqns. (10.44) and (10.45) must obviously be equal, and hence

$$p_{21} = p_{12}$$

Equations of the form (10.42) and (10.43) can be solved for the charges $Q_1, Q_2, \ldots$ Thus

$$Q_1 = \frac{\begin{vmatrix} V_1 & p_{12} & p_{13} & . & . & . \\ V_2 & p_{22} & p_{23} & . & . & . \\ V_3 & p_{32} & . & . & . & . \\ . & & & & & \\ . & & & & & \\ . & & & & & \end{vmatrix}}{\begin{vmatrix} p_{11} & p_{12} & p_{13} & . & . & . \\ p_{21} & p_{22} & p_{23} & . & . & . \\ p_{31} & p_{32} & . & . & . & . \\ . & & & & & \\ . & & & & & \\ . & & & & & \end{vmatrix}} \text{ etc.}$$

These determinants may be reduced to the form

$$\left.\begin{aligned} Q_1 &= C_{11}V_1 + C_{12}(V_1 - V_2) + C_{13}(V_1 - V_3) + \ldots \\ Q_2 &= C_{21}(V_2 - V_1) + C_{22}V_2 + C_{23}(V_2 - V_3) + \ldots \\ Q_3 &= C_{31}(V_3 - V_1) + \ldots \\ &. \\ &. \\ &. \end{aligned}\right\} \quad (10.46)$$

The coefficients C_{mn} are functions of the coefficients p_{mn} and are called the *capacitance coefficients*. C_{mn} is the charge on m per unit voltage difference between m and n. C_{mm} is the charge on m per unit voltage of m with respect to the zero reference of potential. Obviously $C_{mn} = C_{nm}$.

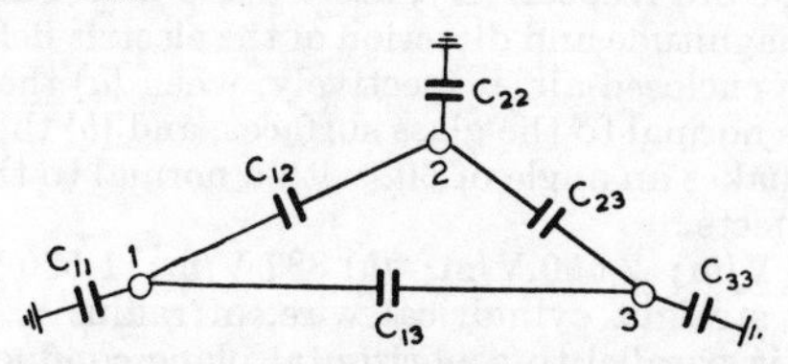

FIG. 10.13. CAPACITANCE COEFFICIENTS

Eqns. (10.46) indicate that any system of charged conductors can be represented by a system of capacitances (C_{mn}) between each pair of conductors, and capacitances C_{mm} from each conductor to earth (self-capacitance) as shown in Fig. 10.13.

Bibliography

Carter, *The Electromagnetic Field in its Engineering Applications* (Longmans).

Jones, *An Introduction to Advanced Electrical Engineering* (English Universities Press).

Shepherd, Morton and Spence, *Higher Electrical Engineering* (Pitman).

Problems

10.1. A capacitor consists of two concentric spheres of radii r_1 and r_2, respectively, where $r_1 < r_2$. The dielectric medium between the spheres has an absolute permittivity ϵ and its electric strength is E_0. Derive an expression for the greatest potential difference between the two spheres, so that the field strength nowhere exceeds the critical value E_0. Assuming r_2 to be constant, for what value of r_1 will this potential difference be a maximum, and what will then be the capacitance?

(*L.U. Part II Theory*, 1958)

(*Ans.* $r_1 = \frac{1}{2}r_2$; $8\pi\epsilon_r$ F.)

10.2. If a wire, of diameter 6 mm and length 2 km, is maintained at a uniform height of 10 m above marshy ground, calculate from first principles the capacitance between the wire and earth in microfarads. Assume the permittivity of free space to be 8·85 pF/m.

(*Ans.* 0·01265 μF.) (*L.U. Part II Theory*, 1960)

10.3. The charging current of a 10-mile single-core cable was observed to be 10·7 A when a voltage of 11 kV at 50 c/s was applied between the core and the sheath. The diameter of the core was 2 cm, and the inner diameter of the sheath was 4 cm. The insulation consisted of two equal thicknesses of dielectric, the permittivity of the layer adjacent to the core being twice that of the layer adjacent to the sheath. Calculate the permittivity of each dielectric, deriving the formula used.

(*Ans.* 3·4; 1·7.) (*I.E.E. Meas.*, 1959)

10.4. Two parallel sheets of plate glass mounted vertically are separated by a uniform air-gap between their inner surfaces. The sheets suitably sealed round the outer edges, are immersed in oil. A uniform electric field in a horizontal direction exists in the oil. The strength of the electric field in the oil is 1,000 V/m and the relative permittivities of the glass and the oil, respectively, are 6·0 and 2·5. Calculate from first principles the magnitude and direction of the electric field strength in the glass and in the enclosed air, respectively, when (*a*) the direction of the field in the oil is normal to the glass surfaces, and (*b*) the direction of the field in the oil makes an angle of 60° with a normal to the glass surfaces. Neglect edge effects. (*A.E.E.*, 1959)

(*Ans.* (*a*) 416 V/m; 2,500 V/m; (*b*) 887 V/m; 1,520 V/m.)

10.5. A long straight cylindrical wire, of radius r, in a medium of permittivity ϵ, is parallel to a horizontal plane conducting sheet. The axis of the wire is at a distance h above the sheet. Derive an expression for the capacitance per unit length between the wire and the plane, stating any assumptions made.

The potential difference between the wire and the sheet is 5 kV, with $r = 0{\cdot}2$ cm and $h = 10$ cm. Calculate the electric stress in the medium at the upper surface of the sheet: (*a*) vertically below the wire; (*b*) at a point 20 cm from the axis of the wire. (*A.E.E.*, 1961)

(*Ans.* 21·8 kV/m; 5·44 kV/m.)

10.6. A transmission line consists of a pair of long, parallel conductors of radius r, with a spacing D between centres in a medium of permittivity ϵ. Show that when D is large compared with r the capacitance of the pair of conductors per unit length is given approximately by

$$C = \frac{\pi\epsilon}{\log_e \dfrac{D}{r}}$$

The conductors A and B of such a line are 1 cm in diameter and spaced 2 m between centres. A point P alongside the conductors is 1 m from the centre of conductor A and $\sqrt{3}$ m from the centre of conductor B. If, at a given instant, the p.d. between A and B is 1,000 V, A having positive polarity with respect to B, calculate the magnitude and direction of the electric field strength at P, and indicate the direction clearly on a diagram. (*A.E.E.*, 1958)

(*Ans.* 96·4 V/m.)

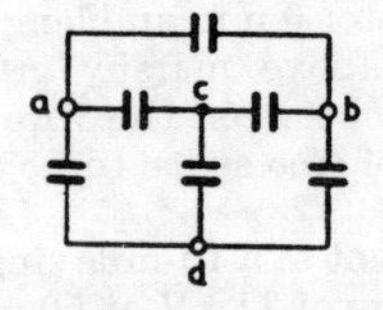

FIG. 10.14

10.7. What is the physical significance of capacitance coefficients in the case of more than two neighbouring insulated conductors with various potential differences between them?

All the capacitance values in the circuit shown in Fig. 10.14, are the same. Show that the net capacitance between terminals a and b is the same as that between terminals c and d.

If a voltage V is applied between terminals a and b, find the charge and stored energy of each capacitor, and hence, by addition, the total energy. Check the result by calculating the energy in terms of the voltage V and the net capacitance between a and b.

(*Ans.* CV^2.) (*L.U. Part III Theory*, 1959)

10.8. An infinitely straight cylindrical wire of small diameter carrying an electric charge of q per unit length is suspended in free space. Derive, using the inverse square law, an expression for the electric field strength at a point at a distance X from the wire which is large compared with the diameter of the wire. Determine the charge per unit length of the wire if the difference of potential between two points situated 1 cm and 10 cm from the wire is 138 V. (*I.E.E. Meas.*, 1956)

(*Ans.* $3{\cdot}33 \times 10^{-9}$ C/m.)

10.9. Show that the outward electric flux from any closed surface is equal to the sum of the charges within the surface. Utilize this theorem to show that, if the component of electric field strength parallel to a given straight line in a charge-free region is independent of the distance from the line, the component perpendicular to this line is proportional to the distance from it.

If the field strength parallel to the line is $10/x^3$ volts per metre, where x is the distance along the line from a point in it, what is the field strength perpendicular to the line at a distance of 1 m from it when $x = 2$ m? (*I.E.E. Meas.*, 1959)

(*Ans.* 0·938 V/m.)

10.10. A capacitor is formed of two concentric spheres. Derive an expression for the capacitance in terms of the radii of the spheres and the permittivity of the medium between them. If the radii are 50 and 10 cm, and there is a steady p.d. of 100 kV between the spheres, calculate the maximum electric stress.

Assuming that the radius of the inner sphere can be varied, while the outer radius and p.d. are fixed, find the smallest value of maximum stress which can be achieved and the radius of the inner sphere for which it occurs. (*A.E.E.*, 1962)

(*Ans.* $1{\cdot}25 \times 10^6$ V/m; 25 cm; 8×10^5 V/m.)

10.11. Derive an expression for the capacitance per unit length between long concentric cylinders.

The working electrode of a standard high-voltage capacitor is a short cylinder mounted concentrically within a longer cylinder. Inner guard-electrodes eliminate end effects from the working electrode. The standard has a capacitance of 100 pF, and the length of the working electrode is 50 cm. When a sinusoidal voltage of 50 kV (r.m.s.) is applied to the capacitor the peak electric stress at the surface of the inner electrode is 15 kV/cm. Calculate the radius of the working electrode, and the inside radius of the outer cylinder. (*A.E.E.*, 1962)

(*Ans.* 17 cm; 22·4 cm.)

CHAPTER 11

Magnetic Fields

MAGNETIC fields are associated with magnets and with current-carrying conductors. Unlike electric charges of opposite polarity, which can exist in isolation, magnetic poles can only exist in pairs. A magnet whose poles are assumed to be concentrated at the ends constitutes a *magnetic dipole*. It is found experimentally that the magnetic effects of such a dipole are equivalent to those produced by a current-carrying loop, and this equivalence forms a possible basis for magnetic field calculations. Before the introduction of the M.K.S. system of units, the fictitious unit magnetic pole was normally used as a starting-point for electromagnetic calculations. This gave the electromagnetic C.G.S. system of units. In the rationalized M.K.S.A. system, the starting-point for defining electrical units is the expression for current in terms of the force between current-carrying conductors. In this system the absolute units, i.e. those obtained from theoretical definitions, are identical with the accepted practical units.

The presence of a magnetic field at any point is demonstrated by placing a compass needle at the point. The direction of the magnetic field determines the orientation of the needle, and this direction is taken as the direction in which the N-pole of the compass points. We can also show the presence of a magnetic field by observing the force on a current-carrying element.

11.1. Magnetic Flux Density and Magnetic Flux

Magnetic dipoles and current-carrying conductors have a *magnetic space flux*, Φ, associated with them, which is independent of the medium in which they are situated. A unit pole is assumed to give rise to unit space flux. The magnetic flux per unit area is called the *magnetic flux density*, B. This is a directed or vector quantity. Flux is measured in webers, and flux density in webers per square metre.

The magnetic condition at a point in space, characterized by a magnetic flux density, is given by the force on a current element $I\,\delta l$ placed at right angles to the direction of the field (i.e. at right angles to the direction in which the N-pole of a compass points)—

$$F = BI\,\delta l \quad \text{newtons} \qquad . \quad . \quad . \quad (11.1)$$

If the element makes an angle θ (Fig. 11.1 (*a*)) with the direction of the field, i.e. with the direction of B, the force is

$$F = BI\,\delta l \sin\theta \quad \text{newtons} \qquad . \qquad . \quad (11.2)$$

The direction of the force is at right angles to the plane containing B and $I\,\delta l$, and is such that the current, field and force form a right-handed Cartesian system (Fig. 11.1 (*b*)). B is measured in webers per square metre when I is amperes and δl in metres.

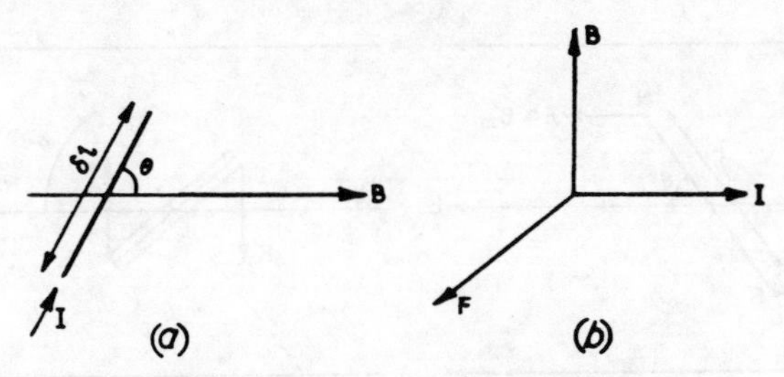

FIG. 11.1. FORCE ON A CURRENT-ELEMENT

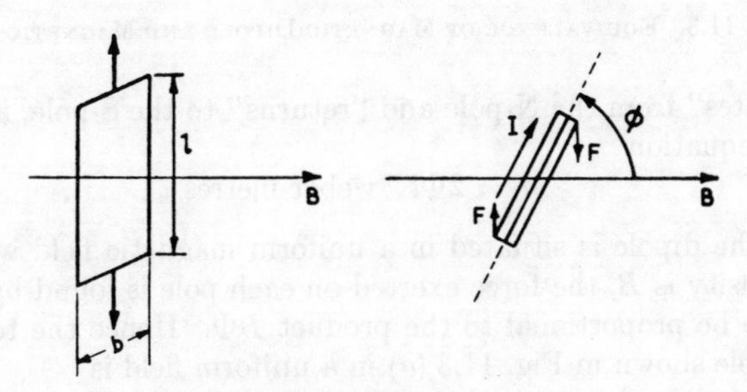

FIG. 11.2. TORQUE ON A COIL

Now consider a rectangular coil with N turns and dimensions $l \times b$, as shown in Fig. 11.2 (*a*), in which the plane of the coil makes an angle ϕ with the direction of the uniform field B. The coil is pivoted about the mid-point of sides b, and carries a current of I amperes. The forces on sides b act in opposite directions towards the centre of the coil and therefore cancel each other. The forces, F, on sides l give rise to a torque about the axis (Fig. 11.2 (*b*)) given by

$$\begin{aligned} T &= Fb\cos\phi \\ &= NBl\,Ib\cos\phi = BAN\,I\cos\phi \quad \text{newton-metres} \; . \quad (11.3) \end{aligned}$$

Magnetic fields may be illustrated by lines delineating the boundaries of tubes of magnetic flux, each tube containing the same amount of flux. The lines are called *lines of magnetic flux*, and their direction is chosen so that the directions of the force on a current-element, the current, and the flux are as illustrated in Fig. 11.1 (*b*).

11.2. The Current Loop and the Magnetic Dipole

The strength of a magnetic pole is measured by the magnetic flux, Φ, which emanates from it. The *magnetic moment*, M, of a magnetic dipole whose length is $2l$ (Fig. 11.3 (*a*)) and whose flux, Φ,

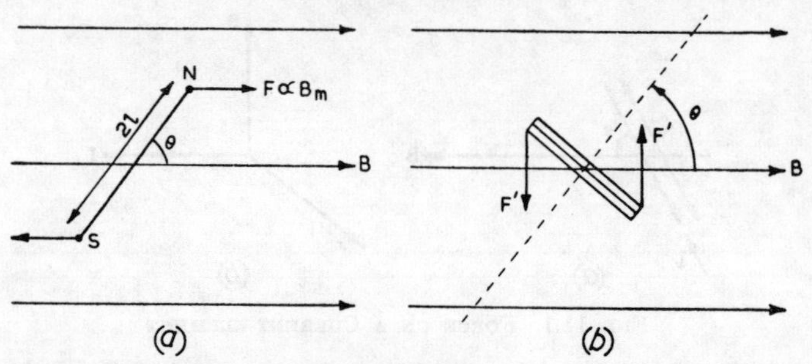

Fig. 11.3. Equivalence of Magnetic Dipole and Magnetic Loop

"emanates" from the N-pole and "returns" to the S-pole, is defined by the equation

$$M = 2\Phi l \quad \text{weber-metres} \qquad (11.4)$$

When the dipole is situated in a uniform magnetic field where the flux density is B, the force exerted on each pole is found by experiment to be proportional to the product $B\Phi$. Hence the torque on the dipole shown in Fig. 11.3 (*a*) in a uniform field is

$$T \propto B\Phi 2l \sin\theta \propto MB \sin\theta \qquad (11.5)$$

The torque on the equivalent single-turn coil, of area A, placed as shown in Fig. 11.3 (*b*) is, from eqn. (11.3),

$$T' = BA\,I \cos(90° - \theta) = BA\,I \sin\theta \qquad (11.6)$$

If these two torques are to be the same, then

$$AI = \text{constant} \times M = \frac{M}{\mu} \quad \text{(say)} \qquad (11.7)$$

where μ is a magnetic constant, known as *permeability*. The torque on the dipole is thus

$$T = \frac{MB \sin\theta}{\mu}$$

so that the force on each pole is $B\Phi/\mu$ newtons. The force per unit pole is B/μ, and this is called the *magnetizing force*, H:

$$B = \mu H \qquad . \quad . \quad . \quad . \quad (11.8)$$

For a vacuum the permeability is given the symbol μ_0. For any magnetic medium, $\mu = \mu_r\mu_0$, where μ_r is the *relative permeability* of the medium.

11.3. Force between Magnetic Poles

If a current-element is placed near a magnetic pole, the force on it varies with the strength of the pole (i.e. with the flux emanating from it) and inversely as the square of the distance between the pole and the current-element.

If a dipole is infinitely long and thin, the flux density on the surface of a sphere of radius r round either pole, of strength Φ_1, is

$$B_r = \frac{\Phi_1}{4\pi r^2} \qquad . \quad . \quad . \quad (11.9)$$

The force, F_m, on a second pole (of another ideal dipole) of strength Φ_2 and at a distance r from the first one, in a non-magnetic medium ($\mu = \mu_0$) will be the product of Φ_2 and the magnetizing force, H_r, due to Φ_1 at r; i.e.

$$\begin{aligned} F_m &= H_r\Phi_2 \\ &= \frac{B_r}{\mu_0}\Phi_2 \\ &= \frac{\Phi_1\Phi_2}{4\pi\mu_0 r^2} \qquad . \quad . \quad . \quad . \quad (11.10) \end{aligned}$$

With Φ_1 and Φ_2 in webers and r in metres, F_m will be in newtons.

The force is directed along the line joining the poles, its direction being given by the rule that like poles repel, and unlike poles attract, one another.

11.4. Magnetic Potential

The *magnetic potential difference*, F, between two points in a magnetic field is analogous to the electric potential difference between two points in an electric field, and is defined as

$$F_{ab} = -\int_b^a H\,\mathrm{d}s \qquad . \quad . \quad . \quad (11.11)$$

Alternatively,

$$H = -\frac{\mathrm{d}F}{\mathrm{d}s} \qquad . \quad . \quad . \quad . \quad (11.12)$$

It is not possible, however, to define magnetic potential uniquely, as it is electric potential. This is because the magnetic fields associated with current-carrying conductors are solenoidal—the magnetic flux forms complete loops round the conductors. If the line integral of H ds is taken round one of these flow lines, the magnetic potential at the starting point (which is also the finishing point) has two values, 0 *and* $\oint H\,ds$. In other words, magnetic potential is a multi-valued function.

It is sometimes helpful to imagine magnetic potential as the work done in moving a unit pole against the magnetizing force of the field in which it is placed. Lines of equal magnetic potential can be drawn at right angles to the direction of magnetizing forces, and therefore to lines of flux.

The *magnetomotive force* (m.m.f.) associated with a magnetic circuit is the work done, or the potential energy acquired, when a unit N-pole is moved once completely round the circuit. It is analogous to e.m.f.

(*a*) *Magnetic Potential due to a Dipole.* In the case of a magnetic dipole, the external flux lines begin and end on the poles, and this case is analogous to that of two electric charges. A dipole of moment $M = 2\Phi l$ is shown in Fig. 11.4 (*a*). The dipole axis makes an angle

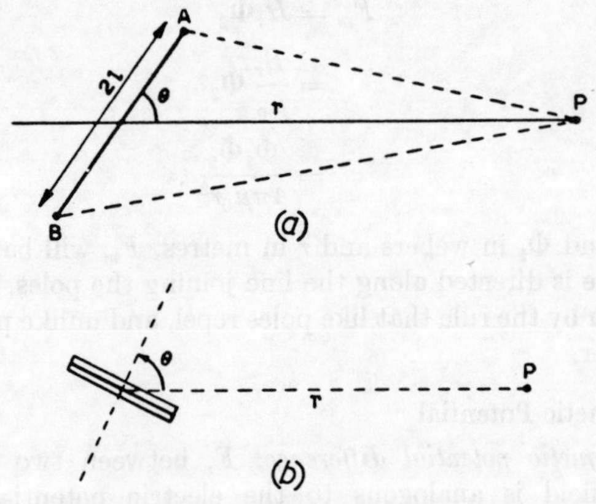

FIG. 11.4. MAGNETIC POTENTIAL DUE TO A DIPOLE, AND ITS EQUIVALENT CURRENT LOOP

θ with the direction to point P at which we wish to calculate the magnetic potential. If $r \gg l$, then to a good approximation the distances of the poles A and B from P are $(r - l\cos\theta)$ and $(r + l\cos\theta)$ respectively. For a pole strength Φ, the flux density at a distance s from A is $B_s = \Phi/4\pi s^2$ (from eqn. (11.10)), and hence $H_s = \Phi/4\pi\mu_0 s^2$. Eqn. (11.11) now gives the potential at P due to A as

$$F_{1P} = -\int_{\infty}^{r-l\cos\theta} H_s\,ds = -\int_{\infty}^{r-l\cos\theta} (\Phi/4\pi\mu_0 s^2)\,ds$$

$$= \frac{\Phi}{4\pi\mu_0(r - l\cos\theta)}$$

the potential at infinity being assumed zero. In the same way, the potential at P due to the opposite pole, B is

$$F_{2P} = \frac{-\Phi}{4\pi\mu_0(r + l\cos\theta)}$$

and the resultant potential (which is a scalar since it is a measure of work) is

$$F_P = F_{1P} + F_{2P} = \frac{\Phi}{4\pi\mu_0}\left(\frac{1}{(r - l\cos\theta)} - \frac{1}{(r + l\cos\theta)}\right)$$

$$= \frac{2\Phi l\cos\theta}{4\pi\mu_0(r^2 - l^2\cos^2\theta)} \simeq \frac{M\cos\theta}{4\pi\mu_0 r^2} \quad . \quad (11.13)$$

(*b*) *Magnetic Potential due to a Current Loop*. The current loop which is equivalent to the dipole of the previous section has $\mu_0 IA = M$ (eqn. (11.7)), and hence the magnetic potential at P due to the current loop is

$$F_P = \frac{IA\cos\theta}{4\pi r^2} = \frac{I\Omega}{4\pi} \quad . \quad . \quad . \quad (11.14)$$

where $\Omega = (A\cos\theta)/r^2$ is the solid angle subtended at P by the loop area A. This equation can obviously be extended to apply to a circuit of large dimensions.

The magnetic potential difference between two points 1 and 2 subtending solid angles Ω_1 and Ω_2 in the field of a current-carrying circuit is

$$F_{12} = \frac{I}{4\pi}(\Omega_1 - \Omega_2) \quad . \quad . \quad . \quad (11.15)$$

Now let us see what happens when we move round a closed path and arrive back at our starting-point. Two cases exist, as shown in Fig. 11.5. Firstly, if the closed path does not link the current circuit (path A), the change in Ω is zero right round the path, and the change in magnetic potential is zero. Mathematically, $-\oint H_s\,ds = 0$, this integral being simply $\Sigma H_s\,ds$ round a closed path. In the second case the path (B) links the electric circuit. At P suppose that the solid angle is zero. At Q, within the current loop, the solid angle is 2π (the solid angle subtended by a plane boundary at a point on the plane within the boundary). Hence the change in solid angle between P and Q is 2π. In the same way the change from Q back to P, continuing round path B, is also 2π, but this represents an *increase* in solid angle from 2π to 4π. This may be seen from Fig. 11.5,

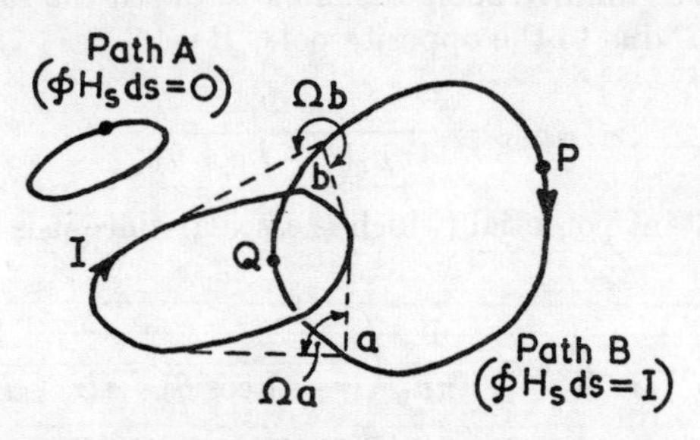

FIG. 11.5. ILLUSTRATING THE WORK LAW

since at a point a in the first half of the path the angle Ω_a is "acute" (i.e. $< 2\pi$), while at b on the return path the angle Ω_b reckoned the same way round as in the first half of the path is "obtuse."

Hence the total change of solid angle is 4π in moving from P once round the path, and the change in magnetic potential is, from eqn. (11.15),

$$F = \frac{I}{4\pi}(4\pi - 0) = I$$

Hence

$$F = |\oint H_s\,ds| = |I|$$

For N turns, this gives

$$F = IN = \oint H_s\,ds \qquad . \quad . \quad . \ (11.16)$$

This is often referred to as the *work law*, since $\oint H_s\,ds$ may be taken to represent the work done in moving a fictitious unit magnetic pole once round a path linking a circuit.

From eqn. (11.16) the unit of magnetic potential (magnetomotive force) is the *ampere-turn*, and it follows from eqn. (11.12) that the unit of H is the *ampere-turn per metre*.

11.5. Field near a Long Straight Wire

The work law may be used to obtain the magnetizing forces near simple idealized circuits. For example, it may be assumed that the flux density, B_r, at a fixed radius r from a long current-carrying

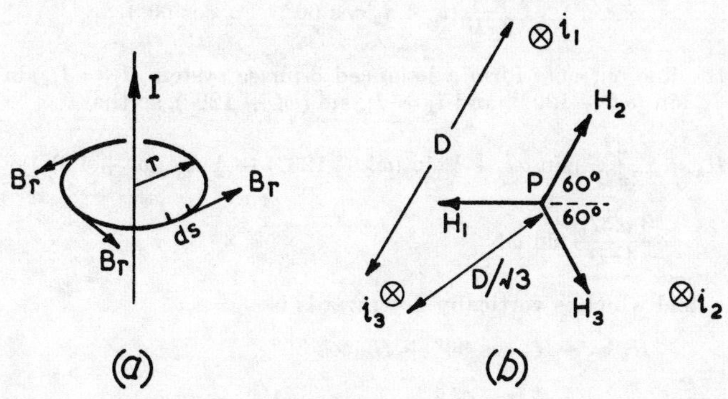

Fig. 11.6. Magnetic Fields near Long Straight Conductors

conductor is constant (Fig. 11.6 (*a*)). In air $H_r = B_r/\mu_0$, and hence H_r is constant, and

$$\oint H_r\,\mathrm{d}s = \Sigma H_r\,\mathrm{d}s = H_r \Sigma \mathrm{d}s = H_r \,.\, 2\pi r$$

By eqn. (11.16), this is equal to the current linked, so that

$$2\pi r H_r = I$$

or

$$H_r = \frac{I}{2\pi r} \quad \text{ampere-turns/metre} \qquad . \qquad . \quad (11.17)$$

Example 11.1. A 3-phase transmission line carrying balanced 3-phase currents consists of three equilaterally-spaced parallel conductors. Show that at a point equidistant from all three conductors there is a pure rotating magnetic field and find an expression for the field strength and for its velocity of angular rotation.

Determine the magnetic field strength when the spacing between conductors is 1 m and the line is carrying 10 MVA at 33 kV line voltage.

(*A.E.E.*, 1957)

The conductor system is shown in Fig. 11.6(*b*). If the spacing between conductors is D, the distance from the centroid, P, to each conductor is $D/\sqrt{3}$. With the current directions shown, the fields at P due to currents i_1, i_2, i_3 will be directed as shown. These fields may be resolved into horizontal and vertical components. From eqn. (11.17), the horizontal field is

$$H_h = H_1 - H_2 \cos 60^\circ - H_3 \cos 60^\circ$$

$$= \frac{i_1}{2\pi D/\sqrt{3}} - \frac{i_2 \cos 60^\circ}{2\pi D/\sqrt{3}} - \frac{i_3 \cos 60^\circ}{2\pi D/\sqrt{3}}$$

$$= \frac{\sqrt{3}}{2\pi D}(i_1 - i_2 \cos 60^\circ - i_3 \cos 60^\circ)$$

If the line currents form a balanced 3-phase system, $i_1 = I_m \sin \omega t$ $i_2 = I_m \sin(\omega t - 120^\circ)$, and $i_3 = I_m \sin(\omega t + 120^\circ)$, so that

$$H_h = \frac{\sqrt{3}I_m}{2\pi D}[\sin \omega t - \tfrac{1}{2}\sin(\omega t - 120^\circ) - \tfrac{1}{2}\sin(\omega t + 120^\circ)]$$

$$= \frac{3\sqrt{3}I_m}{4\pi D}\sin \omega t$$

The field which is vertically downwards is

$$H_v = -H_2 \cos 30^\circ + H_3 \cos 30^\circ$$

$$= \frac{-i_2}{2\pi D/\sqrt{3}}\frac{\sqrt{3}}{2} + \frac{i_3}{2\pi D/\sqrt{3}}\frac{\sqrt{3}}{2}$$

$$= \frac{3I_m}{4\pi D}[-\sin(\omega t - 120^\circ) + \sin(\omega t + 120^\circ)]$$

$$= \frac{3\sqrt{3}I_m}{4\pi D}\cos \omega t$$

The resultant field magnitude at P is

$$H_P = \sqrt{(H_h{}^2 + H_v{}^2)} = \frac{3\sqrt{3}I_m}{4\pi D}\sqrt{(\sin^2 \omega t + \cos^2 \omega t)}$$

$$= \frac{3\sqrt{3}I_m}{4\pi D} \quad \text{ampere-turns/metre}$$

This is constant for a fixed value of I_m.

The instantaneous direction of the field with respect to the downwards vertical axis is given by

$$\phi = \tan^{-1}\frac{H_h}{H_v} = \tan^{-1}(\tan \omega t) = \omega t \quad \text{radians}$$

i.e. ϕ increases linearly with time. The field at P therefore has constant magnitude $3\sqrt{3}I_m/4\pi D$ and rotates at ω radians per second.

With the numerical values given,

$$I_m = \sqrt{2}\,\frac{10 \times 10^6}{\sqrt{3} \times 33 \times 10^3} = 247 \text{ A}$$

Hence $$H_P = \frac{3\sqrt{3} \times 247}{4\pi \times 1} = \underline{\underline{102 \text{ At/m}}}$$

11.6. Force between Two Parallel Wires—the Ampere

Consider two long parallel wires, 1 and 2, separated by d metres and carrying currents I_1 and I_2 respectively. The magnetizing force at wire 2 due to I_1 is, by eqn. (11.17),

$$H_{21} = \frac{I_1}{2\pi d}$$

Hence the flux density at wire 2 due to I_1 is

$$B_{21} = \frac{\mu_0 I_1}{2\pi d}$$

assuming that the medium separating the wires is a vacuum.

From eqn. (11.1) the force on wire 2 per metre length is

$$F_m = \frac{\mu_0 I_1 I_2}{2\pi d} \quad \text{newtons/metre} \qquad . \qquad . \quad (11.18)$$

It is from this equation that the unit of current is defined. One *ampere* is that current which, flowing in two straight, parallel conductors of infinite length, of negligible circular cross-section, and placed 1 metre apart in a vacuum produces a force of 2×10^{-7} newtons per metre length between them. This means that μ_0 must have the value $4\pi \times 10^{-7}$. The unit of μ will be shown later to be the *henry per metre.*

11.7. The Biot–Savart Law (Laplace's Law)

In many cases where it is impracticable to use the simple work law, the *Biot–Savart law* may be employed to determine magnetizing forces near systems of conductors. Consider a current-element $I\,\delta l$ (Fig. 11.7), and let us find the magnetizing force at a point P which is at a distance x from the element. It is convenient to introduce a fictitious unit pole at P, the force on this pole being the magnetizing force at P due to $I\,\delta l$.

The field at δl due to the unit pole at P is, from eqn. (11.10), $B_1 = 1/4\pi x^2$, directed in the x-direction. Hence the force on the current-element is

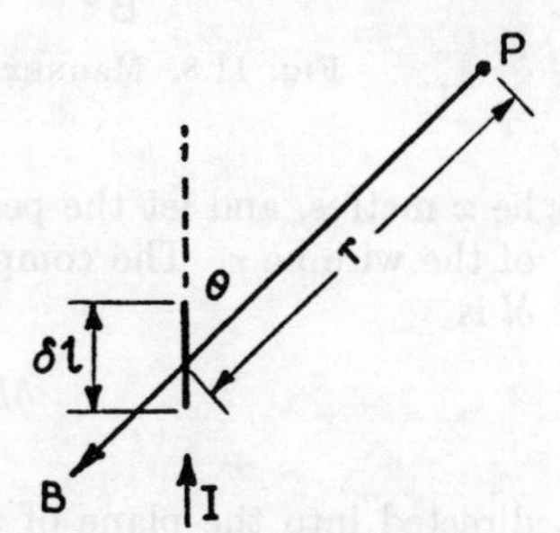

FIG. 11.7. THE BIOT–SAVART LAW GIVES FIELD DUE TO A CURRENT-ELEMENT

$$F_m = \frac{1}{4\pi x^2} I\,\delta l \sin\theta \qquad \text{(from eqn. (11.12))}$$

Since action and reaction are equal and opposite, this force must also be the force on the pole at P due to the current-element, and by definition this is the magnetizing force at P. Hence

$$H_p = \frac{I\,\delta l \sin\theta}{4\pi x^2} \quad \text{ampere-turns/metre} \quad . \qquad . \ (11.19)$$

The direction of H_p is perpendicular to the plane containing δl and x (right-hand screw rule). Eqn. (11.19) is the mathematical expression of the Biot–Savart law, which is sometimes referred to as *Laplace's Law*.

11.8. Field due to a Short Wire

A short length of wire AB carrying I amperes is shown in Fig. 11.8. Let the distance from some point P to an element δl of the wire

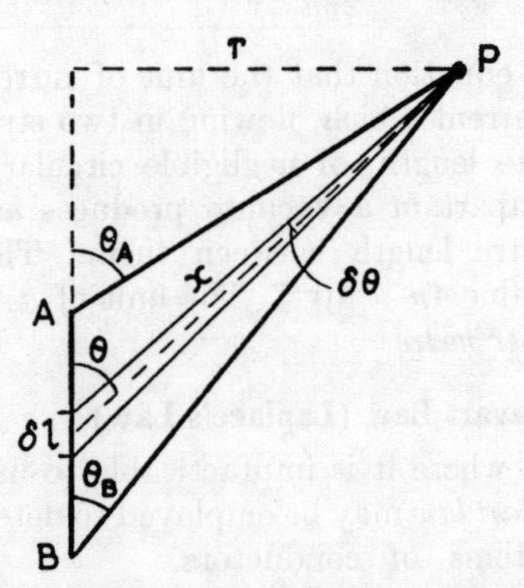

Fig. 11.8. Magnetic Field due to a Short Wire

be x metres, and let the perpendicular distance from P to the axis of the wire be r. The component of magnetizing force at P due to δl is

$$\delta H_p = \frac{I\,\delta l \sin\theta}{4\pi x^2} \qquad \text{(from eqn. (11.19))}$$

directed into the plane of the paper. Now, $x\,\delta\theta = \delta l \sin\theta$, where $\delta\theta$ is the small angle subtended at P by the element δl, and $x = r/\sin\theta$, so that

$$\delta H_p = \frac{I \sin\theta\,\delta\theta}{4\pi r}$$

Each element of AB contributes a field component at P in the same direction; hence the total magnetizing force at P is

$$H_p = \int_B^A \frac{I \sin \theta \, d\theta}{4\pi r} = \frac{I}{4\pi r}(\cos \theta_B - \cos \theta_A) \quad . \quad (11.20)$$

For an infinite wire, $\theta_B = 0$ and $\theta_A = 180°$, so that $H_p = I/2\pi r$, as was obtained in Section 11.5.

11.9. Field on the Axis of a Square Coil

From considerations of symmetry, the field on the axis of a coil will be directed along the axis. Thus, for the square coil shown in Fig. 11.9, the magnetizing forces at P on the axis due to the opposite

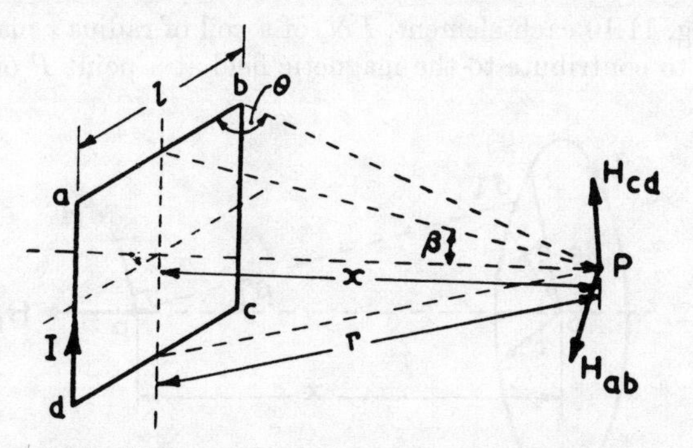

FIG. 11.9. MAGNETIC FIELD ON AXIS OF A SQUARE COIL

sides ab and cd are H_{ab} and H_{cd} directed at right angles to the planes containing P and ab, and P and cd, respectively. H_{ab} and H_{cd} are numerically equal, and hence the components at right angles to the axis will cancel. In the same way, the sides da and bc will also give a resultant axial component only.

If the distance from the centre of the coil to P is x, and the coil side length is l, then from eqn. (11.20),

$$H_{ab} = \frac{I}{4\pi r}[\cos \theta - \cos (180° - \theta)] \quad \text{(where } I \text{ is the coil current)}$$

$$= \frac{I \, . \, 2 \cos \theta}{4\pi\sqrt{(x^2 + l^2/4)}}$$

The axial component of this is

$$H_{ab}{}' = \frac{I \cos \theta}{2\pi\sqrt{(x^2 + l^2/4)}} \sin \beta$$

Since each side will contribute an equal amount to the resultant axial field, the magnetizing force at P is

$$H_p = \frac{2I\cos\theta\sin\beta}{\pi\sqrt{(x^2 + l^2/4)}}$$

$$= \frac{Il^2}{2\pi(x^2 + l^2/4)\sqrt{(x^2 + l^2/2)}} \quad \text{ampere-turns/metre} \quad . \quad (11.21)$$

since $\cos\theta = (l/2)/\sqrt{(x^2 + l^2/2)}$ and $\sin\beta = (l/2)/\sqrt{(x^2 + l^2/4)}$.

11.10. Field on the Axis of a Circular Coil

In Fig. 11.10 each element, $I\,\delta l$, of a coil of radius r may be considered to contribute to the magnetic field at a point P on the axis

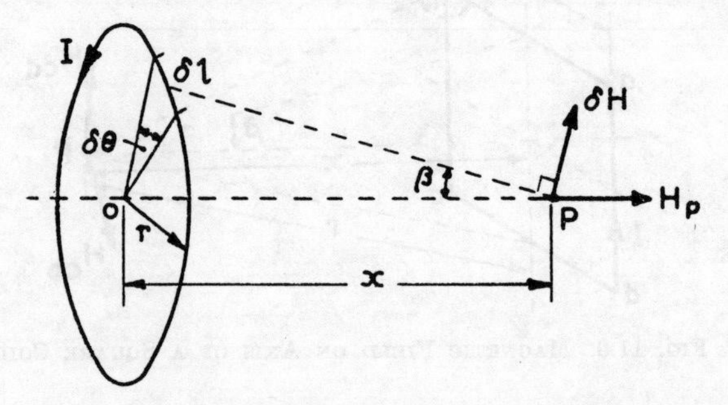

Fig. 11.10. Magnetic Field on Axis of a Circular Turn

of the coil. Diametrically opposite elements $I\,\delta l$ produce magnetizing forces at P whose components at right angles to the axis cancel, so that the resultant field is axial.

If the distance from the centre of the coil to P is x, eqn. (11.19) gives the component of magnetizing force at P:

$$\delta H = \frac{I\,\delta l}{4\pi(x^2 + r^2)}$$

since the distance from δl to P is $\sqrt{(x^2 + r^2)}$ and the angle which this makes with δl is 90°. The axial component is

$$\delta H\sin\beta = \frac{I\,\delta l\sin\beta}{4\pi(x^2 + r^2)}$$

Now, δl subtends an angle $\delta\phi$ at the centre of the coil, so that $\delta l = r\,\delta\phi$, and the resultant axial field at P is

$$H_P = \int_0^{2\pi} \frac{Ir \sin\beta \,\mathrm{d}\phi}{4\pi(x^2 + r^2)} = \frac{Ir \sin\beta}{2(x^2 + r^2)}$$

$$= \frac{Ir^2}{2(x^2 + r^2)^{3/2}} \quad . \quad . \quad . \quad . \quad (11.22)$$

since $\sin\beta = r/\sqrt{(x^2 + r^2)}$.

At the centre of the coil $x = 0$, and

$$H = \frac{I}{2r} \quad \text{ampere-turns/metre} \quad . \quad . \quad (11.23)$$

If the coil has N concentrated turns, each of these expressions is multiplied by N to give

$$H = \frac{IN}{2r} \quad \text{ampere-turns/metre} \quad . \quad . \quad (11.23a)$$

An alternative expression for the axial field may be written in terms of $\sin\beta$ from eqn. (11.22)—

$$H_P = \frac{I \sin^3\beta}{2r} \quad . \quad . \quad . \quad (11.24)$$

11.11. Field on the Axis of a Short Solenoid

The magnetizing force on the axis of a short solenoid may be evaluated by considering that the solenoid is made up of a number of circular elements. If the winding density of the solenoid shown in Fig. 11.11 is n turns per metre, then an element δx wide carries

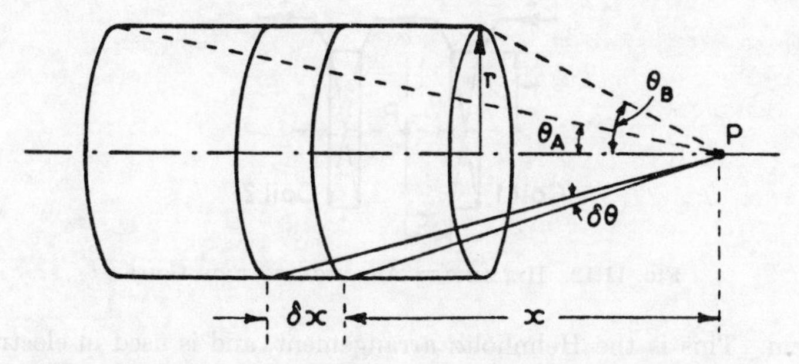

Fig. 11.11. Magnetic Field on Axis of a Short Solenoid

$In\,\delta x$ amperes (where I is the solenoid current), and the axial component of the field at P due to the element is, from eqn. (11.24),

$$\delta H = \frac{In\,\delta x \sin^3\theta}{2r}$$

where θ is the angle between the axis and a line from P to any point on the element δx.

By geometry,

$$\delta x = \frac{\sqrt{(r^2 + x^2)}\,\delta\theta}{\sin\theta}$$

and the total magnetizing force at P is therefore

$$H_P = \int_{\theta_A}^{\theta_B} \frac{In \sin^3\theta}{2r}\,\frac{\sqrt{(r^2+x^2)}}{\sin\theta}\,d\theta$$

$$= \frac{In}{2}\int_{\theta_A}^{\theta_B} \sin\theta\,d\theta \quad \text{since } \sin\theta = \frac{r}{\sqrt{(r^2+x^2)}}$$

Hence

$$H_P = \frac{In}{2}(\cos\theta_A - \cos\theta_B) \quad \text{ampere-turns/metre} \quad . \quad (11.25)$$

At the centre of the solenoid,

$$H = \frac{In}{2}[\cos\theta_A - \cos(180° - \theta_A)] = In\cos\theta_A \quad . \quad (11.26)$$

and for a long solenoid $\theta_A \to 0$, so that

$$H = In \quad \text{ampere-turns/metre} \quad . \quad . \quad (11.27)$$

Helmholtz Arrangement of Coils. Two coils of equal radius may be spaced in such a way as to produce an almost uniform field between

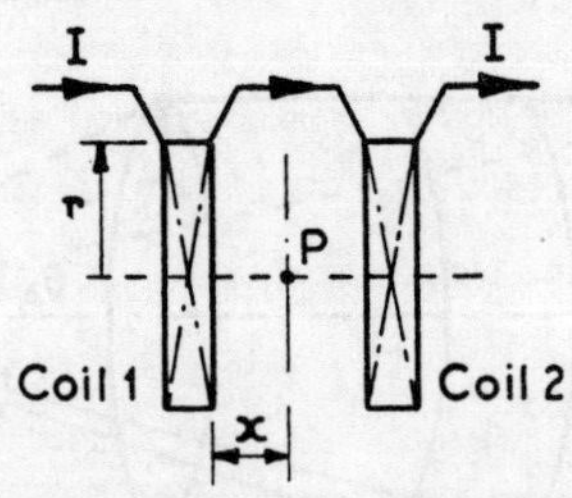

FIG. 11.12. HELMHOLTZ ARRANGEMENT OF COILS

them. This is the Helmholtz arrangement, and is used in electrodynamic instruments. Suppose the coils are of radius r, as shown in

Fig. 11.12. The field due to coil 1 on the axis at P is, from eqn. (11.22),

$$H_{P1} = \frac{Ir^2}{2(x^2 + r^2)^{3/2}} \quad \text{ampere-turns/metre}$$

The rate of change of H_{P1} with distance is

$$\frac{dH_{P1}}{dx} = \frac{Ir^2}{2}\left[-\frac{3}{2}\frac{2x}{(r^2 + x^2)^{5/2}}\right]$$

and there is a point of inflexion when $d^2H_{P1}/dx^2 = 0$; i.e.

$$\frac{3Ir^2}{2}\left[-\frac{1}{(r^2 + x^2)^{5/2}} + \frac{5}{2}\cdot\frac{2x^2}{(r^2 + x^2)^{7/2}}\right] = 0$$

or

$$r^2 + x^2 = 5x^2$$

or

$$x = \frac{r}{2}$$

If the second coil is spaced at a distance r from the first, then midway between the coils the rate of change of field due to coil 1 will be the same as the rate of change of field due to coil 2 but in the opposite sense. The two slopes cancel to give a uniform field at this point.

11.12. Induced E.M.F.

Of fundamental importance in the study of electromagnetism is the proportionality between the rate of change of magnetic flux linking a circuit and the e.m.f. induced in the circuit (Faraday's law). For a coil of N turns, the e.m.f. induced by a changing flux, Φ, is

$$v = \frac{d(\Phi N)}{dt} \quad . \quad . \quad . \quad . \quad (11.28)$$

Lenz's law states that the direction of this induced e.m.f. is such that, if it could cause a current to flow, the current would set up a flux opposing the original change in flux.

Note that $\Phi N = \int v\, dt$, so that the unit of flux (the weber) may also be expressed in *volt-seconds*.

11.13. Inductance

Inductance is a physical constant associated with systems of conductors, expressible in terms of the dimensions and the permeability of the surrounding medium. If the current in a circuit is changed, the flux linking the circuit changes and an e.m.f. is induced opposing the change in current. The e.m.f. is proportional to the rate of change of flux and hence to the rate of change of current. The constant of proportionality is the *self-inductance*, L:

$$v = L\frac{\mathrm{d}i}{\mathrm{d}t} \qquad (11.29)$$

The self-inductance in *henrys* is the e.m.f. in volts induced in a circuit when the current changes at the rate of 1 ampere per second. From eqn. (11.28),

$$v = \frac{\mathrm{d}(\Phi N)}{\mathrm{d}t}$$

and hence

$$L = \frac{\mathrm{d}(\Phi N)}{\mathrm{d}i} \qquad (11.30)$$

Note that if L varies,

$$v = \frac{\mathrm{d}(Li)}{\mathrm{d}t} = i\frac{\mathrm{d}L}{\mathrm{d}t} \quad \text{(for constant } i\text{)} \qquad (11.31)$$

If flux produced by one circuit links a second circuit, the *mutual inductance*, M_{12}, between the circuits is defined by the relation

$$v_2 = M_{12}\frac{\mathrm{d}i_1}{\mathrm{d}t} \qquad (11.32)$$

In the same way a changing current in circuit 2 gives an e.m.f. in circuit 1—

$$v_1 = M_{21}\frac{\mathrm{d}i_2}{\mathrm{d}t} \qquad (11.33)$$

The energy stored in the magnetic field of a current may be deduced by considering an increase δi in the current in δt seconds. The back-e.m.f. induced is $v = L(\delta i/\delta t)$ as $\delta t \to 0$, and hence the energy required from the electric circuit in δt seconds is

$$\delta W = vi\,\delta t = Li\,\delta i$$

The total energy required to increase the current to I is

$$W = \int_0^I Li\,\mathrm{d}i = \tfrac{1}{2}LI^2 \text{ joules} \qquad (11.34)$$

This energy is stored in the magnetic field. In order to evaluate the density of the stored energy (joules per cubic metre), consider two

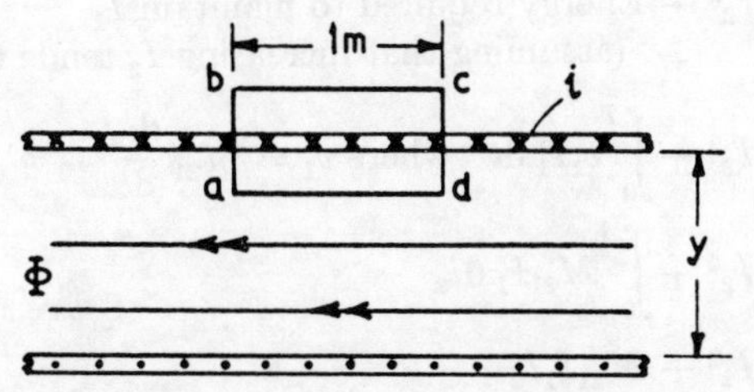

FIG. 11.13. MAGNETIC FIELD DUE TO PARALLEL STRIP CONDUCTORS

plane strip conductors of infinite extent, carrying i amperes perpendicular to the plane of the paper per metre breadth of strip (Fig. 11.13).

If the magnetizing force between the plates is H, then round the closed path $abcd$, $\oint H \, \mathrm{d}s = H \,.\, ad$, assuming H to be zero along bc, ab and cd. Hence $H \,.\, ad = i \,.\, ad$, or $H = i$ (from eqn. (11.16)).

For unit depth of field, the flux is

$$\Phi = \mu H y \,.\, 1 = \mu i y$$

so that

$$\frac{\mathrm{d}\Phi}{\mathrm{d}i} = \mu y = \text{Inductance per unit depth and unit breadth of conductor}$$

Hence,

$$\text{Energy stored} = \tfrac{1}{2}\mu y i^2$$

The volume of the field is $1 \times 1 \times y$, so that the stored energy density is

$$\tfrac{1}{2}\mu i^2 = \tfrac{1}{2}\mu H^2 \quad \text{joules/metre}^3 \qquad . \qquad . \ (11.35)$$

11.14. Mutual Inductance

In Fig. 11.14 two coils of self-inductance L_1 and L_2 are shown. Mutual inductances M_{12} and M_{21} exist between them. Let coil 2 be open-circuited and the current in coil 1 be increased to I_1. From eqn. (11.29), the energy stored is

$$W_1 = \tfrac{1}{2}L_1 I_1^2 \quad \text{joules}$$

FIG. 11.14

If I_1 is kept constant while the current in coil 2 is increased to I_2 in T seconds, the additional energy required is

$$W_2 = \tfrac{1}{2}L_2I_2^2 + \text{Energy required to maintain } I_1$$
(assuming that increasing I_2 tends to reduce I_1)

$$= \tfrac{1}{2}L_2I_2^2 + \int_0^T v_1I_1\,\mathrm{d}t \quad \text{where } v_1 = M_{21}\frac{\mathrm{d}i_2}{\mathrm{d}t}$$

$$= \tfrac{1}{2}L_2I_2^2 + \int_0^{I_2} M_{21}I_1\,\mathrm{d}i_2$$

$$= \tfrac{1}{2}L_2I_2^2 + M_{21}I_1I_2$$

Hence

$$W_1 + W_2 = \tfrac{1}{2}L_1I_1^2 + M_{21}I_1I_2 + \tfrac{1}{2}LI_2^2$$

If the currents had increased in the reverse order the result would have been the same but the middle term would have been $M_{12}I_1I_2$. Since the stored energies cannot be different in the two cases,

$$M_{21} = M_{12} = M \quad . \quad . \quad . \quad (11.36)$$

Note that the energy stored due to the mutual coupling is I_1I_2M joules.

11.15. Inductance of a Long Solenoid

At the centre of a long solenoid, of n turns per metre and carrying i amperes, eqn. (11.27) gives

$$H = in = \frac{iN}{l}$$

where N is the total number of turns and l is the length of the solenoid. If the solenoid is sufficiently long compared with its diameter this magnetizing force will be the same all over the cross-section. Hence

$$\Phi = \mu HA = \frac{\mu iNA}{l}$$

so that

$$L = N\frac{\mathrm{d}\Phi}{\mathrm{d}i} = \frac{\mu AN^2}{l} \quad . \quad . \quad . \quad (11.37)$$

In practice the inductance of a short solenoid is given by $L = K\mu AN^2/l$, where K is *Nagaoka's constant*. K depends on the ratio of length to diameter of the solenoid, and is a tabulated function for single-layer solenoids.

11.16. Inductance of a Toroid

Toroidal coils have the advantage that the external flux is almost zero, so that they may be regarded as self-screening.

Ferrite-cored toroidal coils using both square-loop and linear ferrites are widely used in computer and logic circuits. The cross-sections may be either round or rectangular.

(*a*) *Rectangular Toroid.* In this case it is simplest to use the work law to determine the magnetizing force at any point within the toroid. At radius r in the toroid shown in Fig. 11.15, with inner

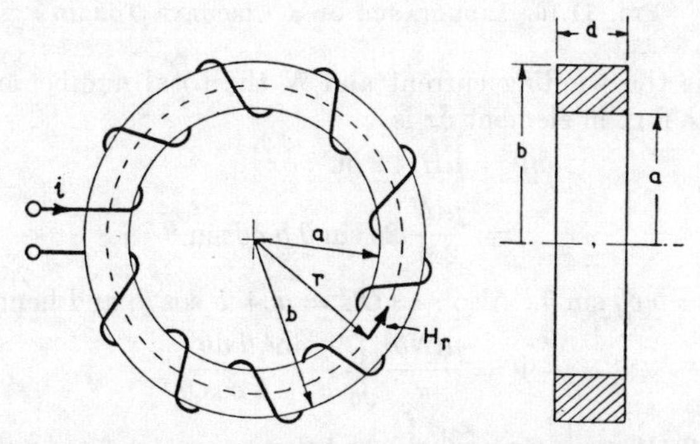

FIG. 11.15. INDUCTANCE OF A RECTANGULAR TOROID

radius a and outer radius b ($a < r < b$), let the magnetizing force be H_r. This will be constant at constant radius. Hence

$$\oint H_r \mathrm{d}r = H_r 2\pi r = iN \qquad \text{(by eqn. (11.16))}$$

where i is the current and N is the total number of turns. The core flux for a toroid of depth d is

$$\Phi = \int_a^b \mu H_r d \, \mathrm{d}r = \mu d \int_a^b \frac{iN}{2\pi r} \mathrm{d}r$$

$$= \frac{\mu d i N}{2\pi} \log_e \frac{b}{a}$$

Hence

$$L = N \frac{\mathrm{d}\Phi}{\mathrm{d}i} = \frac{\mu d N^2}{2\pi} \log_e \frac{b}{a} \quad \text{henrys} \qquad . \quad (11.38)$$

(*b*) *Circular Toroid.* For a toroid whose core radius is b let the mean toroidal radius be a (Fig. 11.16). At radius r from O, H_r is constant and eqn. (11.16) gives

$$\oint H_r \, dr = H_r \,.\, 2\pi r = iN$$

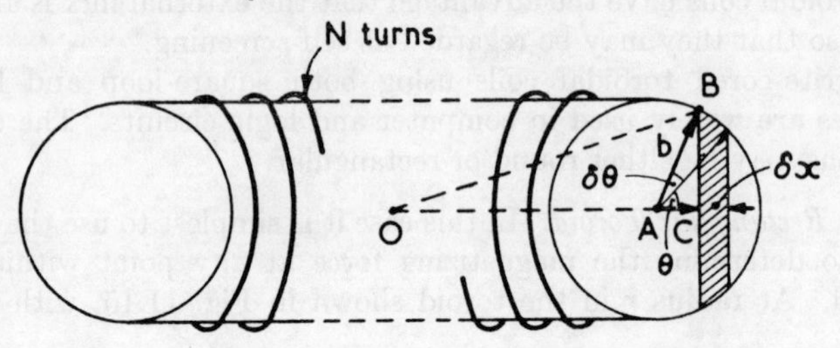

FIG. 11.16. INDUCTANCE OF A CIRCULAR TOROID

where i is the winding current and N the total number of turns. Hence the flux in element δx is

$$\delta\Phi = \mu H_r \,.\, 2BC \,.\, \delta x$$

$$= \frac{\mu i N}{2\pi r} 2b \sin\theta \, b \, \delta\theta \sin\theta$$

since $\delta x = b\,\delta\theta \sin\theta$. Also $r = OC = a + b\cos\theta$, and hence

$$\Phi = \frac{\mu i N b^2}{\pi} \int_0^\pi \frac{\sin^2\theta \, d\theta}{a + b\cos\theta} \qquad . \qquad . \quad (11.39)$$

But $\int_0^\pi f(\theta)\,d\theta = \int_0^\pi f(\pi - \theta)\,d\theta$, and the expression for the flux may therefore be written

$$\Phi = \frac{\mu i N b^2}{\pi} \int_0^\pi \frac{\sin^2\theta \, d\theta}{a - b\cos\theta} \qquad . \qquad . \quad (11.40)$$

Taking half the sum of eqns. (11.39) and (11.40),

$$\Phi = \frac{1}{2}\frac{\mu i N b^2}{\pi} \int_0^\pi \left(\frac{\sin^2\theta}{a + b\cos\theta} + \frac{\sin^2\theta}{a - b\cos\theta} \right) d\theta$$

$$= \frac{\mu i N b^2}{2\pi} \int_0^\pi \frac{2a\sin^2\theta}{a^2 - b^2\cos^2\theta}\,d\theta = \frac{\mu i N b^2 a}{\pi} \int_0^\pi \frac{\cos^2\theta - 1}{b^2\cos^2\theta - a^2}\,d\theta$$

$$= \frac{\mu i N b^2 a}{\pi} \int_0^\pi \left(\frac{1}{b^2} + \frac{(a^2/b^2 - 1)}{b^2\cos^2\theta - a^2} \right) d\theta$$

$$= \mu i N a - \frac{\mu i N (a^2 - b^2) a}{\pi} \int_0^\pi \frac{d\theta}{a^2 - b^2\cos^2\theta}$$

$$= \mu i N \left(a - \frac{2(a^2 - b^2)a}{\pi} \int_0^{\pi/2} \frac{\sec^2\theta}{a^2\sec^2\theta - b^2}\,d\theta \right)$$

Substituting $u = \tan\theta$, $du = \sec^2\theta\, d\theta$, and $\sec^2\theta = 1 + \tan^2\theta$,

$$\Phi = \mu i N \left(a - \frac{2(a^2 - b^2)a}{\pi} \int_0^\infty \frac{du}{(a^2 - b^2) + a^2u^2} \right)$$

$$= \mu i N[a - \sqrt{(a^2 - b^2)}]$$

Therefore $$L = N\frac{d\Phi}{di} = \mu N^2[a - \sqrt{(a^2 - b^2)}] \quad . \quad . \quad (11.41)$$

If $b < 0{\cdot}1a$ this becomes

$$L = \frac{\mu N^2 b^2}{2a} \text{ henrys} \quad . \quad . \quad . \quad (11.42)$$

This last expression is the one obtained by assuming that $H_r = iN/2\pi a$ over the whole cross-section.

11.17. Flux Linkages of a Long Wire

The magnetic field produced by low-frequency or direct current flowing through a long circular conductor will be not only external to the conductor, but, assuming uniform current distribution, will also exist within the conductor. The internal flux will link only a fraction of the current, and external and internal flux linkages must therefore be treated separately. Consider such a conductor of radius a, carrying a current i.

(*a*) *External Flux.* At radius r ($>a$), by eqn. (11.17),

$$H_r = \frac{i}{2\pi r}$$

The flux enclosed by annulus δr (Fig. 11.17 (*a*)) per metre run of

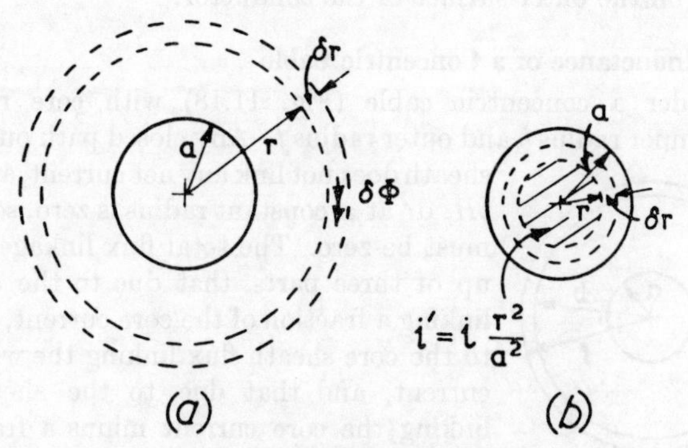

Fig. 11.17. Flux Linkages of a Long Wire

(*a*) External linkages
(*b*) Internal linkages

wire (i.e. for 1 m into the plane of the paper) is

$$\delta\Phi = \mu H_r \, \delta r \times 1$$

Hence the total external flux is

$$\Phi_e = \int_a^R \mu \, H_r \, dr = \frac{\mu i}{2\pi} \int_a^R \frac{dr}{r} = \frac{\mu i}{2\pi} \log_e \frac{R}{a} \quad . \quad (11.43)$$

where R is at a sufficient distance to give zero field. The external flux linkage is thus $\Phi N = \Phi_e$, since there is only one turn.

(*b*) *Internal Flux.* At radius r ($<a$), the magnetizing force, H_r' is due only to the current within the radius r (Fig. 11.17 (*b*)), i.e. to ir^2/a^2.

$$H_r' = \frac{ir^2/a^2}{2\pi r} = \frac{ir}{2\pi a^2}$$

Hence in the element δr the flux per metre run with a core permeability μ_i is

$$\delta\Phi' = \mu_i H_r' \, \delta r \times 1$$

This flux links only r^2/a^2 of one turn so that the flux linkage with the element is

$$\delta\Phi' n = \frac{\mu_i i r}{2\pi a^2} \frac{r^2}{a^2} \, \delta r$$

$$\text{Total internal flux linkage} = \int_0^a \frac{\mu_i i r^3}{2\pi a^4} \, dr = \frac{\mu_i i}{8\pi} \quad . \quad . \quad (11.44)$$

This internal flux linkage rapidly falls to zero as the frequency is increased owing to the skin effect, which tends to concentrate the current on the outer surface of the conductor.

11.18. Inductance of a Concentric Cable

Consider a concentric cable (Fig. 11.18) with core radius a, sheath inner radius b and outer radius f. Any closed path outside the sheath does not link any net current, and hence $\oint H_r \, dr$ at a constant radius is zero, so that H_r must be zero. The total flux linkage is made up of three parts, that due to the core flux linking a fraction of the core current, that due to the core-sheath flux linking the whole core current, and that due to the sheath flux linking the core current minus a fraction of the sheath current. Let the core current be i returning by the sheath.

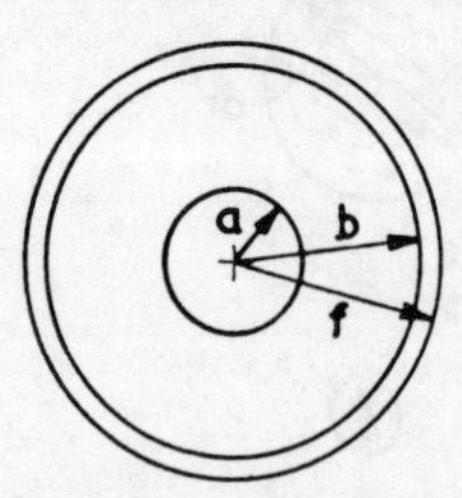

Fig. 11.18. Concentric Cable

(*a*) *Internal Core Flux Linkage.* By eqn. (11.44) the internal core linkage is $\mu_i i/8\pi$, where μ_i is the permeability of the core material

(*b*) *Core–sheath Linkage.* The magnetic field between core and sheath is due entirely to the core current. At radius r $(a < r < b)$

$$H_r = \frac{i}{2\pi r} \qquad \text{(eqn. (11.17))}$$

Hence the flux enclosed between core and sheath, per metre run, is

$$\Phi = \int_a^b \mu_d H_r \,\mathrm{d}r = \frac{\mu_d i}{2\pi} \log_e \frac{b}{a} \quad . \quad . \quad (11.45)$$

where μ_d is the permeability of the dielectric. Since the circuit represents a single turn this is also the core-sheath flux linkage.

(*c*) *Sheath Linkage.* At a radius r within the sheath, the magnetizing force H_r' is constant, so that

$$\begin{aligned} \oint H_r' \,\mathrm{d}r &= 2\pi r H_r' \\ &= \text{Current linked} \\ &= i\left(1 - \frac{r^2 - b^2}{f^2 - b^2}\right) = i\left(\frac{f^2 - r^2}{f^2 - b^2}\right) \end{aligned}$$

Hence
$$H_r' = \frac{i}{2\pi r}\left(\frac{f^2 - r^2}{f^2 - b^2}\right)$$

The flux enclosed in an annulus δr, per metre run, is

$$\delta\Phi' = \mu_s H_r' \,\delta r = \frac{\mu_s i}{2\pi r}\frac{f^2 - r^2}{f^2 - b^2}\,\delta r$$

where μ_s is the permeability of the sheath. This links

$$i[1 - (r^2 - b^2)/(f^2 - b^2)]$$

of the current, so that the linkage with element δr is

$$\frac{\mu_s i}{2\pi r}\frac{(f^2 - r^2)^2}{(f^2 - b^2)^2}\,\delta r = \frac{\mu_s i}{2\pi (f^2 - b^2)^2}\left(\frac{f^4}{r} - 2f^2 r + r^3\right)\delta r$$

The total sheath flux linkage is thus

$$\int_b^f \frac{\mu_s i}{2\pi (f^2 - b^2)^2}\left(\frac{f^4}{r} - 2f^2 r + r^3\right)\mathrm{d}r = \frac{\mu_s i}{2\pi}\left[\frac{f^4 \log_e (f/b)}{(f^2 - b^2)^2} - \frac{f^2}{f^2 - b^2} + \frac{1}{4}\frac{f^2 + b^2}{f^2 - b^2}\right] \quad . \quad (11.46)$$

This flux linkage is usually negligible, since (i) the field is weak and decreasing outwards, (ii) the sheath is usually very thin, and (iii) skin effect concentrates the current on the inner surface of the sheath.

Adding eqns. (11.44), (11.45) and (11.46), and dividing by i, gives the inductance as

$$L = \frac{\mu_i}{8\pi} + \frac{\mu_d}{2\pi}\log_e\frac{b}{a} + \frac{\mu_s}{2\pi}\left[\frac{f^4\log_e(f/b)}{(f^2-b^2)^2} - \frac{f^2}{f^2-b^2} + \frac{1}{4}\frac{f^2+b^2}{f^2-b^2}\right] \quad . \quad . \quad . \quad (11.47)$$

For thin sheaths this becomes

$$L = \frac{\mu_d}{2\pi}\log_e\frac{b}{a} + \frac{\mu_i}{8\pi} \text{ henrys/metre} \quad . \quad . \quad (11.48)$$

and for high frequencies,

$$L = \frac{\mu_d}{2\pi}\log_e\frac{b}{a} \text{ henrys/metre} \quad . \quad . \quad (11.49)$$

Normally μ_i, μ_d and μ_s are equal to μ_0, but if magnetic materials are used for core and sheath μ_i and μ_s may be much larger.

11.19. Skin Effect*

It has been shown in Section 11.17 (*b*) that, when a current flows through a conductor there will be internal flux linkages within the conductor. Hence current at the centre of the conductor will link more flux than that at the surface. If the conductor is considered to be made up of an infinite number of parallel conducting elements, it follows that the inductance of the elements nearer the centre will be greater than that of those near the outside. When the current is alternating the inductive reactance of the elements near the centre will be greater than that of those near the outside, and hence more current will flow in the elements near the outside. A depth of penetration, δ, may be postulated (Fig. 11.19) at which the current density has decreased to 1/e of its value at the outer surface. Within the radius $(a - \delta)$, where a is the conductor radius, the current may be assumed to be zero.

This *skin effect* will obviously be more pronounced the larger the inductive reactance—which varies with the angular frequency of the current, ω, and with the inductance, which in turn varies with

* For a more detailed treatment, *see* p. 393.

the permeability, μ, of the material of the conductor. On the other hand, the greater the resistivity, ρ, of the conductor, the smaller will be the effect of the variation in inductive reactance in causing a non-uniform current distribution. To determine the relationship

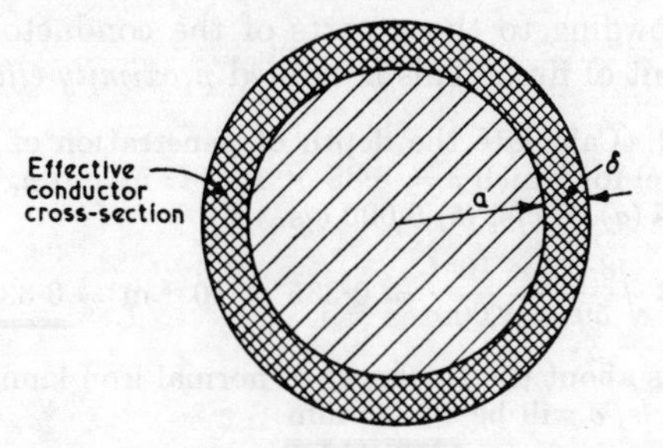

FIG. 11.19. SKIN EFFECT IN A ROUND CONDUCTOR

between the depth of penetration and ω, μ and ρ we may write the dimensional equality

$$[\delta] = \left[\frac{\rho^l}{\omega^n \mu^m}\right]$$

The dimensions of ρ are ohm-metres, of ω are "per second" and of μ are henrys per metre. Now henrys are flux linkages per ampere, or volt-seconds per ampere, and so

$$[\delta] = \frac{[\Omega\text{-metres}]^l}{\left[\dfrac{1}{\text{seconds}}\right]^n \left[\dfrac{\text{volt-seconds}}{\text{ampere-metres}}\right]^m}$$

$$= [\Omega]^{l-m}[\text{metres}]^{l+m}[\text{seconds}]^{n-m} = [\text{metres}]$$

For dimensional equality, $n = m = l = \frac{1}{2}$, since this is the only combination which will give δ in metres. Hence

$$\delta \propto \sqrt{\frac{\rho}{\omega\mu}}$$

In fact it turns out that the constant of proportionality is $\sqrt{2}$, so that

$$\delta = \sqrt{\frac{2\rho}{\omega\mu}} = 503{\cdot}3\sqrt{\frac{\rho}{f\mu_r}} \quad \text{metres} \quad . \quad . \quad (11.50)$$

where $f = \omega/2\pi$.

Similar considerations apply to the penetrations of alternating magnetic flux into conductors, and eqn. (11.50) may be used in this case also, which will then also apply to flat sheets ($a \to \infty$).

When two conductors carrying alternating current are in each other's magnetic fields, there will also be a redistribution of current, the current crowding to those parts of the conductors which link the least amount of flux. This is termed *proximity effect*.

Example 11.2. Calculate the depth of penetration of magnetic flux in a sheet of steel for which $\rho = 8{\cdot}85 \times 10^{-8}$ Ω-m and $\mu_r = 4{,}000$ when the frequency is (*a*) 50 c/s, (*b*) 5,000 c/s.

$$(a)\ \delta = 503{\cdot}3\sqrt{\frac{8{\cdot}85 \times 10^{-8}}{50 \times 4{,}000}} = 0{\cdot}335 \times 10^{-3}\ \text{m} = \underline{\underline{0{\cdot}335\ \text{mm}}}$$

Note that this is about the thickness of normal iron laminations.

(*b*) At 5,000 c/s, δ will be $\underline{\underline{0{\cdot}0335\ \text{mm}}}$

11.20. Inductance of an Isolated Twin Line

The inductance of an isolated twin line may be considered to consist of the self-inductance of each line alone coupled by a mutual inductance which decreases the overall loop inductance (Fig. 11.20 (*a*)). Consider such a line of conductor radius r and spacing d for which $d \gg r$, the lines carrying currents i_A and i_B.

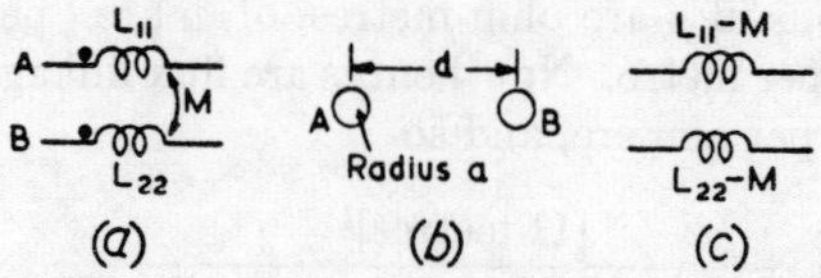

A B L11 L22 M (a) d A B Radius a (b) L11-M L22-M (c)

FIG. 11.20. INDUCTANCE OF A TWIN LINE

By eqn. (11.43), the external self-linkage with A due to its own current is (per metre run)

$$\frac{\mu i_A}{2\pi}\log_e\frac{R}{a} = L_{11}i_A \quad \text{(say)} \qquad . \quad . \quad (11.51)$$

The mutual linkage with A due to i_B is

$$\int_d^R \frac{\mu i_B}{2\pi r}\,\mathrm{d}r = \frac{\mu i_B}{2\pi}\log_e\frac{R}{d} = Mi_B \qquad . \quad . \quad (11.52)$$

where the field is assumed to be zero at the distance R.

The external self-linkage with B due to i_B is

$$\frac{\mu i_A}{2\pi}\log_e\frac{R}{d} = L_{22}i_B \qquad . \quad . \quad . \quad (11.53)$$

and the mutual linkage with B due to i_A is

$$\frac{\mu i_A}{2\pi}\log_e\frac{R}{d} = Mi_A \quad . \quad . \quad . \quad (11.54)$$

If $i_B = -i_A$, the mutual linkage subtracts from the self-linkage, so that the total linkage is

$$L_{11}i_A - Mi_A + L_{22}i_A - Mi_A$$

The inductance per loop metre is

$$L = \frac{\mu}{2\pi}\left(\log_e\frac{R}{a} - \log_e\frac{R}{d} + \log_e\frac{R}{a} - \log_e\frac{R}{d}\right)$$

$$= \frac{\mu}{\pi}\log_e\frac{d}{a} \quad \text{henrys/metre} \quad . \quad . \quad . \quad . \quad (11.55)$$

Including the internal linkage of $\mu_i i_A/8\pi$ per line, the total inductance is

$$L_{eff} = \frac{\mu}{\pi}\log_e\frac{d}{a} + \frac{\mu_i}{4\pi} \quad \text{henrys/metre.} \quad . \quad (11.56)$$

μ_i being the permeability of the conductor material.

Example 11.3. Develop a formula for the inductance per unit length of two parallel, cylindrical wires of equal radius forming the "go" and "return" conductors of a single-phase system.

What is the inductance per unit length of two such parallel wires if each has a diameter of 1 cm and their axes are 5 cm apart? Two cases should be considered: (*a*) when the wires have a relative permeability of unity, and (*b*) when they have a relative permeability of 100. The relative permeability of the surrounding medium is unity in both cases. End effects may be neglected, and the current may be assumed uniformly distributed over a cross-section of the wires. (*I.E.E. Meas.*, 1958)

For an air-spaced line, $\mu = \mu_0$, and if the relative permeability of the conductors is unity, μ_i is also equal to μ_0.

(*a*) From eqn. (11.56), $L = \frac{4\pi \times 10^{-7}}{\pi}\left(\log_e\frac{5}{0.5} + \frac{1}{4}\right) = \underline{\underline{1{\cdot}02\ \mu\text{H/m}}}$

(*b*) $\mu_i = 100\ \mu_0$, and the inductance per metre run is

$$L = \frac{4\pi \times 10^{-7}}{\pi}\left(\log_e\frac{5}{0.5} + \frac{100}{4}\right) = \underline{\underline{10{\cdot}9\ \mu\text{H/m}}}$$

11.21. Inductance of an Isolated Three-phase Line

Maxwell's mesh currents may be used to represent the three line currents i_A, i_B, i_C, in a 3-wire 3-phase system as composed of mesh currents i_1, i_2, i_3, as shown in Fig. 11.21 (*a*), where

$$i_A = i_2 - i_3 \qquad i_B = i_1 - i_2 \qquad i_C = i_3 - i_1 \quad . \ (11.57)$$

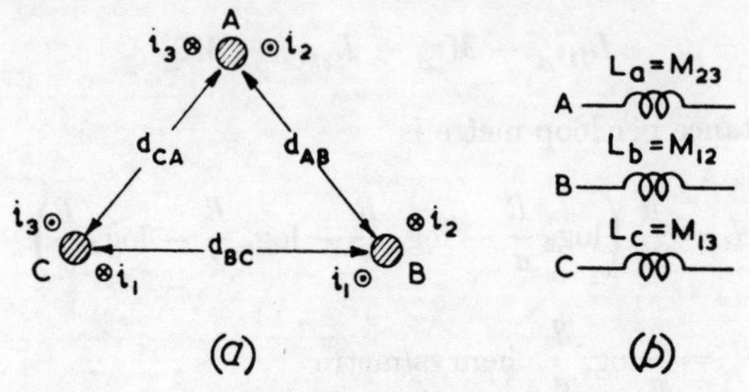

FIG. 11.21. ISOLATED THREE-PHASE LINE

Suppose that the fluxes set up in loops *BC* (loop 1), *AC* (loop 3) and *BA* (loop 2) by a loop current $i_1 = 1$ A (i.e. going down conductor *B* and returning by conductor *C*) are L_{11}, M_{12} and M_{13} respectively. The flux through loop *BC* must pass through either loop *AC* or *BA*. Hence

$$L_{11} = M_{12} + M_{13} \quad . \quad . \quad . \ (11.58)$$

If all three loop currents exist, the total flux (or flux linkage since each loop consists of one turn) through loop 1 is

$$\Phi_1 = L_{11}i_1 - M_{12}i_2 - M_{13}i_3$$

From eqns. (11.57) and (11.58),

$$\Phi_1 = M_{12}(i_1 - i_2) + M_{13}(i_1 - i_3) = M_{12}i_B - M_{13}i_C$$

Hence the induced e.m.f. round loop *BC* is

$$v = \frac{d\Phi_1}{dt} = M_{12}\frac{di_B}{dt} - M_{13}\frac{di_C}{dt} \quad . \quad . \ (11.59)$$

This is the same as the sum of the e.m.f.s induced in a loop consisting of a self-inductance equal to M_{12} carrying i_B amperes and a self-inductance equal to M_{13} carrying i_C amperes, the current directions being as indicated in Fig. 11.21.

Similar considerations apply to the two other loops, and so we may consider the line to be represented by three inductances L_a, L_b, L_c, with no mutual coupling (Fig. 11.21 (*b*)) given by

$$L_b = M_{12} \qquad L_c = M_{13} \qquad L_a = M_{23}$$

These mutual inductances are readily calculated as follows. Suppose that the conductor radius is a, and the spacings are, d_{AB}, d_{BC}, d_{CA}; then, for a 1 m run,

$$\text{Flux linking loop } BC \text{ due to } i_2 \text{ in } B = \int_a^{d_{BC}} \mu H_r \, \mathrm{d}r$$

$$= \frac{\mu i_2}{2\pi} \log_e \frac{d_{BC}}{a}$$

and the flux per ampere is $(\mu/2\pi) \log_e (d_{BC}/a)$.

$$\text{Flux linking loop } BC\text{, due to } -i_2 \text{ in } A = -\int_{d_{AB}}^{d_{AC}} \mu H_r \, \mathrm{d}r$$

$$= -\frac{\mu i_2}{2\pi} \log_e \frac{d_{AC}}{d_{AB}}$$

and the flux per ampere is $-(\mu/2\pi) \log_e (d_{AC}/d_{AB})$.

$$\text{Total flux per ampere} = M_{12} = \frac{\mu}{2\pi}\left(\log_e \frac{d_{BC}}{a} - \log_e \frac{d_{AC}}{d_{AB}}\right)$$

i.e.

$$L_b = \frac{\mu}{2\pi} \log_e \frac{d_{AB} d_{BC}}{a d_{AC}} \quad \text{henrys/metre} \quad . \quad (11.60)$$

In the same way,

$$L_a = \frac{\mu}{2\pi} \log_e \frac{d_{AB} d_{AC}}{a d_{BC}} \quad \text{henrys/metre} \quad . \quad (11.61)$$

and

$$L_c = \frac{\mu}{2\pi} \log_e \frac{d_{AC} d_{BC}}{a d_{AB}} \quad \text{henrys/metre} \quad . \quad (11.62)$$

If the conductors are transposed at regular intervals, so that each occupies each position in turn, the mean inductance of each wire will be

$$L = \frac{1}{3}\frac{\mu}{2\pi}\left(\log_e \frac{d_{AB} d_{AC}}{a d_{BC}} + \log_e \frac{d_{BC} d_{AC}}{a d_{AB}} + \log_e \frac{d_{AB} d_{BC}}{a d_{AC}}\right)$$

$$= \frac{\mu}{2\pi} \log_e \frac{(d_{AB} d_{BC} d_{AC})^{1/3}}{a} \quad \text{henrys/metre} \quad . \quad . \quad (11.63)$$

11.22. Coupling between Transmission Circuits

When alternating currents flow in parallel transmission systems (e.g. a power line in parallel with a communication line), e.m.f.s are induced due to the mutual coupling between the lines. It is particularly important that the e.m.f.s induced by power-line currents in a communication circuit shall be small.

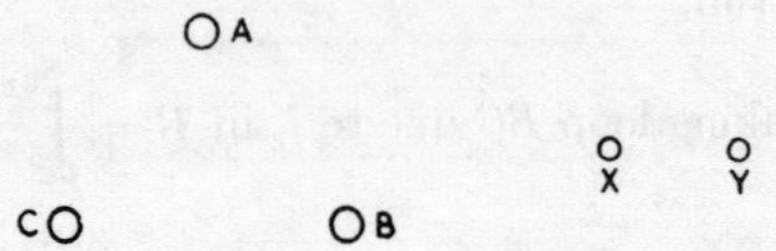

FIG. 11.22. MUTUAL COUPLING BETWEEN TRANSMISSION LINES

Consider a 3-phase line with conductors A, B, C, which runs parallel to a 2-wire circuit XY (Fig. 11.22). The flux linking XY per metre run, due to a current i_A in A is

$$\int_{d_{AX}}^{d_{AY}} \frac{\mu i_A}{2\pi r} \mathrm{d}r = \frac{\mu i_A}{2\pi} \log_e \frac{d_{AY}}{d_{AX}} \quad \text{webers}$$

Hence the mutual flux linkage per ampere is

$$M_{AXY} = \frac{\mu}{2\pi} \log_e \frac{d_{AY}}{d_{AX}}$$

In the same way the mutual linkages per ampere with circuit XY per metre run due to currents i_B in B and i_C in C are

$$M_{BXY} = \frac{\mu}{2\pi} \log_e \frac{d_{BY}}{d_{BX}}$$

and

$$M_{CXY} = \frac{\mu}{2\pi} \log_e \frac{d_{CY}}{d_{CX}}$$

If the currents i_A, i_B and i_C are sinusoidal, with r.m.s. values I_A, I_B, I_C, the resultant e.m.f. induced in XY per metre run is

$$V_{XY} = j\omega I_A M_{AXY} + j\omega I_B \underline{/\phi_B} M_{BXY} + j\omega I_C \underline{/\phi_C} M_{CXY} \quad \text{volts/metre} \quad . \quad (11.64)$$

where I_A is taken as the reference complexor.

If the load on the 3-phase system is balanced, $I_B = I_A \underline{/-120°} = a^2 I_A$ and $I_C = I_A \underline{/120°} = aI_A$. Hence

$$V_{XY} = j\omega I_A(M_{AXY} + a^2 M_{BXY} + aM_{CXY}) \quad \text{volts/metre} \quad (11.65)$$

For regular transposition of the 3-phase line, $M_{AXY} = M_{BXY} = M_{CXY}$, and the average induced e.m.f. in the twin line is

$$V_{XY}' = j\omega I_A(1 + a^2 + a)M_{AXY} = 0$$

Obviously if the twin line is regularly transposed the average induced e.m.f. per metre is zero, irrespective of the balance of the 3-phase system.

11.23. Images in a Perfect Conductor

An alternating flux cannot exist in a perfect conductor, since if it did the e.m.f. it would induce would set up a current which would be limited only by the magnitude required to establish an equal and opposite flux (Lenz's law). Hence an alternating flux can have no component normal to the surface of a perfect conductor, and lines of magnetic flux must therefore be tangential to this surface. If a flat, perfectly conducting plane, is placed below a conductor (Fig.

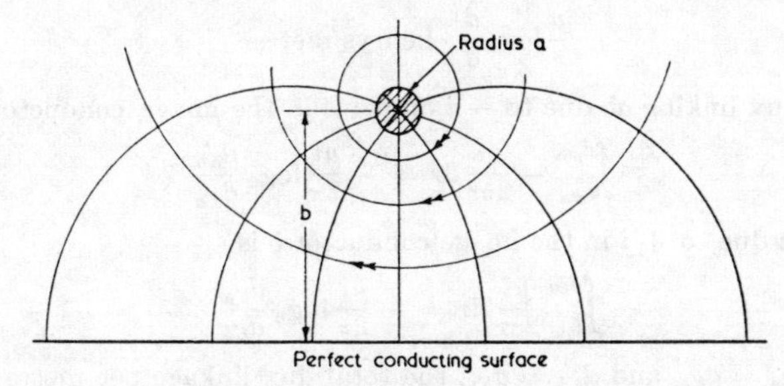

FIG. 11.23. EFFECT OF CONDUCTING EARTH ON A.C. MAGNETIC FIELD

11.23), lines of flux must lie along this plane, giving the field configuration obtained if an image conductor carrying opposite current is placed an equal distance below the plane.

The inductance per metre run of a single wire above a conducting plane is thus half the inductance of the wire and its image, i.e. from eqn. (11.56),

$$L = \frac{\mu}{2\pi}\log_e\frac{2h}{a} + \frac{\mu}{8\pi} \quad \text{henrys/metre at low frequencies} \quad . \quad (11.66)$$

or

$$L = \frac{\mu}{2\pi}\log_e\frac{2h}{a} \quad \text{henrys/metre at high frequencies} \quad . \quad (11.67)$$

Example 11.4. Calculate the inductance per metre of a twin line of conductor of radius 1 cm and spacing 1 m, which runs at a height of 4 m above a perfectly conducting earth. Compare with the inductance if the line were isolated. Neglect internal conductor linkages.

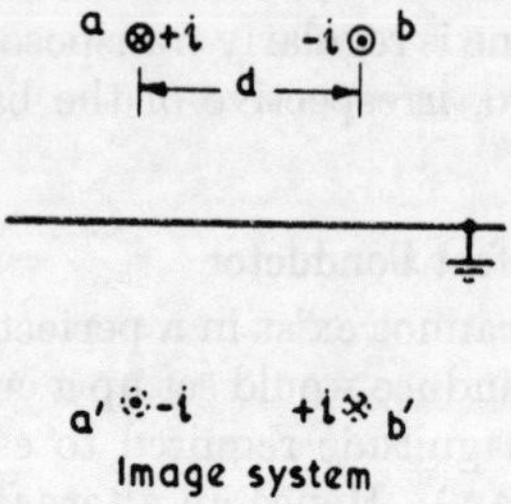

FIG. 11.24

The twin line with its image system is shown in Fig. 11.24. By eqn. (11.55) the self-linkages per ampere for one metre run of the line *ab* is

$$\frac{\mu}{\pi}\log_e\frac{d}{a}\quad \text{henrys/metre}$$

The flux linking *ab* due to $-\,i$ amperes in the image conductor a' is

$$\int_{d_{a'a}}^{d_{a'b}} -\frac{\mu i}{2\pi r}\,dr = -\frac{\mu i}{2\pi}\log_e\frac{d_{a'b}}{d_{a'a}}$$

and that due to $+\,i$ in the image conductor b is

$$\int_{d_{b'a}}^{d_{b'b}} \frac{\mu i}{2\pi r}\,dr = -\frac{\mu i}{2\pi}\log_e\frac{d_{b'a}}{d_{b'b}}$$

Since $d_{a'a} = d_{b'b}$, and $d_{a'b} = d_{b'a}$, the total flux linkage per metre for a current of 1 A (i.e. the self-inductance) is

$$L = \frac{\mu}{\pi}\log_e\frac{d}{a} - \frac{\mu}{\pi}\log_e\frac{d_{a'b}}{d_{a'a}}$$

It will be seen that the presence of the earth plane reduces the effective self-inductance of the circuit by $(\mu/\pi)\log_e(d_{a'b}/d_{a'a})$ henrys/metre compared with the isolated line.

Using the figures given

$$L = 4\times 10^{-7}\left(\log_e\frac{1}{0{\cdot}01} - \log_e\frac{\sqrt{65}}{8}\right)\quad \text{henrys/metre} = 1{\cdot}84\ \mu\text{H/m}$$

For the isolated line, $L = 4\times 10^{-7}\times 4{\cdot}605 = \underline{\underline{1{\cdot}86\ \mu\text{H/m}}}$

11.24. Boundary Conditions

We saw in Section 11.23 that an alternating magnetic field must be parallel to the surface of a perfect conductor (or should be zero at the conductor surface). The magnetizing force, H, is equal to the surface current density (Section 11.13). This is one of the boundary conditions which must be fulfilled by alternating magnetic fields. Static magnetic fields (i.e. those produced by direct currents or permanent magnets) are unaffected by the presence of conductors, provided that the relative permeability of the conductor material is unity. Let us now examine the conditions at the boundary between two materials of permeability μ_1 and μ_2.

Consider the coin-shaped element enclosing such a boundary, and assume that there is no current flowing on either side (Fig. 11.25 (*a*)).

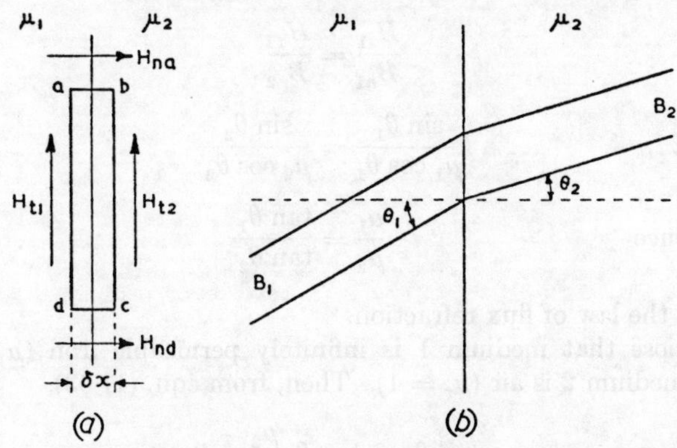

Fig. 11.25. Boundary Conditions for Static Magnetic Fields

Round the closed path *abcd*, $\int H\,\mathrm{d}r = 0$, since no current is linked. Hence

$$H_{t1}da + H_{na}\delta x - H_{t2}bc - H_{nd}\delta x = 0$$

where H_{na} and H_{nd} are the normal magnetizing forces at *a* and *d* respectively.

This gives

$$H_{t1}da - H_{t2}bc = 0 \qquad \text{as } \delta x \to 0$$

or

$$H_{t1} = H_{t2} \quad . \quad . \quad . \quad . \quad (11.68)$$

This is one boundary condition—*that the tangential component of H is continuous across a boundary.*

If the area of the face of the coin-shaped element is δS, and the normal components of flux density are B_{n1} and B_{n2} on each side of the boundary, then $B_{n1}\,\delta S = B_{n2}\,\delta S$, assuming that no magnetic flux emerges from the curved surface as $\delta x \to 0$. Hence

$$B_{n1} = B_{n2} \quad . \quad . \quad . \quad . \quad (11.69)$$

This is the second boundary condition—that *the normal component of flux density is continuous across a boundary.*

Now consider flux lines crossing the boundary as shown in Fig. 11.25 (*b*), with angle of incidence θ_1 and angle of refraction θ_2.

$$H_{t1} = \frac{B_1}{\mu_1} \sin \theta_1 \quad \text{and} \quad H_{t2} = \frac{B_2}{\mu_2} \sin \theta_2 \quad . \quad (11.70)$$

Also $$B_{n1} = B_1 \cos \theta_1 \quad \text{and} \quad B_{n2} = B_2 \cos \theta_2 \quad . \quad (11.71)$$

But, by eqns. (11.68) and (11.69),

$$\frac{H_{t1}}{B_{n1}} = \frac{H_{t2}}{B_{n2}}$$

so that $$\frac{\sin \theta_1}{\mu_1 \cos \theta_1} = \frac{\sin \theta_2}{\mu_2 \cos \theta_2}$$

and hence $$\frac{\mu_1}{\mu_2} = \frac{\tan \theta_1}{\tan \theta_2} \quad . \quad . \quad . \quad (11.72)$$

This is the law of flux refraction.

Suppose that medium 1 is infinitely permeable iron ($\mu_1 \to \infty$) while medium 2 is air ($\mu_r = 1$). Then, from eqn. (11.53),

$$\tan \theta_2 = \tan \theta_1 \frac{\mu_2}{\mu_1} \to 0$$

so that $\theta_2 \to 0$. This shows that magnetic flux emerges into air normal to the surface of infinitely permeable iron.

11.25. Images in Infinitely Permeable Iron

Suppose that a current-carrying conductor is placed parallel to a sheet of infinitely permeable iron (Fig. 11.26 (*a*)). From the previous section we know that at the surface of the iron the magnetic flux

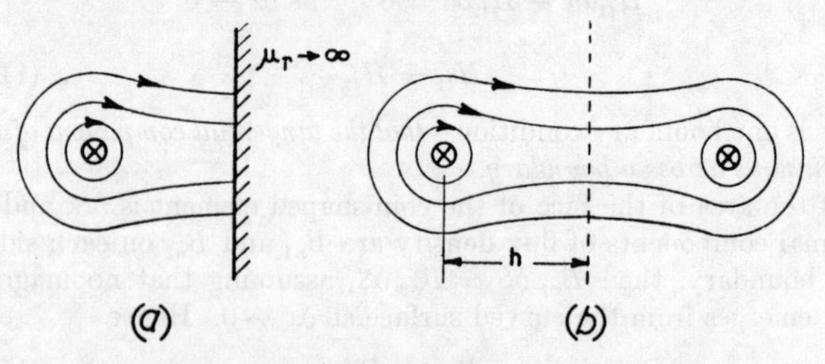

FIG. 11.26. IMAGES IN INFINITELY PERMEABLE IRON

must be normal to the surface. This is the magnetic field configuration produced by the conductor plus an image conductor which carries current in the *same* direction (Fig. 11.26 (*b*)) both currents being direct.

The force between the conductor and the iron must be the same as that which would exist between the conductor and its image. If the conductor carries I amperes, and the perpendicular distance to the plane is h, then, by eqn. (11.18),

$$F = \frac{\mu_0 I^2}{2\pi \times 2h} \text{ newtons/metre} \quad . \quad . \quad (11.73)$$

The above argument may be extended to the case where the conductor lies between intersecting iron surfaces as in Fig. 11.27.

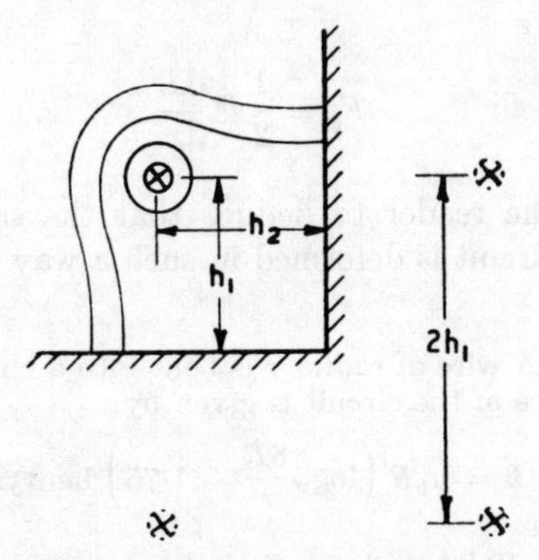

FIG. 11.27. IMAGES IN TWO IRON PLATES

The field pattern will then be the same as that produced by the three image conductors shown, and the force on the conductor will be the vector sum of the forces due to each image conductor.

11.26. Forces Exerted in Inductive Circuits

When the inductance of a circuit is changed by deforming it while it is carrying a current, there is a redistribution of stored energy, and mechanical work is done. At the same time there must be a change in flux linkage, and therefore an induced e.m.f. in the circuit. If the current is held constant, electrical energy must be supplied to or removed from the circuit. Suppose that a circuit of inductance L carries a constant current i, and that this circuit is deformed by a mechanical force F_x acting through a small distance δx in a short time δt. If the deformation reduces the flux linkage of the circuit, the e.m.f., v, induced in the circuit must act, by Lenz's law, in such a direction as to try to maintain the current—hence electrical energy $vi\,\delta t$ is delivered from the circuit to the current source. At the same time the energy stored in the magnetic field is reduced by $\frac{1}{2}i^2\,\delta L$, where δL is the change in inductance, this energy being part of the amount delivered by the circuit. Hence

$$vi\,\delta t - \tfrac{1}{2}i^2\,\delta L = F_x\,\delta x$$

But the rate of change of flux linkage is

$$v = \frac{\delta(Li)}{\delta t} = i\,\frac{\delta L}{\delta t}$$

so that

$$i^2\,\delta L - \tfrac{1}{2}i^2\,\delta L = F_x\,\delta x$$

In the limit

$$F_x = \frac{1}{2}\,i^2\,\frac{\mathrm{d}L}{\mathrm{d}x} \qquad . \quad . \quad . \ (11.74)$$

It is left to the reader to deduce that the same expression is obtained if the circuit is deformed in such a way as to increase the flux linkage.

Example 11.5. A wire of radius r is bent into a circle of mean radius R. The inductance of the circuit is given by

$$L = \mu_0 R\left(\log_e \frac{8R}{r} - 1{\cdot}75\right) \text{ henrys}$$

If the coil carries 10 kA (r.m.s.), find the average tensile force in the conductor for $r = 1$ cm and $R = 100$ cm.

(*A.E.E.*, 1959)

Eqn. (11.74) gives the force in a given direction due to a change of dimensions in that direction. Hence, to find the tensile force we must determine the change of inductance with change in coil circumference, $l\ (= 2\pi R)$. Now,

$$\frac{\mathrm{d}L}{\mathrm{d}l} = \frac{\mathrm{d}L}{\mathrm{d}R}\frac{\mathrm{d}R}{\mathrm{d}l} = \frac{1}{2\pi}\left[\mu_0\left(\log_e\frac{8R}{r} - 1{\cdot}75\right) + \mu_0 R\,\frac{r}{8R}\,\frac{8}{r}\right]$$

$$= 2 \times 10^{-7}\,(\log_e 800 - 1{\cdot}75 + 1) = 2 \times 10^{-7} \times 5{\cdot}94$$

The peak tensile force is therefore (using peak current value)

$$F_m = \tfrac{1}{2}\,10^8(\sqrt{2})^2 \times 2 \times 10^{-7} \times 5{\cdot}94 = \underline{\underline{120\ \mathrm{N}}}$$

The mean tensile force is $120/2 = \underline{\underline{60\ \mathrm{N}}}$

A similar situation arises when two coils carrying constant currents i_1 and i_2 are separated by a force F_x which causes one coil to move a distance δx further from the other in a time δt. The mutual inductance will decrease by some amount δM, and the energy stored in the magnetic field will decrease by $i_1 i_2 \, \delta M$ joules, provided that both i_1 and i_2 enter at "dotted" ends.

The decrease in M gives rise to an induced e.m.f. in coil 1 of $v_1 = \delta M i_2/\delta t$ in such a direction as to attempt to maintain the flux —i.e. by Lenz's law, v_1 acts in the same direction as i_1, and so energy is delivered from the coil. This energy is

$$\frac{\delta M i_2}{\delta t} i_1 \, \delta t$$

In the same way, energy of

$$\frac{\delta M i_1}{\delta t} i_2 \, \delta t$$

is delivered from the second coil. If i_1 and i_2 are held constant, the total energy delivered by both coils is

$$2 i_1 i_2 \, \delta M \quad \text{joules}$$

But $i_1 i_2 \, \delta M$ of this is accounted for by the decrease in stored energy, so that the remainder must represent the mechanical energy, $F_x \, \delta x$, which is supplied. Hence

$$F_x \, \delta x = i_1 i_2 \, \delta M$$

or, in the limit,

$$F_x = i_1 i_2 \frac{\mathrm{d}M}{\mathrm{d}x} \quad \text{newtons} \qquad (11.75)$$

Similar expressions can be derived for torques associated with rotational movements (*see* Sections 14.1 and 14.2)—

$$T_\theta = \frac{1}{2} i^2 \frac{\mathrm{d}L}{\mathrm{d}\theta} \quad \text{for a self-inductance}$$

and

$$T_\theta = i_1 i_2 \frac{\mathrm{d}M}{\mathrm{d}\theta} \quad \text{for a mutual inductance}$$

11.27. Pinch Effect

Recent work on the direct generation of electrical energy from plasmas has emphasized the importance of the forces and pressures which exist in fluid conductors (including gaseous plasmas). Consider a fluid conductor of radius a carrying a current i (Fig. 11.28).

The magnetic flux density at a radius r ($r < a$) is (from Section 11.17 (b))

$$B_r = \frac{\mu i(r^2/a^2)}{2\pi r} = \frac{\mu i r}{2\pi a^2} \quad \text{webers/metre}^2$$

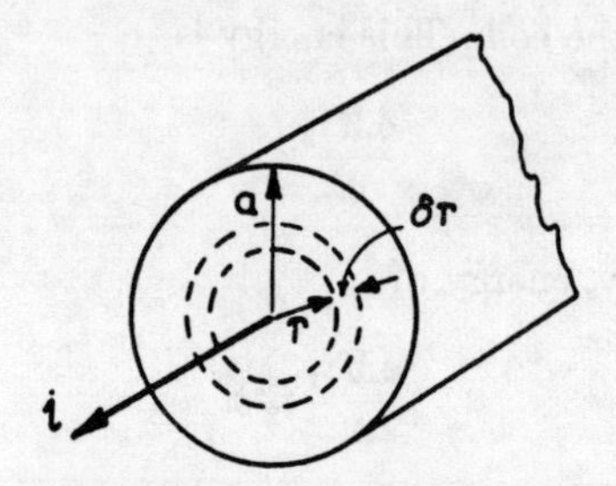

FIG. 11.28. FORCES ON A FLUID CONDUCTOR

The current in the annulus δr is

$$\delta i = \frac{i \,.\, 2\pi r \; \delta r}{\pi a^2} = \frac{2ir \; \delta r}{a^2}$$

Hence the force on δr per metre run of conductor is

$$\delta F = B_r \, \delta i = \frac{\mu i^2 r^2 \; \delta r}{\pi a^4} \quad \text{newtons/metre}$$

This force acts inwards, and the pressure increment at r due to this element is

$$\delta p = \frac{\mu i^2 r^2 \; \delta r}{\pi a^4 2\pi r}$$

The total pressure at r due to all elements outside r is thus

$$p_r = \frac{\mu i^2}{2\pi^2 a^4} \int_r^a r \, \mathrm{d}r = \frac{\mu i^2}{4\pi^2 a^4} (a^2 - r^2) \quad \text{newtons/metre}^2 \quad . \quad (11.76)$$

At the centre of the conductor ($r = 0$) the pressure will be

$$p_0 = \frac{\mu i^2}{4\pi^2 a^2} \quad \text{newtons/metre}^2 \quad . \quad . \quad (11.77)$$

Since pressure in a fluid acts in all directions, the total axial force on the fluid is

$$F_a = \frac{\mu i^2}{4\pi^2 a^4}\int_0^a (a^2 - r^2)2\pi r\, dr = \frac{\mu i^2}{8\pi} \quad \text{newtons} \quad . \quad (11.78)$$

Bibliography

Carter, *The Electromagnetic Field in its Engineering Applications* (Longmans).

Jones, *An Introduction to Advanced Electrical Engineering* (English Universities Press).

Shepherd, Morton and Spence, *Higher Electrical Engineering* (Pitman).

Problems

11.1. A single-turn circular coil of 50 m diameter carries a direct current of 28 × 10^4 A. Assuming Laplace's expression for the magnetizing force due to a current-element, determine the magnetizing force at a point on the axis of the coil and 100 m from the coil. The relative permeability of the space surrounding the coil is unity.

(*Ans.* 79·4 At/m.) (*L.U. Part II Theory*, 1957)

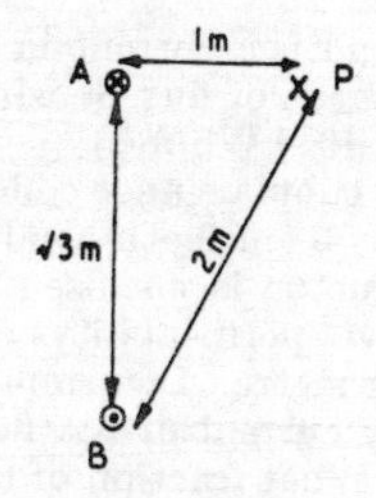

Fig. 11.29

11.2. In Fig. 11.29, *A* and *B* are two long air-insulated conductors of small diameter in a plane normal to that of the paper. The mutual capacitance of the pair of conductors is 6 pF per metre length, and the distance between centres is $\sqrt{3}$ m. The steady potential difference between the conductors is 10 kV, and the direct current carried by the two conductors, in opposite directions, is 100 A. Calculate the magnitude and direction of the electric and magnetic field strengths at point *P*. Conductor *A* is positive with respect to *B*, and carries current downwards into the paper. (*A.E.E.*, 1960)

(*Ans.* 937 V/m; 13·8 At/m.)

11.3. Two large iron plates, one in a horizontal plane and the other in a vertical plane, intersect to form a right-angled corner. A long, straight conductor of small, circular cross-section is mounted in the corner parallel with the plates, at a distance *a* from the vertical plate and *b* from the horizontal plate. If the conductor carries a steady current and the iron is of very high magnetic permeability, make a sketch to show the magnetic field distribution in the space round the conductor and between the plates.

Calculate the magnitude and direction of the force on the conductor per metre length if the conductor carries a current of 1,000 A and if $a = 30$ cm and $b = 40$ cm. (*A.E.E.*, 1958)

(*Ans.* 0·61 N/m, at 42° to horizontal plate.)

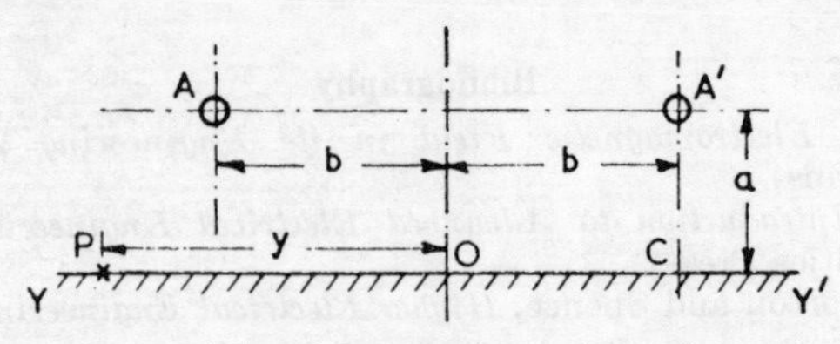

FIG. 11.30

11.4. In Fig. 11.30, A and A' are long parallel conductors which are also parallel to the surface of an infinite plate YOY' of infinitely permeable iron. The conductors carry equal steady currents in opposite directions. Derive an expression for the normal component of magnetic flux density B at any point P on the surface of the plate. Find values of y for which B is zero.

If $a = 4$ cm, $b = 6$ cm, and the current in each conductor is 100 A, calculate the normal component of flux density at the point C.

(*Ans.* $\pm\sqrt{(b^2 + a^2)}$; 3×10^{-4} Wb/m².) (*A.E.E.*, 1961)

11.5. A coaxial line has a tubular inner conductor of radius 1 cm and an outer conductor of radius 5 cm, both conductors being of negligible thickness. The inner conductor is enclosed by a ferrite of constant thickness 1 cm and relative permeability 50. Calculate from first principles the inductance per metre. Determine also the stored magnetic energy per metre if a steady current of 1 A flows in one conductor and returns through the other. What fraction of this energy is stored in the ferrite? (*L.U. Part II Theory*, 1961)

(*Ans.* 7·11 μH/m; 3·55 μJ/m; 97·4 per cent.)

11.6. The three conductors A, B and C of a 3-phase transmission line are equilaterally spaced 1 m apart with conductor B vertically above conductor C. The line carries a balanced load of 10 MVA at 33 kV line voltage. Calculate the mechanical force per metre length, in magnitude and direction, produced magnetically on each of the three conductors at the instant of peak current in conductor A. (*A.E.E.*, 1961)

(*Ans.* $10{\cdot}6 \times 10^{-3}$ N, horizontal; $5{\cdot}3 \times 10^{3}$ N, horizontal; $5{\cdot}3 \times 10^{-3}$ N, horizontal.)

11.7. A long concentric cable consists of a solid central cylindrical conductor of radius a covered with insulation to a radius b over which is a thick annular conductor of inner radius b and outer radius c. The cable carries current in opposite directions in the two conductors, and the current density in each conductor is uniform. Obtain expressions for the magnetic field strength H at any radius (i) inside the central conductor, (ii) in the insulation, (iii) in the outer conductor, (iv) outside the cable.

Sketch the variation of H with radius.

Obtain an expression for the self-inductance of the cable, and calculate

its value per metre length if $a = 0{\cdot}5$ cm, $b = 1{\cdot}0$ cm and $c = \sqrt{2}$ cm. The relative permeability of the conductors and insulation is unity.
(*Ans.* $2{\cdot}17 \times 10^{-7}$ H/m.) (*A.E.E.*, 1960)

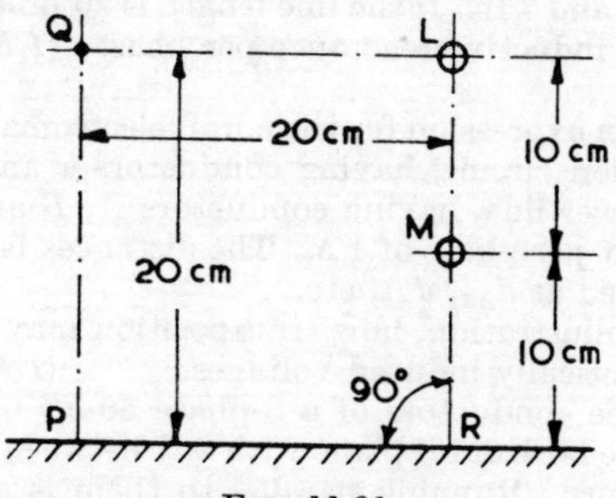

Fig. 11.31

11.8. In Fig. 11.31, L and M represent long, thin parallel conductors, perpendicular to the plane of the paper, and parallel to the surface of an infinite iron plate of high permeability represented by PR. The conductors each carry a steady current of 2,000 A in the directions indicated. Calculate (*a*) the magnetic field strength H in magnitude and direction at the points P and Q, and (*b*) the mechanical force per metre length on conductor L. (*A.E.E.*, 1963)
(*Ans.* 1,500 At/m at 28° to horizontal; 2,390 At/m vertically upwards; 8·67 N/m repulsion.)

11.9. An air-cored solenoid of length $2l$ is uniformly wound with fine wire. There are n turns per unit length, the mean radius of the turns is r and there is a steady current i in the winding. Derive an expression for the magnetic field strength at any point along the axis of the solenoid.

The solenoid is to be used as the primary of a standard mutual inductor. The secondary consists of a short cylinder, wound with 50 turns of fine wire, placed coaxially and centrally inside the solenoid. If $n = 100$ turns/cm and the ratio $l/r = 2$, calculate the required mean radius of the secondary coil to give a mutual inductance $M = 1$ mH.

The secondary winding is now moved from the centre to one end of the primary solenoid. Estimate the new value of M. (*A.E.E.*, 1963)
(*Ans.* 2·38 cm; 0·543 mH.)

11.10. Derive an expression for the magnetic field strength at any point on the axis of a single-turn coil carrying a steady current I, (*a*) when the coil is in the form of a circle of radius r, and (*b*) when it is square with sides of length $2a$. Show that, if the circle and the square have the same area, the respective field strengths at the centres of the coils are in the ratio $\sqrt{(\pi^3/32.)}$ (*A.E.E.*, 1962.)

11.11. A concentric cable consists of two thin-walled tubes, of mean radii r and R. Derive an expression for the inductance of the cable per unit length.

The cable carries a sinusoidal short-circuit current of 3 kA(r.m.s.). The mean radius of the outer tube is 2 cm, and its radial thickness is 2 mm. Assuming uniform current density in the conductors, calculate the peak tensile (hoop) stress in the material of the outer tube.
(*Ans.* 7,160 N/m².) (*A.E.E.*, 1962)

11.12. Explain the technical advantages arising from the transposition of the conductors in an overhead transmission line. A 3-phase 50-c/s overhead line is transposed and has conductors of 0·25 in. radius, at spacings of 4, 5 and 7 ft. If the line length is 20 miles, determine from first principles its inductive reactance per phase. (*I.E.E. Supply*, 1954)
(*Ans.* 13 Ω.)

11.13. Derive an expression for the e.m.f. electromagnetically induced in a communication circuit, having conductors x and y, by a parallel 3-phase 50-c/s power line, having conductors A, B and C and carrying a balanced current per phase of 1 A. The distances between conductors should be expressed as d_{AX}, d_{AB}, etc.

Show, with an illustration, how transposition may be used to reduce such electromagnetically induced voltages. (*I.E.E. Supply*, 1959)

11.14. The three conductors of a 3-phase 50-c/s 66-kV transmission line lie in the same horizontal plane with 9 ft between the centre-lines of adjacent conductors. Running parallel to them is a 2-wire telephone circuit with its two wires vertically below the centre conductor at 15 ft and 17 ft, respectively, from it. If the 3-phase line is carrying a balanced load of 25 MVA, estimate the voltage per loop-mile induced in the telephone circuit. Develop any formulae used.
(*Ans.* 0·7 V/mile.) (*I.E.E. Supply*, 1961)

11.15. A long fluid conductor of circular cross-section, radius R, and magnetic permeability μ, carries an electric current i. Assuming a uniform current density in the conductor, show from first principles that the electromagnetically produced hydrostatic pressure at a point in the conductor distant γ from the axis is

$$\frac{\mu i^2(R^2 - \gamma^2)}{4\pi^2 R^4}$$

A gaseous conductor has a radius of 5 cm when the current is 200 kA. Calculate the pressure at the centre of the conductor. (*A.E.E.*, 1960)
(*Ans.* 510,000 N/m².)

11.16. Give a concise proof that the electromagnetic force F in newtons on any conductor of a circuit of inductance L carrying a current i may be deduced from an expression of the form $F = \frac{1}{2} i^2 (dL/dy)$, where a motion dy of the conductor in the direction of the force increases the circuit inductance by an amount dL.

A rectangular circuit of circular-section conductors of radius r consists of two vertical rods of length y and two horizontal rods of length x. The inductance of the circuit is a function of x, y and r such that

$$\frac{\mathrm{d}L}{\mathrm{d}y} = \frac{\mu_0}{\pi}\left\{\frac{\sqrt{(x^2 + y^2)}}{y} - 0{\cdot}75 + \log_e \frac{2xy}{r[y + \sqrt{(x^2 + y^2)}]}\right\}$$

where $\mathrm{d}L/\mathrm{d}y$ is in henrys per metre.

The circuit, which carries an alternating current of 10 kA(r.m.s.), has the dimensions $x = 1$ m, $y = 2$ m and $r = 1$ cm. Calculate the peak and the average electromagnetic force on the bottom conductor.
(*Ans.* 195 N; 97·5 N.) (*A.E.E.*, 1960)

CHAPTER 12

Time-varying Electromagnetic Fields

WHEN a magnetic field changes, we have already seen that there is an e.m.f. induced in any conductor which links the changing magnetic flux. That is to say, a changing magnetic field is associated with an electric field. It will be shown in this chapter that a changing electric field has an associated magnetic field. Hence alternating fields always have both electric and magnetic components. Radio waves and light waves are examples of alternating electromagnetic fields which radiate in free space.

12.1. Divergence in an Electric Field

Suppose that an electric field characterized by an electric flux density $\boldsymbol{D}$ exists in a co-ordinate space. For the elementary cube $\delta x\, \delta y\, \delta z$ shown in Fig. 12.1, let the charge density in space be ρ

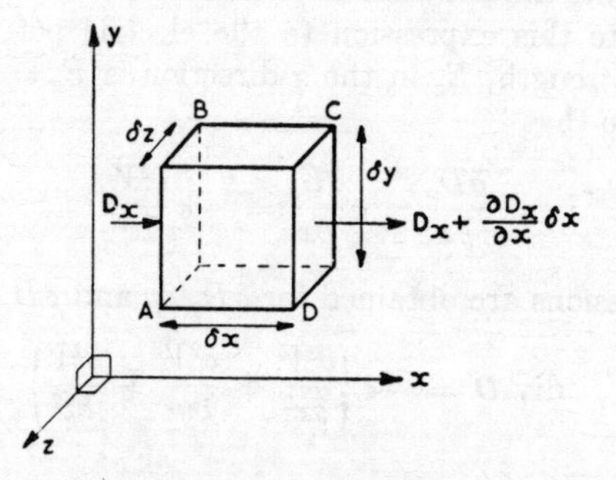

FIG. 12.1. DIVERGENCE OF ELECTRIC FLUX DENSITY

coulombs per cubic metre, so that the total charge enclosed by the element is $\rho\, \delta x\, \delta y\, \delta z$. By Gauss' theorem, this is also the net outward normal electric flux from the element. If D_x is the normal component of electric flux density on face AB, the value over face CD is

$$D_x + \frac{\partial D_x}{\partial x}\, \delta x$$

and we can write

$$\text{Inward normal flux on face } AB = D_x\,\delta y\,\delta z$$

$$\text{Outward normal flux from face } CD = \left(D_x + \frac{\partial D_x}{\partial x}\,\delta x\right)\delta y\,\delta z$$

Hence net outward normal flux in the x-direction is

$$\frac{\partial D_x}{\partial x}\,\delta x\,\delta y\,\delta z$$

Similar expressions will also hold for the net outward normal fluxes in the y- and z-directions, so that

$$\text{Total outward normal flux} = \left(\frac{\partial D_x}{\partial x} + \frac{\partial D_y}{\partial y} + \frac{\partial D_z}{\partial z}\right)\delta x\,\delta y\,\delta z$$

The expression in brackets is called the *divergence* of the vector $\boldsymbol{D}$, or div $\boldsymbol{D}$. Hence

$$\text{div}\,\boldsymbol{D}\,\delta x\,\delta y\,\delta z = \rho\,\delta x\,\delta y\,\delta z$$

or

$$\text{div}\,\boldsymbol{D} = \rho \qquad (12.1)$$

Div $\boldsymbol{D}$ represents the net outflow of flux from a unit volume of field. We may relate this expression to the electric potential since the electric field strength, E_x in the x-direction is $E_x = -\partial V/\partial x$ (from eqn. (10.5)), so that

$$\frac{\partial D_x}{\partial x} = \epsilon\,\frac{\partial E_x}{\partial x} = -\epsilon\,\frac{\partial^2 V}{\partial x^2}$$

Similar expressions are obtained for $\partial D_y/\partial y$ and $\partial D_z/\partial z$. Hence

$$\text{div}\,\boldsymbol{D} = -\epsilon\left\{\frac{\partial^2 V}{\partial x^2} + \frac{\partial^2 V}{\partial y^2} + \frac{\partial^2 V}{\partial z^2}\right\}$$

and

$$\frac{\partial^2 V}{\partial x^2} + \frac{\partial^2 V}{\partial y^2} + \frac{\partial^2 V}{\partial z^2} = -\frac{\rho}{\epsilon} \qquad (12.2)$$

This is known as *Poisson's equation.* If there is no distributed charge in the space considered, $\rho = 0$, and Laplace's equation is obtained, i.e.

$$\frac{\partial^2 V}{\partial x^2} + \frac{\partial^2 V}{\partial y^2} + \frac{\partial^2 V}{\partial z^2} = 0 \qquad (12.3)$$

12.2. Divergence of Magnetic Flux

In the magnetic field produced by a current, the flux paths form closed loops, so that there can be no net outflow of magnetic flux from any element of volume. By similar reasoning to that used in Section 12.1, the net outflow from a volume element $\delta x\ \delta y\ \delta z$ is $\operatorname{div} \boldsymbol{B}\ \delta x\ \delta y\ \delta z$, and since this must be zero

$$\operatorname{div} \boldsymbol{B} = 0 \qquad . \quad . \quad . \quad . \quad (12.4)$$

12.3. Divergence of Current

Suppose that a current flows in the co-ordinate space shown in Fig. 12.1. If the current density on face AB is J_x, the total outward current from the cube $\delta x\ \delta y\ \delta z$ in the x-direction is

$$(\partial J_x/\partial x)\ \delta x\ \delta y\ \delta z$$

by similar reasoning to that used in Section 12.1. The total outward current in all directions is therefore

$$\left(\frac{\partial J_x}{\partial x} + \frac{\partial J_y}{\partial y} + \frac{\partial J_z}{\partial z}\right) \delta x\ \delta y\ \delta z = \operatorname{div} \boldsymbol{J}\ \delta x\ \delta y\ \delta z$$

But the outward current is equal to the rate of change of enclosed charge, so that

$$\operatorname{div} \boldsymbol{J}\ \delta x\ \delta y\ \delta z = \frac{\mathrm{d}}{\mathrm{d}t}(\rho\ \delta x\ \delta y\ \delta z)$$

where ρ is the charge density of the enclosed charge. Hence

$$\operatorname{div} \boldsymbol{J} = \frac{\mathrm{d}\rho}{\mathrm{d}t} \quad . \quad . \quad . \quad . \quad (12.5)$$

12.4. Displacement Current

When an alternating voltage is applied to a capacitor, an ammeter indicates that a current flows in the connecting wires. No conduction current can exist in the dielectric between the plates (assuming the

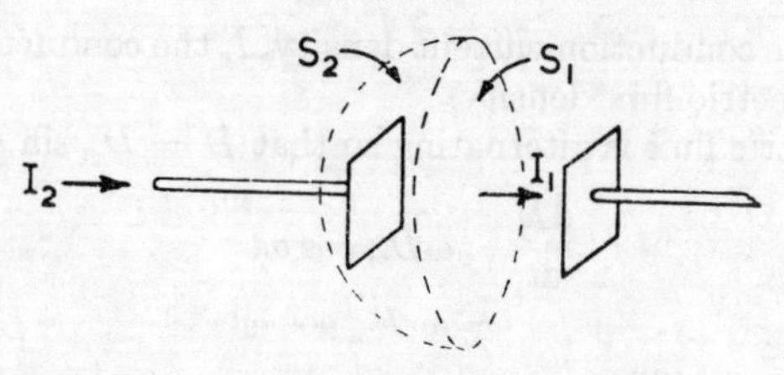

FIG. 12.2. DISPLACEMENT CURRENT

dielectric to be perfect), but Maxwell postulated that the alternating field causes a displacement current, which may be thought of as due to the rate of change of charge on the capacitor plates, and thus numerically equal to the rate of change of electric flux between them.

Consider the two plates shown in Fig. 12.2. The surfaces S_1 and S_2 are traversed by the conduction currents I_1 and I_2, and the rate of increase of charge on the enclosed plate is

$$\frac{\mathrm{d}q}{\mathrm{d}t} = I_2 - I_1$$

If the electric fluxes across S_1 and S_2 are Ψ_1 and Ψ_2, then by Gauss' theorem,

$$q = \Psi_1 - \Psi_2$$

Hence

$$\frac{\mathrm{d}q}{\mathrm{d}t} = \frac{\mathrm{d}\Psi_1}{\mathrm{d}t} - \frac{\mathrm{d}\Psi_2}{\mathrm{d}t}$$

so that

$$I_2 - I_1 = \frac{\mathrm{d}\Psi_1}{\mathrm{d}t} - \frac{\mathrm{d}\Psi_2}{\mathrm{d}t}$$

and

$$I_1 + \frac{\mathrm{d}\Psi_1}{\mathrm{d}t} = I_2 + \frac{\mathrm{d}\Psi_2}{\mathrm{d}t} = I_T \qquad . \qquad . \quad (12.6)$$

In this equation I_1 and I_2 are the normal conduction currents, while $\mathrm{d}\Psi_1/\mathrm{d}t$ and $\mathrm{d}\Psi_2/\mathrm{d}t$ are called the *displacement currents*. The *total current*, I_T, is the sum of the conduction and displacement currents. In conductors the displacement current is negligible, while in insulators the conduction current is usually negligible.

It is often convenient to write eqn. (12.6) in terms of current densities, by dividing throughout by the area, A. Thus

$$J_T = \frac{I_T}{A} = \frac{I_c}{A} + \frac{\mathrm{d}\left(\frac{\Psi}{A}\right)}{\mathrm{d}t} = J_c + \frac{\mathrm{d}D}{\mathrm{d}t} \quad . \qquad . \quad (12.7)$$

where J_c is the conduction current density, I_c the conduction current, and D the electric flux density.

If the electric flux is alternating so that $D = D_m \sin \omega t$, then

$$\begin{aligned}\frac{\mathrm{d}D}{\mathrm{d}t} &= \omega D_m \cos \omega t \\ &= \omega \epsilon E_m \cos \omega t \\ &= \omega \epsilon \rho J_c \cos \omega t \qquad . \quad (12.8)\end{aligned}$$

This follows from $V/I = R = \rho l/A$, where ρ is the resistivity of the material. Hence

$$E = \frac{V}{l} = \rho \frac{I}{A} = \rho J$$

For a copper conductor at 15°C, $\rho = 0{\cdot}017\ \mu\Omega$-m, and $\epsilon = \epsilon_0 = 1/(36\pi \times 10^9)$ so that $2\pi\epsilon_0\rho \simeq 10^{-18}$, and hence the displacement current density will be negligible compared to the conduction current density up to some 10^{18} c/s.

The introduction of displacement current affects the magnetic circuit law of eqn. (11.16), which must now be modified to read

$$\oint H\,\mathrm{d}s = I_T = I_c + \frac{\mathrm{d}\Psi}{\mathrm{d}t} \qquad . \quad . \quad . \quad (12.9)$$

12.5. Relation between *H* and *J* for Distributed Currents

Let us suppose that at a point P in space, the *total* current density, J_T, has components J_{Tx}, J_{Ty}, J_{Tz} in the mutually perpendicular x-, y- and z-directions (Fig. 12.3), and that the magnetizing force of the associated magnetic field has components H_x, H_y, H_z in these

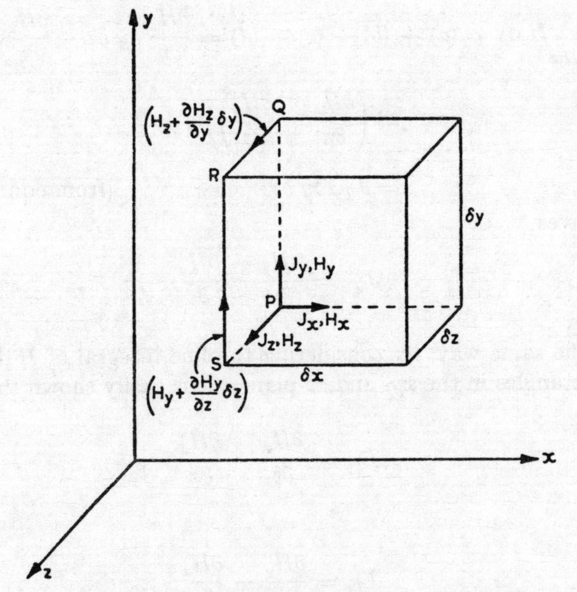

FIG. 12.3. DISTRIBUTED CURRENTS AND FIELDS

same directions. An elementary cube with sides δx, δy, δz is now constructed with P at one corner. For the loop $PQRS$, by eqn. (12.9),

$$\oint \boldsymbol{H}\, \mathrm{d}\boldsymbol{s} = I_{Tx} = J_{Tx}\, \delta y\, \delta z \quad . \quad . \quad . \quad (12.10)$$

If each side of this loop is considered separately, we obtain

(*a*) For side PQ, $\displaystyle\int_P^Q \boldsymbol{H}\, \mathrm{d}\boldsymbol{s} = H_y\, \delta y$

(*b*) For side QR, $\displaystyle\int_Q^R \boldsymbol{H}\, \mathrm{d}\boldsymbol{s} = \left(H_z + \frac{\partial H_z}{\partial y}\, \delta y\right) \delta z$

(*c*) For side RS, $\displaystyle\int_R^S \boldsymbol{H}\, \mathrm{d}\boldsymbol{s} = -\left(H_y + \frac{\partial H_y}{\partial z}\, \delta z\right) \delta y$

(*d*) For side SP, $\displaystyle\int_S^P \boldsymbol{H}\, \mathrm{d}\boldsymbol{s} = -H_z\, \delta z$

since the magnetizing force on side QR is $[H_z + (\partial H_z/\partial y)\, \delta y]$ and on side RS is $-[H_y + (\partial H_y/\partial z)\, \delta z]$.

Hence, adding the components round the loop,

$$\oint_{PQRS} \boldsymbol{H}\, \mathrm{d}\boldsymbol{s} = (a) + (b) + (c) + (d) = \frac{\partial H_z}{\partial y}\, \delta y\, \delta z - \frac{\partial H_y}{\partial z}\, \delta y\, \delta z$$

$$= \left(\frac{\partial H_z}{\partial y} - \frac{\partial H_y}{\partial z}\right) \delta y\, \delta z$$

$$= J_{Tx}\, \delta y\, \delta z \qquad \text{(from eqn. (12.10))}$$

This gives

$$J_{Tx} = \frac{\partial H_z}{\partial y} - \frac{\partial H_y}{\partial z} \quad . \quad . \quad . \quad (12.11a)$$

In the same way, by considering the line integral of $\boldsymbol{H}\, \mathrm{d}\boldsymbol{s}$ round the rectangles in the xy- and xz-planes, it is easily shown that

$$J_{Ty} = \frac{\partial H_x}{\partial z} - \frac{\partial H_z}{\partial x} \quad . \quad . \quad . \quad (12.11b)$$

and

$$J_{Tz} = \frac{\partial H_y}{\partial x} - \frac{\partial H_x}{\partial y} \quad . \quad . \quad . \quad (12.11c)$$

In free space there will be no conduction current density and the above equations will become (from eqn. (12.7))

$$\frac{\partial H_z}{\partial y} - \frac{\partial H_y}{\partial z} = \frac{\mathrm{d}D_x}{\mathrm{d}t} = \epsilon_0 \frac{\mathrm{d}E_x}{\mathrm{d}t} \quad . \quad . \quad (12.12a)$$

$$\frac{\partial H_x}{\partial z} - \frac{\partial H_z}{\partial x} = \epsilon_0 \frac{\mathrm{d}E_y}{\mathrm{d}t} \quad . \quad . \quad . \quad (12.12b)$$

$$\frac{\partial H_y}{\partial x} - \frac{\partial H_x}{\partial y} = \epsilon_0 \frac{\mathrm{d}E_z}{\mathrm{d}t} \quad . \quad . \quad . \quad (12.12c)$$

These equations show that the space rate of change of a magnetic field at a point in free space depends on the time rate of change of the electric field at the same point.

12.6. Faraday's Law—a Second Look

When a single-turn conductor is linked by a changing magnetic flux, the e.m.f. induced in it is given by

$$v = \frac{\mathrm{d}\Phi}{\mathrm{d}t}$$

and the direction of v is found by Lenz's law, as illustrated in Fig. 12.4 (*a*).

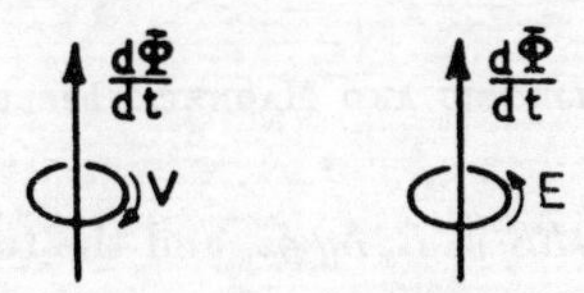

Fig. 12.4. E.M.F. and Potential Gradient due to a Changing Magnetic Flux

Even if there is no conductor, the changing flux sets up a potential gradient, $\boldsymbol{E}$, such that $\oint \boldsymbol{E}\,\mathrm{d}\boldsymbol{s}$ round a closed loop is equal to v (Fig. 12.4 (*b*)). In vector analysis the conventional positive directions of an axis and a rotation are related by the right-hand screw rule. Since the direction of $\boldsymbol{E}$ in Fig. 12.4 (*b*) is physically opposite to this in relation to the axis of $\mathrm{d}\Phi/\mathrm{d}t$, a negative sign is included in the vector relation, so that we write

$$\oint \boldsymbol{E}\,\mathrm{d}\boldsymbol{s} = -\frac{\mathrm{d}\Phi}{\mathrm{d}t} \quad . \quad . \quad . \quad (12.13)$$

Consider, now, a point P in space where the components of a varying magnetic flux density on three mutually perpendicular directions are B_x, B_y, B_z (Fig. 12.5). The components of potential gradient associated with this changing flux are E_x, E_y, E_z. If an elementary cube is now constructed with one corner at P, the total

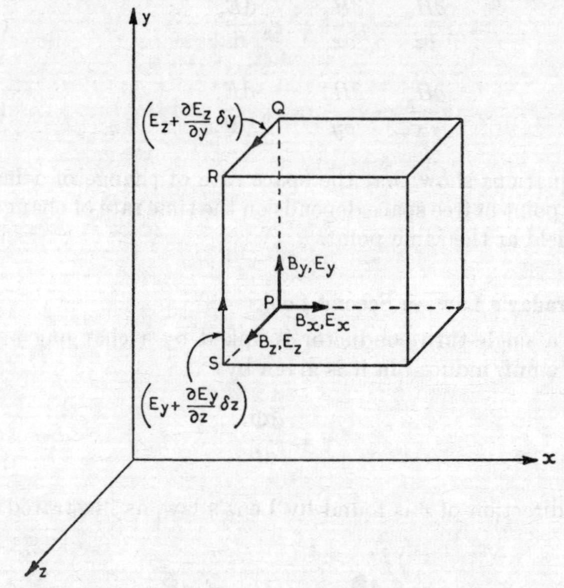

FIG. 12.5. ELECTRIC AND MAGNETIC FIELDS IN SPACE

flux through face $PQRS$ is $B_x \, \delta y \, \delta z$, and the total induced voltage round the loop is

$$v = -E_y \,.\, PQ - \left(E_z + \frac{\partial E_z}{\partial y}\delta y\right).\, QR + \left(E_y + \frac{\partial E_y}{\partial z}\delta z\right).\, RS + E_z \,.\, SP$$

$$= -\frac{\partial E_z}{\partial y}\delta y\, \delta z + \frac{\partial E_y}{\partial z}\delta y\, \delta z$$

But $v = \dfrac{\partial \Phi_x}{\partial t} = \dfrac{\partial B_x}{\partial t}\delta x\, \delta y$ for loop $PQRS$. Hence

$$\frac{\partial E_z}{\partial y} - \frac{\partial E_y}{\partial z} = \frac{\partial B_x}{\partial t} \qquad . \quad . \quad . \quad (12.14a)$$

In the same way,

$$\frac{\partial E_x}{\partial z} - \frac{\partial E_z}{\partial x} = -\frac{\partial B_y}{\partial t} \quad . \quad . \quad . \quad (12.14b)$$

and

$$\frac{\partial E_y}{\partial x} - \frac{\partial E_x}{\partial y} = -\frac{\partial B_z}{\partial t} \quad . \quad . \quad . \quad (12.14c)$$

These equations give the relation between the time variation of a magnetic field and the space variation of the associated electric field.

Eqns. (12.1) and (12.4) (i.e. div $\boldsymbol{D} = \rho$ and div $\boldsymbol{B} = 0$), together with eqns. (12.11) and (12.14) are known as *Maxwell's equations.*

12.7. Solution of Maxwell's Equations for a Plane Wave in Free Space

The simplest form of electromagnetic field in free space is one in which the electric field is in one direction only. The variation of this electric field in space may be represented by a wave as shown in

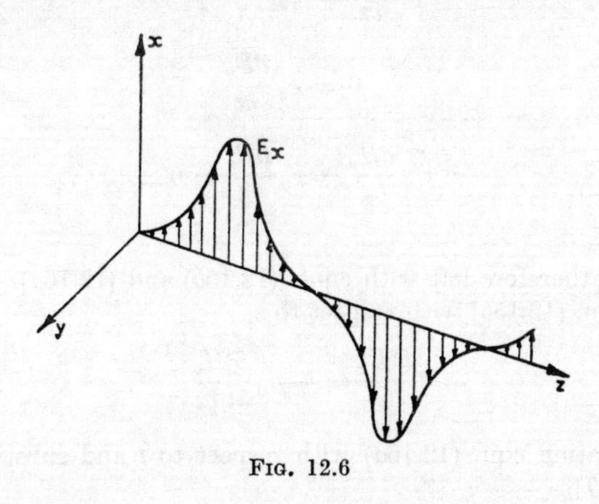

Fig. 12.6

Fig. 12.6. The field is directed parallel to the x-direction and varies in magnitude along the z-direction, but not along the y-direction. This is called a *plane wave.* The components E_y and E_z are zero, and so

$$\frac{\partial E_y}{\partial z} = \frac{\partial E_y}{\partial x} = \frac{\partial E_z}{\partial y} = \frac{\partial E_z}{\partial x} = \frac{\partial E_x}{\partial y} = 0$$

Only $\partial E_x/\partial z$ has any value; hence eqns. (12.14) become

$$-\frac{\partial B_x}{\partial t} = 0 = -\mu_0 \frac{\partial H_x}{\partial t} \quad . \quad . \quad . \quad (12.15a)$$

$$-\frac{\partial B_y}{\partial t} = \frac{\partial E_x}{\partial z} = -\mu_0 \frac{\partial H_y}{\partial t} \quad . \quad . \quad (12.15b)$$

$$-\frac{\partial B_z}{\partial t} = 0 = -\mu_0 \frac{\partial H_z}{\partial t} \quad . \quad . \quad . \quad (12.15c)$$

It follows that H_x and H_z must either be constant or zero, and since we are considering alternating fields only, it further follows that $H_x = H_z = 0$. The magnetic field has components in the y-direction only, and so it, too, is a plane wave. It is thus obvious that

$$\frac{\partial H_z}{\partial y} = \frac{\partial H_z}{\partial x} = \frac{\partial H_x}{\partial y} = \frac{\partial H_x}{\partial z} = 0$$

Eqns. (12.12) can now be reduced to

$$-\frac{\partial H_y}{\partial z} = \epsilon_0 \frac{\partial E_x}{\partial t} \quad . \quad . \quad . \quad . \quad (12.16a)$$

$$0 = \epsilon_0 \frac{\partial E_y}{\partial t} \quad . \quad . \quad . \quad . \quad (12.16b)$$

$$\frac{\partial H_y}{\partial x} = \epsilon_0 \frac{\partial E_z}{\partial t} = 0 \quad . \quad . \quad . \quad (12.16c)$$

since $E_z = 0$.

We are therefore left with eqns. (12.15*b*) and (12.16*a*). Differentiating eqn. (12.15*b*) with respect to z,

$$\frac{\partial^2 E_x}{\partial z^2} = -\mu_0 \frac{\partial^2 H_y}{\partial t\, \partial z} \quad . \quad . \quad . \quad (12.17)$$

Differentiating eqn. (12.16*a*) with respect to t and substituting in eqn. (12.17),

$$\frac{\partial^2 E_x}{\partial z^2} = \mu_0 \epsilon_0 \frac{\partial^2 E_x}{\partial t^2} \quad . \quad . \quad . \quad (12.18)$$

Solving for E_x,

$$E_x = Af(t - z/c) + Bf(t + z/c) \quad . \quad . \quad (12.19)$$

where A and B are constants, and $c = 1/\sqrt{(\mu_0 \epsilon_0)}$. This result may be checked by substituting back in eqn. (12.18). Note that this is also the general solution of the loss-free transmission line equation in Chapter 7.

Consider the first part of the solution for E_x, namely $E_{x1} = A\,f(t - z/c)$. At some time δt seconds later, there must be an increment δz of z for which

$$t - \frac{z}{c} = t + \delta t - \frac{z + \delta z}{c} \quad . \quad . \quad . \quad (12.20)$$

This means that the curve of E_{x1} is repeated at a distance δz further in the positive z-direction after δt seconds. From eqn. (12.20),

$$\delta t = \frac{\delta z}{c} \qquad \text{or} \qquad \frac{\delta z}{\delta t} = c$$

But $\delta z/\delta t$ is the velocity of the wave in the z-direction, and so $A\,f(t - z/c)$ represents a wave travelling at a velocity $c = 1/\sqrt{(\mu_0\epsilon_0)}$ in the positive z-direction. $A\,f(t - z/c)$ is called a *retarded function.*

In the same way, $B\,f(t + z/c)$ represents a wave motion travelling at velocity c in the negative z-direction.

12.8. Plane Sinusoidal Waves in Free Space

Let us suppose that the time variation of a plane electric field in free space is sinusoidal so that we may write for the forward wave

$$E_{x1} = E_{xm} \cos \omega(t - z/c) = E_{xm} \cos (\omega t - \omega z/c)$$

Notice that the time variation of E_x for a given value of z is also the space variation of E_x for a given value of t. Also since $f\lambda = c$, where λ is the wavelength,

$$\frac{\omega}{c} = \frac{2\pi}{\lambda} \quad . \quad . \quad . \quad . \quad (12.21)$$

This is the phase-change coefficient β, in radians per metre, as derived for transmission lines. Thus

$$\beta = \omega/c = \omega\sqrt{(\mu_0\epsilon_0)} \quad \text{radians/metre} \quad . \quad (12.21a)$$

The magnetic field associated with this electric field is given by eqn. (12.15*b*), since if

$$\frac{\partial E_x}{\partial z} = -\mu_0 \frac{\partial H_y}{\partial t}$$

then

$$H_y = -\frac{1}{\mu_0} \int \frac{\partial E_x}{\partial z}\, dt$$

The constant of integration is zero, since there is no d.c. component of magnetic field, and hence

$$H_y = -\frac{1}{\mu_0}\int E_{xm}\frac{\omega}{c}\sin\left(\omega t - \frac{\omega z}{c}\right)\mathrm{d}t$$

$$= \frac{1}{\mu_0 c}E_{xm}\cos\left(\omega t - \frac{\omega z}{c}\right) \quad . \quad . \quad . \quad (12.22)$$

The waves are shown in Fig. 12.7 (*a*). The magnetic-field wave is a retarded function, travelling along the z-axis at the velocity of light. Peak values of E_x and H_y occur at the same values of z.

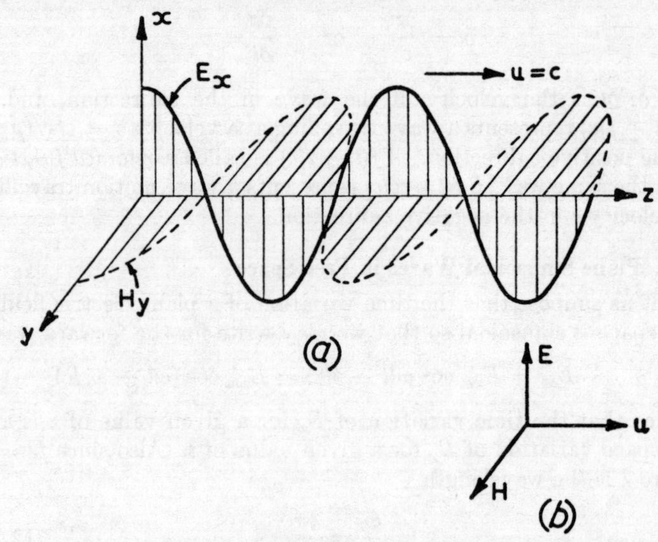

FIG. 12.7. SINUSOIDAL PLANE ELECTROMAGNETIC WAVES IN FREE SPACE

The relationship between peak electric and magnetic field strengths is obtained from eqn. (12.22) as

$$H_{ym} = \frac{1}{\mu_0 c}E_{xm} \qquad \text{or} \qquad E_{xm} = cB_{ym} \quad . \quad (12.23)$$

The quotient E_x/H_y has the dimensions of an impedance (volts per metre divided by amperes per metre = ohms), and is called the *wave impedance*, Z_w. For free space,

$$Z_w = \frac{E_x}{H_y} = \mu_0 c = \sqrt{\frac{\mu_0}{\epsilon_0}} \simeq 4\pi \times 10^{-7} \times 3 \times 10^8 = 120\pi \text{ ohms} \quad . \quad . \quad (12.24)$$

since $c = 1/\sqrt{(\mu_0\epsilon_0)}$.

If the backward electric field wave is considered,

$$E_{xb} = E_{xm} \cos \omega(t + z/c)$$

and hence

$$H_{yb} = -\frac{1}{\mu_0} \int \frac{\partial E_x}{\partial z} \, dt = -\frac{E_{xm}}{\mu_0 c} \cos\left(\omega t + \frac{\omega z}{c}\right) \quad . \quad (12.25)$$

Note that, for the backward wave,

$$\frac{E_{xb}}{H_{yb}} = -Z_w$$

For both the forward and backward waves the directions of c, E, and H form a right-handed orthogonal system as shown in Fig. 12.7 (*b*). The plane of polarization of the waves is taken as the plane in which the E-vector lies—thus the wave shown in Fig. 12.7 (*a*) is vertically polarized.

If the medium in which the electromagnetic wave propagates is a dielectric with relative permittivity ϵ_r, the analysis which has been carried out for free-space propagation remains valid, but in this case the velocity of propagation will be reduced to

$$u = 1/\sqrt{(\mu_0 \epsilon_0 \epsilon_r)} = c/\sqrt{(\epsilon_r)} \quad \text{metres/second} \quad . \quad (12.26)$$

and the derivation of H_y from E_x gives the relation

$$H_{ym} = \frac{1}{\mu_0 u} E_{xm} = \sqrt{\frac{\epsilon_0 \epsilon_r}{\mu_0}} E_{xm} = \frac{\sqrt{\epsilon_r}}{\mu_0 c} E_{xm} \quad . \quad (12.27)$$

The wave impedance will therefore be

$$Z_w' = \frac{E_{xm}}{H_{ym}} = \frac{\mu_0 c}{\sqrt{\epsilon_r}} = \sqrt{\frac{\mu_0}{\epsilon_0 \epsilon_r}} = \frac{120\pi}{\sqrt{\epsilon_r}} . \quad . \quad (12.28)$$

Both the velocity and the wave impedance are reduced by a factor $\sqrt{\epsilon_r}$ from their free space values.

12.9. Poynting's Power Flow Vector

The electric and magnetic field patterns of a plane wave move with a velocity u, and hence the stored energy associated with them is moving through the medium with this same velocity. This represents a flow of power through the medium. Consider one square metre of wavefront in a plane wave moving in the z-direction with components E_x and H_y (Fig. 12.8 (*a*)): the energy stored in the element δz thick is

$$\delta W = \tfrac{1}{2}\epsilon E_x^2 \, \delta z + \tfrac{1}{2}\mu H_y^2 \, \partial z \quad \text{joules/metre}^3$$

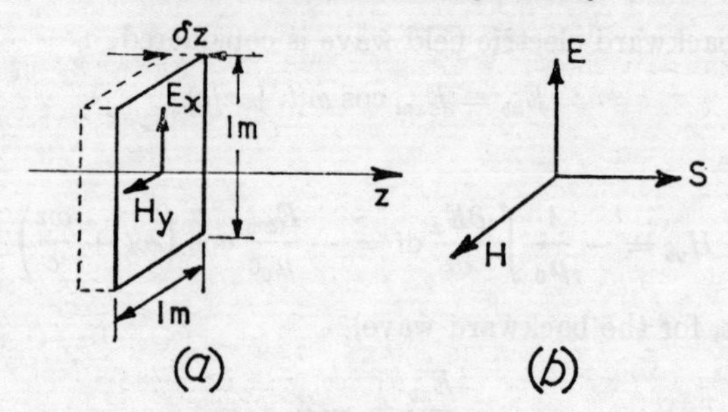

FIG. 12.8. POWER FLOW

The energy passing per second is

$$\frac{\delta W}{\delta t} = \frac{1}{2}\,\epsilon E_x{}^2\,\frac{\delta z}{\delta t} + \frac{1}{2}\,\mu H_y{}^2\,\frac{\delta z}{\delta t} \quad \text{joules/sec-metre}^2\text{, or watts/metre}^2$$

where δt is the time taken for the wave to move a distance δz. This must represent the instantaneous power flow, $\boldsymbol{S_i}$, through the area of one square metre. Thus

$$\boldsymbol{S_i} = \tfrac{1}{2}\epsilon E_x{}^2 u + \tfrac{1}{2}\mu H_y{}^2 u \quad \text{watts/metre}$$

$$= \left(\frac{1}{2}\,\epsilon u E_x \times H_y\sqrt{\frac{\mu}{\epsilon}}\right) + \left(\frac{1}{2}\,\mu u H_y \times E_x\sqrt{\frac{\epsilon}{\mu}}\right) \text{ (from eqn. (12.24))}$$

Hence

$$\boldsymbol{S_i} = \tfrac{1}{2}E_x H_y + \tfrac{1}{2}E_x H_y = E_x H_y$$

since $u = 1/\sqrt{(\mu\epsilon)}$.

The mean power flow is

$$\boldsymbol{S} = \boldsymbol{E_{rms}} \times \boldsymbol{H_{rms}} \text{ watts/metre}^2 \qquad (12.29)$$

This is the *Poynting vector*. It is directed in the same sense as the velocity, u, so that S, E and H form a right-handed orthogonal system (Fig. 12.8 (*b*)).

Example 12.1. A sinusoidal plane wave is transmitted through a medium whose breakdown strength is 30 kV/m and whose relative permittivity is 4. Determine the maximum possible r.m.s. power flow density, and the peak value of the associated magnetizing force.

For the plane wave, from eqn. (12.28),

Hence $\quad Z'_w = E/H = 120\,\pi/\sqrt{\epsilon_r} = 60\pi$, since $\epsilon_r = 4$

$$S = E_{rms} \times H_{rms} = E^2{}_{rms}/60\pi = (9 \times 10^8)/2 \times 60\pi = \underline{\underline{2{,}390 \text{ kW/m}^2}}$$

Also $\quad H_{max} = (2{,}390/E_{rms})\sqrt{2} = \underline{\underline{0{\cdot}112 \text{ A/m}}}$

12.10. Plane Waves in a Lossy Medium

Let us consider the field between the conductors of an infinitely wide strip transmission line. As shown in Fig. 12.9, this will be in the form of a plane wave.

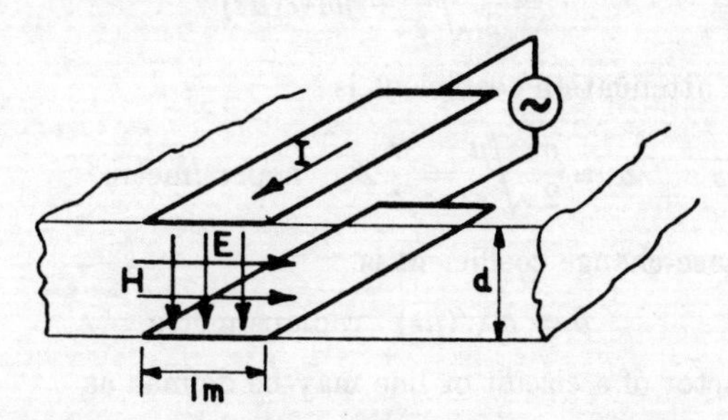

FIG. 12.9. FIELD BETWEEN CONDUCTORS OF A STRIP LINE

The capacitance per unit width and length of the line is

$$C = \frac{\epsilon}{d} \text{ farads/metre}^2$$

where d is the separation in metres.

If A is the *linear current density*, i.e. the current per metre width of the line, the inductance is

$$L = \Phi/A = B(1 \times d)/A = \mu Hd/A$$

But $H = A$, so that

$$L = \mu d \quad \text{henrys}$$

If the conductivity of the dielectric between the lines is σ, the conductance between the two conductors per metre length and width is

$$G = \frac{\sigma}{d} \quad \text{mhos/metre}^2$$

From transmission line theory (eqn. (7.11) with $R = 0$), the characteristic impedance per unit breadth is

$$Z_0 = \sqrt{\frac{j\omega\mu d}{\sigma/d + j\omega\epsilon/d}} = d\sqrt{\frac{j\omega\mu}{\sigma + j\omega\epsilon}} \quad \text{ohms} \qquad . \ (12.30)$$

and, from eqn. (7.7) with $R = 0$, the propagation coefficient is

$$\gamma = \sqrt{[j\omega\mu d(\sigma/d + j\omega\epsilon/d)]} = \sqrt{[j\omega\mu(\sigma + j\omega\epsilon)]} \quad \text{per metre}$$

If $\sigma \ll \omega\epsilon$, we may use the binomial theorem to evaluate γ:

$$\gamma \simeq j\omega\sqrt{(\mu\epsilon)}\left(1 + \frac{\sigma}{j2\omega\epsilon}\right)$$

$$= \frac{\sigma}{2}\sqrt{\frac{\mu}{\epsilon}} + j\omega\sqrt{(\mu\epsilon)} \quad . \quad . \quad . \quad (12.31)$$

so that the attenuation coefficient is

$$\alpha = \frac{\sigma}{2}\sqrt{\frac{\mu}{\epsilon}} = \frac{\sigma}{2} Z_w \quad \text{nepers/metre}$$

and the phase-change coefficient is

$$\beta = \omega\sqrt{(\mu\epsilon)} \quad \text{radians/metre}$$

The Q-factor of a circuit or line may be defined as

$$Q = 2\pi \frac{\text{Peak energy stored in electric or magnetic field}}{\text{Energy dissipated per cycle}}$$

$$= 2\pi f \frac{\text{Peak energy stored in electric field}}{\text{Mean power dissipation}}$$

since equal energies are stored in the electric and magnetic fields, and since the mean power, or the energy dissipated per second, is the energy dissipated per cycle multiplied by the number of cycles per second.

The peak stored energy density in the electric field is $\frac{1}{2}D_m E_m$ joules per cubic metre, where D_m is the peak electric flux density, and E_m is the peak electric field strength. Hence the peak energy stored per metre width and length of dielectric, of thickness d, is

$$W = \tfrac{1}{2}D_m E_m d \quad \text{joules}$$

Also the mean power dissipation for a peak sinusoidal voltage V_m ($= E_m d$) applied to a conductor of conductance G is

$$P = \tfrac{1}{2}G V_m{}^2 = \tfrac{1}{2}G(E_m d)^2$$

$$= \frac{1}{2}\frac{\sigma}{d} E_m{}^2 d^2 = \frac{1}{2}\sigma d E_m{}^2$$

so that

$$Q = 2\pi f \frac{W}{P} = 2\pi f \frac{\frac{1}{2}D_m E_m d}{\frac{1}{2}\sigma d E_m{}^2}$$

$$= 2\pi f \frac{\epsilon E_m{}^2}{\sigma E_m{}^2} = 2\pi f \frac{\epsilon}{\sigma}$$

$$= \omega\frac{\epsilon}{\sigma} \quad . \quad . \quad . \quad . \quad . \quad . \quad (12.32)$$

This corresponds to the Q-factor of a circuit with a capacitance C in parallel with a conductance G ($Q = \omega C/G$).

If the medium in which the plane wave propagates is a good conductor, then $\sigma \gg \omega\epsilon$, and the propagation coefficient becomes,

$$\gamma \simeq \sqrt{(j\omega\mu\sigma)}$$

$$= \sqrt{\frac{\omega\mu\sigma}{2}} + j\sqrt{\frac{\omega\mu\sigma}{2}} \text{ per metre} \quad . \quad . \quad (12.33)$$

The length δ, for which the wave is attenuated to $1/e$ of its initial value is such that $\sqrt{(\omega\mu\sigma/2)}\,\delta = 1$, so that

$$\delta = \sqrt{\frac{2}{\omega\mu\sigma}} \quad . \quad . \quad . \quad . \quad (12.34)$$

This is the depth of penetration (or skin depth) of alternating currents into conductors. Eqn. (12.33) also shows that as the wave penetrates the conductor there is a phase change of $\beta = \sqrt{(\omega\mu\sigma/2)}$ radians per metre.

12.11. Reflection from a Conductor

When a plane wave is incident on a perfect conductor, no energy can be absorbed by the conductor and hence all the energy of the incident wave must be reradiated in a reflected wave. At the conductor surface the boundary conditions for both electric and magnetic fields must be satisfied (Sections 10.8 and 11.24).

(*a*) *Normal Incidence—Perfect Conductor*. A normally incident plane wave has E and H vectors perpendicular to the conductor surface (Fig. 12.10). Since the tangential component of E is continuous across the boundary, and there can be no electric stress in a perfect conductor, the resultant E-field just outside the conductor must be zero. Hence the reflected electric field, E_r, must be equal and opposite to the incident electric field, E_i. The magnetic field gives rise to induced currents in the conductor which support both incident and reflected H-fields, and applying the right-hand rule for the directions of u, E_r and H_r, it is seen that the reflected wave is directed away from the conductor. Incident and reflected plane waves interfere to give a standing-wave pattern just as in a short-circuited transmission line.

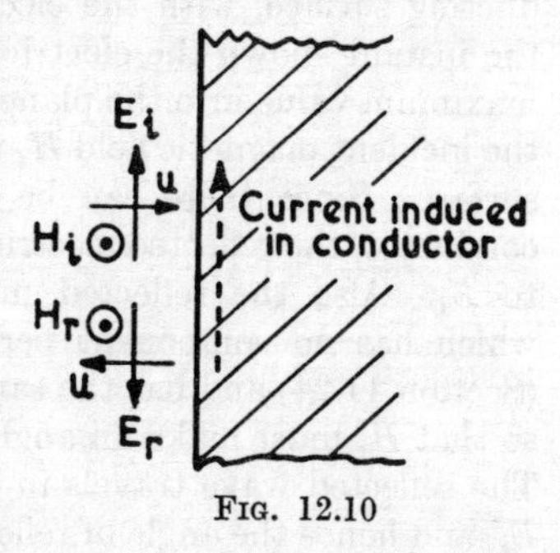

Fig. 12.10

For the incident wave,

$$Z_{wi} = \frac{E_i}{H_i} \quad \text{(from eqn. (12.28))}$$

For the reflected wave, $E_r = -E_i$ and $H_r = H_i$ and so we may write

$$Z_{wr} = -\frac{E_r}{H_r} \quad . \quad . \quad . \quad (12.35)$$

(*b*) *Oblique Incidence—Horizontal Polarization*. Fig. 12.11 shows

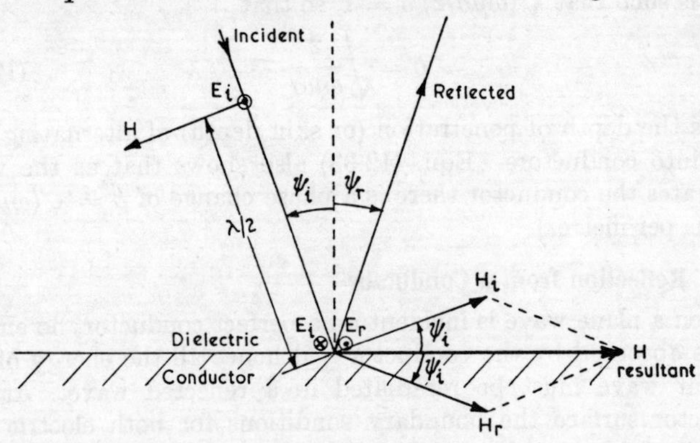

FIG. 12.11

a wave incident at an angle Ψ_i to the normal to a perfectly conducting surface, with the electric field parallel to the surface. At the instant shown the electric field of the incident wave, E_i, has a maximum value into the plane of the diagram at the conductor, and the incident magnetic field H_i makes an angle Ψ_i with the conductor surface. Since there can be no component of E parallel to the conductor, the reflected electric field, E_r, must be equal and opposite to E_i. Also the reflected magnetic field must give a resultant which has no component perpendicular to the conductor surface (Section 11.24) and has the same component parallel to the surface, so that H_r must make an angle of $-\Psi_i$ with the surface, as shown. The reflected wave travels in a direction at right angles to H_r and E_r, and hence the angle of reflection, Ψ_r, must by geometry be equal to the angle of incidence, Ψ_i.

(*c*) *Oblique Incidence—Vertical Polarization*. If the plane wave incident on a perfectly conducting plane has its H-field parallel to the conductor surface, and has an angle of incidence Ψ_i, then by geometry E_i makes an angle Ψ_i with the surface (Fig. 12.12). Since there can be no component of resultant electric field parallel to the conductor surface, the reflected E-field must be such as to cancel the horizontal component of E_i and have the same vertical component. It is therefore directed as shown at the instant when E_i is a

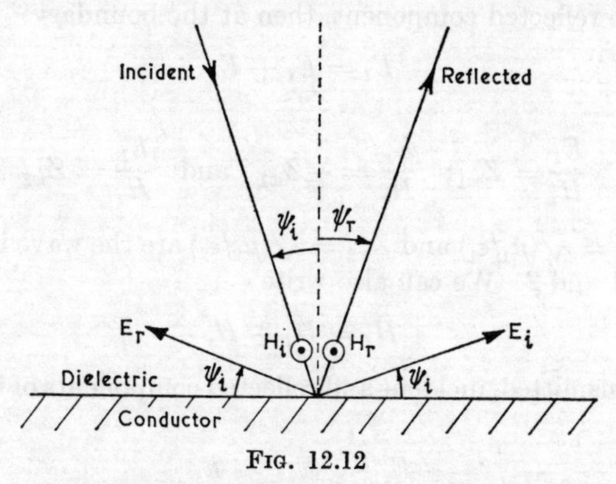

FIG. 12.12

maximum at the conductor. The reflected H-field is the same as the incident H-field, since it is supported by the same current, and hence the reflected wave (at right angles to the plane of E_r and H_r) makes an angle of reflection, Ψ_r, equal to Ψ_i.

12.12. Transmission through a Dielectric

When a uniform plane wave is normally incident on a dielectric, reflection and transmission take place in a manner analogous to the

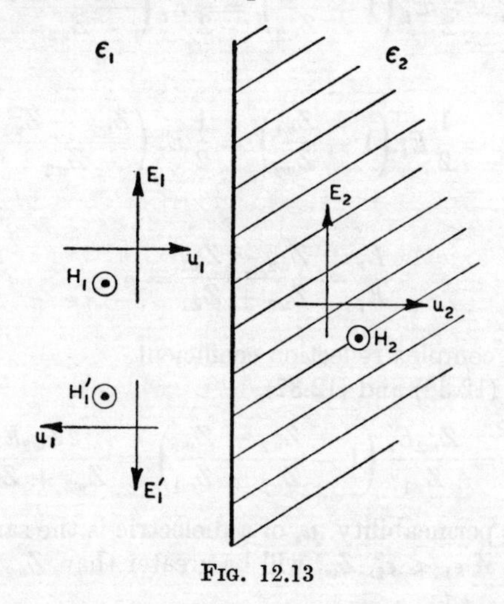

FIG. 12.13

processes which occur at a discontinuity in a transmission line. Consider the wave entering a loss-free dielectric of permittivity ϵ_2 from a medium of permittivity ϵ_1 (Fig. 12.13). If E_t is the transmitted component of the electric field, E_i the incident component, and E_r the reflected component, then at the boundary

$$E_t = E_i + E_r \quad . \quad . \quad . \quad (12.36)$$

Also

$$\frac{E_i}{H_i} = Z_{w1}; \quad \frac{E_r}{H_r} = -Z_{w1} \quad \text{and} \quad \frac{E_t}{H_t} = Z_{w2}$$

where $Z_{w1} = \sqrt{(\mu_1/\epsilon_1)}$ and $Z_{w2} = \sqrt{(\mu_2/\epsilon_2)}$ are the wave impedances of media 1 and 2. We can also write

$$H_t = H_i + H_r$$

for the transmitted, incident and reflected components of the H-field. Hence

$$\frac{E_t}{Z_{w2}} = \frac{E_i}{Z_{w1}} - \frac{E_r}{Z_{w1}}$$

and so

$$\frac{E_t Z_{w1}}{Z_{w2}} = E_i - E_r \quad . \quad . \quad . \quad (12.37)$$

Adding eqns. (12.36) and 12.37),

$$E_i = \frac{1}{2} E_t \left(1 + \frac{Z_{w1}}{Z_{w2}}\right) = \frac{1}{2} E_t \left(\frac{Z_{w2} + Z_{w1}}{Z_{w2}}\right)$$

Subtracting,

$$E_r = \frac{1}{2} E_t \left(1 - \frac{Z_{w1}}{Z_{w2}}\right) = \frac{1}{2} E_t \left(\frac{Z_{w2} - Z_{w1}}{Z_{w2}}\right)$$

Hence

$$\frac{E_r}{E_i} = \frac{Z_{w2} - Z_{w1}}{Z_{w2} + Z_{w1}} = \rho \quad . \quad . \quad . \quad (12.38)$$

where ρ is the complex reflection coefficient.

From eqns. (12.38) and (12.37),

$$E_t = \frac{Z_{w2} E_i}{Z_{w1}} \left(1 - \frac{Z_{w2} - Z_{w1}}{Z_{w2} + Z_{w1}}\right) = \frac{2 Z_{w2} E_i}{Z_{w2} + Z_{w1}} \quad . \quad (12.39)$$

Usually the permeability, μ, of a dielectric is the same as for free space, and so, if $\epsilon_1 < \epsilon_2$, Z_{w1} will be greater than Z_{w2} and ρ will be negative. If, however, $\epsilon_1 > \epsilon_2$ then $Z_{w1} < Z_{w2}$ and ρ will be positive. The reflected E-field will then have the same sense as the incident E-field.

The velocity u_1 in medium 1 is $1/\sqrt{(\mu_1\epsilon_1)}$, and in medium 2 is $1/\sqrt{(\mu_2\epsilon_2)}$.

If medium 2 has a finite conductivity, σ, from eqn. (12.31) the wave will be attenuated as it travels through the medium, and at a distance x from the surface the r.m.s. value of the electric field will be

$$E = E_{2\,rms} \exp\left(-\frac{\sigma}{2}\sqrt{\frac{\mu_2}{\epsilon_2}}\,x\right) \quad . \quad . \quad (12.40)$$

Example 12.2. A sinusoidally distributed plane wave in air has a peak magnetizing force of 1·5 A/m. It is incident normally on a dielectric of relative permittivity 4. Calculate the peak values of the incident, reflected and transmitted electric and magnetic fields.

From eqn. (12.23), the incident electric field strength is

$$E_{im} = cB_m = c\mu_0 H_{im} = 3 \times 10^8 \times 4\pi \times 10^{-7} \times 1{\cdot}5 = \underline{\underline{565{\cdot}6 \text{ V/m}}}$$

For air, the wave impedance is $Z_{w1} = 120\pi = 377$ (eqn. (12.24)).
For the dielectric $Z_{w2} = \sqrt{(\mu_0/\epsilon_2)} = Z_{w1}/\sqrt{\epsilon_r} = 189\ \Omega$

Hence, from eqn. (12.38), the reflected wave has a peak electric field strength, E_{rm}, of

$$E_{rm} = \frac{189 - 377}{189 + 377} E_{im} = -1/3 \times 565{\cdot}6 = \underline{\underline{-188{\cdot}9 \text{ V/m}}}$$

From eqn. (12.39), the transmitted wave has a peak electric field strength of

$$E_{tm} = \frac{2 \times 189}{189 + 377} E_{im} = 2/3 \times 565{\cdot}6 = \underline{\underline{376{\cdot}7 \text{ V/m}}}$$

From eqn. (12.35),

$$H_{rm} = -\frac{E_{rm}}{Z_{w1}} = \frac{188{\cdot}9}{377} = \underline{\underline{0{\cdot}5 \text{ A/m}}}$$

From eqn. (12.28),

$$H_{tm} = \frac{E_{tm}}{Z_{w2}} = \frac{376{\cdot}7}{189} = \underline{\underline{2 \text{ A/m}}}$$

Note that $H_{tm} = H_{im} + H_{rm}$, where H_{rm} is the peak value of the reflected H-field.

12.13. Refraction

Since light is an electromagnetic radiation, we would expect that the laws of reflection and refraction which are well known for light waves would also apply to the plane-polarized waves which we have been considering. This has already been shown for reflection from a perfect conductor. Consider now a plane-polarized wave making oblique incidence with a perfect dielectric (Fig. 12.14).

The wave is said to be horizontally polarized if the E-field vector is parallel to the boundary (Fig. 12.14 (a)), and vertically polarized if the H-field vector is parallel to the boundary (Fig. 12.14 (b)). For both conditions, the tangential component of E must be continuous across the boundary for all points on the z-axis, where this axis is taken to correspond with the boundary. This means that the velocity with which the wavefront (which is perpendicular to the direction of propagation) cuts the z-axis is the same for the incident, reflected and refracted waves. This velocity is called the *phase velocity* of the wave in the z-direction, and is given by

$$u_{pa} = \frac{u}{\sin \Psi_i} = \frac{u}{\sin \Psi_r} = \frac{u_t}{\sin \Psi_t}$$

where Ψ_t is the angle of refraction. Hence $\Psi_i = \Psi_r$, and

$$\frac{u}{u_t} = \frac{\sin \Psi_i}{\sin \Psi_t}$$

Now $u = 1/\sqrt{(\mu_1 \epsilon_1)}$ and $u_t = 1/\sqrt{(\mu_2 \epsilon_2)}$ are the velocities in media 1 and 2, and we may assume that $\mu_1 = \mu_2 = \mu_0$, so that

$$\frac{\sin \Psi_i}{\sin \Psi_t} = \sqrt{\frac{\epsilon_2}{\epsilon_1}} = \sqrt{\frac{\epsilon_{2r}}{\epsilon_{1r}}} \quad . \quad . \quad . \quad (12.41)$$

where ϵ_{1r} and ϵ_{2r} are relative permittivities.

The magnitudes of the reflected and refracted components of the incident wave are different for the cases of horizontal and vertical polarization, and may be found by considering the normal components of the waves. For the horizontally polarized wave, the components of E and H for the wave component in the x-direction are E and $H \cos \Psi$. Thus the wave impedances for this component are

$$Z_{w1x} = \frac{E_i}{H_i \cos \Psi_i} = \frac{Z_{w1}}{\cos \Psi_i} \quad \text{for the incident wave}$$

$$Z_{w2x} = \frac{E_t}{H_t \cos \Psi_t} = \frac{Z_{w2}}{\cos \Psi_t} \quad \text{for the transmitted wave}$$

since $E_i/H_i = Z_{w1}$ and $E_t/H_t = Z_{w2}$. Hence, from eqn. (12.39),

$$E_t = \frac{2Z_{w2x}}{Z_{w2x} + Z_{w1x}} E_i \quad . \quad . \quad (12.42)$$

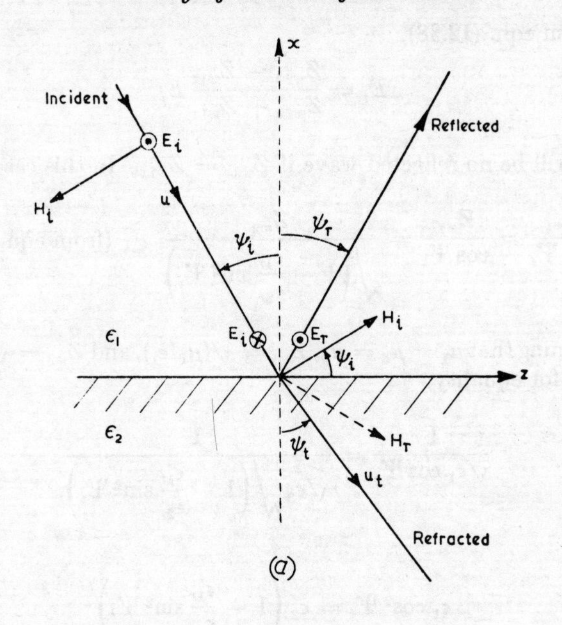
x
Incident
E_i
H_i
u
Reflected
ψ_r
ψ_i
H_i
ε_1
E_i
E_r
ψ_i
z
ε_2
H_r
ψ_t
u_t
Refracted
(a)

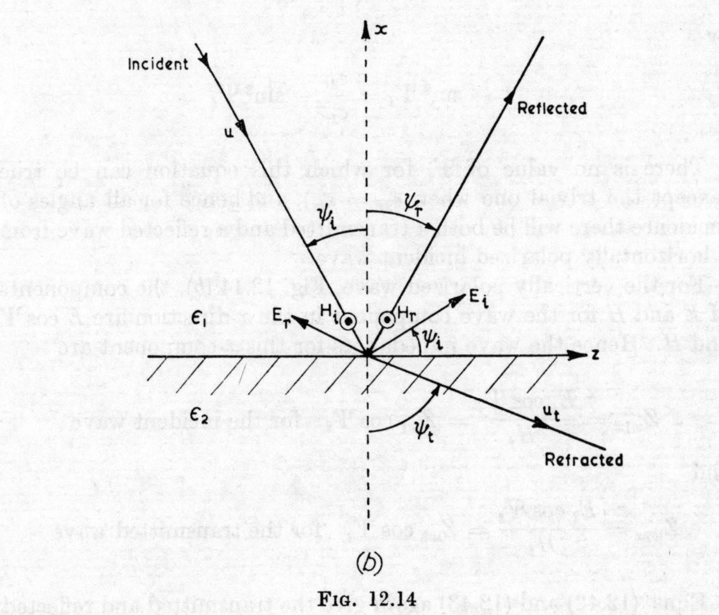
x
Incident
u
Reflected
ψ_i
ψ_r
E_i
ε_1
E_r
H_i
H_r
ψ_i
z
ε_2
u_t
ψ_t
Refracted
(b)

FIG. 12.14

and from eqn. (12.38),

$$E_r = \frac{Z_{w2x} - Z_{w1x}}{Z_{w2x} + Z_{w1x}} E_i \quad . \quad . \quad . \quad (12.43)$$

There will be no reflected wave if $Z_{w2x} = Z_{w1x}$. In this case

$$\frac{Z_{w1}}{\cos \Psi_i} = \frac{Z_{w2}}{\cos \Psi_t} = \frac{Z_{w2}}{\sqrt{\left(1 - \frac{\epsilon_{1r}}{\epsilon_{2r}} \sin^2 \Psi_i\right)}} \qquad \text{(from eqn. (12.41))}$$

Assuming that $\mu_1 = \mu_2 = \mu_0$, $Z_{w1} = \sqrt{(\mu_0/\epsilon_1)}$, and $Z_{w2} = \sqrt{(\mu_0/\epsilon_2)}$, so that for equality,

$$\frac{1}{\sqrt{\epsilon_1} \cos \Psi_i} = \frac{1}{\sqrt{\epsilon_2} \sqrt{\left(1 - \frac{\epsilon_{1r}}{\epsilon_{2r}} \sin^2 \Psi_i\right)}}$$

Hence

$$\epsilon_1 \cos^2 \Psi_i = \epsilon_2 \left(1 - \frac{\epsilon_{1r}}{\epsilon_{2r}} \sin^2 \Psi_i\right)$$

or

$$1 - \sin^2 \Psi_i = \frac{\epsilon_{2r}}{\epsilon_{1r}} - \sin^2 \Psi_i$$

There is no value of Ψ_i for which this equation can be true (except the trivial one when $\epsilon_{2r} = \epsilon_{1r}$), and hence for all angles of incidence there will be both a transmitted and a reflected wave from a horizontally polarized incident wave.

For the vertically polarized wave, Fig. 12.14 (*b*), the components of E and H for the wave component in the x-direction are $E \cos \Psi$ and H. Hence the wave impedances for this x-component are

$$Z_{w1x} = \frac{E_i \cos \Psi_i}{H_i} = Z_{w1} \cos \Psi_i \quad \text{for the incident wave}$$

and

$$Z_{w2x} = \frac{E_t \cos \Psi_t}{H_t} = Z_{w2} \cos \Psi_t \quad \text{for the transmitted wave}$$

Eqns. (12.42) and (12.43) again give the transmitted and reflected waves.

In this case there will be no reflection at an angle of incidence Ψ_i' which makes $Z_{w1x} = Z_{w2x}$; i.e. if

$$\sqrt{\frac{\mu_0}{\epsilon_1}} \cos \Psi_i' = \sqrt{\frac{\mu_0}{\epsilon_2}} \sqrt{\left(1 - \frac{\epsilon_{1r}}{\epsilon_{2r}} \sin^2 \Psi_i'\right)}$$

or if

$$1 - \sin^2 \Psi_i' = \frac{\epsilon_{1r}}{\epsilon_{2r}}\left(1 - \frac{\epsilon_{1r}}{\epsilon_{2r}} \sin^2 \Psi_i'\right)$$

This yields

$$\sin \Psi_i' = \sqrt{\frac{\dfrac{\epsilon_{1r}}{\epsilon_{2r}} - 1}{\dfrac{\epsilon_{1r}^2}{\epsilon_{2r}^2} - 1}} = \sqrt{\frac{\epsilon_{2r}}{\epsilon_{1r} + \epsilon_{2r}}} \quad . \quad . \quad (12.44)$$

This is a real angle, and is called the *Brewster angle*. At this particular angle of incidence of a vertically polarized wave, all the energy of the incident wave is transmitted through the dielectric.

Example 12.3. A vertically polarized uniform plane electromagnetic wave is incident on a dielectric plate of relative permittivity 3. Determine the Brewster angle, and the corresponding angle of refraction.

From eqn. (12.44), the Brewster angle is given by

$$\sin \Psi_i' = \sqrt{[3/(1 + 3)]} = 0{\cdot}866$$

Hence $\Psi_i' = \underline{\underline{60°}}$

The angle of refraction, Ψ_t is found from eqn. (12.41): $\sin \Psi_i'/\sin \Psi = \sqrt{(3/1)}$, so that $\sin \Psi_t = 0{\cdot}866/\sqrt{3}$.

Hence $\Psi_t = \underline{\underline{30°}}$

Note that a plane wave in the dielectric incident on the surface at this angle also gives no internal reflection, so that this is the Brewster angle for the dielectric–air boundary.

12.14. Radiation from a Current Element

It has been seen in Section 11.7 that the field strength at a point P due to a current-element $I\,dl$ (Fig. 12.15) is given by eqn. (11.19) as

$$H_P = \frac{I\,dl \sin\theta}{4\pi r^2}$$

Hence from eqn. (11.11) the magnetic potential at P is

$$F_P = -\int_0^r H\,dr = \frac{I\,dl \sin\theta}{4\pi r}$$

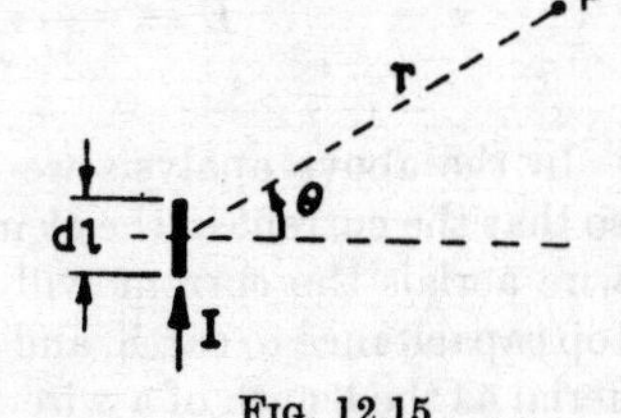

FIG. 12.15

If the current I is alternating, the field too is alternating, and the magnetic potential is a retarded function, since the effect of any change in I will not appear at P until the alternating field has travelled the distance r at the velocity of propagation u, in the medium. Hence the expression for the magnetic potential at P (from Section 12.8) becomes

$$F_P' = \frac{I_m \, \mathrm{d}l \sin \theta}{4\pi r} \sin\left(\omega t - \frac{\omega r}{u}\right)$$

where I_m is the peak alternating current in the element dl.

The magnetizing force at P is thus

$$H_P' = -\frac{\partial F_P'}{\partial r} = \frac{I_m \, \mathrm{d}l \sin \theta}{4\pi}\left[\frac{\sin(\omega t - \omega r/u)}{r^2} + \frac{\omega \cos(\omega t - \omega r/u)}{ur}\right] \quad . \quad . \quad (12.45)$$

The first term in this expression is called the *induction field*, and since it is inversely proportional to r^2 it will rapidly decrease as r increases. The second term represents the *radiation field*, and since it is inversely proportional to r it will not decay nearly as rapidly as the induction field. Notice, however, that the radiation field is proportional to frequency, so that at low frequencies we may expect it to be very small.

At a sufficient distance from the current-element for the induction field to be negligible, the r.m.s. value of the magnetic radiation field will be

$$H = \frac{\omega I \, \mathrm{d}l \sin \theta}{4\pi ur} = \frac{I \, \mathrm{d}l \sin \theta}{2\lambda r} \text{ amperes/metre} \quad . \quad (12.46)$$

since, from eqn. (12.21), $\omega/u = 2\pi/\lambda$.

For an air dielectric, eqn. (12.24) gives $E/H = 120\pi$, so that the electric field associated with the magnetic field given by eqn. (12.46) is

$$E = \frac{60\pi I \, \mathrm{d}l \sin \theta}{\lambda r} \text{ volts/metre} \quad . \quad . \quad (12.47)$$

In the above analysis we have assumed that the length d$l \ll \lambda$, so that the current in the element is uniform. For practical radiating wire aerials the current will not be uniform unless there is a large top capacitance to earth, and we may define the effective length of an aerial as the length of a wire carrying a uniform current equal to the input current of the actual aerial, and producing the same field strength on the equatorial plane ($\theta = 0$).

Bibliography

Carter, *The Electromagnetic Field in its Engineering Applications* (Longmans).

Glazier and Lamont, *The Services Textbook of Radio, Vol.* 5, *Transmission and Propagation* (H.M. Stationery Office).

CHAPTER 13

Electron Ballistics

THE electron is the most important negatively charged particle of matter. It has a mass of $9{\cdot}1 \times 10^{-31}$ kg when at rest, and a negative charge of $1{\cdot}6 \times 10^{-19}$ C. In this chapter the behaviour of electrons in free space under the influence of electric and magnetic fields will be studied. It is assumed that the reader is familiar with the usual method of producing free electrons in a vacuum by thermionic emission from a heated cathode. Although the thermionic valve is now largely being superseded by the transistor, electron ballistics is still of the greatest importance in many fields, including the cathode-ray tube, high-power microwave valves and particle accelerators. In the case of particle accelerators we are interested in other charged particles as well as electrons, as will be seen later.

13.1. Electron Acceleration in a Uniform Electric Field

Consider two large flat parallel plates l metres apart (Fig. 13.1), with potentials V_1 and V_2 (or a potential difference $V = V_1 - V_2$

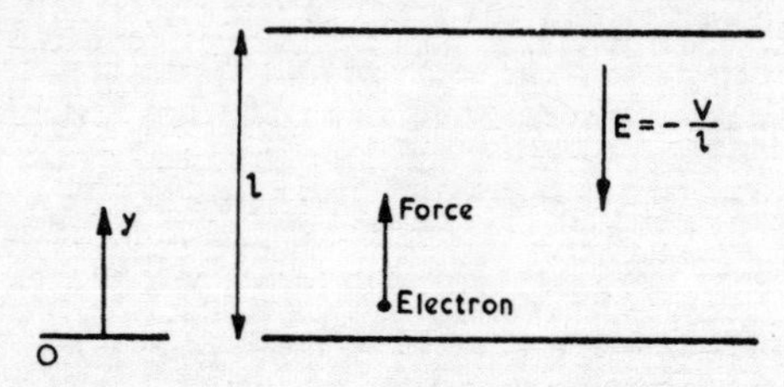

FIG. 13.1. ACCELERATION OF A FREE ELECTRON TOWARDS THE POSITIVE POTENTIAL IN A UNIFORM ELECTRIC FIELD

volts between them), and let the lower of the two plates be negative with respect to the upper. Then the electric field strength, E, between the plates is uniform, and is directed towards the lower plate. The value of E is

$$E = -\frac{\partial V}{\partial l} = -\frac{V}{l} \quad \text{volts/metre}$$

The negative sign indicates that the direction of E is the opposite of the positive y-direction in Fig. 13.1.

From the definition of E in Section 10.2, the force on an electron of charge $-e$ coulombs, in this field, is

$$F = -eE \quad \text{newtons} \qquad (13.1)$$

The negative sign shows that the direction of the force is opposite to that of the field strength, E.

Now suppose that the lower plate is a thermionic emitter, and gives a supply of electrons of zero velocity at its surface. The force, F, on these electrons accelerates them towards the positive electrode, with an acceleration given by the equation

$$F = \text{Rate of change of momentum} = \frac{\mathrm{d}(mu)}{\mathrm{d}t} = ma$$

where m is the mass of the electron (assumed constant) and $a = \mathrm{d}u/\mathrm{d}$ is its acceleration. Hence

$$ma = -eE$$

or
$$a = -\frac{eE}{m} \quad \text{metres/second} \qquad (13.2)$$

The negative sign indicates that the acceleration is oppositely directed to E.

The work done when an electron moves a distance $+\delta l$ (i.e. in the direction of the force F) is

$$W = -eE\,\delta l \quad \text{newton-metres}$$

Hence the total work done when the electron moves between the two plates is

$$W = -\int_0^l eE\,\mathrm{d}l = eV \qquad (13.3)$$

since $\int_0^l E\,\mathrm{d}l$ is, by definition, $-V$. This work represents an increase in the kinetic energy of the moving electron. If the velocity attained at the top electrode is u, the kinetic energy is

$$\tfrac{1}{2}mu^2 = eV \quad \text{joules}$$

Hence
$$u = \sqrt{\frac{2eV}{m}} \quad \text{metres/second} \qquad (13.4)$$

$$= 5{\cdot}93 \times 10^5\sqrt{V} \quad \text{metres/second} \qquad (13.4a)$$

Note that we have not assumed a uniform field, but only that the electron has moved through a total potential difference of V volts from rest. Eqn. (13.4) shows that, under these conditions, the final velocity depends on the potential difference V only, and not on E. If the electron has an initial velocity u_0, its final velocity will be $(u_0 + u)$, since eqn. (13.2) shows that the acceleration is independent of u_0.

It is often convenient to use as a unit of energy the kinetic energy gained when an electron is accelerated through a potential difference of 1 V. This is called the *electron-volt* (eV).

$$\begin{aligned} 1 \text{ eV} &= \text{Work done in accelerating an electron through 1 V} \\ &= eV \text{ joules} = 1{\cdot}6 \times 10^{-19} \text{ J} \qquad (13.5) \end{aligned}$$

since $e = 1{\cdot}6 \times 10^{-19}$ C.

In the above derivation we have assumed that the mass of the electron remains constant. This is true up to velocities approaching that of light, as will be shown in the next section.

13.2. Mass and Energy—Rest Mass—Effective Mass

Einstein has postulated that there is a fixed relationship between the energy of a body and its mass when at rest, m_0, given by

$$W_0 = c^2 m_0 \qquad (13.6)$$

where c is the velocity of light.

When a body is moving it has kinetic energy, and if we assume that eqn. (13.6) still applies, we may define an effective mass, m_e, for the moving body given by

$$W_0 + \text{kinetic energy} = c^2 m_e \qquad (13.7)$$

For an electron of rest mass m_0, moved through a potential difference of V volts, the kinetic energy gained is eV joules, and therefore

$$c^2 m_0 + eV = c^2 m_e$$

Differentiating with respect to time, and since both c and m_0 are constant,

$$\frac{\mathrm{d}(eV)}{\mathrm{d}t} = \frac{\mathrm{d}(c^2 m_e)}{\mathrm{d}t} \qquad (13.8)$$

This rate of change of energy must be the power input to the electron to maintain its acceleration. If the accelerating force is F, this instantaneous power input is Fu watts, or newton-metres per second. But $F = \mathrm{d}(m_e u)/\mathrm{d}t$, and hence

$$\text{Instantaneous power} = u\,\frac{\mathrm{d}(m_e u)}{\mathrm{d}t}$$

Comparing this expression with eqn. (13.8), we see that

$$u\,\frac{\mathrm{d}(m_e u)}{\mathrm{d}t} = c^2\,\frac{\mathrm{d}m_e}{\mathrm{d}t}$$

It follows that

$$\mathrm{d}m_e = \frac{u}{c^2}\,\mathrm{d}(m_e u) = \frac{u}{c}\,\mathrm{d}\left(m_e\frac{u}{c}\right) = \left(\frac{u^2}{c^2}\right)\mathrm{d}m_e + m_e\frac{u}{c}\,\mathrm{d}\left(\frac{u}{c}\right)$$

whence

$$\frac{\mathrm{d}m_e}{m_e} = \frac{\dfrac{u}{c}\,\mathrm{d}\left(\dfrac{u}{c}\right)}{1 - \dfrac{u^2}{c^2}} = \frac{\frac{1}{2}\,\mathrm{d}\left(\dfrac{u^2}{c^2}\right)}{1 - \dfrac{u^2}{c^2}}$$

Integrating both sides,

$$\log_e m_e = \frac{1}{2}\left[-\log_e\left(1 - \frac{u^2}{c^2}\right)\right] + \log_e A$$

$$= \log_e \frac{A}{\sqrt{\left(1 - \dfrac{u^2}{c^2}\right)}}$$

where A is a constant.

Hence
$$m_e = \frac{A}{\sqrt{\left(1 - \dfrac{u^2}{c^2}\right)}}$$

But when $u = 0$ the mass is the rest mass, m_0, and hence $m_0 = A$, so that,

$$m_e = \frac{m_0}{\sqrt{\left(1 - \dfrac{u^2}{c^2}\right)}} \quad . \quad . \quad . \quad (13.9)$$

This important relationship shows that, if any body is accelerated to 0·9 of the velocity of light, its relativistic mass, or *effective mass*, is

$$m_e = \frac{m_0}{\sqrt{(1 - 0{\cdot}81)}} = 2{\cdot}35 m_0$$

whereas at $0{\cdot}2c$ the effective mass is

$$m_e = \frac{m_0}{\sqrt{(1 - 0{\cdot}04)}} \simeq 1{\cdot}02 m_0$$

Generally, up to about 20 per cent of the velocity of light the effective mass may be taken to be the same as the rest mass. This will apply to electrons accelerated from rest through potentials of V given by

$$0{\cdot}2 \times 3 \times 10^8 = 5{\cdot}93 \times 10^5 \sqrt{V}$$

or

$$V = 10 \text{ kV}$$

For larger accelerating voltages the change in effective mass must be taken into account.

Example 13.1. Electrons leave a cylindrical cathode of radius 0·5 cm at zero velocity, and are attracted towards a concentric cylindrical anode of radius 1 cm whose potential is 200 V above that of the cathode. Determine (*a*) the velocity with which the electrons arrive at the anode, and (*b*) the radius at which the velocity is half the final velocity.

Since the final velocity depends only upon the p.d. between the electrodes (the initial velocity being zero), it is given by eqn. (13.4) as

$$u_b = 5{\cdot}93 \times 10^5 \sqrt{200} = \underline{\underline{83{\cdot}8 \times 10^5 \text{ m/sec}}}$$

From Section 10.13 the potential at any radius x between the electrodes is

$$V_x = -\int_x^a \frac{V}{r \log_e (b/a)} \, dr = V \frac{\log_e (x/a)}{\log_e (b/a)}$$

where V is the p.d. between the electrodes of radii a (inner) and b (outer).

Now,

$$\frac{u_x}{u_b} = \sqrt{\frac{V_x}{V}}$$

and for $u_x = 0{\cdot}5\, u_b$,

$$4 = \frac{V_x}{V} = \frac{\log_e (x/a)}{\log_e (b/a)}$$

Hence

$$\frac{x}{a} = \left(\frac{b}{a}\right)^{1/4} \quad \text{or} \quad x = (b/a^3)^{1/4} = (1 \times 0{\cdot}5^3)^{1/4} = \underline{\underline{0{\cdot}594 \text{ cm}}}$$

3.3. The Electron Gun

The *electron gun* is a means of producing a sharply defined beam of electrons in a vacuum. A typical electron gun is shown in Fig. 13.2. The electron source is an indirectly heated cathode in the form of a nickel cylinder, the end cap of which has an oxide coating. Surrounding and extending beyond the cathode is another cylinder (the *grid*) which is maintained at a negative potential with respect to the cathode. A constriction with a central hole at the end of the grid serves to concentrate the electrons into a rough beam before they pass through the first *accelerating anode* (A_1). If the grid is made negative enough the electron beam is cut off, and hence the grid potential control serves as a means of varying the beam intensity.

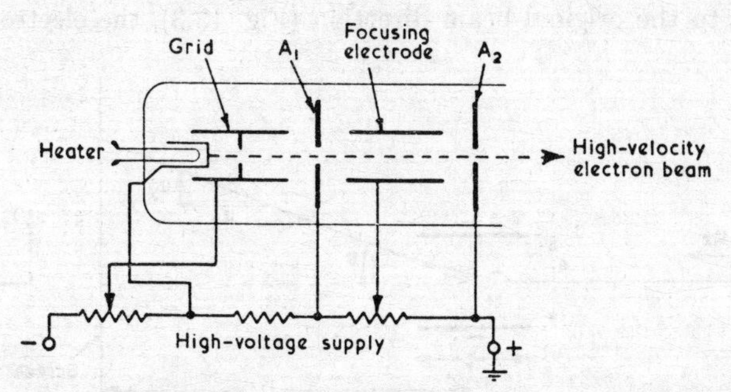

FIG. 13.2. ELECTRON GUN, IN WHICH A BEAM OF HIGH-VELOCITY ELECTRONS IS PRODUCED

The anodes A_1 and A_2 are in the form of discs with small central holes through which the electrons pass. Between them is a third cylinder which serves to focus the electrons into a narrow pencil beam, this arrangement being called an *electron lens*. In a cathode-ray oscilloscope, the final anode A_2 is normally held at earth potential, so that the deflecting plates and screen may also be at about zero voltage. This means that the cathode must be maintained at a considerable negative potential with respect to earth, and must therefore be well insulated from earth. Voltages of from 2 to 4 kV are common, but much higher voltages are also used.

Example 13.2. For an electron gun such as is shown in Fig. 13.2, the anode A_2 is maintained at zero potential, while the anode A_1 is at −1,250 V, and the cathode is at −2,000 V. Determine the velocity of an axial electron (*a*) at A_1, and (*b*) in the beam emerging from A_2.

Since the p.d. between the cathode and A_1 is 750 V, the velocity at A_1 is (eqn. (13.4))

$$u_1 = 5{\cdot}93 \times 10^5\sqrt{750} = \underline{\underline{162 \times 10^5 \text{ m/sec}}}$$

Also, since the p.d. between the cathode and A_2 is 2,000 V, the final beam velocity is

$$u_2 = 5{\cdot}93 \times 10^5\sqrt{2{,}000} = \underline{\underline{447 \times 10^5 \text{ m/sec}}}$$

These answers assume that the initial velocity at the cathode is zero. Also, since both velocities are well below 20 per cent of the velocity of light, the error due to effective mass will be negligible.

13.4. Electrostatic Deflexion of an Electron Beam

The velocity *in vacuo* of a beam of electrons from an electron gun will remain constant if the electrons move through a field-free space. If such a beam of moving electrons passes between two parallel charged deflecting plates which give a uniform electric field at right angles to the original beam direction (Fig. 13.3), the electrons will

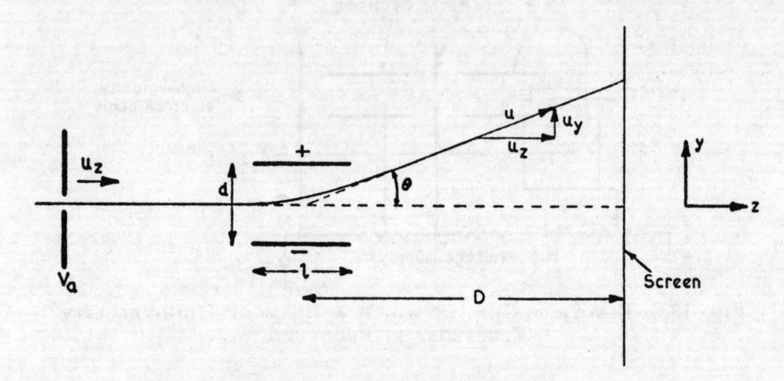

FIG. 13.3. AN ELECTRON BEAM CAN BE DEFLECTED BY A PERPENDICULAR ELECTRIC FIELD

experience an acceleration at right angles to their original direction and will be deflected. The velocity in the original direction will, however, remain constant.

If the accelerating voltage of the electron gun is V_A, the electrons will leave with a velocity given by

$$u_z = \sqrt{\frac{2eV_A}{m}} \quad \text{metres/second}$$

in the z-direction (Fig. 13.3).

For a p.d. of V volts between the parallel deflecting plates, the vertical acceleration (which is unaffected by u_z) is given by eqn. (13.2) as

$$a_y = \frac{e}{m}\frac{V}{d}$$

where d is the distance between the plates.

If the length of the plates is l metres, the time which each electron spends in the vertical deflecting field is $t = l/u_z$ seconds and hence the final vertical velocity is

$$u_y = a_y t = \frac{e}{m}\frac{V}{d}\frac{l}{u_z} \quad \text{metres/second}$$

and the resultant velocity, u, after deflexion is

$$u = \sqrt{(u_y^2 + u_z^2)} \qquad . \quad . \quad . \quad (13.10)$$

The final trajectory of the electron makes an angle $\theta = \tan^{-1}(u_y/u_z)$ with the z-direction. Although the actual electron path between the deflecting plates is parabolic, there is no loss of accuracy in assuming that the beam suddenly deflects by an angle θ at the centre of the plates.

If a flat fluorescent screen is placed a distance D from the centre of the deflecting plates, the deflexion, y, produced by a plate p.d. V is

$$y = D \tan \theta = D \frac{u_y}{u_z}$$

$$= D \frac{eVl}{mdu_z^2} = \frac{1}{2} D \frac{l}{d} \frac{V}{V_a} \qquad . \quad . \quad . \quad (13.11)$$

and the deflexion sensitivity is

$$\frac{V}{y} = 2 \frac{d}{l} \frac{V_A}{D} \quad \text{volts/metre} \qquad . \quad . \quad . \quad (13.12)$$

The maximum deflexion occurs when the electron beam just misses the edge of the deflecting plate, i.e. when $\theta = \tan^{-1}(d/l)$.

Deflexion in the x-direction can obviously be obtained by employing a second pair of deflecting plates at right angles to the y-deflexion plates. In the cathode-ray tube for an oscilloscope, the y-plates are placed nearer the electron gun, since this arrangement gives the best signal sensitivity. The maximum "writing speed" of such a tube depends on the type of phosphor used for the screen, and on the energy of the electrons on arrival. High writing speeds demand a large electron-gun accelerating voltage, and hence the time spent between the deflecting plates is reduced, with consequent decrease in deflexion sensitivity. To maintain signal sensitivity combined with high writing speeds the electrons may be given a further acceleration after passing through the deflecting system. This is called *post-deflexion acceleration* (p.d.a.).

13.5. Deflexion of a Moving Electron in a Uniform Magnetic Field

An electron moving at a velocity u is equivalent to a current-carrying element. For a spacing between electrons of δl, the electron transit time is $\delta l/u$ seconds, and the equivalent current is therefore

$$I = \frac{-e}{\delta l/u} = \frac{-eu}{\delta l} \quad \text{amperes} \qquad . \quad . \quad . \quad (13.13)$$

Hence $$I\,\delta l = -eu \qquad (13.13a)$$

The negative sign indicates that the conventional direction of I is opposite to the direction of u.

If the electron beam passes through a magnetic field of flux density B webers per square metre which is perpendicular to the electron velocity, the electron will experience a force, F, which is perpendicular to both u and B and is given by

$$F = BI\,\delta l = -Beu \text{ newtons} \qquad (13.14)$$

as illustrated in Fig. 13.4 (*a*).

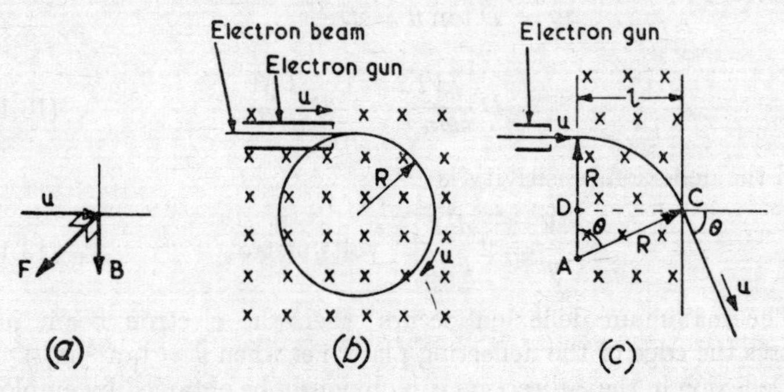

FIG. 13.4. A BEAM OF ELECTRONS BENDS IN A MAGNETIC FIELD

(*a*) *Circular Orbit.* If an electron gun injects a beam of electrons with velocity u at right angles to a magnetic field which is extensive enough or strong enough, then, since the force is always at right angles to both u and B, the electron will describe a circular orbit of radius R, given by

$$Beu = \frac{mu^2}{R}$$

or $$R = \frac{mu}{Be} \text{ metres} \qquad (13.15)$$

This is illustrated in Fig. 13.4 (*b*).

(*b*) *Magnetic Deflexion.* If the magnetic field into which the electron beam is injected is as illustrated in Fig. 13.4 (*c*), the electrons will emerge at an angle θ to their original direction, given by

$$\sin\theta = \frac{l}{R} = \frac{Ble}{mu} \qquad (13.16)$$

Since the linear velocity of the electrons is unchanged by a force which is always at right angles to the motion, the exit velocity will be the same as that of the injected beam.

Magnetic deflexion is sometimes used where a wide angle of deflexion is required (for instance, in television tubes). The magnetic field is produced by current-carrying coils which are mounted externally round the neck of the cathode-ray tube. The deflexion sensitivity is then usually expressed as coil current per unit deflexion.

13.6. Deflexion in Combined Fields

When an electron moves in combined electric and magnetic fields its resultant motion is dependent upon the relative strengths of the fields and on the electron velocity. One method of measuring the velocity of an electron beam is to inject the electrons at right angles to mutually perpendicular electric and magnetic fields as shown in Fig. 13.5 (*a*).

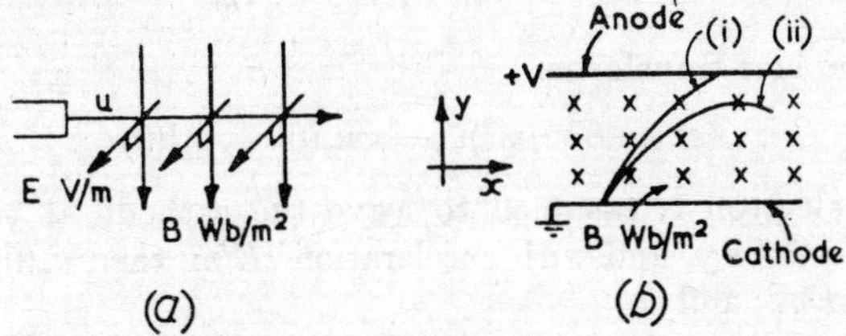

Fig. 13.5. Electrons in Combined Electric and Magnetic Fields

The force due to the motion in the magnetic field is given by eqn. (13.14) as

$$|F_m| = Beu$$

acting out of the plane of the paper, and that due to the electric field is

$$|F_e| = eE \qquad \text{(eqn. (13.1))}$$

acting into the plane of the paper. If these two forces are equal, there will be no resultant deflexion of the beam, and hence, for this condition,

$$u = E/B \quad \text{metres/second}$$

A second example of the use of crossed fields is the *magnetron effect*, illustrated in Fig. 13.5 (*b*). Electrons from the thermionic cathode are attracted towards the anode. A magnetic field of B webers per square metre is arranged as shown in the cathode–anode space. As the electron accelerates, its motion in the magnetic field causes it to trace a curved path (e.g. path (i)). If the magnetic field is strong enough, the electrons will bend completely round and will fail to reach the anode (path (ii)).

If the x- and y-co-ordinates of the electron velocity at any point are u_x, u_y, the acceleration in the x-direction is given by

$$m \frac{\mathrm{d}u_x}{\mathrm{d}t} = Beu_y \quad . \quad . \quad . \quad (13.17)$$

and in the y-direction by

$$m \frac{\mathrm{d}u_y}{\mathrm{d}t} = eE - Beu_x \quad . \quad . \quad . \quad (13.18)$$

where E is the electric field strength and B the magnetic flux density. From these equations,

$$\frac{\mathrm{d}^2u_y}{\mathrm{d}t^2} = - \frac{Be}{m} \frac{\mathrm{d}u_x}{\mathrm{d}t} = - \frac{B^2e^2}{m^2} u_y$$

Taking Laplace transforms,

$$(s^2 + B^2e^2/m^2)\bar{u}_y = su_y(0) + u_y'(0)$$

If the electron is assumed to leave the cathode at time $t = 0$ with zero velocity, and with acceleration eE/m, then $u_y(0) = 0$ and $u_y'(0) = eEm/$, and so

$$\bar{u}_y = \frac{eE}{m(s^2 + B^2e^2/m^2)}$$

or

$$u_y = \frac{E}{B} \sin (Be/m)t \quad . \quad . \quad . \quad (13.19)$$

From eqn. (13.18),

$$u_x = \frac{E}{B} - \frac{E}{B} \cos (Be/m)t \quad . \quad . \quad (13.20)$$

These equations represent cycloidal motion. The y-co-ordinate at any time t is given by

$$y = \int_0^t \frac{E}{B} \sin (Be/m) \, \mathrm{d}t = - \frac{mE}{eB^2} \cos (Be/m)t + \frac{mE}{B^2e}$$

The maximum value of y occurs when $t = \pi m/Be$ and is

$$y_{max} = 2 \frac{mE}{B^2e} \quad . \quad . \quad . \quad (13.21)$$

If this is less than the distance between cathode and anode, the electrons will never reach the anode, but will return to the cathode space charge.

13.7. Electron Entering a Magnetic Field at an Angle to the Normal

When a current-carrying conductor lies along the direction of a magnetic field it experiences no force. In the same way the electrons in an electron beam which lies in the direction of the lines of magnetic flux experience no force. If an electron beam enters a magnetic field at some angle θ to the direction of the field, the force experienced by the electrons will be due only to their normal component of velocity. The motion along the field will be quite unaffected. In Fig. 13.6 an electron gun injects electrons with velocity u at an

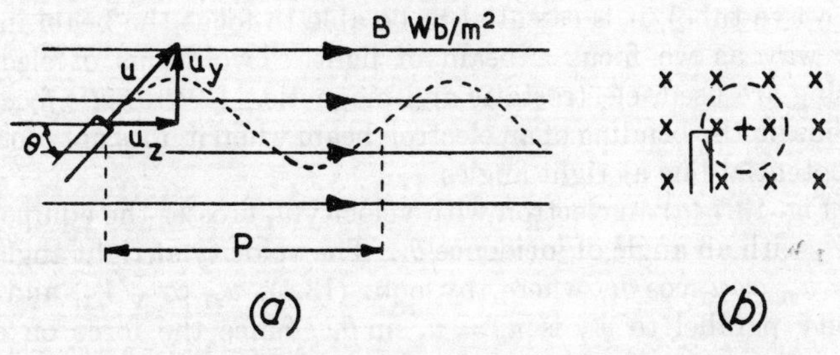

FIG. 13.6. AN ELECTRON SPIRALS IN A MAGNETIC FIELD IF IT ENTERS AT AN ANGLE

angle θ to the axial direction of a magnetic field, which may be considered to be due to a current-carrying coil.

The axial component of velocity is $u_z = u \cos \theta$, and the electrons move from left to right at this rate. The normal component of velocity is $u_y = u \sin \theta$ at the electron gun, so that the force at right angles to both u_y and B is

$$F = Beu \sin \theta \quad \text{newtons}$$

If the electron moved under this force alone its path would be a circular orbit of radius

$$R = \frac{mu \sin \theta}{Be} \qquad \text{(from eqn. (13.15))}$$

as shown in Fig. 13.6 (*b*).

The actual movement is, of course, a combination of the circular and the axial velocity, so that the path is a helix. The pitch of the helix, P, is given by the product of the axial velocity, u, and the time, T, which would be required for one circular orbit. Hence

$$P = u_z T = u_z \frac{2\pi R}{u_y}$$

$$= 2\pi \frac{mu \cos \theta}{Be} \quad . \quad . \quad . \quad (13.22)$$

This spiralling effect can be used to focus an electron beam, electrons entering an axial magnetic field at different angles being spiralled round to meet at one particular point.

13.8. Electrostatic Focusing

In order to produce a sharp spot on a cathode-ray-tube screen, or to keep electrons on a straight and narrow path (as in some forms of microwave tube), it is essential to be able to focus the beam in the same way as we focus a beam of light. Two forms of electron focusing are used, electrostatic and magnetic. *Electrostatic focusing* depends on the bending of an electron beam when it does not cross an equipotential line at right angles.

In Fig. 13.7 (a) an electron with velocity u_1 crosses the equipotential V_1 with an angle of incidence θ_1. The velocity at right angles to V_1 is $u_{z1} = u_1 \cos \theta_1$, where, by eqn. (13.4), $u_{z1} \propto \sqrt{V_1}$, and the velocity parallel to V_1 is $u_y = u_1 \sin \theta_1$. Since the force on each electron is in the direction of the electric field, the y-velocity at V_2 is the same as at V_1. The z-velocity, however, increases owing to the electric field in the z-direction, and its value at the equipotential V_2 is proportional to $\sqrt{V_2}$ (eqn. (13.4)). In the velocity triangles AFG and CDE, we therefore have

$$FG = ED = u_y$$

$$FA \propto \sqrt{V_1} \quad \text{and} \quad CD \propto \sqrt{V_2}$$

Hence

$$\frac{\tan \theta_1}{\tan \theta_2} = \frac{FG/FA}{ED/CD} = \sqrt{\frac{V_2}{V_1}} \quad . \quad . \quad . \quad (13.23)$$

The path of the electrons between the equipotentials may be shown to be parabolic, as follows. If the equipotentials V_1 and V_2 are close enough for the electric field strength, E, between them to be considered uniform and equal to $(V_2 - V_1)/\delta l$ (δl being the distance between the equipotentials), the acceleration in the z-direction may be written

$$a_z = \frac{e}{m} \frac{(V_2 - V_1)}{\delta l} \qquad \text{(from eqn. (13.2))}$$

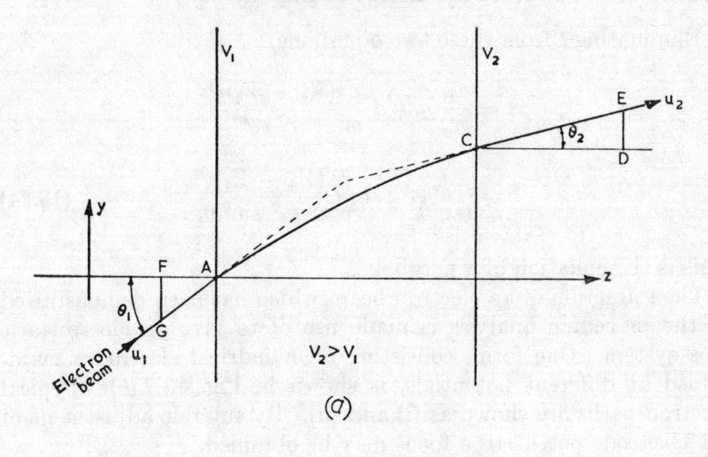

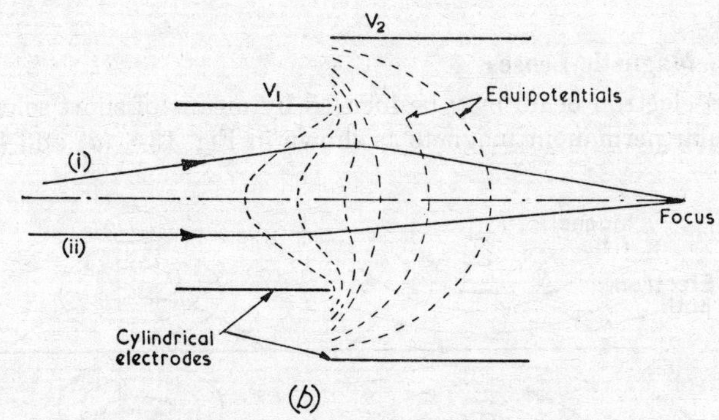

FIG. 13.7. REFRACTION OF AN ELECTRON BEAM, AND A CYLINDRICAL ELECTRON-LENS SYSTEM

Hence the z-component of velocity at V_2 is

$$u_{z2} = u_{z1} + \frac{e}{m}\frac{(V_2 - V_1)}{\delta l}\,t$$

where t is the transit time between V_1 and V_2.

The distance travelled along the z-axis in this time is

$$z = u_{z1}t + \tfrac{1}{2}\frac{e}{m}\frac{(V_2 - V_1)}{\delta l}\,t^2$$

while the distance travelled in the y-direction is

$$y = u_y t$$

Eliminating t from these two equations,

$$z = \frac{u_{z1}}{u_y} y + \frac{1}{2} \frac{e}{m} \frac{(V_2 - V_1)y^2}{\delta l \, v_y^2}$$

$$= \frac{y}{\tan \theta_1} + \frac{1}{2} \frac{e}{m} \frac{(V_2 - V_1)y^2}{\delta l \, v_1^2 \sin^2 \theta_1} \qquad . \qquad . \quad (13.24)$$

This is the equation of a parabola.

The refraction of an electron beam which has been demonstrated in the preceding analysis is made use of to give an electrostatic lens system. One form, consisting of cylindrical electrodes maintained at different potentials, is shown in Fig. 13.7 (*b*). Typical electron paths are shown as (i) and (ii). By suitable adjustment of the electrode potentials a focus may be obtained.

13.9. Magnetic Lenses

An electron beam may be focused by means of short solenoids or annular permanent magnets as shown in Fig. 13.8 (*a*) and (*b*). The

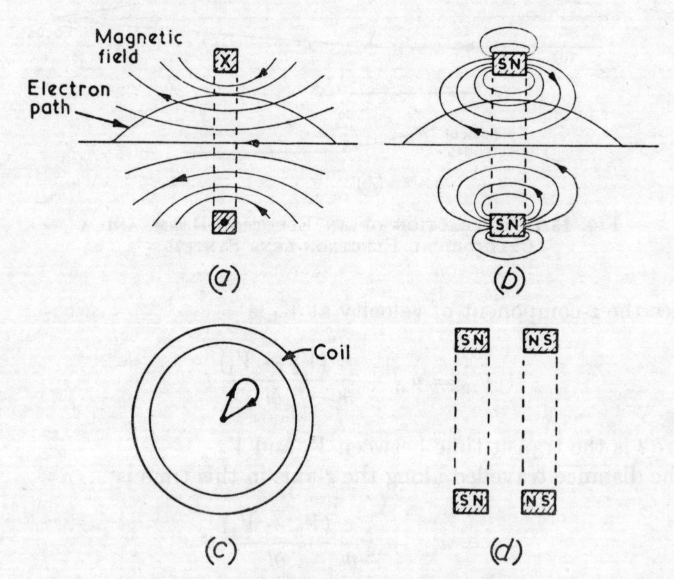

FIG. 13.8. MAGNETIC-LENS SYSTEM

(*a*) Short solenoid
(*b*) Annular permanent magnet
(*c*) End view of electron path
(*d*) Double-magnet lens

radial component of the magnetic field causes bending and rotation of the electron beam, and focusing results. An end view of a typical electron path is shown in Fig. 13.8 (*c*). Since electrons farther from the axis are rotated more than those near the axis, a form of aberration may exist, which is particularly unwanted in television applications. For this reason, magnetic lenses are usually constructed in pairs of opposite polarity, in which case focusing is not affected, but image rotations take place in opposite directions in the two parts of the lens and therefore cancel (Fig. 13.8 (*d*)). Focusing is then achieved by moving one annular magnet relative to the other.

13.10. Particle Accelerators

High-energy sub-atomic particles are required for many applications, including the study of atomic structure, the production of over thirty types of elementary particles, the creation of radioactive isotopes, nuclear physics, medical therapy, and the production of high-energy X-rays for medical deep therapy and industrial radiography. A few particles (e.g. the electron and the proton) are stable and have an electric charge so that they can be accelerated by high voltages. This direct acceleration raises difficulties as the energies required are increased, and particle accelerators generally use lower accelerating voltages which are applied cyclically to give the final particle energy required.

Particle accelerators are designed mainly to accelerate β-particles (electrons), protons, deuterons (proton plus neutron) and α-particles (helium nuclei). High-energy neutrons, which have no electric charge, may be produced by bombarding a beryllium target with protons or deuterons. Very "hard" (i.e. high-energy) X-rays are produced by bombarding a target with high-energy electrons.

Since the non-relativistic kinetic energy of a moving particle is $\frac{1}{2}mu^2$ ($= QV$, where Q is the particle charge, and V is the effective total accelerating voltage), it follows that the heavier the particle, the larger is the energy it will acquire in accelerating to a given velocity. For this reason, protons, deuterons and α-particles can achieve much higher energies than electrons. Heavier ions are not commonly used, owing to their poor penetration of targets.

Particle accelerators fall into two main categories, those employing a linear mode of acceleration, and those employing a circular mode.

13.11. Linear Accelerators

The simplest form of *linear accelerator* consists of a pair of electrodes in an evacuated tube. A high direct voltage is applied between the electrodes, and particles introduced into the electric field between them are accelerated, to attain a final energy of QV electron-volts, where Q is the particle charge and V the total accelerating potential (eqn. (13.3)). The required accelerating voltage may be obtained from a Van de Graaff generator (Fig. 13.9). A plastic or rubber belt, A, is driven by a motor. A pointed electrode, B (maintained at a potential of several kilovolts) causes an induced positive charge on the belt, which is carried up to a second pointed electrode, D, fixed on the inside of the conducting sphere, E. A negative charge is drawn from D to neutralize the charge on the belt, and the outer surface of the sphere thus acquires a positive charge. Potentials of several megavolts may be induced on the upper sphere, but insulation difficulties usually limit the range of application of this technique to 2 or 3 MV.

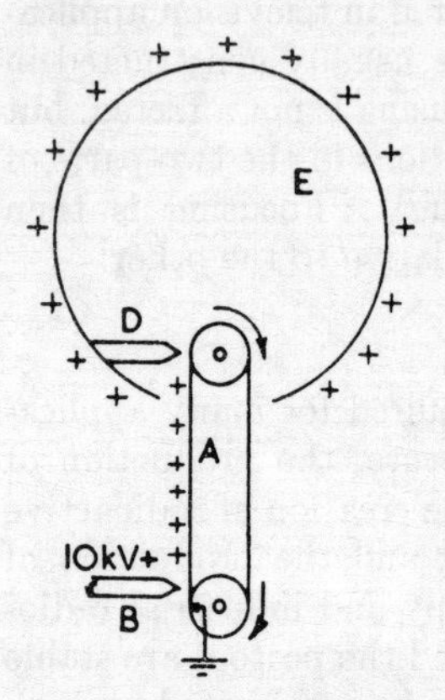

FIG. 13.9. THE VAN DE GRAAFF GENERATOR

For higher energies the linear accelerator shown in Fig. 13.10 may be used. Particles are produced by the ion source, at one end of an evacuated envelope, which contains several tubular electrodes and a target, as shown. The ion source could be a heated filament maintained at some 200 V negative with respect to an exciting anode. A small amount of gas introduced into the evacuated envelope becomes ionized by collision with the electrons between the

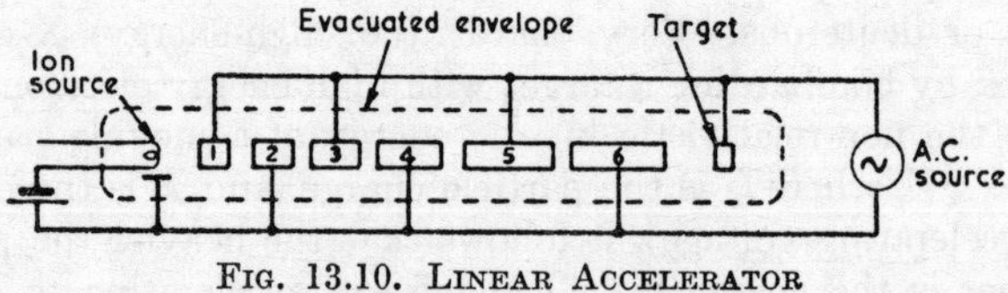

FIG. 13.10. LINEAR ACCELERATOR

filament and the excitation anode (with hydrogen gas, protons are produced; with deuterium, deuterons; with helium, α-particles). The positive ions are accelerated towards the first cylindrical electrode (1) at the instant when the a.c. source makes this of negative polarity. They then drift at a constant velocity through electrode 1 (since once they are inside the cylinder, changes in the electrode potential cannot affect their velocity), and provided that the electrode length and a.c. source voltage are correctly chosen, they arrive at the gap between electrode 1 and electrode 2 at the instant when electrode 2 is negative. The ions are therefore accelerated across this gap, and travel through electrode 2 at an increased velocity, to arrive at the gap between electrodes 2 and 3 at the instant when electrode 3 is negative. Thus at each gap the particles are accelerated, and are emitted from the final cylinder towards the target with extremely high energies. Since the drift spaces provided by each electrode must be successively larger with increasing numbers of accelerations (in order that the particles shall arrive at the next gap at the correct instant in the a.c. source cycle), the linear accelerator must be extremely long for particles with very high energies.

If the peak alternating accelerating voltage is V_m, and is the same for each gap, then neglecting relativitistic effects, the energy acquired by a particle of charge Q at the nth electrode is

$$nQV_m = \tfrac{1}{2}mu_n^2 \qquad (13.25)$$

where m is the particle mass, and u_n the velocity when entering the nth electrode.

To maintain the acceleration, the time, t, taken to travel through the nth electrode must be the time for one half-cycle of the a.c. supply, so that

$$t = \frac{1}{2f} = \frac{l_n}{u_n} \qquad (13.26)$$

where l_n is the length of the nth electrode. From eqns. (13.25) and (13.26),

$$l_n = \frac{u_n}{2f} = \sqrt{\frac{nQV_m}{2f^2m}}$$

The overall accelerator length for N electrodes is therefore

$$L = \sqrt{\left(\frac{QV_m}{2f^2m}\right)} \sum_{n=1}^{N} n^{1/2} \qquad (13.27)$$

and the final energy at the target is

$$W = (N+1)QV_m \qquad (13.28)$$

Modern linear accelerators for electrons and light positive particles use microwave frequencies and "slow wave" techniques to produce energies of tens and hundreds of mega-electron-volts.

13.12. Circular Accelerators—the Cyclotron

Circular accelerators employ magnetic fields to give orbital motion to the particles, and it is arranged that energy is acquired with each orbit. From eqn. (13.15), the momentum of a particle of charge Q and mass m, moving at velocity u in an orbit at right angles to a magnetic field in which the flux density is B is given by

$$mu = BQR \qquad (13.29)$$

where R is the radius of the orbit. The orbital frequency (*cyclotron resonance frequency*) is

$$f_o = \frac{\omega_0}{2\pi} = \frac{u}{2\pi R} \qquad (13.30)$$

Hence, from eqn. (13.29),

$$f_o = \frac{BQ}{2\pi m} \text{ cycles/second} \qquad (13.30a)$$

This equation shows that as long as m is constant (i.e. the velocity does not approach the velocity of light), the orbital frequency is

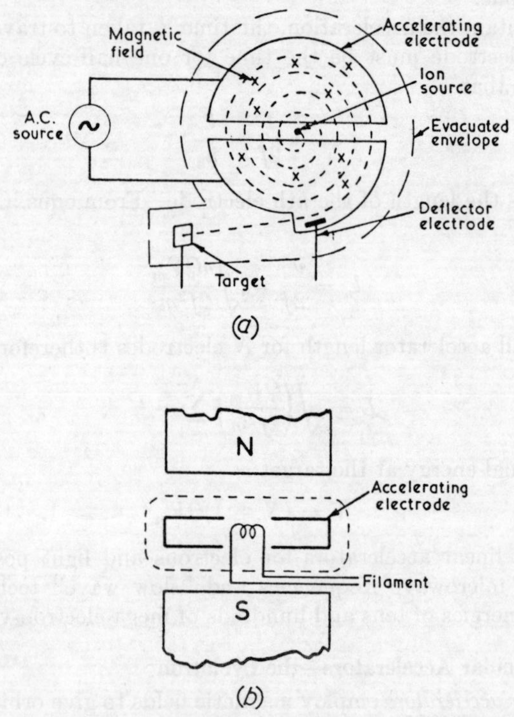

Fig. 13.11. Schematic Views of a Cyclotron

independent of the radius of the orbit, while eqn. (13.30) shows that u increases linearly with R. This is the principle of the *cyclotron*, whose schematic form is shown in Fig. 13.11.

Two hollow D-shaped electrodes are mounted in an evacuated envelope between the poles of a powerful electromagnet. An ion source is located at the centre of the cyclotron between the electrodes, which are connected to a high-frequency generator. The positive ions are accelerated towards the negative electrode, and under the influence of the magnetic field they follow a circular orbit within the electrode. If the source frequency is chosen so that it corresponds to the cyclotron resonance frequency, the ions after one half-circle will arrive back at the accelerating gap at the instant when the second electrode has its maximum negative potential, and will be further accelerated across the gap. They then follow a circular path of increased radius (eqn. (13.39)) inside the second electrode, and again arrive at the gap when the first electrode is once more negative. Thus the ion path is an outward spiral, there being an increase of velocity and of orbital radius of $\sqrt{2}$ at each crossing of the accelerating gap. The ions are deflected out of the electrodes when they reach their outermost orbit by means of a deflecting electrode, and strike the required target.

The power of the cyclotron is limited by the maximum size of the magnetic field which can be produced (some 5 m diameter), and by the fact that as the particles approach the velocity of light the energy acquired on crossing the gap goes to increase the effective mass rather than the velocity. The particles then fall out of resonance with the accelerating field.

If R_m is the maximum radius, the particle energy on leaving the electrodes is

$$W = \tfrac{1}{2}m_0u^2 = \tfrac{1}{2}m_0\,\frac{B^2Q^2R_m{}^2}{m_0{}^2} \quad \text{(from eqn. (13.29))}$$

$$= \tfrac{1}{2}\,\frac{B^2Q^2R_m{}^2}{m_0} \qquad . \quad . \quad . \quad (13.31)$$

Energies of some 20 MeV have been achieved using deuterons.

Owing to their small mass, electrons are not used in cyclotrons, since the orbital frequencies given by eqn. (13.30*a*) are impracticably large. Thus with a flux density of 1 Wb/m², the cyclotron resonance frequency for an electron is

$$f_o = \frac{1 \times 1{\cdot}6 \times 10^{-19}}{2\pi \times 9{\cdot}1 \times 10^{-31}} = 28 \times 10^9 \text{ c/s}$$

while for a deuteron, whose rest mass is 3,680 times that of an electron,

$$f_o = \frac{28 \times 10^9}{3{,}680} = 7{\cdot}62 \times 10^6 \text{ c/s}$$

13.13. High-energy Circular Accelerators

When very high energies are required, the particles are accelerated up to almost the velocity of light, and increased energy is obtained through an increase in effective mass, as well as through an increase in velocity. The effect of the effective mass increase on resonance frequency, orbital radius and particle energy can be deduced as follows.

From eqn. (13.9) for the effective mass, we can write

$$m_0 = m_e \sqrt{\frac{c^2 - u^2}{c^2}} \quad \text{or} \quad m_e{}^2u^2 = m_e{}^2c^2 - m_0{}^2c^2$$

This, in turn, yields

$$m_e u = \sqrt{(m_e{}^2c^2 - m_0{}^2c^2)} = \frac{\sqrt{(m_e{}^2c^4 - m_0{}^2c^4)}}{c}$$

$$= \frac{\sqrt{(W^2 - W_0{}^2)}}{c} \quad \text{(from eqns. (13.6) and (13.7))}$$

where $W = m_e c^2 = W_0 + K$, W_0 being the rest energy, and K the particle kinetic energy, in joules.

Hence,

$$m_e u = \frac{\sqrt{[K(2W_0 + K)]}}{c}$$

But eqn. (13.29) gives the momentum as BQR, and hence the orbit radius becomes

$$R = \frac{\sqrt{[K(2W_0 + K)]}}{cBQ} \quad . \quad . \quad . \quad (13.32)$$

If the kinetic energy is large compared with W_0, we can write

$$K \simeq cBQR \quad . \quad . \quad . \quad (13.33)$$

Otherwise eqn. (13.32) must be solved for K.

Example 13.3. Determine the stable orbital radius for (*a*) an electron, (*b*) a deuteron with energy of (i) 1 MeV, (ii) 1,000 MeV in a steady magnetic field of 1 Wb/m². The rest energy of an electron is 0·51 MeV, and of a deuteron, 1,877 MeV.

(*a*) For the electron $Q = e = 1{\cdot}6 \times 10^{-19}$ C, and hence for an energy of 1 MeV ($= 1 \times 10^{6} \times 1{\cdot}6 \times 10^{-19}$ J) eqn. (13.32) gives

$$R = \frac{1{\cdot}6 \times 10^{-13}\sqrt{[(2 \times 0{\cdot}51 + 1)]}}{3 \times 10^{8} \times 1 \times 1{\cdot}6 \times 10^{-19}} = 0{\cdot}00474\ \text{m}$$

For 1,000 MeV,

$$R = \frac{1{\cdot}6 \times 10^{-13}\sqrt{[(2 \times 0{\cdot}51 + 1{,}000)\ 1{,}000]}}{3 \times 1{\cdot}6 \times 10^{-11}} = 3{\cdot}33\ \text{m}$$

(*b*) For the deuteron, $Q = e$, and hence eqn. (13.32) gives

(i) For 1 MeV

$$R = \frac{1{\cdot}6 \times 10^{-13}\sqrt{[(2 \times 1{,}877 + 1)]}}{3 \times 10^{8} \times 1 \times 1{\cdot}6 \times 10^{-19}} = 0{\cdot}204\ \text{m}$$

(ii) For 1,000 MeV

$$R = \frac{1{\cdot}6 \times 10^{-13}\sqrt{[(3{,}754 + 1{,}000)\ 1{,}000)]}}{6 \times 1{\cdot}6 \times 10^{-11}} = 7{\cdot}26\ \text{m}$$

This example gives some idea of the dimensions required in circular accelerators.

The orbital frequency required to maintain a stable orbit when relativistic mass changes are taken into account is given by replacing m_0 in eqn. (13.30*a*) by its relativistic equivalent. Thus

$$f = \frac{BQ\sqrt{(1 - u^2/c^2)}}{2\pi m_0} \qquad . \quad . \quad . \ (13.34)$$

We may express this in terms of energies, by writing

$$f = \frac{BQ}{2\pi m_0}\sqrt{\frac{c^2 - u^2}{c^2}} = f_o\sqrt{\frac{m_e c^2 - m_e u^2}{m_e c^2}}$$

$$= f_o\sqrt{\frac{W_0}{W}} = f_o\sqrt{\frac{W_0}{W_0 + K}} \qquad . \quad . \quad . \ (13.35)$$

This equation shows that as K increases the orbital frequency must be reduced to maintain the particles in the "resonant" condition. This is the principle of the *synchrocyclotron*, in which, after the particles have been accelerated by cyclotron action, the supply frequency is slowly reduced. The particles will then remain "in phase" with the supply and will gain energy and increase in orbit radius at a rate which is set by the rate of change of frequency. If the frequency is changed too rapidly, however, the particles will not acquire sufficient energy on each crossing of the gap to remain in the correct phase relation with the supply. Deuterons with energies of over 600 MeV have been produced by synchrocyclotrons. Owing to the large cyclotron resonance frequency, and hence to the large frequency changes which would be involved, electrons are not used in synchrocyclotrons.

13.14. Electron Acceleration—the Betatron

The *betatron* (Fig. 13.12) is a commonly used accelerator for electrons. A highly evacuated tube is placed round an iron core,

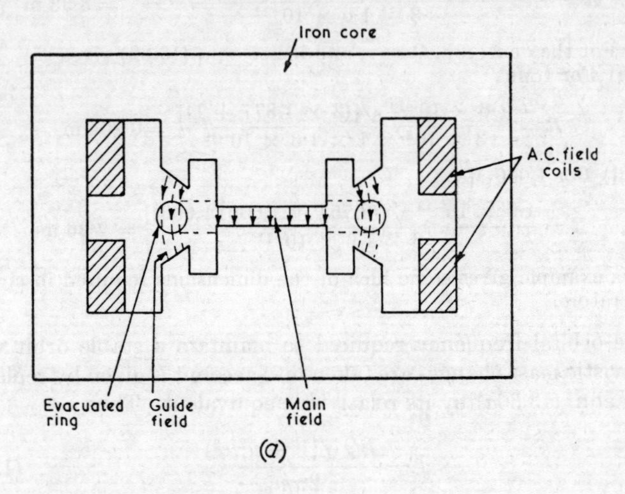

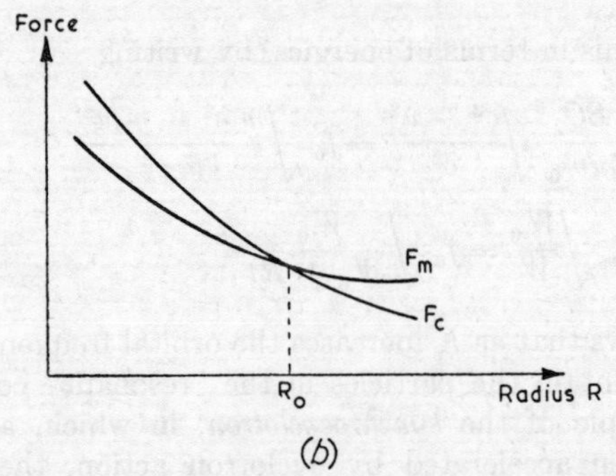

Fig. 13.12. Schematic View of a Betatron

whose magnetic field may be varied from zero to some suitable maximum value by a.c. field coils, usually at power frequencies. High-velocity electrons from an electron gun are injected into the tube at a given instant in the a.c. field cycle. The guide field causes the electrons to orbit in the evacuated tube. Increasing the magnetic field gives rise to a voltage gradient round the tube, which accelerates the electrons which are orbiting there, and so increases their energy. At the same time, the guide field must increase in order to preserve the same orbital radius. Thus an electron of velocity u in a magnetic field where the flux density is B describes an orbit of radius R given by eqn. (13.15) as

$$R = \frac{mu}{Be} \text{ metres}$$

so that if u/B is constant, the radius will be constant at some value, R_0.

For an enclosed flux Φ which is changing at the rate of $\mathrm{d}\Phi/\mathrm{d}t$ webers per second, the electric field strength, E_R, at radius R is

$$E_R = \frac{-1}{2\pi R}\frac{\mathrm{d}\Phi}{\mathrm{d}t} \text{ volts/metre}$$

since the voltage induced in a single turn is $V = \mathrm{d}\Phi/\mathrm{d}t = -2\pi RE_R$.

The electron velocity is related to the changing flux by the equation

Force on electron = Rate of change of momentum

Hence

$$eE_{R0} = \frac{\mathrm{d}(mu)}{\mathrm{d}t} = \frac{-e}{2\pi R_0}\frac{\mathrm{d}\Phi}{\mathrm{d}t}$$

where E_{R0} is the field strength at radius R_0, and the negative sign gives the vector relation between the direction of E and that of the flux axis.

Integrating and taking magnitudes,

$$mu = \frac{e}{2\pi R_0}(\Phi - \Phi_0)$$

where Φ is the enclosed flux after $\mathrm{d}t$ seconds and Φ_0 is the initial flux. But, for a stable orbit, $mu = BeR_0$ (eqn. (13.15)), and hence

$$BeR_0 = \frac{e}{2\pi R_0}(\Phi - \Phi_0)$$

or

$$B = \frac{1}{2\pi R_0^2}(\Phi - \Phi_0) \quad . \quad . \quad . \quad (13.36)$$

The actual change of flux linking the electron orbit for a stable orbit must therefore be

$$\Phi - \Phi_0 = 2\pi R_0^2 B \quad . \quad . \quad . \quad (13.37)$$

This is twice the change in flux ($\pi R_0^2 B$) which would take place if the guide-field density, B, were uniform over the area enclosed by the ring, and gives the betatron "2 to 1" rule. The required condition is achieved by having the main-field flux density much larger than that of the guide field.

A further condition for radial stability of the electrons is that, if any electrons leave the orbit (owing, for example, to collisions with residual gas molecules), the forces acting on them are such that they are driven back into the orbit. This is achieved, provided that the rate of change with radius of the magnetic force F_m ($= Beu$) acting towards the centre is less than the rate of change with radius of centrifugal force, F_c ($= mu^2/R$), which acts outwards and requires correct shaping of the guide-field pole faces to give a decreasing value of B with increasing radius. Fig. 13.12 (*b*) shows how F_c and F_m should vary with R. For $R < R_0$, $F_m < F_c$ and electrons are forced outwards towards R_0. For $R > R_0$, $F_m > F_c$ and electrons are forced inwards towards R_0.

The electrons are emitted from the betatron when they have acquired sufficient energy, by suitably disturbing the guide field by means of a further auxiliary field. They then hit the required target or are ejected through a thin window. Energies of over 400 MeV have been achieved.

13.15. The Synchrotron

The *synchrotron* combines the action of the betatron and the synchrocyclotron to give the highest energy particles so far achieved. In the *electron synchrotron*, electrons are accelerated in a circular orbit by betatron action up to almost the velocity of light (energy of 2–4 MeV). The radio-frequency accelerating system is then switched on to increase the electron energy to several hundred mega-electron-volts. In the *proton synchrotron*, since the particle velocities do not approach the velocity of light, the radio-frequency field must be suitably frequency modulated in order to maintain stable orbital motion, as in the synchrocyclotron. Energies in excess of 25 GeV have been achieved (1 GeV $= 10^3$ MeV).

Bibliography

Amos and Birkinshaw, *Television Engineering, Vol. I* (Iliffe).
Ed. Lovell, *Electronics* (Pilot Press).
Livingstone, *High Energy Accelerators* (Interscience, New York.)

Problems

(The ratio of charge to mass for an electron at rest is $1{\cdot}76 \times 10^{11}$ C/kg.)

13.1. A high-vacuum diode has a cylindrical anode of diameter 1 cm. The cathode, of very small diameter, is on the axis of the cylinder. The anode is maintained at a positive potential of 800 V relative to the

cathode. What value of uniform axial magnetic field is required just to cause the anode current to be zero? Derive any necessary formulae and state clearly any assumptions made. (*I.E.E. Eln.*, 1956)
(*Ans.* $3\cdot8 \times 10^{-2}$ Wb/m^2.)

13.2. Given that mass and energy are related by the equation $W = mc^2$, show that the mass of an electron having a rest mass m_0 and travelling with a velocity v is given by

$$m = \frac{m_0}{(1 - v^2/c^2)^{1/2}}$$

(*L.U. Part III Eln.*, 1957)

13.3. The spacing between the anode and the cathode in a planar diode is a metres and the potential of the anode is V volts positive relative to the cathode. A magnetic field, of flux density B webers per square metre, is directed parallel to the electrode surfaces. For what value of the flux density will an electron emitted from the cathode with zero velocity just reach the anode? What is the transit time from cathode to anode under these conditions? (*L.U. Part III Eln.*, 1958)

13.4. In a cathode-ray tube a beam of electrons is accelerated through a potential difference V_1 before passing between a pair of parallel deflecting plates. The distance between the plates is S and the effective axial length of the plates is l. If a steady potential difference V_2 is maintained between the plates, derive an expression for the angular deflexion of the electron beam. It may be assumed that the electric field between the plates is constant over the effective length. Hence calculate the angular deflexion of an electron beam when $V_1 = 3{,}000$ V, $V_2 = 150$ V, $l = 6$ cm, and $S = 2$ cm. (*A.E.E.*, 1957)
(*Ans.* $4\cdot3°$.)

13.5. Electrons are accelerated through a potential difference V_0 to a velocity u. They form two long, thin and parallel cylindrical beams at a distance x apart. A current I is carried by each beam. Derive an expression for the total force per unit length between the beams. If $V_0 = 20$ kV, $x = 1$ cm, and $I = 20$ mA calculate the force acting on each beam per metre length and indicate the direction of the force. [*Hint.* Consider both attractive and repulsive force.] (*A.E.E.*, 1958)
(*Ans.* 94×10^{-9} N/m.)

13.6. Derive an expression for the electric field strength in the annular space bounded by two concentric cylinders when there is a potential difference between them.

An electron is injected with a certain velocity and at a certain radius into the evacuated space between the cylinders in a tangential direction. Determine the relation which must exist between electron velocity, cylinder radii and potential difference if the electron is to follow a concentric circular orbit. Calculate the p.d. required to give a circular orbit if the electron velocity is 10^7 m/sec and the relevant cylinder radii are 2 cm and 6 cm respectively. (*A.E.E.*, 1957)
(*Ans.* 625 V.)

13.7. A cathode-ray oscillograph has a final-anode voltage of $+2\cdot0$ kV with respect to the cathode. Calculate the beam velocity.

Parallel deflecting plates are provided, $1\cdot5$ cm long and $0\cdot5$ cm apart, their centre being 50 cm from the screen. (*a*) Find the deflexion sensitivity in volts applied to the deflecting plates per millimetre deflexion at the

screen. (*b*) Find the density of a magnetic cross-field, extending over 5 cm of the beam path and distant 40 cm from the screen, that will give a deflexion at the screen of 1 cm. Prove all the formulae used.
(*Ans.* 2·67 V/mm; $0{\cdot}76 \times 10^{-4}$ Wb/m^2.) (*A.E.E.*, 1956)

13.8. A charged particle is projected with velocity u into a uniform magnetic field of flux density B. Derive from first principles the subsequent trajectory of the particle, and hence discuss with the aid of diagrams either (*a*) the measurement of current by means of a cathode-ray tube, or (*b*) the principle of operation of the cyclotron.
(*A.E.E.*, 1954)

13.9. An electron is accelerated through a potential difference V and then enters a uniform magnetic field of magnetic flux density B in a direction perpendicular to the direction of the field. Show that the electron will describe a circular path, and derive an expression for the radius of the circle.

If the flux density, B, is 10^{-3} Wb/m^2, and the electron leaves the magnetic field 0·0005 μsec after entering it, calculate the angle between the direction of entry of the electron and its direction of leaving the field.
(*Ans.* 5°.) (*A.E.E.*, 1958)

13.10. An electron accelerated through a potential difference V enters a uniform magnetic field, of flux density B, in a direction perpendicular to the direction of the field. Show that the electron will describe a circular path, and derive expressions for the radius of the circle and the angular velocity of the electron.

The magnetic flux density between a pair of plane, parallel, circular pole faces is 0·01 Wb/m^2. The flux density is assumed to be uniform with negligible edge effects, the effective radius of the pole faces being 3 cm. An electron accelerated through a potential difference of 1 kV enters the field in a radial direction. Calculate the interval of time during which the electron is between the pole faces, and the direction in which it leaves the field relative to the direction of entry. (*A.E.E.*, 1960)
(*Ans.* $1{\cdot}4 \times 10^{-9}$ sec.; 141°.)

13.11. The anode and cathode of a diode are coaxial cylinders of radii R and r, respectively, with $R > r$. A potential difference V is maintained between them. An electron leaves the cathode radially with negligible velocity. Derive an expression for the subsequent velocity of the electron as a function of the radius x from the axis of the cylinders. Show that the velocity reaches half of its final value when $x = (Rr^3)^{1/4}$. The effects of space charge are to be neglected.
(*A.E.E.*, 1961)

13.12. Two long, thin, parallel beams of electrons each constitute a current I in the same direction. The distance between the beams is x and the velocity of the electrons in each beam is u. Derive an expression for the electric force and for the magnetic force per unit length between the beams. Hence show that the resultant force must always be one of repulsion.

If $x = 4$ mm, $I = 100$ mA and the beams were formed initially by accelerating the electrons through a potential difference of 10 kV, calculate the resultant force per metre length on the beams.
(*Ans.* $12{\cdot}3 \times 10^{-6}$ N/m, repulsion.) (*A.E.E.*, 1963)

CHAPTER 14

Units, Dimensions and Standards

An essential part of the theory of any science or technology consists in the development of expressions relating the various quantities which are used in that particular discipline. The units chosen for each quantity should be such that simple and logical relations exist between them, since otherwise awkward and useless conversion factors must be used. A coherent system of units for the physical sciences has been adopted by the General Conference on Weights and Measures and is called the International System of Units. The units are called S.I. units (from the French, *Système Internationale d'Unités*).

There are six basic S.I. units, from which all others in the system are derived. The basic units are the metre, kilogramme, second, ampere, degree Kelvin, and candela. In electrical engineering we are normally concerned with the basic units of mass, length, time and current, and this part of the international system is called the M.K.S.A. system. The magnitudes of the basic units are realized in standards.

In Britain the foot, pound and second have long been established as the fundamental units of length, mass and time in mechanical engineering (along with the degree Fahrenheit and the British thermal unit) to give the F.P.S. system. In physics it is still usual to employ the centimetre, gramme and second (C.G.S. system). Units in these systems obviously bear a fixed relation to those in the M.K.S.A. system.

Units whose magnitudes are directly related to the six basic units are called absolute units. Since the introduction of the M.K.S.A. system and due to advances in the techniques of measurement, these absolute units are also the practical units in which, for instance, indicating instruments are calibrated. In order to ensure national and international uniformity, reference and absolute standards are maintained in many national laboratories (e.g. the National Physical Laboratory in Britain, and the National Bureau of Standards in the United States). Absolute standards are those whose values can be directly determined in terms of the six fundamental standards.

Reference standards have their values determined by the use of absolute standards. These are checked against each other, and against standards held in other national standards laboratories.

Previous to the introduction of the M.K.S.A. system, a system of electrical comparison standards was used which gave a set of international units (abandoned in 1948, and not to be confused with the present ones). Thus the old international ohm was the resistance at 0°C of a column of mercury of mass 14·4521 g, uniform cross-section, and length 106·300 cm. There is a difference of some 0·015 per cent between the old and the new international units.

14.1. The Fundamental Standards for the M.K.S.A. System

The only basic S.I. unit which is represented by a prototype is the unit of mass. The kilogramme is the mass of a particular piece of metal which is carefully preserved in the Bureau International des Poids et Mesures (B.I.P.M.) near Paris. Nominally it is 1,000 times the mass of 1 cm^2 of water at maximum density.

The unit of length is nominally the distance between two marks on a metal bar kept at the B.I.P.M., but is now defined as 1,650,763·73 wavelengths *in vacuo* of the radiation corresponding to the transition between the energy levels P_{10} and $5D_5$ of the krypton-86 atom. Obviously such a standard of length was not possible until accurate methods of measuring such wavelengths had been developed.

The unit of time is the second of ephemeris time, or the ephemeris second. This is defined as 1/31,556,925·9747 of the tropical year for 1900. The exact year must be specified, since the earth's rotation is gradually slowing down, and hence the length of each year is gradually increasing in terms of ephemeris time—i.e. of time reckoned in seconds relative to the second in 1900. Molecular clocks can be constructed and calibrated in ephemeris seconds. In the caesium standard clock, atomic resonance in caesium is stimulated at 9,192,631,770 $\pm$ 20 c/s in terms of ephemeris time. An alternative using the electronuclear properties of rubidium vapour, which produces an effect at 6,834,682,614 c/s with an annual stability of 5 parts in 10^{11} has also been constructed. Using such clocks it is possible to detect variations in the rotational velocity of the earth.

In some cases the second of universal time, or 1/8,640 of the mean solar day, is taken as a convenient measure of time.

The fourth fundamental unit in the M.K.S.A. system is the ampere. This is defined in such a way that the practical electrical units which were in use before the adoption of the M.K.S.A. system will also be the units used in theoretical work. The ampere is thus defined as the current which, flowing through two straight parallel conductors of infinite length, of negligible circular cross-section, and placed 1 metre apart in a vacuum, gives rise to a force of 2×10^{-7} newton per metre length between them.

14.2. Dimensions

The dimensions of any derived unit give the relation between the derived unit and the fundamental units of length [L], mass [M], time [T], and (in electrical engineering) current [I]. The "dimensional equations" giving these relations are conventionally written in square brackets in order to differentiate between them and the equations relating the quantities to which they refer.

Thus velocity, u, is length travelled per unit time, so that the dimensions of velocity are

$$[u] = [\mathrm{LT^{-1}}] \qquad (14.1)$$

Similarly acceleration, a, is change in velocity per unit time so that dimensionally

$$[a] = [u][\mathrm{T^{-1}}] = [\mathrm{LT^{-2}}] \qquad (14.2)$$

Other mechanical units follow—

Force = Mass × Acceleration $\quad [F] = [\mathrm{LMT^{-2}}] \qquad (14.3)$

Momentum = Mass × Velocity $\quad [p] = [\mathrm{LMT^{-1}}] \qquad (14.4)$

Work = Force × Distance $\quad [W] = [\mathrm{L^2MT^{-2}}] \qquad (14.5)$

Power = Rate of working $\quad [P] = [\mathrm{L^2MT^{-3}}] \qquad (14.6)$

In the M.K.S.A. system the unit of force is the newton, which is the force required to give a mass of 1 kg an acceleration of 1 m/sec^2; the unit of work or energy is the joule (the work done when a force of 1 N moves its point of application 1 m); and the unit of power is the watt (a rate of working of 1 J/sec). Note that the joule is also the unit of heat (heat produced when a dissipation of 1 W is continued for 1 sec). Calories are obsolescent units.

The dimensions of electrical units in the M.K.S.A. system may be deduced as follows.

Charge = Current × Time $\quad [Q] = [\mathrm{TI}] \qquad (14.7)$

Force between charges, $F = \dfrac{Q_1Q_2}{4\pi\epsilon r^2}$, so that

Permittivity, $\epsilon = \dfrac{Q_1Q_2}{4\pi F r^2} \quad [\epsilon] = [Q^2F^{-1}\mathrm{L^{-2}}]$

$$= [\mathrm{L^{-3}M^{-1}T^4I^2}] \qquad (14.8)$$

Potential difference $= \dfrac{\text{Work}}{\text{Charge}}$ $\qquad [V] = [WQ^{-1}]$

$$= [\mathrm{L^2MT^{-3}I^{-1}}] \quad . \quad (14.9)$$

Resistance = P.D./Current $\qquad [R] = [VI^{-1}]$

$$= [\mathrm{L^2MT^{-3}I^{-2}}] \quad . \quad (14.10)$$

Since $V = \mathrm{d}\Phi/\mathrm{d}t$,

Magnetic flux = Voltage × Time $\qquad [\Phi] = [VT]$

$$= [\mathrm{L^2MT^{-2}I^{-1}}] \quad . \quad (14.11)$$

Magnetic flux density

= Magnetic flux/Area $[B] = [\Phi\mathrm{L^{-2}}]$

$$= [\mathrm{MT^{-2}I^{-1}}] \quad . \quad (14.12)$$

Magnetizing force

= Current/Length $[H] = [\mathrm{L^{-1}I}]$ . . (14.13)

Permeability, $\mu = B/H$ $\qquad [\mu] = [\mathrm{LMT^{-2}I^{-2}}]$. (14.14)

Since $V = L(\mathrm{d}i/\mathrm{d}t)$,

Inductance $= \dfrac{\text{Voltage} \times \text{Time}}{\text{Current}}$ $\qquad [L] = [VT\mathrm{I^{-1}}]$

$$= [\mathrm{L^2MT^{-2}I^{-2}}] \quad . \quad (14.15)$$

$$= [\mu\mathrm{L}] \quad . \quad (14.16)$$

Capacitance = Charge/P.D. $\qquad [C] = [QV^{-1}]$

$$= [\mathrm{L^{-2}M^{-1}T^4I^2}] \quad . \quad (14.17)$$

$$= [\epsilon\mathrm{L}] \quad . \quad . \quad (14.18)$$

Electric flux density = Charge/Area $[D] = [Q\mathrm{L^{-2}}]$

$$= [\mathrm{L^{-2}TI}]. \quad . \quad (14.19)$$

Electric field strength

= P.D./Length $[E] = [V\mathrm{L^{-1}}]$

$$= [\mathrm{LMT^{-3}I^{-1}}] \quad . \quad (14.20)$$

Note that $[\mu\epsilon] = [\mathrm{L^{-2}T^2}] = 1/[u]^2$, i.e. $1/\sqrt{(\mu\epsilon)}$ has the dimensions of a velocity—confirming the results obtained in Chapter 7. A further example of the use of dimensional analysis is found in the derivation of skin depth in Section 11.19.

14.3. Realization of the Absolute Ampere

Since it is inconvenient to measure the force between two long parallel wires, the absolute measurement of current is carried out by means of a current balance. In the National Physical Laboratory the *Ayrton-Jones current balance* is used.

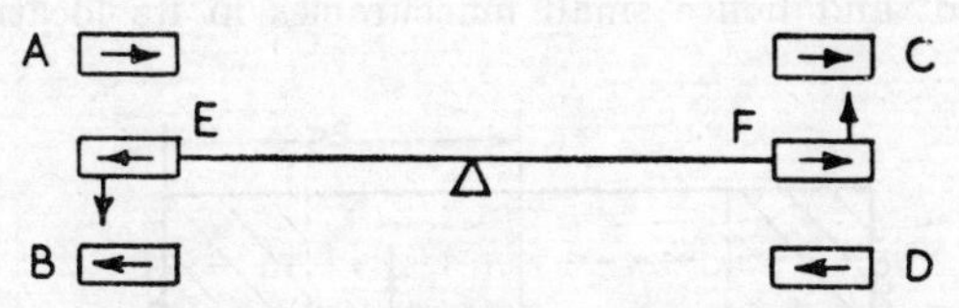

FIG. 14.1. CURRENT BALANCE

A schematic arrangement of a current balance is shown in Fig. 14.1. There are two sets of fixed coils (A,B and C,D), between which are suspended the moving coils E and F. This arrangement eliminates any effect of convection air currents due to coil heating, since the torques due to such currents cancel. The force of attraction and repulsion between the fixed and moving coils is given by eqn. (11.56) as

$$F = I^2 \frac{\mathrm{d}M}{\mathrm{d}x}$$

since all coils carry the same current I.

The quantity $\mathrm{d}M/\mathrm{d}x$ can be calculated from the dimensions of the coils, and the arrangement is such that small inaccuracies in these dimensions cause only minor errors in the final result. The beam is balanced by a sliding weight to bring it to equilibrium, and hence the value of I^2 can be determined. To prevent inaccuracy due to the effects of the fields of coils A and B on coil F, and of coils C and D on coil E, the balance is repeated with all the currents on one side of the balance reversed, and a mean is taken. The ultimate accuracy is limited by the determination of the gravitational constant, and the calculation of $\mathrm{d}M/\mathrm{d}x$ to some 1 part in 10^5, through results reproducible to within 2 parts in 10^6 have been obtained.

14.4. Absolute Determination of Inductance

It was seen in Section 14.2 that the dimensions of inductance are the same as those of the product of μ and a length. The value of $\mu = \mu_0$ for air is accurately known, and hence inductance can be measured in terms of a mechanical length. The reference standard of inductance at the National Physical Laboratory is the mutual inductance shown in Fig. 14.2. The primary consists of two single-layer coils of bare copper wire wound in two sections under tension in grooves which are machined in a marble former. The marble has a temperature coefficient of expansion of some -5×10^{-6}, so that changes in dimensions with temperature are extremely small. With the dimensions shown, the secondary is located at a position of zero field, and hence small inaccuracies in its location or the

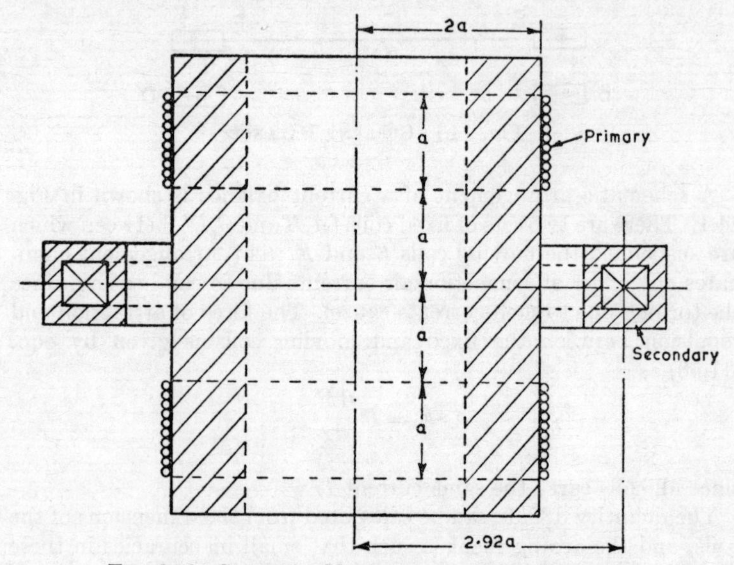

FIG. 14.2. CAMPBELL MUTUAL-INDUCTANCE STANDARD

measurement of its dimensions produce negligible errors in the calculated value of mutual inductance. The secondary is a multi-turn coil, also wound on a marble former. The standard has a nominal value of 10 mH and is known to an accuracy of at least 1 part in 10^5, the value being determined by measurements of the dimensions of the arrangement, and the known value of μ_0.

14.5. Absolute Determination of Resistance

The dimensions of resistance given in Section 14.2 are

$$[L^2MT^{-3}I^{-2}] = [\mu][LT^{-1}] = [\mu][\text{velocity}] \quad (\text{or } [\text{inductance}][T^{-1}])$$

Hence resistance may be measured in absolute terms as the product of the constant μ and a velocity, or in terms of an inductance and time. A method called the Lorenz method gives resistance in terms of inductance and time. The principle of the method is illustrated in Fig. 14.3.

A phosphor-bronze disc is driven at constant speed in the field of a coil which is wound in grooves on a marble former, the mutual inductance, M, between the coil and the disc being accurately calculable. The e.m.f. induced between the centre and circumference of the disc depends on the speed of the disc, and is balanced against the voltage drop due to the coil current flowing through R.

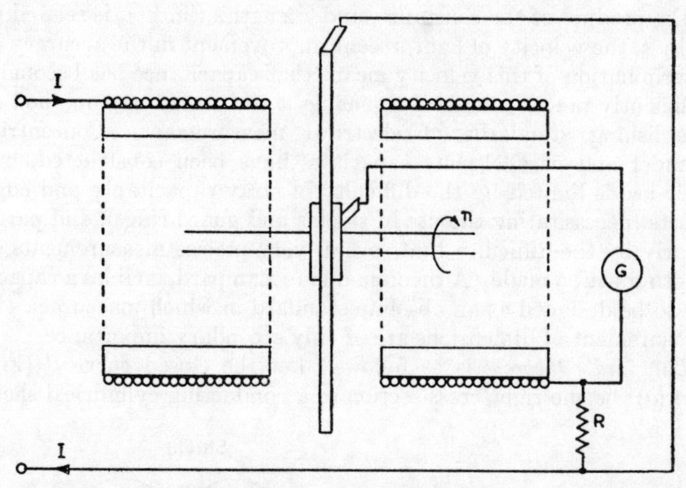

FIG. 14.3. LORENZ DISC METHOD FOR DETERMINATION OF RESISTANCE

Thus, the flux cutting the disc is MI webers, and the induced e.m.f. is MIn volts, where n is the disc speed in revolutions per second. At balance, there is no current in the galvanometer, G; hence

$$MIn = IR$$

or

$$R = Mn \qquad (14.21)$$

Both M and n can be obtained to an accuracy of better than 1 part in 10^5, so that an accuracy of 1 part in 10^4 is readily obtainable in the determination of R.

The effect of the earth's magnetic field is minimized by locating the disc in the plane of the field, and is eliminated by taking a second measurement with the direction of the current reversed. Thermoelectric e.m.f.s are eliminated by arranging two similar discs and solenoids in such a way that the e.m.f.s at the brush contacts cancel.

Advanced measurement methods enable zero-frequency resistance to be determined to within 2 parts in 10^7.

14.6. Absolute Determination of Capacitance—Lampard's Theorem

From the dimensions of capacitance given in Section 14.2,

$$[\text{capacitance}] = [L^{-2}M^{-1}T^4I^2] = [\epsilon][L]$$

it will be seen that we can determine capacitance in absolute terms as the product of the constant ϵ and a length. Since ϵ_0 is related to μ_0 by c, the velocity of light, recent improvement in the accuracy of determination of this velocity means that capacitance has become a sufficiently precise standard to enable it to be used as a method of establishing standards of electrical measurement. Concentric-cylinder and parallel-plate capacitors have been constructed, but their use is limited by the difficulty of stray capacitance and edge effects (necessitating the use of shields and guard rings), and particularly by the difficulty that several very precise measurements of length must be made. A theorem due to Lampard, enables a capacitor to be designed as an absolute standard in which inaccuracies in measurement of dimensions are of only secondary importance.

Lampard's theorem is as follows. Let the closed curve S (Fig. 14.4 (a)) be the right cross-section of a conducting cylindrical shell,

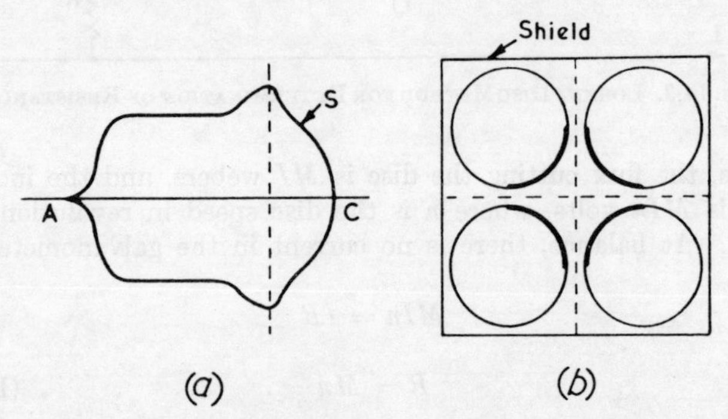

Fig. 14.4. Capacitance between Any Two Opposing Segments of Cylindrical Shells

the cross-section having one axis of symmetry AC, but otherwise being arbitrary in shape. If the shell is divided into four parts by two planes at right angles, the line of intersection of the planes being parallel to the generators of the cylinder, and one of the planes containing the line AC, then the direct capacitance per unit length of the cylinder between either pair of opposing parts of the shell is a constant

$$C_0 = \frac{\epsilon \log_e 2}{\pi} \text{ farads per metre length} \quad . \quad (14.22)$$

where ϵ is the permittivity of the dielectric between the conducting surfaces.

The cross-section of a practical capacitor is shown in Fig. 14.4 (*b*). This arrangement gives a result which is highly insensitive to the width of the gaps between the segments, whose active surfaces are shown by the heavy lines. Guard rings are used to eliminate end effects. The capacitance between any pair of opposite segments is $(\epsilon_0/\pi) \log_e 2$ for an air dielectric.

14.7. Reference Standards

The National Physical Laboratory maintains a range of reference standards of resistance, e.m.f., inductance and capacitance. The e.m.f. standard is that of the Weston standard cadmium cell, which is 1·01859 V at 20°C. Laboratory standards with accuracies to 1 part in 10^4 may be constructed and checked by potentiometer or bridge methods against the National Physical Laboratory reference standards.

For work where accuracy to 1 part in 10^3 is acceptable, standard instruments are used whose scale indication and reading accuracy are within the required limits. In a.c. instrument calibration, transfer standard instruments are used. These are of the electrodynamic or electrostatic type and may be calibrated by means of a potentiometer on direct current, and used on low-frequency alternating current as comparison standards.

14.8. The C.G.S. Systems

The M.K.S.A. system of units described in this chapter is a rationalized system. This means that unit charges and unit magnetic poles are assumed to emit unit fluxes. It follows that expressions involving spherical symmetry normally contain a factor 4π (e.g. field strength near a unit charge), expressions involving circular symmetry have a factor 2π (e.g. potential near a charged line), while expressions involving plane symmetry have neither. Before the introduction of the M.K.S.A. system two electrical systems of theoretical units were in use, based on the centimetre, gramme and second. The two C.G.S. systems are generally used in an unrationalized form (unit charges and poles emit 4π units of flux), and differ both from each other and from the practical units used for normal measurement.

(*a*) *The C.G.S. Electromagnetic System.* This system is based on the assumption that the force between two unit magnetic poles 1 cm apart *in vacuo* is 1 dyne ($= 10^{-5}$ N). Hence the permeability of free space, or of air, is unity and $B = H$ in such media. The electromagnetic unit (emu) of current is then defined as the current which, when flowing through a 1-cm arc of a circle 1 cm in radius produces a force of 1 dyne on a unit pole at the centre. An emu of e.m.f. is induced in a single-turn coil when 1 emu of magnetic flux changes in one second. Hence the units for all electrical quantities in emus can be built up.

The relation between these units and the M.K.S.A. units can be found by considering the equivalent relations. Thus the pole strength in webers required to produce 10^{-5} N at a separation of 1 cm between two magnetic poles is

$$\Phi = \sqrt{(F \times 4\pi\mu_0 r^2)} = 4\pi \times 10^{-8} \text{ Wb (from eqn. (11.10))}$$

so that

$$1 \text{ emu of pole strength} = 4\pi \times 10^{-8} \text{ Wb}$$

Also in the unrationalized system a unit pole emits 4π units of flux, so that

$$1 \text{ emu of magnetic flux} = 10^{-8} \text{ Wb}$$

and hence

$$1 \text{ emu of voltage} = 10^{-8} \text{ V}$$

The relation between current units is found by considering what current in amperes flowing in a circle of 1 cm radius produces a force of $2\pi \times 10^{-5}$ N on a pole of strength $4\pi \times 10^{-8}$ Wb at the centre. Now, the field strength at the centre of a coil of 1 cm radius carrying I amperes is $I/(2 \times 10^{-2})$ ampere-turns per metre, so that the force on a pole of strength $4\pi \times 10^{-8}$ Wb placed here is $2\pi \times 10^{-6} I$ newtons. Hence $2\pi \times 10^{-6} I = 2\pi \times 10^{-5}$ so that $I = 10$ A; i.e.

$$1 \text{ emu of current} = 10 \text{ A}$$

The relationships between other electrical units in the two systems follow.

(*b*) *The C.G.S. Electrostatic System.* This system is based on the assumption that the force between two unit charges 1 cm apart *in vacuo* is 1 dyne. Hence the permittivity of free space, and approximately of air, is unity and $D = E$ in such media. The charge in coulombs which gives the above result is

$$Q = \sqrt{(4\pi\epsilon \times 10^{-4} \times 10^{-5})} = \tfrac{1}{3} \times 10^{-9}\ \mathrm{C}$$

Hence

$$1\ \text{esu of charge} = \tfrac{1}{3} \times 10^{-9}\ \mathrm{C}$$

The electric flux from 1 electrostatic unit (esu) of charge is 4π esu, so that the electric field strength at radius r from this charge is $1/r^2$ esu. In the M.K.S.A. system the electric field strength at this radius from a charge $\frac{1}{3} \times 10^{-9}$ C is

$$E = \frac{\frac{1}{3} \times 10^{-9}}{4\pi\epsilon_0 r^2 \times 10^{-4}} = \frac{3 \times 10^4}{r^2}$$

where r is in centimetres. Hence

$$1\ \text{esu of electric field strength} = 3 \times 10^4\ \mathrm{V/m}$$

The esu of potential is the work done in ergs in moving 1 esu of charge 1 cm in a field of strength 1 esu. In the M.K.S.A. system this gives a p.d. of $3 \times 10^4 \times 10^{-2}$ V, i.e.

$$1\ \text{esu of voltage} = 300\ \mathrm{V}$$

Note that 1 erg is the work done when a force of 1 dyne moves its point of application 1 cm (i.e. 1 erg $= 10^{-7}$ J).

Other relations can be deduced in a similar way.

Bibliography

Buckingham and Price, *Principles of Electrical Measurements* (English Universities Press).

Golding and Widdis, *Electrical Measurements and Measuring Instruments* (Pitman).

Lampard, A new theorem in electrostatics with applications to calculable standards of capacitance, *Proc. Instn Elect. Engrs.*, **104C**, p. 271 (1957).

CHAPTER 15

Single-phase Measuring Networks

THE accurate measurement of circuit-element values by alternating currents may be carried out by inserting the unknown elements in a bridge network containing known variable standards and a sensitive alternating-current detector. The variable elements are adjusted until a null is indicated on the detector, and analysis of the network in this "balanced" condition then gives the relation between the known and unknown elements. The principal advantage of this type of measurement (null measurement) is that it is independent of any instrument calibration, and hence the accuracy depends only on the accuracy of the standards used and on the sensitivity (but not the calibration) of the detector. The Wheatstone bridge network may be adapted for power- and audio-frequency measurements, but since both the a.c. source and the detector cannot be earthed, difficulties may arise at radio frequencies. At these frequencies (up to some 250 Mc/s), bridged- and parallel-T networks, and networks based on the differential transformer give accurate results.

At power frequencies the a.c. potentiometer may be used to determine the magnitude and phase of alternating voltages, but the accuracy is low compared with the d.c. potentiometer, since the slide-wire currents can be standardized only by using a calibrated instrument. Electronic instruments giving accurate phase measurements are now replacing the a.c. potentiometer in many applications.

Commercial instruments are available employing bridge, T, and differential-transformer networks for the measurement of component values over wide ranges.

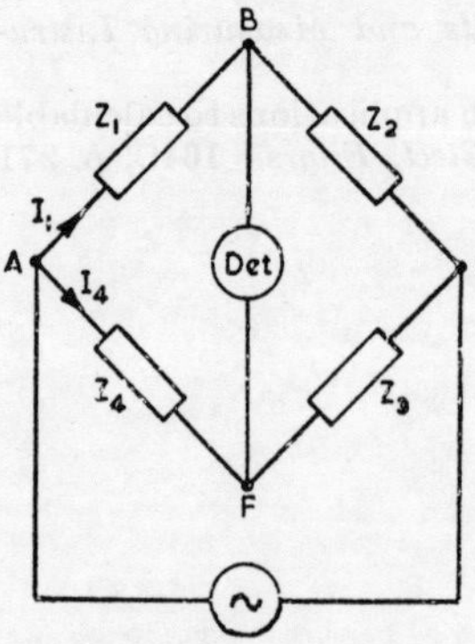

FIG. 15.1. SIMPLE A.C. BRIDGE

15.1. The Simple A.C. Bridge

The general circuit configuration of a simple a.c. bridge network without mutual couplings is shown in Fig. 15.1.

If no current flows through the detector, then I_1 flows through Z_2, and I_4 through Z_3. Also the potential of B must be equal to that of F, so that

$$I_1Z_1 = I_4Z_4 \quad \text{and} \quad I_1Z_2 = I_4Z_3$$

From these equations,

$$\frac{Z_1}{Z_2} = \frac{Z_4}{Z_3} \quad \text{or} \quad Z_1 = \frac{Z_2Z_4}{Z_3} \qquad . \quad . \quad (15.1)$$

This equation represents the balanced condition of the bridge, and enables the value of an unknown, Z_1, to be found in terms of known variable standards, Z_2, Z_3 and Z_4. In general, Z_1 will be complex, so that eqn. (15.1) may be rewritten as

$$Z_1 = R_1 \pm jX_1 \qquad . \quad . \quad . \quad (15.1a)$$

where R_1 and X_1 depend on the values of Z_2, Z_3 and Z_4. Since the unknown impedance will generally consist of a resistive and a reactive part, two independent variables must be used to obtain balance.

If the bridge is to balance rapidly it should be arranged that varying one known standard varies R_1 only, to give the resistive balance, while varying a second known standard varies X_1 only, to give the reactive balance. In practice successive adjustment of the two variable standards is performed, while the detector sensitivity is increased as balance is approached.

The balance condition may be written

$$Z_1 = \frac{Z_2}{Z_3}(R_4 \pm jX_4) \qquad . \quad . \quad . \quad (15.2)$$

When Z_2 and Z_3 are fixed standards, while Z_4 contains both the variable elements, the network is known as a *ratio bridge.* Eqn. (15.2) shows that only if the ratio Z_2/Z_3 is a reference number or is entirely quadrate will the resistive and reactive balance conditions be independent. If the ratio is complex, balance will only be achieved by many successive adjustments, and the bridge is said to be slow to converge. Obviously, if Z_2/Z_3 is a reference number, the known variable reactance must be of the same kind as the unknown. If Z_2/Z_3 is quadrate, its sign must be the same as that of the unknown reactive term, while the sign of the reactive component of Z_4 must be such that it will give a positive reference number when multiplied by Z_2/Z_3. In practice, only capacitors give pure quadrate impedances, and this limits the types of ratio bridge.

Product bridges are formed when Z_2 and Z_4 are the fixed elements and Z_3 contains the variables. The balance equation can be expressed in the form

$$Z_1 = (Z_2Z_4)Y_3 \quad . \quad . \quad . \quad . \quad (15.3)$$

As in the case of the ratio bridge the product Z_2Z_4 must be either a reference or a quadrate number if the bridge is to converge quickly. Frequently the impedance Z_3 may consist of two variable standards in parallel, so that $Y_3 = G_3 \pm jB_3$ and a rapid convergence is possible, by successive adjustments of G_3 and B_3. If the product Z_2Z_4 is a reference number then B_3 must have a sign opposite to that of X_1. If Z_2Z_4 is quadrate its sign must be negative since in practice pure inductance is not possible, and the sign of B_3 must then be such that it will give a positive reference number when multiplied by Z_2Z_4; i.e. B_3 must be a positive susceptance. Hence

$$Z_1 = -jA(G_3 + jB_3)$$

where $-jA = Z_2Z_4$. It follows that the reactive component of Z_1 must be negative (capacitive). Only one bridge network fulfils this condition—the *Schering bridge* discussed in Section 15.3 (*e*).

In the measurement of capacitance it is often convenient to have the answer in terms of the equivalent parallel combination of C_p and R_p. Eqn. (15.1) will then be written

$$Y_1 = G_1 + jB_1 = \frac{Z_3}{Z_2Z_4} \quad . \quad . \quad . \quad (15.4)$$

where $G_1 = 1/R_p$, and $C_p = B_1/\omega$.

15.2. Owen Bridge for Inductance

The circuit for the *Owen bridge* is shown in Fig. 15.2 (*a*). The derivation of the balance conditions and complexor diagram are

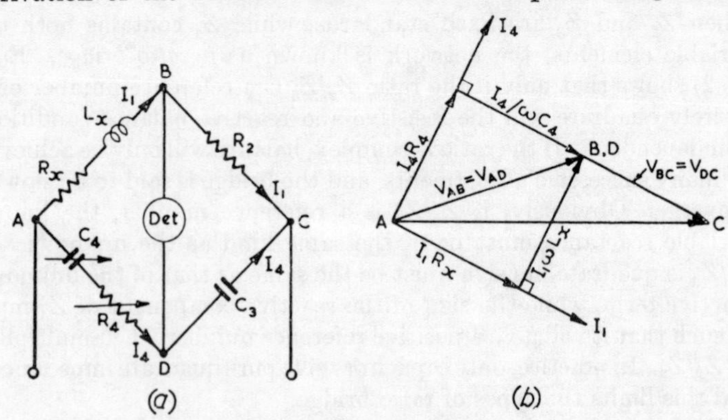

FIG. 15.2. OWEN BRIDGE

given as an illustration of the method employed. The bridge is used for the measurement of a wide range of inductance. It is a ratio bridge in which $Z_2/Z_3 = j\omega C_3 R_2$. The balance equation,

$$Z_1 = \frac{Z_2}{Z_3} Z_4$$

gives

$$R_x + j\omega L_x = j\omega C_3 R_2 \left(R_4 - \frac{j}{\omega C_4}\right)$$

$$= \frac{R_2 C_3}{C_4} + j\omega C_3 R_2 R_4$$

Hence, equating reference and quadrate terms,

$$R_x = \frac{R_2 C_3}{C_4} \quad \text{and} \quad L_x = C_3 R_2 R_4 \quad . \quad . \quad (15.5)$$

Obviously, if the variables are R_4 and C_4 the bridge balance will be rapidly convergent.

The complexor diagram for the balanced bridge is shown in Fig. 15.2 (*b*). The supply voltage is V_{AC}. The current I_1 lags behind this, and the voltage V_{AB} is made up of the components $I_1 R_x$ in phase with I_1 and $I_1 \omega L_x$ leading I_1 by 90°. The point B must be such that the voltage V_{BC} $(= I_1 R_2)$ is in phase with I_1. Now, at balance $V_{BC} = V_{DC} = -jI_4/\omega C_3$, so that I_4 must be perpendicular to V_{BC}. The voltage drops across R_4 (in phase with I_4) and C_4 (lagging I_4 by 90°) must then be as shown, to give a total voltage drop $V_{AD} = V_{AB}$.

15.3. Typical Ratio and Product Bridges

Some of the more common ratio and product bridges are shown in Fig. 15.3. It is left as an exercise for the reader to verify the balance conditions and draw typical complexor diagrams at balance. Although the more common uses of the bridges are indicated, it should be remembered that they may be used to compare the elements in any arm with those in the other arms.

(*a*) *Hay Bridge* (Fig. 15.3 (*a*)). The Hay bridge is a product bridge, and using eqn. (15.4) the balance conditions are

$$R_x = \frac{R_2 R_4}{R_3} \quad \text{and} \quad L_x = C_3 R_2 R_4 \quad . \quad . \quad (15.6)$$

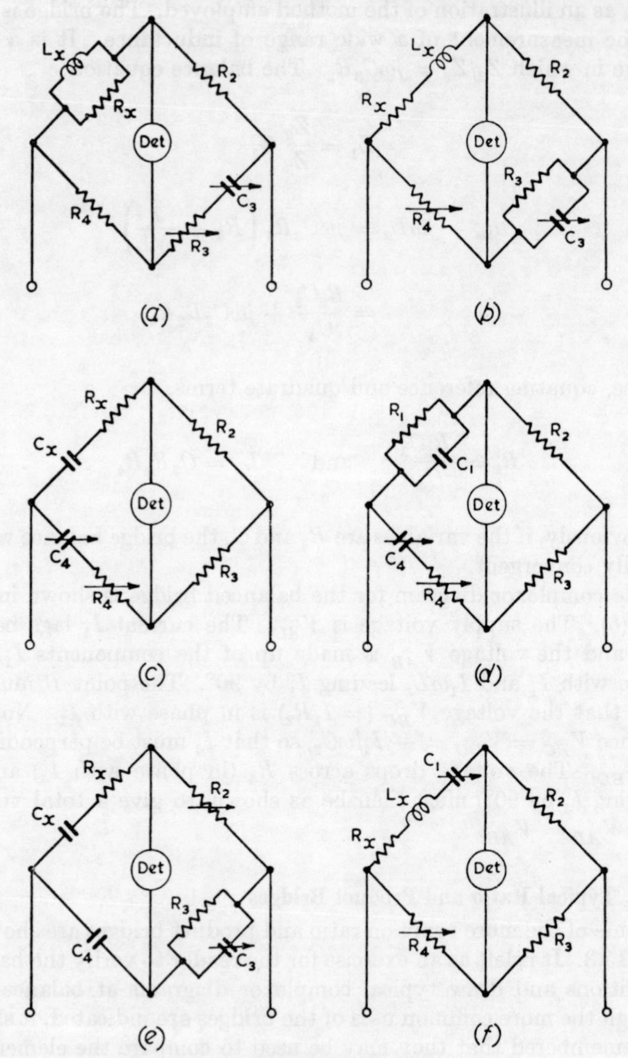

FIG. 15.3. RATIO AND PRODUCT BRIDGES

(*a*) Hay bridge (product)
(*b*) Maxwell bridge (product)
(*c*) Wien series bridge (ratio)
(*d*) Wien parallel bridge (ratio)
(*e*) Schering bridge (product)
(*f*) Resonance bridge

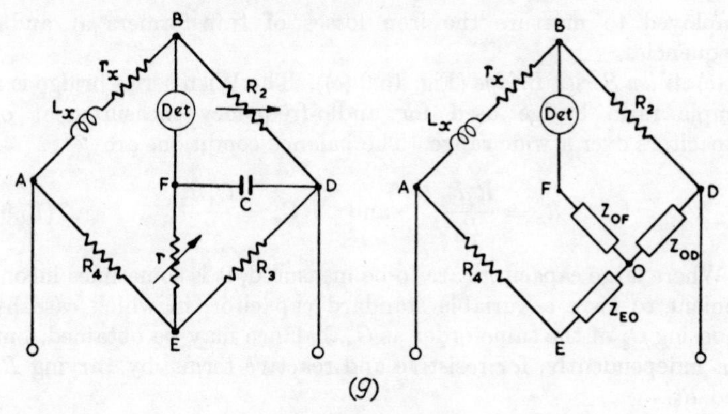

FIG. 15.3. (*Contd.*)
(*g*) Anderson bridge

The equivalent series combination for L_x and R_x may readily be evaluated, since the equivalent series circuit is given by

$$Z_x = R_s + j\omega L_s = \frac{1}{Y_x} = \frac{1}{1/R_x - j/\omega L_x}$$

$$= \frac{R_x\omega^2 L_x{}^2}{R_x{}^2 + \omega^2 L_x{}^2} + \frac{j\omega L_x R_x{}^2}{R_x{}^2 + \omega^2 L_x{}^2} \quad . \quad (15.7)$$

Normally $R_x \gg \omega L_x$; this expression then becomes

$$Z_x = \frac{\omega^2 L_x{}^2}{R_x} + j\omega L_x$$

so that $$R_s = \frac{\omega^2 L_x{}^2}{R_x} \quad \text{and} \quad L_s = L_x \quad . \quad . \ (15.7a)$$

It should be noticed that the solution is not now independent of frequency.

(*b*) *Maxwell Bridge* (Fig. 15.3 (*b*)). The Maxwell bridge is a product bridge for the measurement of inductance in terms of capacitance. From eqn. (15.3) the balance conditions are

$$R_x = \frac{R_4 R_2}{R_3} \quad \text{and} \quad L_x = R_2 R_4 C_3 \quad . \quad . \ (15.8)$$

Resistive and reactive terms balance independently, and the conditions are independent of frequency. This bridge is sometimes employed to measure the iron losses of transformers at audio frequencies.

(*c*) *Wien Series Bridge* (Fig. 15.3 (*c*)). The Wien series bridge is a simple ratio bridge used for audio-frequency measurement of capacitors over a wide range. The balance conditions are

$$R_x = \frac{R_2 R_4}{R_3} \quad \text{and} \quad C_x = \frac{C_4 R_3}{R_2} \quad . \quad . \quad (15.9)$$

Where large capacitors are to be measured, it is sometimes inconvenient to have a variable standard capacitor, in which case by choosing C_4 of the same order as C_x, balance may be obtained, but not independently, for resistive and reactive terms, by varying R_3 and R_4.

(*d*) *Wien Parallel Bridge* (Fig. 15.3 (*d*)). The Wien parallel bridge is a ratio bridge used mainly as the feedback network in wide-range audio-frequency *RC* oscillators. It may be used (as can any network whose balance conditions are frequency dependent) to measure audio frequencies, though modern digital frequency meters now give accuracies to better than 1 part in 10^6 and are to be preferred for frequency measurement. The solution for the balance conditions is

$$\omega^2 = \frac{1}{C_1 C_4 R_1 R_4} \quad . \quad . \quad . \quad (15.10)$$

(*e*) *Schering Bridge* (Fig. 15.3 (*e*)). The Schering bridge is the only product bridge configuration which is feasible in which the product is quadrate. This may be seen by writing eqn. (15.3) in the form

$$Z_1 = \pm jK(G_3 \pm jB_3) \quad . \quad . \quad . \quad (15.11)$$

Since the only practical configuration for arm 3 is a pure capacitor in parallel with a pure resistor (pure inductance not being possible unless superconducting conditions prevail), then Y_3 must be of the form $(G_3 + jB_3)$. Hence the sign of the product term must be negative to give a resultant positive reference component. This negative sign is possible only if Z_4 is a pure capacitance and Z_2 a pure resistance, or vice versa, giving the Schering bridge configuration.

The main application of the Schering bridge is in the measurement of capacitance and equivalent series loss resistance of high-voltage cables and capacitors. The solution of eqn. (15.11) is

$$R_x = R_2 \frac{C_3}{C_4} \quad \text{and} \quad C_x = C_4 \frac{R_3}{R_2} \quad . \quad . \quad (15.11a)$$

The bridge is essentially safe to operate, since the impedances Z_x and Z_4 are very much higher than Z_2 and Z_3. Thus the voltage across the detector and across the arms containing the variable elements will be only a small fraction of the high supply voltage (which may exceed 100 kV). The common connexion of the variables can be earthed, and these elements may have an earthed screen for safety, so ensuring that only a relatively low voltage exists between any control which the operator has to adjust, or the detector, and earth.

The *loss angle*, δ, of the unknown capacitor is the angle between the actual capacitor current and the complexor which leads the voltage across the capacitor by 90°. If R_x and C_x are the equivalent series resistance and capacitance,

$$\delta = \tan^{-1} \omega C_x R_x$$

Substituting from eqn. (15.11),

$$\delta = \tan^{-1} \omega C_3 R_3 \qquad . \quad . \quad . \ (15.12)$$

If the capacitor is represented by an equivalent parallel circuit of R_p and C_p, then

$$\delta = \tan^{-1} (1/\omega C_p R_p) \qquad . \quad . \quad . \ (15.12a)$$

In this case the series and parallel circuit representations of the lossy capacitor will be equivalent if

$$Z = R_x + \frac{1}{j\omega C_x} = \frac{R_p}{1 + j\omega C_p R_p}$$

$$= \frac{R_p}{1 + \omega^2 C_p{}^2 R_p{}^2} - \frac{j\omega C_p R_p{}^2}{1 + \omega^2 C_p{}^2 R_p{}^2} \qquad . \ (15.13)$$

Normally $\omega C_p R_p \gg 1$ so that

$$R_x \simeq 1/\omega^2 C_p{}^2 R_p \text{ and } C_x \simeq C_p \qquad . \ .(15.13a)$$

(*f*) *Resonance Bridge* (Fig. 15.3 (*f*)). In the resonance bridge the reactive elements in arm 1 are balanced out by tuning to series resonance. The resistance element is then balanced by varying R_4. This bridge may be used to measure frequency in terms of L_x and C_1; inductance in terms of the variable capacitance; and frequency or capacitance in terms of a variable inductance and frequency.

(*g*) *Anderson Bridge* (Fig. 15.3 (*g*)). The balance conditions for the Anderson bridge are most easily obtained by converting the mesh of impedances r, C and R_3 to a star, by the delta-star transformation (eqns. (1.21)). Thus

$$Z_{EO} = rR_s/(r + R_3 + 1/j\omega C)$$

and

$$Z_{OD} = (R_3/j\omega C)/(r + R_3 + 1/j\omega C) = Z_3$$

From this $Z_4 = R_4 + Z_{EO}$, and, solving for balance,

$$R_x = \frac{R_2 R_4}{R_3} \quad \text{and} \quad L_x = R_2 C\left(\frac{R_4 r}{R_3} + R_4 + r\right) \tag{15.14}$$

Balance conditions are not independent for each variable, but the bridge gives excellent results, and is used for the measurement of inductance over the range from a few microhenrys to several henrys.

(*h*) *General-purpose Commercial Bridges.* These are generally of the type in which the unknown impedance is balanced by variable resistances and capacitive reactances. The ratio or product is usually variable in decade steps, so that a very wide range of unknowns can be measured.

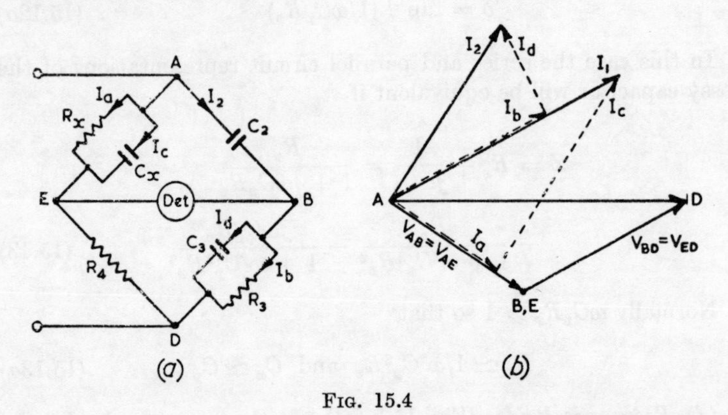

FIG. 15.4

Example 15.1. The bridge shown in Fig. 15.4(*a*) is used to measure a capacitor represented by a capacitance C_x and a dielectric loss resistance R_x.

Derive, from the balance equations for the bridge, an expression for C_x in terms of the known elements. Determine C_x and the loss angle if balance is obtained at 400 c/s with $C_2 = C_3 = 0{\cdot}01\ \mu\text{F}$, $R_3 = 4\ \text{k}\Omega$, and $R_4 = 11{\cdot}5\ \text{k}\Omega$. Draw a complexor diagram for the bridge in its balanced condition. (*A.E.E.*, 1961)

From the balance equation (15.4),

$$Y_x = \frac{Z_3}{Z_2 Z_4}$$

Hence
$$\frac{1}{R_x} + j\omega C_x = \frac{\dfrac{R_3}{1 + j\omega C_3 R_3}}{R_4/j\omega C_2}$$

$$= \frac{R_3 j\omega C_2}{R_4(1 + j\omega C_3 R_3)} = \frac{\omega^2 C_2 C_3 R_3^2 + j\omega C_2 R_3}{R_4(1 + \omega^2 C_3^2 R_3^2)}$$

This gives
$$R_x = \frac{R_4(1 + \omega^2 C_3^2 R_3^2)}{\omega^2 C_2 C_3 R_3^2}$$

and
$$C_x = \frac{C_2 R_3}{R_4(1 + \omega^2 C_3^2 R_3^2)}$$

With the numerical values given, $\omega^2 C_3^2 R_3^2 = 1{\cdot}01 \times 10^{-2}$,

$$\underline{\underline{R_x = 11{\cdot}5 \times 10^5\ \Omega}} \quad \text{and} \quad \underline{\underline{C_x = 0{\cdot}344 \times 10^{-8}\ \text{F}}}$$

Notice that the bridge in Example 15.1 is essentially a Schering bridge, with the unknown represented by its equivalent parallel capacitance and resistance.

The complexor diagram is shown in Fig. 15.4 (*b*). The voltage V_{AB} is given by

$$V_{AB} = \frac{V_{AD}}{\dfrac{1}{j\omega C_2} + \dfrac{R_3}{1 + j\omega C_3 R_3}} \frac{1}{j\omega C_2}$$

$$= \frac{V_{AD}(1 + j\omega C_3 R_3)}{1 + 2j\omega C_3 R_3}$$

since $C_2 = C_3$ in the given problem.

Substituting numerical values, V_{AB} ($= V_{AE}$) may be drawn to scale lagging behind V_{AD} by the appropriate phase angle. The voltage V_{BD} ($= V_{ED}$) must be such that it completes the triangle *ABD*. I_2 can now be drawn leading V_{AB} by 90°, and the components I_b and I_d add vectorially to give I_2, with I_b parallel to V_{BD}, and I_d leading V_{BD} by 90°. The current I_1 will be in phase with V_{ED}, while its components I_a and I_c will be in phase and in quadrature respectively with V_{AE}.

15.4. Bridges with Mutual-inductance Elements

The balance conditions of bridges which include mutual-inductance elements are usually most readily obtained by equating voltage drops across the bridge arms. A selection of such bridges will be considered.

(*a*) *Felici Bridge* (Fig. 15.5). Any bridge suitable for measuring self-inductance may be used to measure mutual inductance by obtaining the self-inductance of the two coils connected (*a*) in series aiding, and (*b*) in series opposing. Then, since $L_a = L_1 + L_2 + 2M$ and $L_b = L_1 + L_2 - 2M$, we obtain $M = (L_a - L_b)/4$.

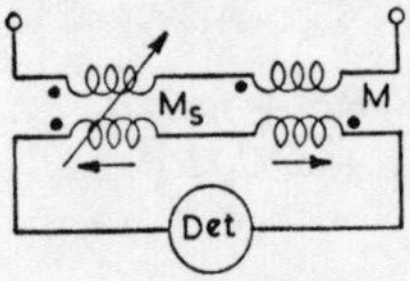

FIG. 15.5. FELICI METHOD FOR MUTUAL INDUCTANCE

A simple bridge for the direct comparison of an unknown mutual inductance with a laboratory standard mutual inductometer is the *Felici bridge*, illustrated in Fig. 15.5. When the detector indicates zero, the secondary e.m.f.s cancel and $M = M_s$. Obviously the unknown, M, must have a value within the range of the standard, M_s. Difficulties may be encountered at high frequencies owing to the effects of the coil self-capacitances and of the phase defect of the mutual inductances.

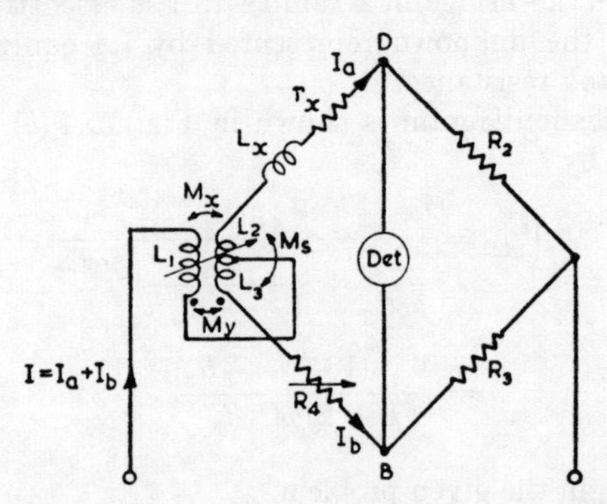

FIG. 15.6. CAMPBELL–HEAVISIDE EQUAL-RATIO BRIDGE

(*b*) *Campbell–Heaviside Equal-ratio Bridge* (Fig. 15.6). The Campbell–Heaviside bridge is used to measure small inductances in terms of a mutual inductometer reading.

In the equal-ratio bridge $R_2 = R_3$ and hence $I_a = I_b = \frac{1}{2}I$. Equating voltage drops from the mid-point of the mutual-inductometer secondary to D and B yields (for the winding directions indicated by the dot notation),

$$\tfrac{1}{2}I(r_x + j\omega L_x) + \tfrac{1}{2}Ij\omega L_2 - \tfrac{1}{2}Ij\omega M_s - Ij\omega M_x$$
$$= \tfrac{1}{2}I(R_4 + j\omega L_3) - \tfrac{1}{2}Ij\omega M_s + Ij\omega M_y$$

Equating reference terms,

$$r_x = R_4 \qquad . \qquad . \qquad . \quad (15.15)$$

Equating quadrate terms,

$$L_x + L_2 - L_3 = 2(M_x + M_y) \qquad . \qquad .(15.15a)$$

$(M_x + M_y)$ is the dial reading on the mutual inductometer, and L_2 and L_3 are known accurately.

If $L_2 = L_3$ the expression reduces to

$$L_x = 2 \times \text{Dial reading}$$

Since the bridge is employed for small inductance measurements, the residual inductance of the leads may be important. When this is so, the bridge is balanced initially with the unknown terminals short-circuited and then with the unknown inserted. The difference reading gives the value of the unknown.

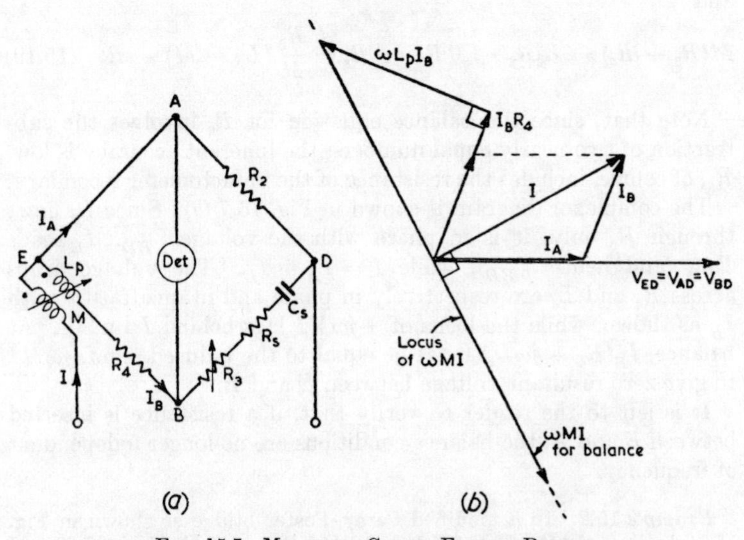

Fig. 15.7. Modified Carey–Foster Bridge

(c) *The Modified Carey–Foster Bridge* (Fig. 15.7). The modified Carey–Foster bridge is used to measure capacitance in terms of mutual inductance, or vice versa. At balance, $V_{EB} = 0$, since $V_{EA} = 0$, and hence for the mutual inductometer winding directions indicated by the dot notation on the diagram, and for dial reading M

$$(I_A + I_B)j\omega M = I_B(R_4 + j\omega L_p)$$

or

$$I_A j\omega M = I_B(R_4 + j\omega L_p - j\omega M) \quad . \quad . \quad (15.16)$$

where L_p is the known primary self-inductance of the inductometer.

Also, since $V_{AD} = V_{BD}$,

$$I_A R_2 = I_B(R_3 + R_s + 1/j\omega C_s) \quad . \quad . \quad (15.17)$$

Dividing eqn. (15.16) by eqn. (15.17) and cross-multiplying,

$$j\omega M(R_3 + R_s + 1/j\omega C_s) = R_2(R_4 + j\omega L_p - j\omega M)$$

and hence

$$\frac{M}{C_s} = R_2 R_4 \quad \text{or} \quad C_s = \frac{M}{R_2 R_4} \quad . \quad . \quad (15.18)$$

and

$$M(R_3 + R_s) = L_p R_2 - M R_2 \text{ or } R_s = \frac{R_2}{M}(L_p - M) - R_3 \quad (15.19)$$

Note that, since the balance equation for R_s involves the subtraction of two nearly equal numbers, the inherent accuracy is low. R_4, of course, includes the resistance of the inductometer secondary.

The complexor diagram is shown in Fig. 15.7 (*b*). Since I_A flows through R_2 only, it is in phase with the voltage V_{ED}. I_B leads V_{BD} (and hence V_{ED}), while $I = I_A + I_B$. The voltage drops across R_4 and L_s are respectively in phase and in quadrature with I_B as shown, while the locus of $-j\omega MI$ lags behind I by 90°. At balance, $I_B(R_4 + j\omega L_s)$ must be equal to the induced e.m.f. $j\omega MI$ to give zero resultant voltage between E and B.

It is left to the reader to verify that, if a resistance is inserted between E and A, the balance conditions are no longer independent of frequency.

Example 15.2. In a modified Carey–Foster bridge as shown in Fig. 15.7(*a*), $R_2 = 10\ \Omega$, $R_3 = 25{\cdot}1 \pm 0{\cdot}1\ \Omega$, $R_4 = 100\ \Omega$, $L = 22{\cdot}82$ mH and $M = 4{,}920 \pm 10\ \mu$H. Find the unknown capacitance C_5 and the series loss resistance R_5. Give the limits of accuracy determined by the limits of the balance measurements indicated.

From eqn. (15.18),

$$C_s = \frac{4{,}920 \pm 10}{10 \times 100} = \underline{\underline{4{\cdot}92 \pm 0{\cdot}01\ \mu\text{F}}} \text{ (i.e. } \pm 0{\cdot}2 \text{ per cent)}$$

From eqn. (15.19),

$$\begin{aligned} R_s &= \frac{R_2 L}{M} - R_2 - R_3 \\ &= \frac{10 \times 22{\cdot}82 \times 10^{-3}}{(4{,}920 \pm 10) \times 10^{-6}} - 10 - 25{\cdot}1 \pm 0{\cdot}1 \\ &= \frac{22{\cdot}8 \times 10^4}{4{,}920\left(1 \pm \frac{1}{492}\right)} = 35{\cdot}1 \pm 0{\cdot}1 \\ &= 46{\cdot}5\left(1 \mp \frac{1}{492}\right) - 35{\cdot}1 \pm 0{\cdot}1 \text{ (by the binomial theorem)} \\ &= 46{\cdot}5 - 35 \mp 0{\cdot}095 \pm 0{\cdot}1 \\ &= \underline{\underline{11{\cdot}4 \pm 0{\cdot}195\ \Omega}} \text{ (or } \pm\ 1{\cdot}71 \text{ per cent)} \end{aligned}$$

Note that the maximum possible limits of error are always given, so that, although one part of the answer has a possible error of $\mp 0{\cdot}095$, and a second part has a possible error of $\pm 0{\cdot}1$, the possible error of the sum or difference of the two parts is $\pm 0{\cdot}195$.

15.5. Inaccuracy in Bridges

The precision of 4-arm bridges so far described depends on achieving a voltage balance. The accuracy with which any measurement can be made will be governed by (*a*) the voltage sensitivity of the detector, (*b*) the accuracy with which the standard elements are known, and (*c*) the effect of stray couplings between components and stray couplings to earth.

Provided that the standard elements used are reliable to better than 1 part in 10^4, bridge measurements should be accurate to about 1 part in 10^3 or 10^4. The limits of uncertainty of the measurement may readily be observed by noting the range of variation over which there is no observable detector current. This will depend on the sensitivity of the detector, and on the minimum possible change in the standards.

If the unknown is computed from an expression involving products only, the inaccuracy of the measurement can be found as follows.

Suppose that

$$X = \frac{AB}{C}$$

Then

$$\log_e X = \log_e A + \log_e B - \log_e C$$

and hence

$$\frac{dX}{X} = \frac{dA}{A} + \frac{dB}{B} - \frac{dC}{C}$$

or, for small changes,

$$\frac{\delta X}{X} = \pm\left(\frac{\delta A}{A} + \frac{\delta B}{B} + \frac{\delta C}{C}\right) \quad . \quad . \quad (15.20)$$

Note that, since we are interested only in the maximum possible error, the fractional error, $\delta X/X$, will be the *sum* of all the individual errors. This is because positive errors in A and B may well occur when the error in C is negative.

If the unknown is computed as a sum or difference, the overall error is the sum of the *actual* errors in each term. The *fractional* error in the result must then be calculated from the actual error found. Thus if

$$X = A - B \qquad \text{then} \qquad \delta X = \pm(\delta A + \delta B)$$

where $\pm\delta A$ is the actual error in A, and $\pm\delta B$ is the actual error in B. The fractional error is then

$$\frac{\delta X}{X} = \frac{\pm(\delta A + \delta B)}{A - B} \quad . \quad . \quad (15.21)$$

Note that, for a sum $X = A + B$, the fractional error is $\delta X/X = \pm(\delta A + \delta B)/(A + B)$, which will be much smaller than that obtained in eqn. (15.21) for a difference. This means that the inaccuracy possible when two nearly equal quantities are subtracted may be considerable.

In addition to the inaccuracy of measurement, the effects of stray electromagnetic and capacitive couplings between bridge elements and to earth may become important, particularly at high frequencies and for small-valued components. For this reason, components should be adequately screened, in which case the capacitive coupling to the screen is fixed and may be measured and allowed for. Stray electromagnetic coupling may be minimized by spacing the components and twisting together the connecting leads, although this will increase the stray capacitive coupling.

The stray capacitance of the detector to earth may cause a detector current at "balance" (head effect) unless the detector is then at earth potential. In some cases special screened detector transformers must be used to couple the detector to the bridge.

Fig. 15.8 (*a*) shows diagrammatically the possible stray earth capacitances of an a.c. bridge network (*see* also Section 1.9). If one end of the detector is connected directly to earth, then the capacitors C_a and C_c appear in shunt across Z_4 and Z_3, and will therefore introduce unknown effects in the balance conditions. This difficulty may be overcome by the *Wagner earthing device* shown in Fig. 15.8 (*b*).

In this device the additional variable elements Z_5 and Z_6 (whose actual values need not be known) are chosen so that a balance with Z_1 and Z_2 is possible with the key in position 2. The centre-point of Z_5 and Z_6 is earthed. The detector current is first reduced to a

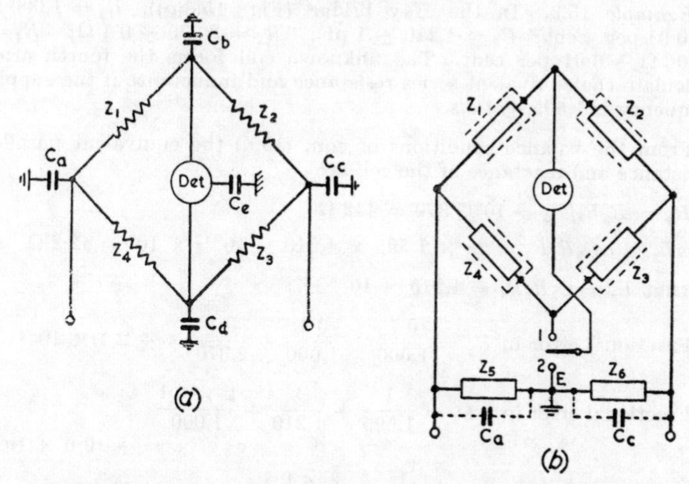

FIG. 15.8. STRAY CAPACITANCE AND THE WAGNER EARTH

minimum with the key in position 1 by adjustments of Z_3 and/or Z_4, then by adjusting Z_5 and Z_6 when the key is in position 2. Successive rebalancing with the key in alternate positions leads to a zero with the key in position 1 and the detector at earth potential. The stray capacitances C_a and C_c are accounted for when balance is achieved with the key in position 2, and will not now upset the main balance, since they may be considered as permanently connected across Z_5 and Z_6. The stray capacitances C_b and C_d are short-circuited when balance is achieved. The Wagner earth is particularly useful when one or more of the bridge arms contain a small capacitance.

In many cases increased accuracy is achieved by the use of substitution methods. The most direct substitution method is to balance the bridge with the unknown inserted. The unknown is then replaced by a variable standard which is adjusted without altering the other bridge arms until balance is once more achieved. An obvious disadvantage of the method is the provision of suitable variable standards. A second method (illustrated by Problem 15.4) is to balance the bridge with a fixed standard of the same order as the unknown, and then to rebalance with the unknown. The difference between the balance conditions then gives the unknown, and stray couplings and residuals will have only secondary effects on the accuracy. Notice, however, that the limits of accuracy of the standards still limit the accuracy of the measurement.

Example 15.3. In the Hay bridge (Fig. 15.3(*a*)), $R_2 = 1{,}000\ \Omega \pm 0{\cdot}01$ per cent; $C_3 = 4{,}210 \pm 1$ pF; $R_3 = 2{,}370 \pm 0{\cdot}1\ \Omega$; $R_4 = 1{,}000\ \Omega \pm 0{\cdot}01$ per cent. The unknown coil forms the fourth arm. Calculate the equivalent series resistance and inductance if the supply frequency is $1{,}595 \pm 1$ c/s.

From the balance conditions of eqn. (15.6) the equivalent parallel resistance and reactance of the coil are

$$R_p = R_2R_4/R_3 = 10^6/2{,}370 = 442\ \Omega$$

$$\omega L_p = \omega C_3R_2R_4 = 2\pi \times 1{,}595 \times 4{,}210 \times 10^{-12} \times 10^6 = 42{\cdot}2\ \Omega$$

so that $L_p = C_3R_2R_4 = 4{,}210 \times 10^{-6}$ H.

$$\text{Fractional error in } R_p = \frac{0{\cdot}1}{1{,}000} + \frac{0{\cdot}1}{1{,}000} + \frac{0{\cdot}1}{2{,}370} = \pm\, 2{\cdot}4 \times 10^{-4}$$

$$\text{Fractional error in } \omega L_p = \frac{1}{1{,}595} + \frac{1}{4{,}210} + \frac{2 \times 0{\cdot}1}{1{,}000}$$

$$= \pm\, 10{\cdot}6 \times 10^{-4}$$

$$\text{Fractional error in } L_p = \frac{1}{4{,}210} + \frac{2 \times 0{\cdot}1}{1{,}000} = \pm\, 4{\cdot}4 \times 10^{-4}$$

From eqn. (15.7) the equivalent series resistance is

$$R_s = \frac{\omega^2 L_p{}^2 R_p}{R_p{}^2 + \omega^2 L_p{}^2} = \frac{42{\cdot}2^2 \times 442}{197{,}200} = 4{\cdot}000\ \Omega$$

where $R_p{}^2 + \omega^2 L_p{}^2 = 195{,}400(1 \pm 2 \times 2{\cdot}4 \times 10^{-4})$

$$+\ 1{,}800\ (1 \pm 2 \times 10{\cdot}6 \times 10^{-4})$$

$$= 197{,}200 \pm 100$$

Hence,

$$\text{Fractional error in } R_s = (2 \times 10{\cdot}6 + 2{\cdot}4 + 100/19{\cdot}72) \times 10^{-4}$$

$$= \pm\, 3{\cdot}1 \times 10^{-4}$$

so that

$$R_s = 4{\cdot}000(1 \pm 3{\cdot}1 \times 10^{-4}) = \underline{\underline{4{\cdot}00 \pm 0{\cdot}01\ \Omega}}$$

Similarly,

$$L_s = \frac{L_pR_p{}^2}{R_p{}^2 + \omega^2 L_p{}^2} = \underline{\underline{4{\cdot}172 \pm 0{\cdot}006\ \text{mH}}}$$

15.6. Residuals

Circuit-elements cannot be constructed to have only resistance, inductance or capacitance. Each practical element will contain small residual amounts of the others, and these residuals may materially affect the measured values, particularly at high frequencies. Equivalent circuits showing the residuals are depicted in Fig. 15.9. The equivalent circuit configurations for resistors and inductors are similar, and show the residual shunt capacitance together with series inductance or resistance. For capacitors the connexions will introduce residual series inductance. By careful construction the residuals are reduced to low values, but circuit-elements should not normally be operated at frequencies for which the residual effects are marked.

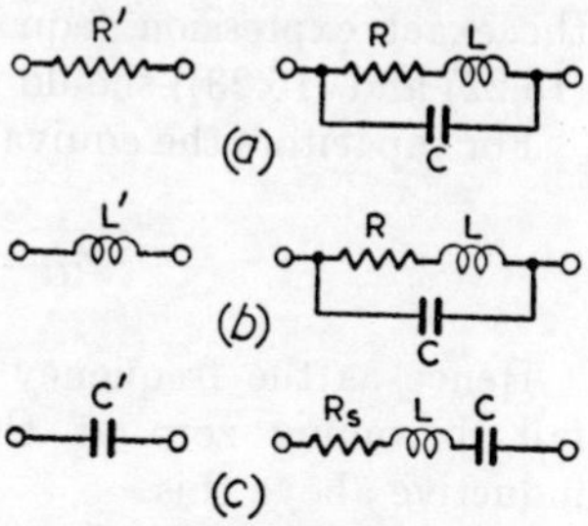

FIG. 15.9. EQUIVALENT CIRCUITS OF RESISTORS, INDUCTORS AND CAPACITORS

The equivalent impedance of a resistor or an inductor is

$$Z = \frac{(R + j\omega L)/j\omega C}{R + j\omega L + 1/j\omega C} = \frac{R + j\omega L}{1 - \omega^2 LC + j\omega CR}$$

$$= \frac{(R + j\omega L)(1 - \omega^2 LC - j\omega CR)}{(1 - \omega^2 LC)^2 + \omega^2 C^2 R^2}$$

$$= \frac{R}{(1 - \omega^2 LC)^2 + \omega^2 C^2 R^2} + j\omega \frac{(L(1 - \omega^2 LC) - CR^2)}{(1 - \omega^2 LC)^2 + \omega^2 C^2 R^2} \quad . \quad (15.21)$$

For resistors operated well below the self-resonant frequency ($f_0 = 1/2\pi\sqrt{(LC)}$), $\omega^2 LC \ll 1$, and if, further, $\omega^2 C^2 R^2 \ll 1$, the equivalent impedance is

$$Z_R \simeq R + j\omega(L - CR^2) \quad . \quad . \quad (15.22)$$

Thus for physically identical resistors the residual reactance will depend on the relationship between L and CR^2. If $R < \sqrt{(L/C)}$ the reactance is inductive, if $R = \sqrt{(L/C)}$ there is no reactance, and if $R > \sqrt{(L/C)}$ the reactance is capacitive. Obviously the reactance will increase with frequency.

For inductors, R will normally be small, and the terms CR^2 and $\omega^2 C^2 R^2$ may often be neglected in eqn. (15.21). Hence the equivalent impedance is

$$Z_L \simeq \frac{R}{(1 - \omega^2 LC)^2} + \frac{j\omega L}{(1 - \omega^2 LC)} \quad . \quad . \quad (15.23)$$

As the frequency increases, the effective reactance will increase towards infinity at the self-resonant frequency. Thereafter it will become capacitive. This neglects the increase in R due to high-frequency effects. It should be noted that the self-resonant frequency of large iron-cored coils may be relatively low, and may well be in the audio-frequency range. A simple calculation will show whether the exact expression (eqn. (15.21)) or the approximations (eqns. (15.22) and (15.23)) should be used.

For capacitors the equivalent impedance, including residuals, is

$$Z_C = R_s + \frac{1 - \omega^2 LC}{j\omega C} \quad . \quad . \quad . \quad (15.24)$$

Hence as the frequency increases the capacitive reactance will fall, becoming zero at the self-resonant frequency, and being inductive above this.

Note that eqn. (15.13a) indicates that R_s varies with frequency if it is assumed that the equivalent parallel loss resistance, R_p, is constant.

15.7. Parallel and Bridged-T Networks—Balance Conditions

These networks have the advantage at radio frequencies that both generator and detector may be solidly earthed. This reduces the effects of stray-capacitance couplings, simplifies the shielding of components, and enables high accuracy to be obtained, provided that the detector is suitably screened from the signal source.

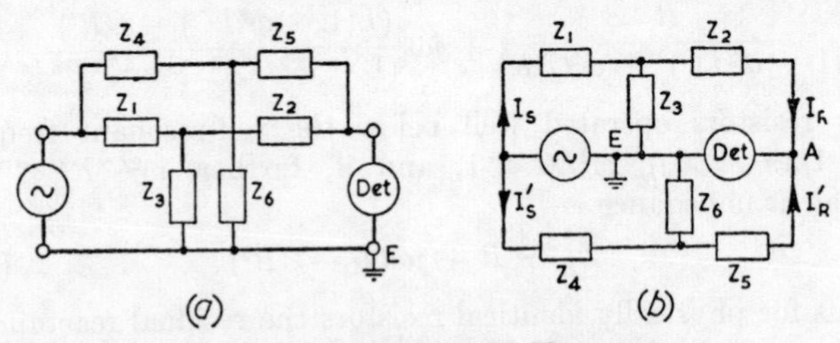

FIG. 15.10. PARALLEL-T NETWORK

(*a*) *Parallel-T Network.* The conventional circuit for a general parallel-T network is shown in Fig. 15.10 (*a*), and is redrawn at (*b*) in a form which facilitates analysis. At balance there is no current in the detector, and hence A and E may be considered to be joined. The input impedance of the top T is then

$$Z_s = Z_1 + \frac{Z_2 Z_3}{Z_2 + Z_3}$$

Also
$$I_s = \frac{V_s}{Z_s}$$

and hence
$$I_R = \frac{V_s}{Z_s}\frac{Z_3}{Z_2 + Z_3}$$

In the same way, if Z_s' is the input impedance of the lower T $[Z_s' = Z_4 + Z_5Z_6/(Z_5 + Z_6)]$, then
$$I_s' = \frac{V_s}{Z_s'} \quad \text{and} \quad I_R' = \frac{V_s}{Z_s'}\frac{Z_6}{Z_5 + Z_6}$$

Now, if the detector current is zero, $I_R = -I_R'$, and hence
$$\frac{V_s}{Z_s}\frac{Z_3}{(Z_2 + Z_3)} = -\frac{V_s}{Z_s'}\frac{Z_6}{(Z_5 + Z_6)}$$

Therefore
$$\frac{Z_3}{Z_1Z_2 + Z_1Z_3 + Z_2Z_3} = \frac{-Z_6}{Z_4Z_5 + Z_4Z_6 + Z_5Z_6}$$

Inverting and dividing out,
$$Z_1 + Z_2 + \frac{Z_1Z_2}{Z_3} = -Z_4 - Z_5 - \frac{Z_4Z_5}{Z_6}$$

or
$$Z_1 + Z_2 + \frac{Z_1Z_2}{Z_3} + Z_4 + Z_5 + \frac{Z_4Z_5}{Z_6} = 0 \quad . \; (15.25)$$

This relation represents the balance condition of the parallel-T network.

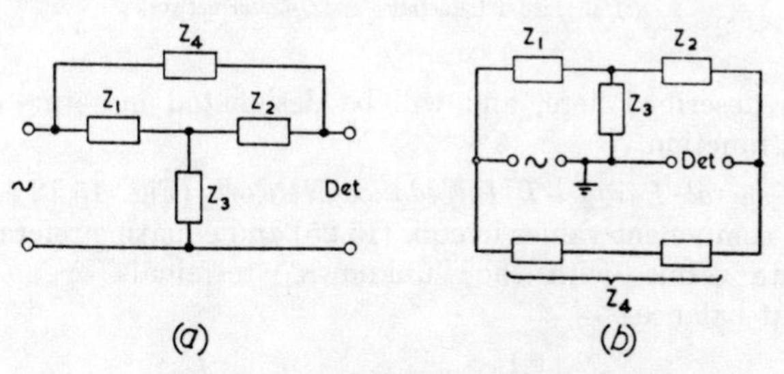

FIG. 15.11. BRIDGED-T NETWORK

(*b*) *Bridged-T Network.* The bridged-T network is shown in its usual form in Fig. 15.11 (*a*) and is redrawn at (*b*), from which it can be seen that it is equivalent to a parallel-T in which Z_4 is the sum of the top impedance of the T, and $Z_6 = \infty$. It follows from eqn. (15.25) that the balance condition for the bridged-T circuit is

$$Z_1 + Z_2 + \frac{Z_1 Z_2}{Z_3} + Z_4 = 0 \qquad . \qquad . \ (15.26)$$

15.8. Typical Parallel-T and Bridged-T Circuits

Bridged-T and parallel-T networks are not graced with the names of their originators as are bridge networks. Some typical networks

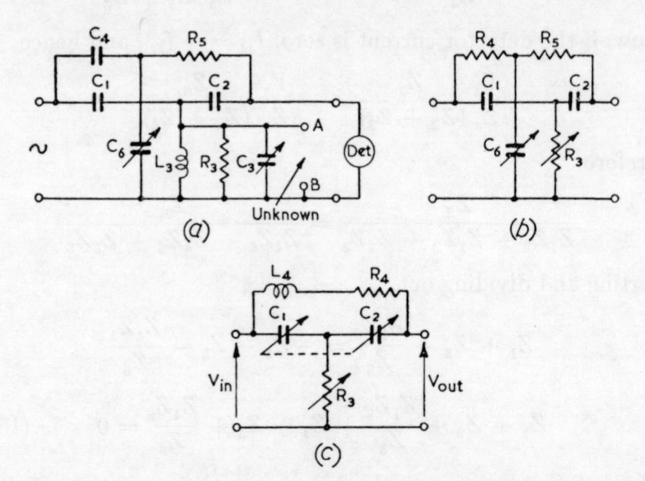

Fig. 15.12. Typical Parallel-T and Bridged-T Networks

(*a*) General parallel-T impedance network
(*b*) Parallel-T frequency network
(*c*) Bridged-T inductance and *Q*-factor network

will be described here, and will be designated in terms of their normal function.

(*a*) *General Parallel-T Impedance Network* (Fig. 15.12 (*a*)). Inserting component values in eqn. (15.25) and equating reference and quadrate terms, with the "unknown" terminals open-circuited yields at balance

$$G_3 = \frac{1}{R_3} = \omega^2 R_5 C_1 C_2 \left(1 + \frac{C_6}{C_4}\right) \qquad . \qquad . \ (15.27)$$

and

$$B_3 = \frac{1}{\omega L_3} = \omega \left[C_3 + C_1 C_2 \left(\frac{1}{C_1} + \frac{1}{C_2} + \frac{1}{C_4}\right)\right] \quad . \ (15.28)$$

The unknown impedance, Z, represented by a resistance R_p in parallel with a reactance ωL_p, is now inserted and a second balance is obtained with C_3 altered to C_3' and C_6 to C_6'. Hence

$$G_3 + \frac{1}{R_p} = \omega^2 R_s C_1 C_2 \left(1 + \frac{C_6'}{C_4}\right)$$

Subtracting eqn. (15.27),

$$\frac{1}{R_p} = \frac{\omega^2 R_5 C_1 C_2}{C_4}(C_6' - C_6) . \qquad . \quad (15.29)$$

In the same way,

$$\frac{1}{\omega L_3} + \frac{1}{\omega L_p} = \omega \left[C_3' + C_1 C_2 \left(\frac{1}{C_1} + \frac{1}{C_2} - \frac{1}{C_4} \right) \right]$$

Subtracting eqn. (15.28).

$$\frac{1}{\omega L_p} = \omega(C_3' - C_3)$$

Hence

$$\text{Reactance of unknown} = \frac{1}{\omega(C_3' - C_3)} . \qquad . \quad (15.30)$$

If the unknown reactance is capacitive, then $C_3' < C_3$ and a negative answer will be obtained in eqn. (15.30).

The bridge may be used up to some 30 Mc/s. The method of substitution largely eliminates the effect of stray couplings from the result. Note that the network balance conditions give independent balancing for resistive and reactive terms by using C_6 and C_3 as the variables.

(*b*) *Parallel-T Frequency Network* (Fig. 15.12 (*b*)). This is equivalent to a Wien bridge, and as the balance conditions depend on frequency, the network could be used to measure frequency over a wide range with suitably chosen components. Also, since the detector current is zero at one particular frequency, the network may be used as a rejection circuit. Owing to the shape of its output/frequency characteristic, this circuit is sometimes called a *notch filter*. It is in these applications that it is most often used.

Balance conditions yield,

(i) equating reference terms in eqn. (15.25),

$$\omega^2 = \frac{1}{C_1 C_2 R_3 (R_4 + R_5)}$$

(ii) equating quadrate terms,

$$\omega^2 = \frac{C_1 + C_2}{C_1 C_2 C_6 R_4 R_5}$$

If now $C_1 = C_2$, and $R_4 = R_5$, we obtain

$$\omega^2 = \frac{1}{2C_1{}^2 R_3 R_4} = \frac{2}{C_1 C_6 R_4{}^2} \quad . \quad . \quad (15.31)$$

Since the variables R_3 and C_6 give independent balance for reference and quadrate terms the balance will be quickly achieved.

(*c*) *Bridged-T Inductance and Q-factor Network* (Fig. 15.12 (*c*)). The unknown inductance is inserted as the bridge arm (L_4, R_4). Solving the balance equation,

$$\omega L_4 = \frac{1}{\omega C_1} + \frac{1}{\omega C_2} \quad \text{and} \quad R_4 = \frac{1}{\omega^2 C_1 C_2 R_3} \quad . \quad (15.32)$$

If C_1 and C_2 are two equal ganged variable capacitors, $C_1 = C_2 = C$, and eqns. (15.32) reduce to

$$\omega L_4 = \frac{2}{\omega C} \quad \text{and} \quad R_4 = \frac{1}{\omega^2 C^2 R_3} \quad . \quad . \quad (15.33)$$

Hence

$$Q = \frac{\omega L_4}{R_4} = 2\omega C R_3 \quad . \quad . \quad (15.34)$$

It should be noted that the balance conditions of many of these networks are frequency dependent, but since modern digital frequency meters can have accuracies to better than 1 part in 10^5, this measurement of frequency presents no problems up to several megacycles per second.

The circuit of Fig. 15.12 (*c*) can also be used to measure very high resistances at radio frequencies, provided that the unknown $R_4 \gg R_3$. In this case, since $L_4 \rightarrow 0$, $2/\omega^2 C \rightarrow 0$ (eqn. (15.33)), and hence $\omega^2 C \gg 1$. If this condition is fulfilled, the expression for R_4 shows that CR_3 must be very much less than unity. The "balance" is not a true one, but gives satisfactory practical results.

15.9. Transformer Ratio-arm Bridge

This type of bridge is capable of measuring admittances over a very wide range of frequencies, from 50 c/s up to some 250 Mc/s. Stray capacitance effects are reduced to a minimum. Accuracies to

1 part in 10^3 can be achieved in single-frequency commercial instruments employing this method, while wide-frequency-range bridges for very high frequencies have accuracies to 1 or 2 per cent.

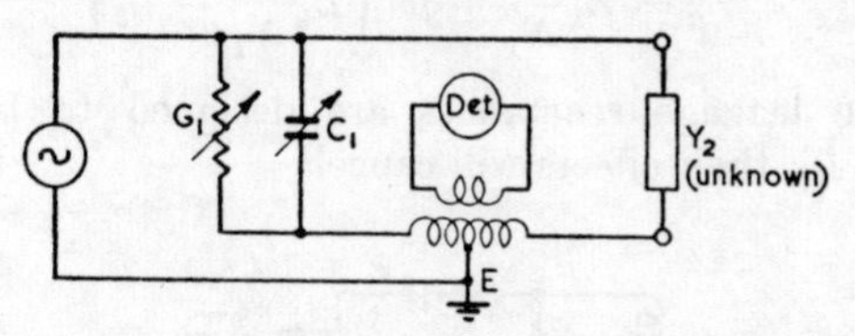

FIG. 15.13. PRINCIPLE OF DIFFERENTIAL-TRANSFORMER BRIDGE

The basic principle of operation is illustrated in Fig. 15.13. The detector will indicate zero when the ampere-turns in the two halves of the primary of the differential transformer balance, that is when

$$I_1N_1 = I_2N_2$$

But

$$I_1 = VY_1 \quad \text{and} \quad I_2 = VY_2$$

so that

$$N_1Y_1 = N_2Y_2$$

Hence

$$Y_2 = Y_1\frac{N_1}{N_2} \quad \text{or} \quad G_2 \pm jB_2 = (G_1 \pm jB_1)\frac{N_1}{N_2} \tag{15.35}$$

If Y_1 is a standard admittance consisting of a variable resistance in parallel with a variable susceptance, balance is obtained only when the two independent conditions

$$G_2 = G_1\frac{N_1}{N_2} \quad \text{and} \quad B_2 = B_1\frac{N_1}{N_2} \tag{15.36}$$

are satisfied.

At balance there will be no resultant core ampere-turns, so that the ends of Y_1 and Y_2 which are connected to the differential transformer are effectively at earth potential.

The influence of the transformer leakage reactance can be seen by expressing the balance condition in impedance terms, and including the leakage reactance. Thus

$$Z_{2\,total} = Z_{1\,total}\frac{N_2}{N_1}$$

Therefore

$$Z_2 + j\omega L_2 = (Z_1 + j\omega L_1)\frac{N_2}{N_1}$$

where ωL_2 is the leakage reactance of the right-hand side of the primary, and ωL_1 that of the left-hand side. Hence

$$Z_2 = Z_1 \frac{N_2}{N_1} + j\omega \left(L_1 \frac{N_2}{N_1} - L_2 \right)$$

Thus if the leakage reactances are designed to be such that $L_1 N_2 / N_1 = L_2$, their effects will cancel.

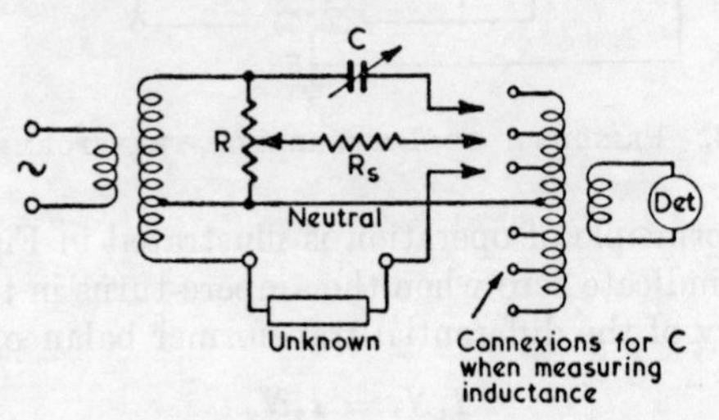

FIG. 15.14. PRACTICAL FORM OF DIFFERENTIAL-TRANSFORMER BRIDGE

A practical form of the differential-transformer admittance bridge is shown in Fig. 15.14. The bridge is direct reading in terms of parallel conductance and capacitance. In addition to the standard resistor and capacitor, the standard arm contains uncalibrated balancing elements, which are not shown. These are used to obtain a balance when the unknown is not connected and the standards are set to zero. This eliminates internal and external stray effects.

Hence, for example, a small capacitor at the end of long leads may be accurately measured by balancing out the stray effects when the capacitor is open-circuited, and then measuring the admittance on the standard dials when the capacitor is connected.

The admittance range of the bridge is increased by having tappings on the differential transformer. Inductive admittances can be measured by connecting the standard capacitor to the end of the differential transformer opposite to the resistive element. The value of the unknown parallel inductance is then obtained by equating ωL and $1/\omega C$, where C is the indicated value of the standard capacitor. In this bridge the standard and the unknown are fed in anti-phase from the input transformer to give balance when the outputs are connected in the same sense to the output transformer.

One advantage of the differential-transformer bridge is that it may be used to measure the transfer characteristics of 3-terminal networks. This is illustrated in Fig. 15.15. The network is connected in series with the terminating resistance required—at balance the transformer end of this resistance is at the neutral potential. The output current can be calculated from the values of the standards, the range multipliers required for balance, and the input voltage, and hence the transfer admittance may be found. Active 3-terminal networks can have their current or admittance transfer characteristic measured in the same manner.

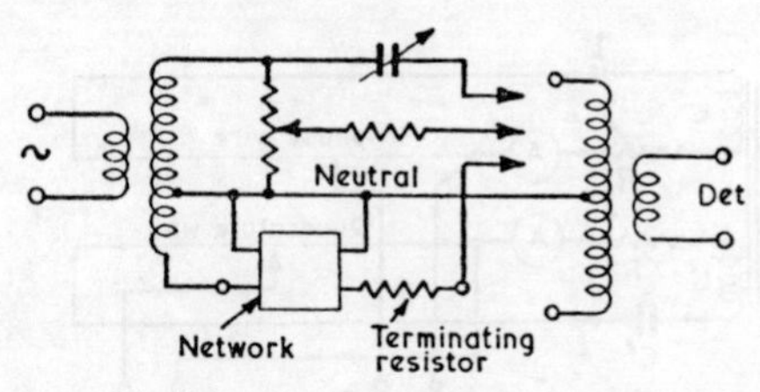

FIG. 15.15. MEASUREMENT OF THREE-TERMINAL NETWORK PARAMETERS

A further advantage is that components may often be measured *in situ*. Thus a delta of capacitors can be connected with the desired element across the "unknown" terminals, while the third terminal is connected to the "neutral" of the bridge. The two capacitors whose value is not required are thus connected across the input and output of the transformers, and do not affect the balance conditions (since at balance there is no voltage across the output transformer).

15.10. The A.C. Potentiometer

This instrument forms a convenient method of measuring the magnitude and phase of small alternating voltages at power frequencies. It is, however, rather difficult to calibrate accurately, and normally depends on the calibration accuracy of indicating electrodynamic ammeters. For this reason care must be taken to prevent harmonics from being present. The frequency range is limited by mutual coupling effects. Generally modern electronic phase-sensitive voltmeters are now used for the measurements for which the a.c. potentiometer is suited.

(*a*) *Rectangular Co-ordinate A.C. Potentiometer*. A schematic of the Gall rectangular co-ordinate potentiometer is shown in Fig. 15.16. The two slide-wires are fed with equal currents in phase quadrature from a common a.c. source. This quadrature is achieved provided that

$$I(R + j\omega L) = jI(R' + j\omega L' + 1/j\omega C')$$

Hence

$$R' = \omega L \quad \text{and} \quad R = \frac{1}{\omega C'} - \omega L'$$

The potential gradient along each wire is usually 1 V/m when the standard currents of 100 mA (as measured on the electrodynamic ammeters) flow through them. The detector, which is normally a vibration galvanometer, will indicate a null only if the reference and quadrature components of the unknown voltage are balanced by

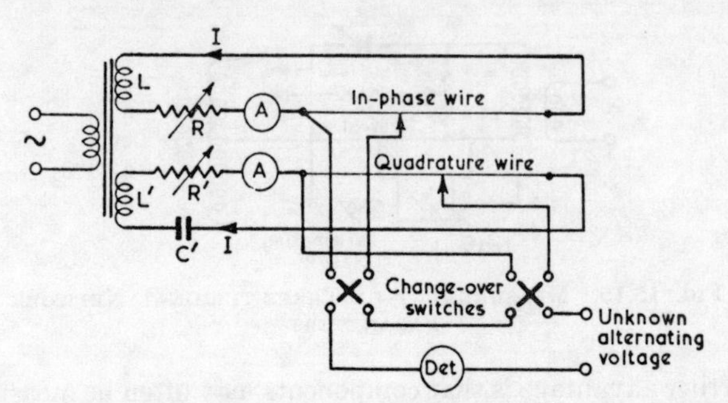

Fig. 15.16. Gall Rectangular Co-ordinate Potentiometer

equal and opposite reference and quadrature voltage drops along the two slide-wires, and hence the rectangular co-ordinates of the unknown are found in terms of the slider positions when balance is achieved. The change-over switches enable measurements to be made over the full 360° of phase angle.

The accuracy of the quadrature relationship between the two slide-wires can be checked by connecting a standard mutual inductometer in series with the quadrature wire and measuring the secondary voltage on the in-phase wire. With standard wire currents of I_1 and I_2 the inductometer secondary e.m.f. is

$$v = j\omega MI_2 = -\omega MI_1 \qquad \text{if } I_1 = I_2$$

This voltage is now set on the reference wire by setting the slider at a position to give the calculated voltage (Fig. 15.17). Then, if the wire currents are not in quadrature, the detector will show a deflexion.

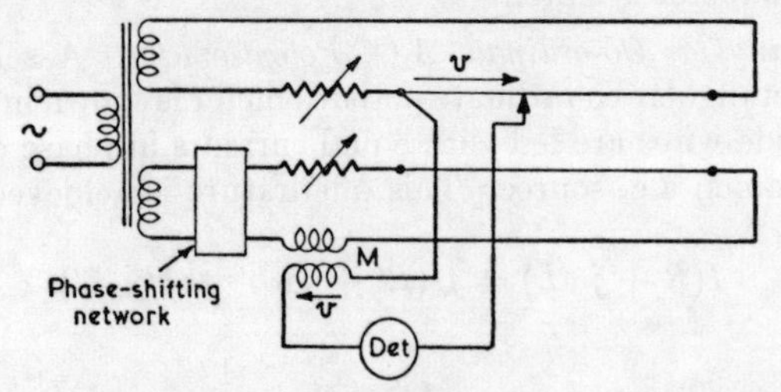

Fig. 15.17. Calibration of Rectangular Co-ordinate Potentiometer

A phase-shifting network in the quadrature-wire circuit can now be adjusted to give balance.

(*b*) *Polar Co-ordinate A.C. Potentiometer*. In this form only one wire is required. It is supplied through a calibrated phase-shifting transformer, and carries the standard current of 100 mA. Balance is obtained by varying the slider position and the setting of the phase-shifting transformer, which consists basically of a stator wound with a 2-phase winding. This gives a rotating field in the air-gap. The rotor winding will have an induced e.m.f. of constant magnitude and a phase that will vary with the angular position of the rotor. The 2-phase supply for the stator is obtained from a single-phase supply by means of a phase-shifting network.

15.11. Detectors

The sensitivity of null measuring networks depends largely on that of the detectors used. With a given detector, maximum sensitivity is achieved when the source impedance matches the bridge input impedance, and in a.c. bridges this consideration may determine the relative positions of source and detector. Networks where the source and detector may both be earthed avoid the use of screened detector transformers.

At power frequencies the vibration galvanometer is commonly used. This instrument is tuned mechanically to the source frequency, and is not affected by harmonics of small amplitude, but requires a stable supply frequency.

Headphones may be used at audio frequencies. With experience, balance can be determined for the fundamental even in the presence of harmonics, but this will depend largely on the operator. Head effect may be serious, and a Wagner earth may have to be employed in bridge circuits.

At higher frequencies the most convenient form of detector is the electronic heterodyne detector. This is tuned to give a 1-kc/s beat note with the source frequency. This note may be amplified and fed to headphones or to an indicating instrument. The heterodyne detector is unaffected by harmonics, and may have a high sensitivity.

Valve voltmeters and cathode-ray oscillographs are sometimes used as detectors, but care is required since these instruments normally have one input terminal earthed.

Bibliography

Hague, *A.C. Bridge Methods* (Pitman).

Golding and Widdis, *Electrical Measurements and Measuring Instruments* (Pitman).

Calvert, *The Transformer Ratio-arm Bridge* (Wayne Kerr Monograph).

Problems

15.1. The a.c. bridge network shown in Fig. 15.18 is used to measure the mutual inductance M. Derive the conditions required for balance. What modification is required to make the balance independent of frequency? Draw the vector [complexor] diagram for the bridge so modified, in its condition of balance. (*A.E.E.*, 1961)

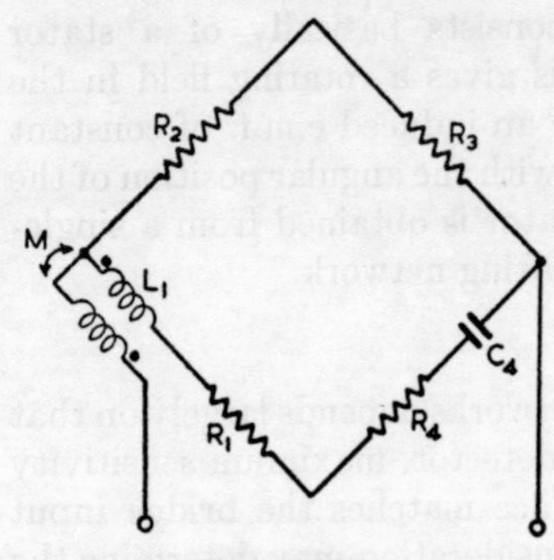

Fig. 15.18

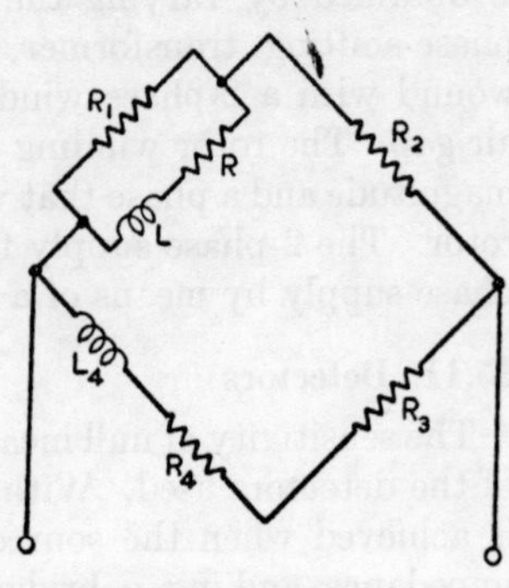

Fig. 15.19

15.2. The bridge shown in Fig. 15.19, is used to measure the value of the unknown inductance L. Derive the condition required for balance when $R_2 = R_3$.

Balance is obtained at a frequency of 400 c/s with R_1, R and R_4 respectively equal to 100, 25·6 and 50 Ω and the ratio R_2/R_3 equal to unity. Determine the unknown inductance. (*A.E.E.*, 1960)

(*Ans.* 38·3 mH.)

15.3. The bridge shown in Fig. 15.20, is used to measure the capacitance C_x and the resistance R_x. Derive the balance equations of the bridge, and determine C_x and R_x when $R_2 = R_3$, $C_1 = 0{\cdot}1\ \mu$F, $R_1 = 2{,}100\ \Omega$ and the frequency is 1,000 c/s. Draw a complexor diagram for the balanced condition of the bridge. (*A.E.E.*, 1958)

(*Ans.* 0·0365 μF; 3,310 Ω.)

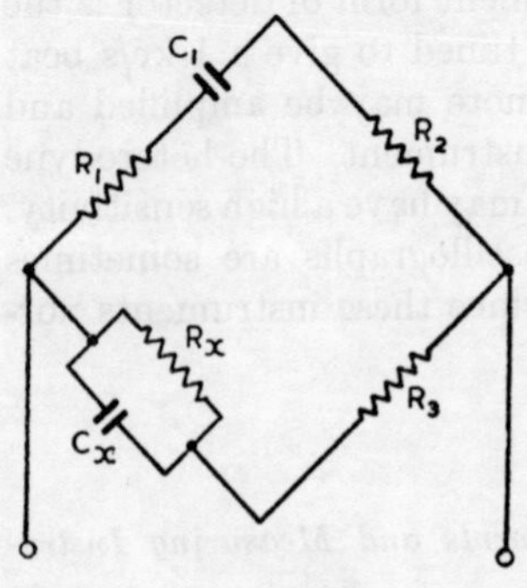

Fig. 15.20

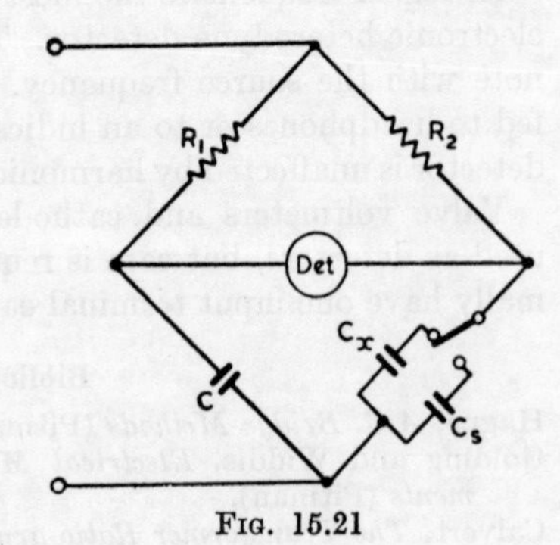

Fig. 15.21

15.4. Discuss the advantages of the substitution method applied to an a.c. bridge.

The bridge shown in Fig. 15.21 is used to measure the unknown capacitance C_x. The standard capacitor C_s is capable of step adjustment. A balance with C_x in circuit is achieved with $R_1 = 1{,}000\ \Omega$. When C_s, set at 0·1 μF, is inserted in place of C_x, balance is restored by increasing R_1 to 1,030 Ω. Calculate the value of C_x.

If the value of R_1 is known to within 0·1 per cent, what will be the uncertainty in the value determined for C_x? (*A.E.E.*, 1963)

(*Ans.* 0·0971 $\pm$ 0·0002 μF).

15.5. A wire-wound resistor for use in a resistance box for audio frequencies has a resistance $R = 10{,}000\ \Omega$, a residual self-inductance $L = 2$ mH and a residual self-capacitance equivalent to a capacitance $C = 30$ pF connected across the terminals of the resistor. Derive approximate expressions in terms of R, L, C and the angular frequency ω for the impedance and time-constant of the resistor. Justify the approximations made.

Assuming that the ohmic resistance of the wire is independent of frequency, calculate the effective resistance and the phase-angle error of the resistor when the frequency is 100 kc/s. (*A.E.E.*, 1963)

(*Ans.* $Z \simeq R/(1 + \omega^2C^2R^2) + j\omega\,(L - CR^2)/(1 + \omega^2C^2R^2)$; $\tau = R/[\omega^2(CR - L)]$; 9,670 Ω, 3·8° leading.)

15.6. The components of a 4-arm bridge when balanced have the following values—

Arm AB, 0·5 μF.
Arm AC, 1,000 Ω (non-reactive).
Arm BD, 1 μF in series with a decade resistance box set at 146·4 Ω.
Arm CD, a coil in series with a decade resistance box set at 495·1 Ω.

Derive the equations of balance, and calculate the resistance and inductance of the coil. If the individual resistances of the decade boxes are limited in accuracy to 0·2 per cent, what is the uncertainty in the determination of the coil resistance? (*L.U. Part II Theory*, 1957)

(*Ans.* $R = 4{\cdot}9 \pm 1{\cdot}0\ \Omega$, $L = 73{\cdot}2$ mH.)

15.7. An a.c. bridge $PQRS$ consists of the following components—

PQ, a 2,000-Ω non-reactive resistor.
QR, a 120,000-Ω non-reactive resistor shunted by a 0·005-μF loss-free capacitor.
RS, an imperfect capacitor.
SP, a 138-Ω non-reactive resistor.

Determine from first principles the constants of the arm RS, in both series and parallel form, if the bridge balances at a frequency of 8,000 c/s. (*L.U. Part II Theory*, 1958)

(*Ans.* $C_p = 0{\cdot}0725\ \mu$F, $R_p = 8{,}280\ \Omega$; $C_s = 0{\cdot}0725\ \mu$F, $R_s = 9{\cdot}2\ \Omega$.)

15.8. Derive for the network shown in Fig. 15.12(c) on p. 462, an expression for the frequency at which the attenuation is infinite.

Calculate the resistance R_4 and the inductance L_4, when each of the capacitances is 0·1 μF and the resistance R_3 is 2,100 Ω, the output being zero when the frequency is 2,000 c/s. Derive any formula used.

(*Ans.* 302 Ω, 0·127 H.) (*A.E.E.*, 1958)

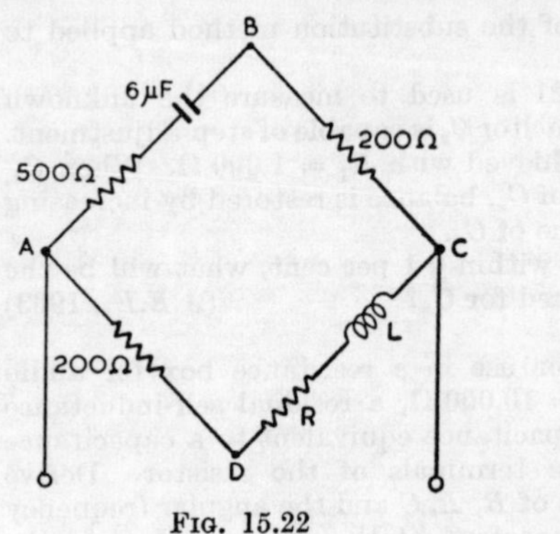

Fig. 15.22

Fig. 15.23

15.9. An a.c. bridge is shown in Fig. 15.22. The supply frequency is 50 c/s. Determine from first principles the values of R and L when the detector shows a null. Sketch a vector [complexor], diagram for the balanced condition, identifying all vectors [complexors] with reference to a corresponding circuit diagram. (*L.U. Part II Theory*, 1961)
(*Ans.* $R = 37{\cdot}3\ \Omega$, $L = 127$ mH.)

15.10. Describe a method of determining the self-capacitance of a radio-frequency coil.

The self-capacitance C_0 of a coil is found to be 20 pF, and when the coil is measured at 1 kc/s on the Maxwell bridge shown in Fig. 15.23 a balance is obtained with $R = 10\ \Omega$ and $C = 1\ \mu$F. Calculate the effective coil series resistance and inductance at a frequency of 2 Mc/s, assuming that skin effect causes the resistance to be increased 1·2 times from 1 kc/s to 2 Mc/s. (*I.E.E. Radio*, 1959)
(*Ans.* 25·6 Ω, 146 μH.)

15.11. A balanced bridge has the following components connected between its five nodes, A, B, C, D and E—

Between A and B, 1,000 Ω resistance.
Between B and C, 1,000 Ω resistance.
Between C and D, an inductor.
Between D and A, 218 Ω resistance.
Between A and E, 469 Ω resistance.
Between E and B, 10 μF capacitance.
Between E and C, a detector.
Between B and D, a power supply (a.c.).

Derive the equations of balance, and hence deduce the resistance and inductance of the inductor. (*L.U. Part III Theory*, 1957)
(*Ans.* 218 Ω, 7·89 mH.)

15.12. A 4-node bridge, balanced at $10/2\pi$ kc/s, has the following components connected between nodes—

Between A and B, a 98·7-Ω resistor having an inductance of 1·90 mH, in parallel with a 1·024-μF loss-free capacitor.
Between B and C, a 1,000-Ω non-reactive resistor.

Between C and D, an inductor having across its terminals a 0·0310-μF loss-free capacitor.

Between D and A, a 1,000-Ω non-reactive resistor.

Calculate the inductance and resistance of the inductor.

(*Ans.* 442 Ω, 270 mH.) (*L.U. Part III Theory*, 1959)

15.13. A 4-arm unbalanced a.c. bridge is supplied from a source having negligible impedance. The bridge has non-reactive resistors of equal resistance, R, in adjacent arms. The third arm has an inductor of resistance R and reactance X, where X is numerically equal to R. The fourth arm has a variable non-reactive resistor. The detector is connected between the junction of the first and second arms and the junction of the third and fourth arms, and has a resistance R and negligible reactance. Determine the value of the variable resistor when the detector current is in quadrature with the supply current.

(*Ans.* 1·37 R.) (*L.U. Part III Theory*, 1960)

15.14. The arms of a 4-arm bridge network have the following components: arm AB, a coil in series with a variable resistance P; arm BC, a capacitance C, in series with a variable resistance Q; arm CD, a capacitance C_2; arm DA a resistance R.

An a.c. source is connected across AC and a detector across BD. Derive the equations of balance, and calculate the inductance and resistance of the coil if the values of the components at balance are $P = 47\ \Omega$, $Q = 300\ \Omega$, $R = 100\ \Omega$, $C_1 = 1\ \mu$F and $C_2 = 0{\cdot}5\ \mu$F.

Explain how a Wagner earth would be arranged for use with the above bridge. (*L.U. Part III Theory*, 1962)

(*Ans.* 15 mH, 3 Ω.)

15.15. A current of 10 A at a frequency of 50 c/s was passed through the primary of a mutual inductor having a negligible phase defect. The voltages at the primary and secondary terminals were measured on a co-ordinate potentiometer and are given below.

With secondary winding open-circuited,

Secondary volts . . $-2{\cdot}72 + j1{\cdot}57$
Primary volts . . $-0{\cdot}211 + j0{\cdot}352$.

With secondary winding short-circuited,

Primary volts . . $-0{\cdot}051 + j0{\cdot}329$.

The phase of the primary current relative to the potentiometer current was the same in both tests.

Determine the self- and mutual inductances of the inductor.

[The phase defect of a mutual inductance is the angle between the secondary e.m.f. and the complexor which leads the primary current by 90°. *Hint.* Do not neglect the primary and secondary resistances.] (*L.U. Part III Theory*, 1960)

(*Ans.* 0·1145 mH, 18·1 mH, 1 mH.)

15.16. Derive an expression for the ratio V_{out}/V_{in} for the network of Fig. 15.12(c) on p. 462, and show that at a certain frequency the output voltage is zero. Determine this frequency if $C_1 = C_2 = 0{\cdot}05\ \mu$F and the inductance of the coil is 150 mH. (*A.E.E.*, 1961)

(*Ans.* 2,600 c/s.)

15.17. Derive the balance conditions for the twin-T network shown in Fig 15.12(a) on p. 462, solving for R_3 and L_3.

The bridge is balanced with $C_3 = 210$ pF, $R_5 = 50\ \Omega$, $C_1 = C_2 = 25$ pF, $C_4 = 0{\cdot}01\ \mu$F and $C_6 = 350$ pF. The frequency is 25 Mc/s. When a capacitor C_x is connected across AB, at balance C_3 becomes 103 pF and $C_6 = 483$ pF. Find the value of C_x and its equivalent parallel loss resistance. (*R.R.E.*)

(*Ans.* 107 pF, 97·5 kΩ.)

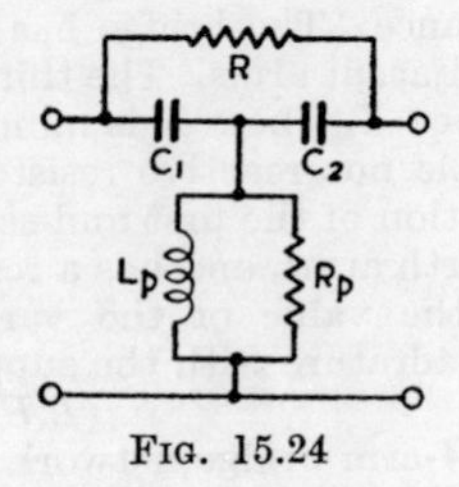

Fig. 15.24

15.18. Derive the balance conditions for the bridged-T network shown in Fig. 15.24, solving for L_p and R_p. If $C_1 = C_2 = 102$ pF, $R = 122\ \Omega$ and $f = 1$ Mc/s, find L and R, and hence determine the equivalent series resistance of the coil.

[*Hint.* Express the L_pR_p parallel circuit as an admittance.]

(*Ans.* 0·124 mH, 20 kΩ, 30·3 Ω.)

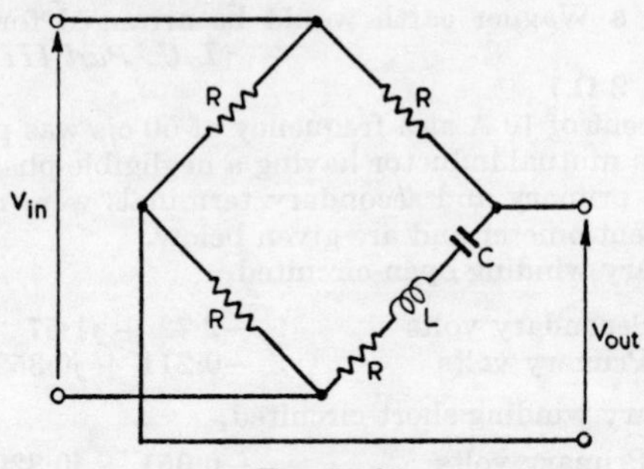

Fig. 15.25

15.19. If X is the reactance of the LC combination in the bridge network shown in Fig. 15.25, derive expressions (in terms of X and R) for (*a*) the magnitude and (*b*) the phase of the voltage transfer function, K, of the network.

If the bridge balances at the angular frequency $\omega_0 = 1/\sqrt{(LC)}$, and if $\mathrm{d}|K|/\mathrm{d}X$ is then equal to $1/4R$, show that

$$\left(\frac{\mathrm{d}|K|}{\mathrm{d}\omega}\right)_{\omega_0=0} = \frac{L}{2R}$$

(*A.E.E.*, 1964)

CHAPTER 16

Instrument Theory

THE D'Arsonval moving-coil principle is widely used for direct current and voltage measurement. Moving-coil instruments combine sensitivity with accuracy. It is assumed that the reader is familiar with the construction and theory of such instruments.

The instruments used for the measurement of alternating quantities can be grouped into three main divisions. In the first category are rectifier instruments, which employ a moving-coil movement the torque on which is proportional to the mean current passing through it. These instruments can be calibrated in r.m.s. alternating current or voltage, with sinusoidal waveforms assumed, and with suitable rectifiers they can be used at high frequencies. The second category includes instruments whose indication depends on the square of the operating current, and which therefore should be accurate in measuring r.m.s. values (either a.c. or d.c.). Moving-iron and electrodynamic instruments belong to this category. The frequency range is limited by stray capacitance and by the impedance of the instrument (especially voltmeters), and they are normally employed only at power frequencies. Thermal and electrostatic instruments also belong to this category, but have a much larger frequency range; thermal instruments are available for current measurement at radio frequencies, and for voltage measurement up to 20 kc/s. The third category includes those instruments which are suitable for alternating current only. In this class are the instruments operating on the induction principle, which are most commonly used as energy meters.

16.1. Moving-iron Instruments

In both the attraction type of moving-iron instrument and the more usual repulsion type, an iron armature which is attached to a spindle moves relative to a fixed coil carrying the operating current. The movement takes place against the restoring torque of a control spring, and is dependent on the operating current. Any movement of the iron causes a change in the inductance of the coil, and alters the energy stored in the system.

Consider a coil of inductance L henrys carrying a current i amperes, which causes an angular rotation, θ, of the spindle against the restoring torque, T, of the control spring. The stored energy is $W_1 = \frac{1}{2}Li^2$ joules. Suppose now that the current changes by a small amount δi, causing a further rotation $\delta\theta$ of the spindle, and an increase in the inductance of δL. The stored energy is now

$$W_2 = \tfrac{1}{2}(L + \delta L)(i + \delta i)^2$$
$$\simeq \tfrac{1}{2}Li^2 + \tfrac{1}{2}i^2\,\delta L + Li\,\delta i \quad \text{(neglecting second-order terms)}$$

Hence the change in stored energy is

$$W_2 - W_1 \simeq \tfrac{1}{2}i^2\,\delta L + Li\,\delta i$$

and the work done is $T\,\delta\theta$ joules.

During the change of position an e.m.f. is induced in the coil, given by

$$v = \frac{\mathrm{d}(Li)}{\mathrm{d}t} = L\frac{\mathrm{d}i}{\mathrm{d}t} + i\frac{\mathrm{d}L}{\mathrm{d}t}$$

since both L and i are changing.

The electrical energy supplied in δt seconds is $vi\,\delta t$ joules—

$$vi\,\delta t = iL\frac{\mathrm{d}i}{\mathrm{d}t}\,\delta t + i^2\frac{\mathrm{d}L}{\mathrm{d}t}\,\delta t$$
$$= iL\,\delta i + i^2\,\delta L$$

Hence

$$iL\,\delta i + i^2\,\delta L = \tfrac{1}{2}i^2\,\delta L + Li\,\delta i + T\,\delta\theta$$

so that

$$T = \frac{1}{2}i^2\frac{\delta L}{\delta\theta} \quad \text{newton-metres}$$

In the limit as $\delta\theta \to 0$, this becomes

$$T = \frac{1}{2}i^2\frac{\mathrm{d}L}{\mathrm{d}\theta} \quad \text{newton-metres} \qquad (16.1)$$

For a linear control spring the deflexion, θ, is directly proportional to the torque. If $\mathrm{d}L/\mathrm{d}\theta$ is constant the deflexion is therefore proportional to the square of the current. For alternating currents the deflexion will be proportional to the mean square current, and hence the scale may be calibrated in r.m.s. values. This will give non-uniform graduations (square-law scale), but the scale shape may be altered to some extent by having changing values of $\mathrm{d}L/\mathrm{d}\theta$ with θ.

16.2. The Electrostatic Voltmeter

In an electrostatic voltmeter a set of fixed plates is interleaved with a set of plates attached to a spindle which may rotate against the restoring torque of a control spring. If an instantaneous voltage, v, exists between the fixed and moving plates, charges will be set up on the plates, and the force on the edge of the movable plates will pull them into the field of the fixed plates. Consider a small increase δv in the voltage between the plates, causing them to move by a small angle $\delta\theta$, and so to give a change in the capacitance between the plates of δC. The change in charge due to this will be $\delta q = C\,\delta v + v\,\delta C$ coulombs.

The work done against the restoring torque, T, of the control spring is $T\,\delta\theta$ joules.

The increase in stored energy in the electric field is $\frac{1}{2}v^2\,\delta C + vC\,\delta v$ joules. The energy taken from the supply is, $v\,\delta q = vC\,\delta v + v^2\,\delta C$ joules. Hence

$$vC\,\delta v + v^2\,\delta C = \tfrac{1}{2}v^2\,\delta C + vC\,\delta v + T\,\delta\theta$$

The instantaneous torque, T, is

$$T = \frac{1}{2}\,v^2\,\frac{\delta C}{\delta\theta}$$

In the limit, as $\delta\theta \to 0$,

$$T = \frac{1}{2}\,v^2\,\frac{\mathrm{d}C}{\mathrm{d}\theta} \quad \text{newton-metres} \qquad . \quad . \quad (16.2)$$

Hence the deflexion, θ, with a linear control spring will be proportional to $\frac{1}{2}v^2(\mathrm{d}C/\mathrm{d}\theta)$. With alternating current the instrument deflexion is proportional to the mean square voltage (if $\mathrm{d}C/\mathrm{d}\theta$ is constant), and the scale can be calibrated in r.m.s. values. Note that with direct voltage there is no steady operating current, and with alternating voltage there is only a small capacitive current, provided that the frequency is not too high.

16.3. Electrodynamic Instruments

In the electrodynamic (or dynamometer) instrument one current-carrying coil is free to rotate in the field of a second fixed current-carrying coil. Usually the fixed coil is arranged in two parts (Helmholtz arrangement) with the moving coil between them. If the fixed coil carries an instantaneous current i_1 amperes and has an inductance L_1, the moving coil carries i_2 amperes and has an inductance

L_2, and there is a mutual inductance M between the coils, the instantaneous stored energy in the magnetic fields is

$$W_1 = \tfrac{1}{2}L_1 i_1^2 + \tfrac{1}{2}L_2 i_2^2 + i_1 i_2 M$$

As the moving coil rotates, M will change. Consider what happens if i_1 and i_2 are held constant and the moving coil is rotated an amount $\delta\theta$ against the restoring torque, T, of a control spring. Then

$$\text{Work done} = T\,\delta\theta$$

and

Change in stored energy $= i_1 i_2\,\delta M$ (since L_1 and L_2 are constant)

The electrical energy supplied from the external circuits is

$$i_1 \frac{\mathrm{d}Mi_2}{\mathrm{d}t}\,\delta t + i_2 \frac{\mathrm{d}Mi_1}{\mathrm{d}t}\,\delta t$$

if the movement takes δt seconds. This equals $i_1 i_2\,\delta M + i_1 i_2\,\delta M$ since i_1 and i_2 are fixed. Hence

$$2i_1 i_2\,\delta M = T\,\delta\theta + i_1 i_2\,\delta M$$

or

$$T = i_1 i_2 \frac{\delta M}{\delta\theta}$$

In the limit the instantaneous torque is

$$T = i_1 i_2 \frac{\mathrm{d}M}{\mathrm{d}\theta} \qquad . \quad . \quad . \ (16.3)$$

In the electrodynamic ammeter or voltmeter the fixed and moving coils are connected in series so that $i_1 = i_2 = i$, say. With direct current the operating torque (and therefore the deflexion) is proportional to i^2 if $\mathrm{d}M/\mathrm{d}\theta$ is constant. With alternating current the torque and deflexion will be proportional to the mean square current, so that a scale of r.m.s. values can be used.

When the instrument is used as a wattmeter the fixed coils carry the load current, i, while the moving coil carries a current proportional to the operating voltage, v ($i_2 = v/R$, where R is the moving-coil circuit resistance, this normally being external to the actual coil).

Hence the mean torque, T_{av}, is

$$T_{av} = \frac{1}{\tau}\int_0^\tau i i_2 \frac{\mathrm{d}M}{\mathrm{d}\theta}\,\mathrm{d}t$$

$$= II_2 \frac{\mathrm{d}M}{\mathrm{d}\theta} \quad \text{for direct currents } I \text{ and } I_2 . \qquad . \ (16.4)$$

If the currents are alternating, so that $i = I_m \sin(\omega t - \phi)$ and $i_2 = I_{2m} \sin \omega t$, the mean torque over the period, τ, is

$$T_{av} = \frac{\mathrm{d}M/\mathrm{d}\theta}{\tau} \int_0^\tau I_m I_{2m} \sin(\omega t - \phi) \sin \omega t \, \mathrm{d}t$$

$$= \frac{\mathrm{d}M}{\mathrm{d}\theta} \frac{I_m I_{2m}}{\tau} \int_0^\tau \frac{1}{2} (\cos\phi - \cos(2\omega t - \phi) \, \mathrm{d}t)$$

$$= II_2 \cos\phi \frac{\mathrm{d}M}{\mathrm{d}\theta} \quad . \quad . \quad . \quad . \quad (16.4a)$$

where I and I_2 are r.m.s. values and ϕ is the phase angle between them. If the voltage coil has no reactance, $I_2 = V/R$, whence

$$T_{av} = \frac{VI\cos\phi}{R} \frac{\mathrm{d}M}{\mathrm{d}\theta}$$

$$= \text{constant} \times VI\cos\phi \quad . \quad . \quad . \quad (16.5)$$

if $\mathrm{d}M/\mathrm{d}\theta$ is constant. In this case ϕ is the phase angle between V and I, and hence the deflexion is proportional to the mean circuit power. This, of course, applies to d.c. power as well. Note also that the mean circuit power is

$$\frac{1}{\tau} \int_0^\tau vi \, \mathrm{d}t$$

Effect of Inductance of Voltage Coil. The circuit of an electrodynamic wattmeter is shown in Fig. 16.1 (*a*). Suppose that the

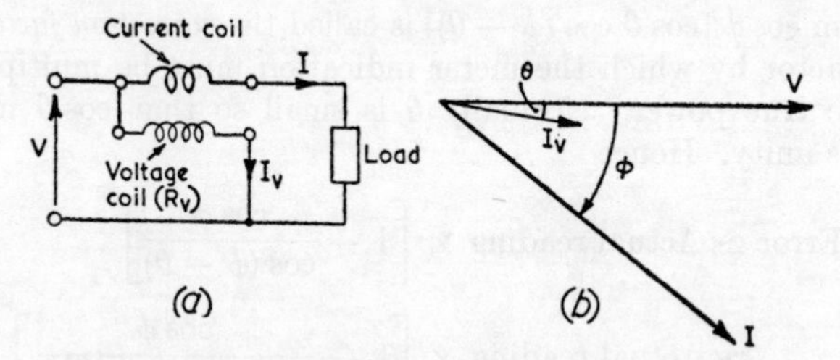

FIG. 16.1. ELECTRODYNAMIC WATTMETER CONNEXIONS

inductance of the voltage coil causes the voltage-coil current, I_v, to lag behind the circuit voltage, V, by an angle θ. Then the instrument deflexion is

$$\text{Deflexion} \propto II_v \cos(\phi - \theta) \quad . \quad . \quad (16.6)$$

for a lagging load power factor, as illustrated in the complexor qiagram of Fig. 16.1 (*b*).

If the load power factor is leading,

$$\text{Deflexion} \propto II_v \cos(\phi + \theta) \qquad . \qquad . \quad (16.7)$$

Hence, with a lagging load power factor, eqn. (16.6) shows that the instrument will read more than the true load power, while with a leading load power factor the instrument will read low.

For a lagging power factor,

$$\text{Instrument reading} = \frac{KIV}{Z}\cos(\phi - \theta)$$

where K is a constant and Z is the impedance of the voltage coil. Hence

$$\text{Actual reading} = \frac{KIV}{(R/\cos\theta)}\cos(\phi - \theta)$$

since $Z = R/\cos\theta$.

The true load power which should be indicated is $(KIV/R)\cos\phi$, so that the error is

$$\frac{KIV}{R}\left[\cos\theta\cos(\phi - \theta) - \cos\phi\right]$$

$$= \frac{KIV}{R}\cos\theta\cos(\phi - \theta)\left[1 - \frac{\cos\phi}{\cos\theta\cos(\phi - \theta)}\right]$$

$$= \text{Actual reading} \times \left[1 - \frac{\cos\phi}{\cos\theta\cos(\phi - \theta)}\right]$$

The term $\cos\phi/[\cos\theta\cos(\phi - \theta)]$ is called the *correction factor*, and is the factor by which the meter indication must be multiplied to give the true power. Normally θ is small so that $\cos\theta$ may be taken as unity. Hence

$$\text{Error} \simeq \text{Actual reading} \times \left[1 - \frac{\cos\phi}{\cos(\phi - \theta)}\right]$$

$$\simeq \text{Actual reading} \times \left[1 - \frac{\cos\phi}{\cos\phi + \sin\theta\sin\phi}\right]$$

$$\simeq \text{Actual reading} \times \left[\frac{\sin\theta\sin\phi}{\cos\phi + \sin\theta\sin\phi}\right]$$

$$\simeq \text{Actual reading} \times \left[\frac{\sin\theta}{\cot\phi + \sin\theta}\right] \qquad . \qquad . \quad (16.8)$$

In the case of a leading power factor we readily obtain

$$\text{Error} \simeq \text{Actual reading} \times \left[\frac{-\sin\theta\sin\phi}{\cos\phi - \sin\theta\sin\phi}\right] \quad . \; (16.9)$$

Connexion Error. In an uncompensated electrodynamic wattmeter there are obviously two ways in which the voltage coil can be connected: either (i) before or (ii) after the current coil. In case (i) the instrument measures load power plus current-coil losses, but gives negligible errors with high-impedance loads. In case (ii) the voltage-coil losses are measured by the instrument, but the error is negligible with low-impedance loads.

When the power in low-power-factor loads is being measured, the effect of the power absorbed in losses in the instrument may be important. One way of overcoming this is to have a compensating winding which carries the voltage-coil current, wound with the fixed coils as illustrated in Fig. 16.2. The effective current in the fixed coils is then the load current only, and the voltage-coil current is effectively proportional to the load voltage. There is no torque due to the voltage-coil current flowing in the current coil.

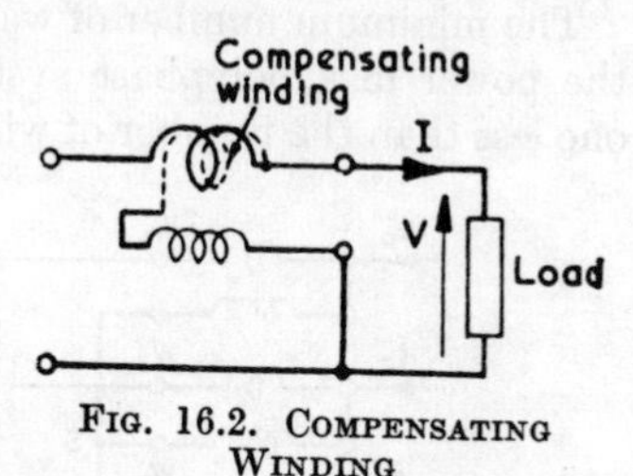

FIG. 16.2. COMPENSATING WINDING

Example 16.1. Derive an expression for the error in the power indication resulting from the inductance of the voltage coil of an electrodynamic wattmeter.

The current coil of a wattmeter is connected in series with an ammeter and a capacitive load. The voltage circuit of the wattmeter and a voltmeter are connected across the 3,000-c/s supply. The ammeter reading is 7·5 A, while the voltmeter and wattmeter readings are respectively 240 V and 45 W. The inductance of the voltage coil is 1·4 mH ($= L_v$) and the resistance of the voltage circuit is 5,000 Ω ($= R_v$). Calculate the actual power dissipated by the load. Neglect the voltage drops across the ammeter and the current coil. (*A.E.E.*, 1960)

For a capacitive load of power factor cos ϕ, the true power is greater than the actual reading and the error is given by eqn. (16.9).

The phase angle, θ, of the voltage coil circuit is

$$\theta = \tan^{-1}\frac{\omega L_v}{R_v} = \tan^{-1}\frac{2\pi \times 3{,}000 \times 1\cdot4 \times 10^{-3}}{5{,}000}$$

$$\simeq \frac{6\pi \times 1\cdot4}{5{,}000} \text{ rad, since } \theta \text{ is very small.}$$

$$\simeq 18\cdot1'$$

The actual wattmeter reading is

$$45 = 240 \times 7{\cdot}5 \cos(\phi + \theta)$$

so that $\cos(\phi + \theta) = 0{\cdot}025$, and $\phi = 88° 16'$. Hence

$$\text{Error} = 45 \times \frac{\sin 18{\cdot}1'}{\cot 88° 16' - \sin 18{\cdot}1'} = 9{\cdot}3$$

and

$$\text{True power} = 45 + 9{\cdot}3 = \underline{\underline{54{\cdot}3 \text{ W}}}$$

As a check, $240 \times 7{\cdot}5 \times \cos 88° 16' = 54{\cdot}4$ W.

16.4. Three-phase Power Measurements

The minimum number of wattmeter readings required to measure the power in a polyphase system is given by *Blondel's theorem* as one less than the number of wires in the system. This is because the

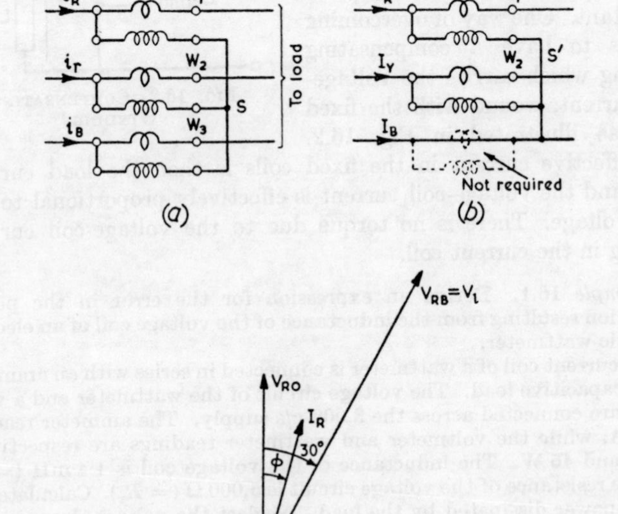

Fig. 16.3. Power Measurement in Three-phase Three-wire System

(*a*) With three wattmeters
(*b*) With two wattmeters

sum of the line currents is zero. Hence in a 3-phase 4-wire system a minimum of three wattmeters is required, while for a 3-phase 3-wire system the minimum number is two. It is obvious that the power in a 3-phase 3-wire load can be measured as the sum of the readings of three wattmeters connected as shown in Fig. 16.3 (*a*).

If 0 is the system neutral point, and S is the star point of the wattmeters, the sum of the wattmeter readings is

$$P_1 + P_2 + P_3 = \frac{1}{T}\int_0^T (i_R v_{RS} + i_Y v_{YS} + i_B v_{BS})\,\mathrm{d}t$$

$$= \frac{1}{T}\int_0^T [i_R(v_{R0} - v_{0S}) + i_Y(v_{Y0} - v_{0S}) + i_B(v_{B0} - v_{0S})]\,\mathrm{d}t$$

$$= \frac{1}{T}\int_0^T [i_R v_{R0} + i_Y v_{Y0} + i_B v_{B0} - (i_R + i_Y + i_B)v_{0S}]\,\mathrm{d}t$$

$$= \frac{1}{T}\int_0^T (i_R v_{R0} + i_Y v_{Y0} + i_B v_{B0})\,\mathrm{d}t$$

$$= \text{Mean power delivered to load} \quad . \quad . \quad (16.10)$$

since $i_R + i_Y + i_B = 0$.

Notice that the wattmeters need not be indentical. It is also apparent that the potential of the star point S does not affect the result, and hence if S is connected to one line, the wattmeters will still read the total delivered power. In this case the wattmeter whose current coil is connected to the line which is acting as the common voltage reference will read zero, and may be removed, giving the two-wattmeter connexion of Fig. 16.3 (*b*). Note also that no assumptions need be made regarding balance of load, phase sequence or waveform.

For a balanced load and sinusoidal waveform the two-wattmeter method gives additional power-factor information. Thus for the connexion shown in Fig. 16.3 (*b*), and for a lagging load phase angle, ϕ (with respect to phase voltages), the reading on W_1 is

$$P_1 = V_l I_l \cos(\phi - 30°)$$

and on W_2,

$$P_2 = V_l I_l \cos(\phi + 30°)$$

as can be seen from the complexor diagram, Fig. 16.3 (*c*), where V_l and I_l are line values. Hence

$$P_1 + P_2 = V_l I_l [\cos(\phi - 30°) + \cos(\phi + 30°)]$$
$$= \sqrt{(3)}\, V_l I_l \cos\phi \qquad (16.11)$$
$$= \text{Circuit power}$$

In the same way,

$$P_1 - P_2 = V_l I_l [\cos(\phi - 30°) - \cos(\phi + 30°)] = V_l I_l \sin\phi \quad (16.12)$$

so that
$$\tan\phi = \sqrt{3}\left(\frac{P_1 - P_2}{P_1 + P_2}\right) \qquad (16.13)$$

For balanced loads a single wattmeter can be used with an artificial star (Fig. 16.4 (*a*)) whose limbs are equal to the wattmeter

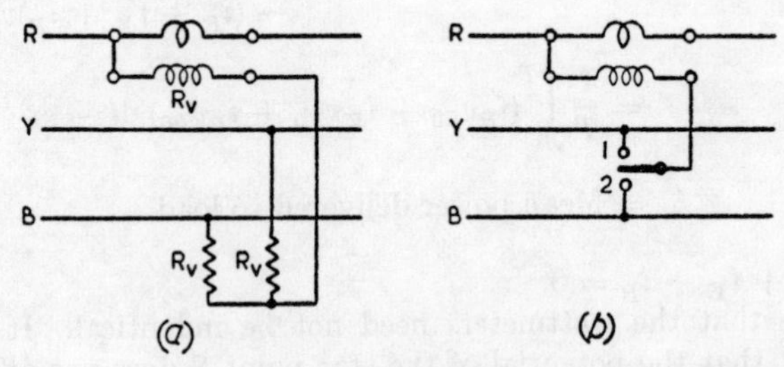

FIG. 16.4. USE OF A SINGLE WATTMETER FOR MEASUREMENT OF BALANCED THREE-PHASE POWER
(*a*) With artificial star
(*b*) With switched voltage coil

voltage-coil resistance. The total power is then three times the wattmeter reading. Alternatively a single wattmeter with switched voltage coil can be used (Fig. 16.4 (*b*)). In this case it is left to the reader to verify that the circuit power is the sum of the wattmeter readings with the switch in position 1 and then in position 2.

16.5. The Induction-type Movement

An induction movement is one in which a torque is produced on a conducting disc (which is pivoted and free to rotate) by the interaction of alternating fluxes and the currents which they induce in the disc. Instruments working on the induction principle are suitable only for a.c. operation.

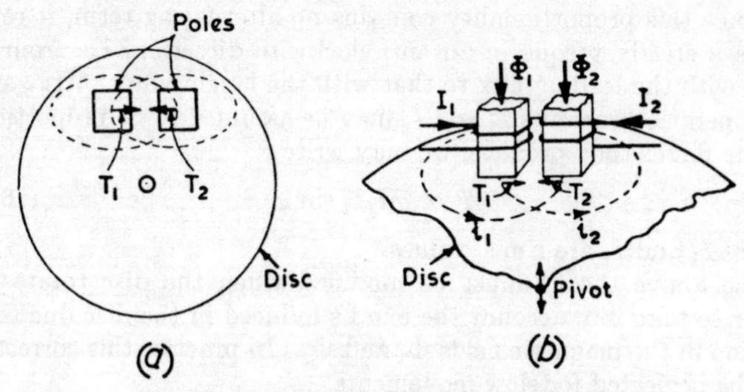

FIG. 16.5. THE INDUCTION PRINCIPLE

Fig. 16.5 gives a diagrammatic view of a section of a pivoted disc with two operating coils, carrying alternating currents I_1 and I_2, respectively, and producing fluxes Φ_1 and Φ_2 threading the disc. If the phase angle between the currents is ψ, the fluxes may be written

$$\Phi_1 = \Phi_{1m} \sin \omega t \quad \text{and} \quad \Phi_2 = \Phi_{2m} \sin (\omega t + \psi)$$

The alternating flux Φ_1 causes an induced e.m.f. in the disc which is proportional to $\mathrm{d}\Phi_1/\mathrm{d}t$, and hence an induced current i_1 which is also proportional to $\mathrm{d}\Phi_1/\mathrm{d}t$. Similarly the flux Φ_2 gives rise to an induced alternating current i_2 in the disc which is proportional to $\mathrm{d}\Phi_2/\mathrm{d}t$. The instantaneous torque, T_1, produced by the interaction of i_2 with Φ_1, and acting in a clockwise direction (left-hand rule), is given by

$$\begin{aligned} T_1 &\propto \Phi_1 i_2 \\ &\propto \Phi_{1m} \sin \omega t \frac{\mathrm{d}\Phi_2}{\mathrm{d}t} \\ &\propto \omega \Phi_{1m} \Phi_{2m} \sin \omega t \cos (\omega t + \psi) \end{aligned}$$

Similarly the instantaneous torque, T_2, produced by the interaction of Φ_2 with i_1, and acting in an anti-clockwise direction, is given by

$$\begin{aligned} T_2 &\propto \Phi_2 i_1 \\ &\propto \omega \Phi_{1m} \Phi_{2m} \sin (\omega t + \psi) \cos \omega t \end{aligned}$$

The resultant instantaneous torque T acting in an anti-clockwise direction is

$$\begin{aligned} T &= T_2 - T_1 \\ &\propto \omega \Phi_{1m} \Phi_{2m} [\sin (\omega t + \psi) \cos \omega t - \sin \omega t \cos (\omega t + \psi)] \\ &\propto \omega \Phi_{1m} \Phi_{2m} \sin \psi \qquad (16.14) \end{aligned}$$

Since this proportionality contains no alternating term, it represents a steady torque in an anti-clockwise direction, i.e. from the pole with the leading flux to that with the lagging flux. Also, since the operating currents I_1 and I_2 may be assumed to be proportional to the fluxes they produce, we may write

$$T \propto \omega I_1 I_2 \sin \psi \quad . \quad . \quad . \quad (16.15)$$

where I_1 and I_2 are r.m.s. values.

The above theory must be modified when the disc rotates, in order to take into account the e.m.f.s induced in the disc due to its motion in the magnetic fields Φ_1 and Φ_2. In practice this correction can be neglected for slow movements.

From eqns. (16.14) and (16.15) we can see that if Φ_1 and Φ_2 are in phase there will be no resultant torque, while maximum torque occurs when $\psi = 90°$. With a single-phase supply, rotation of an induction disc can be achieved by arranging that the power factors of the two exciting coil windings are different (e.g. by connecting a large capacitor in series with one of them), or by a pole-shading winding.

The Shaded-pole Principle. In the shaded-pole motor, only one exciting winding is required. The pole is split as shown in Fig. 16.6 (*a*), and a winding is placed round one half of the split core.

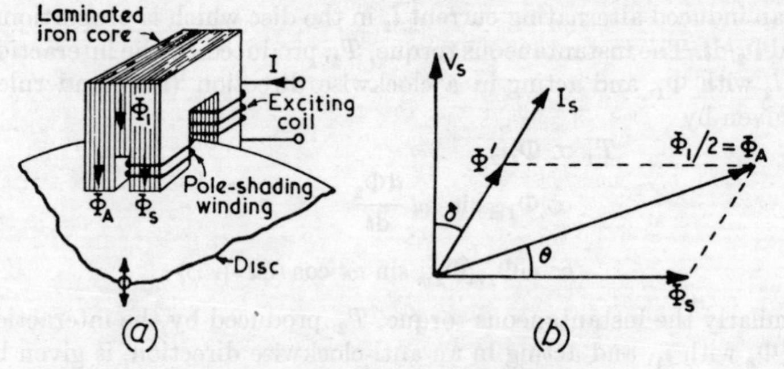

FIG. 16.6. THE SHADED-POLE PRINCIPLE

This winding, which may be a single short-circuited turn, is called a *pole-shading winding*. The alternating flux in the unshaded half of the pole is produced by the exciting ampere-turns alone. The flux through the shaded half is due to the sum of the exciting ampere-turns and the shading-winding ampere-turns. If Φ_s is the flux in the shaded portion, the e.m.f. induced in the shading winding is V_s, leading Φ_s by 90° (Fig. 16.6 (*b*)). The current in this winding is I_s, lagging behind V_s by an angle δ, and the flux produced by this current alone would be Φ'. Since Φ' opposes the exciting flux Φ_1, we may write

$$\frac{\Phi_1}{2} - \Phi' = \Phi_s \quad \text{or} \quad \frac{\Phi_1}{2} = \Phi_s + \Phi'$$

Since half of Φ_1 passes through the unshaded pole, the unshaded-pole flux, Φ_A, is given by

$$\Phi_A = \frac{\Phi_1}{2}$$

The complexor diagram shows that a phase displacement, θ, between Φ_A and Φ_s has been achieved, and so there will be a torque tending to rotate the disc from the unshaded towards the shaded pole.

16.6. The Induction Wattmeter and Energy Meter

(*a*) *Induction Wattmeter.* In the induction wattmeter, the disc rotates against the restoring torque of a control spring. One flux, Φ_1, is produced by a low-resistance low-inductance winding which carries the line current. The second flux, Φ_2, is obtained by means of a highly reactive winding connected across the supply, so that this flux lags behind the supply voltage by an angle, β, which is nearly 90°. In practice a secondary closed winding on the voltage coil core ensures that the phase angle of the flux is exactly 90°. The arrangement is shown in Fig. 16.7.

$$\Phi_1 \propto I \quad \text{and is in phase with } I$$

$$\Phi_2 \propto I_v \left(= \frac{V}{\omega L}\right) \quad \text{and lags behind } V \text{ by } 90°$$

where I is the load current; V is the load voltage; I lags behind V by ϕ; and the voltage-coil inductance is L.

From eqn. (16.14), and assuming that $\beta = 90°$,

$$\begin{aligned} \text{Torque} &\propto \omega\Phi_1\Phi_2 \sin \underline{/\Phi_1, \Phi_2} \\ &\propto \omega \frac{IV}{\omega L} \sin (90° - \phi) \\ &\propto VI \cos \phi \qquad . \qquad . \qquad . \qquad (16.16) \end{aligned}$$

Hence the deflexion against the control spring is proportional to the load power.

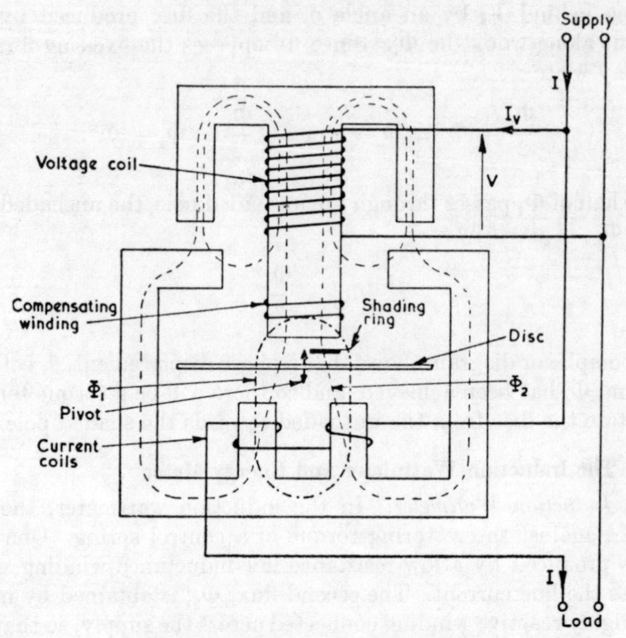

FIG. 16.7. INDUCTION WATTMETER

The induction wattmeter has a uniform scale of power and is largely independent of frequency. A long scale is possible (300° rotation). It is inherently less accurate than the electrodynamic instrument and has a larger power consumption. Also, owing to the variation in resistance of the disc with temperature, it is temperature dependent unless compensation is provided.

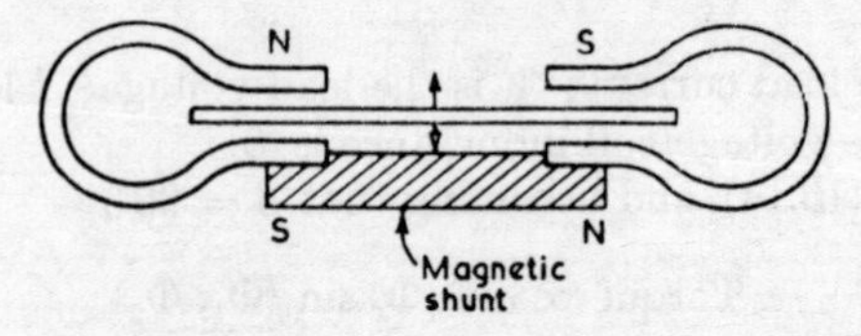

FIG. 16.8. METHOD OF DAMPING AN INDUCTION-DISC MOVEMENT

Electromagnetic damping is provided by arranging permanent magnets across the disc as shown in Fig. 16.8. The e.m.f. induced in the disc due to its motion at velocity u past the permanent-magnet field of flux density B is

$$V \propto Bu$$

This gives an induced current, I', which is also proportional to u. The current I', reacts with the magnetic field to give a braking torque, $T \propto BI' \propto I' \propto u$. The actual braking torque can be controlled by altering the radial position of the braking magnets.

(*b*) *Induction Energy Meter.* A construction similar to that indicated in Fig. 16.7 may be used for an induction-type energy meter. A braking magnet is provided, but there is no control spring, and the disc is allowed to rotate continuously and drive an indicating gear train.

$$\text{Driving torque} \propto VI \cos\phi \qquad \text{(eqn. (16.16))}$$

$$\text{Braking torque} \propto u \propto n$$

where n is the speed of the disc.

For a steady disc speed these two torques are equal, so that

$$\text{Power} \propto \text{Disc speed}$$

$$\begin{aligned}\text{Total revolutions of disc} &\propto \int n\,\mathrm{d}t \\ &\propto \int VI\cos\phi\,\mathrm{d}t \\ &\propto \text{Energy consumed} \quad . \quad (16.17)\end{aligned}$$

Since the accuracy of the energy meter depends on having a constant braking torque, the effect of temperature in reducing the strength of the permanent braking magnets must be compensated for. This is achieved by having a magnetic shunt between the poles of the braking magnets as shown in Fig. 16.8. The permeability of this magnetic shunt is chosen to decrease linearly with temperature, so that, as the ambient temperature rises, a smaller proportion of the permanent-magnet flux passes through the shunt and hence a larger proportion passes through the disc. By suitable design the actual flux passing through the disc can be made practically independent of temperature.

A second error which must be overcome is that due to static friction, or stiction. Without some form of compensation under low-load conditions the torque on the disc may not be sufficient to overcome stiction. Compensation is achieved by incorporating a "shading ring" over part of the voltage-coil pole face. By the shaded-pole principle, the torque produced due to the shading is arranged to be just sufficient to overcome stiction.

A third error which may arise in induction energy meters is that due to the fact that the voltage-coil flux may not lag exactly 90° behind the applied voltage. This is overcome by having a compensating winding round the tip of the voltage-coil pole. By adjusting the resistance of this compensating winding, it is possible to arrange that the voltage-coil flux lags exactly 90° behind the applied voltage.

16.7. The Current Transformer

The ranges of d.c. instruments may be extended by the use of shunts and multipliers. Under a.c. conditions the inductance of the shunt leads may seriously affect the accuracy, and current and voltage transformers are to be preferred. In addition, current and voltage transformers have the advantages that the transformation ratio is largely independent of the instrument used with them, and that they isolate the instrument from the circuit under test (this being particularly advantageous in high-voltage circuits).

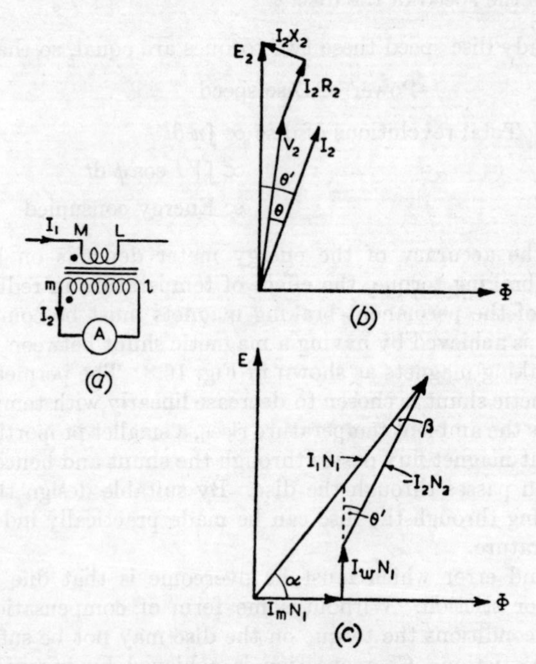

Fig. 16.9. Current Transformer

The connexion diagram for a current transformer is shown in Fig. 16.9 (*a*). The core is normally of a high-permeability steel, whose iron loss must be small and whose magnetizing current is low. Very often a Berry type of construction is employed (circular stampings with windings all the way round.)

The complexor diagram for the secondary is shown in Fig. 16.9 (*b*). The induced e.m.f., E_2, leads the core flux by 90°, and the secondary terminal voltage V_2 is less than this by the impedance drops in the windings. The phase angle of the secondary burden (i.e. load) is θ, and the angle between secondary current, I_2, and E_2 is θ'. It should be realized that the secondary burden is virtually a short-circuit (actually the low impedance of the ammeter), and that the current transformer operates under short-circuit conditions, with a fixed primary m.m.f. which is set by the external circuit. This primary m.m.f., I_1N_1, must have a magnetizing component, I_mN_1, and a core-loss component, I_wN_1, and must balance out the secondary m.m.f., $-I_2N_2$. The primary complexor diagram is shown in Fig. 16.9 (*c*).

Although, for clarity, the complexor diagrams exaggerate the magnetizing and core-loss components, it is evident from them that the actual current ratio I_1/I_2, will differ slightly from the nominal ratio, N_2/N_1, and that there will also be a small phase error, β, between primary and secondary. If the components I_mN_1, I_wN_1 and $-I_2N_2$ are projected on the complexor I_1N_1, we obtain

$$I_1N_1 = I_2N_2 \cos \beta + I_mN_1 \cos \alpha + I_wN_1 \sin \alpha$$

where $\alpha = 90 - (\theta' + \beta)$, by the geometry of the diagram. Hence

$$\frac{I_1}{I_2} = \frac{N_2}{N_1} \cos \beta + \frac{I_m \sin (\theta' + \beta) + I_w \cos (\theta' + \beta)}{I_2}$$

Since the primary e.m.f., E_1, is small, both I_m and I_w will be small, and hence β will be only a few minutes of arc in good-quality transformers, and at the most around a degree for poor transformers, so that $\cos \beta \simeq 1$ and we can write

$$\frac{I_1}{I_2} \simeq \frac{N_2}{N_1} + \frac{I_m \sin \theta' + I_w \cos \theta'}{I_2} \quad . \quad . \quad (16.18)$$

Usually the phase angle of the secondary burden is small, so that $\sin \theta' \to 0$ and $\cos \theta' \to 1$, and we can write, very approximately,

$$\frac{I_1}{I_2} \simeq \frac{N_2}{N_1} + \frac{I_w}{I_2} . \quad . \quad . \quad . \quad (16.19)$$

The *ratio error* is defined by

$$\text{Ratio error} = \frac{\text{Nominal ratio} - \text{Actual ratio}}{\text{Actual ratio}} \quad . \quad (16.20)$$

When the current I_2 (and hence also I_1) is reduced the core flux decreases in proportion. The magnetizing and core-loss components of current, however, fall by a relatively smaller amount. Hence the term I_w/I_2 in eqn. (16.19) increases, and the ratio error must therefore also increase for low operating currents.

The phase error, β, is obtained from

$$\tan\beta = \frac{\text{(Projection of } I_m N_1 - \text{Projection of } I_w N_1 \text{ on perpendicular to } I_1 N_1)}{\text{Projection of } I_2 N_2 \text{ on } I_1 N_1}$$

$$= \frac{I_m N_1 \cos(\theta' + \beta) - I_w N_1 \sin(\theta' + \beta)}{I_2 N_2 \cos\beta} \quad . \quad . \quad (16.21)$$

i.e. $$\beta = \frac{N_1}{N_2}\frac{I_m \cos\theta' - I_w \sin\theta'}{I_2} \quad . \quad . \quad (16.22)$$

since β is very small. For $\theta' \to 0$ this gives

$$\beta \simeq \frac{N_1}{N_2}\frac{I_m}{I_2} \quad . \quad . \quad . \quad . \quad (16.23)$$

The phase error is of importance when current transformers are used with wattmeters.

Precautions in Use of a Current Transformer. If the secondary of a current transformer is allowed to become open-circuited during use, there will be no secondary m.m.f., and all of the constant primary m.m.f. will be used in magnetizing the core. The core flux increases to a very large value, with a consequent large increase in iron losses. The resulting heat may damage the insulation. In addition, the e.m.f. induced in the secondary may become dangerously high ($N_2 \gg N_1$). A further effect is that the current ratio may alter owing to the saturation of the iron core. It is, of course, quite permissible to short-circuit the secondary in operation—the transformer normally operates in almost this condition. Should an open-circuit occur, the core must be demagnetized before the transformer is used again.

Transformer Specifications. Current transformer ratings and maximum permissible errors are specified in B.S. 81. This gives a range of classifications from class D with a ratio error of 0·05 and unspecified phase error, through class A with a ratio error of 0·005 and a phase error of 35′ (at 60–120 per cent of rated current), to a laboratory standard class AL with a ratio error of 0·0015 and a phase error (at 60–120 per cent of rated current) of 3′.

Ratings of secondary burdens vary from 2·5 VA to 30 VA.

Compensation. Eqn. (16.18) shows that the actual current ratio of a current transformer will always be somewhat larger than the nominal turns ratio. If the number of secondary turns is reduced by some 1 per cent, this partially compensates for the added term due to magnetizing and iron-loss current, and can give a ratio error which varies from negative to positive as the measured current is decreased. In this way closer tolerances can be achieved, the nominal ratio not now being the turns ratio.

Effect of Power Factor of Secondary Burden. The effect of the power factor of the secondary burden on the actual ratio of a current transformer is shown by eqn. (16.18). Since normally $I_m > I_w$, increasing the phase angle of the burden increases the ratio error. If, however, the secondary burden has a leading power factor, the ratio error decreases to zero at the phase angle θ' for which $I_m \sin \theta' = I_w \cos \theta'$. Thereafter it increases with reversed sign.

Eqn. (16.21) shows how the phase error, β, changes with the phase angle, θ', of the secondary burden. For lagging secondary power factors β decreases with increasing θ' and is zero when $I_m \cos \theta' = I_w \sin \theta'$. With leading secondary power factors β reaches a maximum when $\tan \theta' = I_w/I_m$ and thereafter decreases.

Example 16.2. A compensated current transformer of nominal ratio 20:1 has 960 secondary and 50 primary turns. The core requires magnetizing and core-loss m.m.f.s of 60 At and 30 At respectively when the secondary current is 5 A. Determine the primary current and the ratio and phase-angle errors when the secondary current is 5 A at a power factor of 0·8 lagging.

The secondary current, I_2, is given with reference to the secondary voltage as

$$I_2 = 5\underline{/-36{\cdot}9^\circ} = 4 - j3$$

The magnetizing component of the primary current is 60/50 = 1·2 A lagging behind the e.m.f. by 90°.

The iron-loss component of the primary current is 30/50 = 0·6 A in phase with the e.m.f.

Hence the primary current ratio is given by eqn. (16.18) as

$$\frac{I_1}{I_2} = \frac{960}{50} + \frac{(1{\cdot}2 \times 0{\cdot}6) + (0{\cdot}6 \times 0{\cdot}8)}{5} = 19{\cdot}44$$

and so $$I_1 = 97{\cdot}2 \text{ A}$$

The ratio error (eqn. (16.20)) is $\dfrac{20 - 19\cdot44}{19\cdot44} = \underline{\underline{+0\cdot0288}}$

The phase error (or phase defect) is given by eqn. (16.22) as

$$\beta = \frac{50}{960}\,\frac{1\cdot2 \times 0\cdot8 - 0\cdot6 \times 0\cdot6}{5} = 0\cdot00625 \text{ rad}$$

$$= \underline{\underline{0\cdot358^\circ}}$$

16.8. The Voltage Transformer

The connexion and complexor diagrams for the voltage transformer are shown in Fig. 16.10. The secondary burden is the high

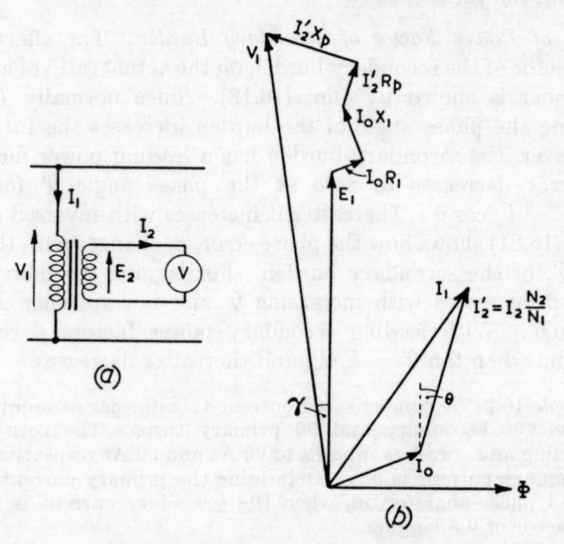

FIG. 16.10. VOLTAGE TRANSFORMER
θ = Phase angle of burden

impedance of a voltmeter, and so the voltage transformer operates effectively on open-circuit. The primary current will be of the same order as the open-circuit current I_0.

If all impedances are referred to the primary, and if $I_2' = I_2N_2/N_1$ = reflected secondary current, then the following complexor equations hold—

$$I_1 = I_2' + I_0$$

and

$$E_1 = V_1 - I_2'X_p - I_2'R_p - I_0X_1 - I_0R_1$$

as shown in Fig. 16.8 (*b*), where R_p and X_p are the winding resistances and reactances referred to the primary.

$$R_p = R_1 + R_2(N_1/N_2)^2 \qquad \text{and} \qquad X_p = X_1 + X_2(N_1/N_2)^2$$

where subscript 1 refers to the primary and subscript 2 to the secondary winding. It follows that the secondary terminal voltage, V_2, is found from

$$V_2 = E_1 \frac{N_2}{N_1}$$

If numerical values are given it is relatively easy to determine the actual ratio V_1/V_2, the ratio error (as defined for current transformers), and the phase defect.

It can be seen from the complexor diagram that the actual secondary voltage will tend, under normal conditions, to be slightly lower than is given by the nominal ratio, N_1/N_2. The ratio error can be reduced by increasing the secondary turns slightly (without changing the "nominal" ratio). The transformer is then said to be compensated.

B.S. 81 gives a range of permissible ratio errors and phase defects for different classes of transformer. These range from a ratio error of ±0·0025 and phase defect of ±10′ from 80 to 110 per cent of rated voltage for class AL, down to a ratio error of 0·05 and an unspecified phase defect for class D. The rated burden for class AL is 10 VA.

16.9. Testing Current Transformers—the Biffi Test

Several effective methods are available for determining the ratio and phase-angle errors of a current transformer. One method which involves readily available components is the *Biffi test*, the circuit for which is shown in Fig. 16.11.

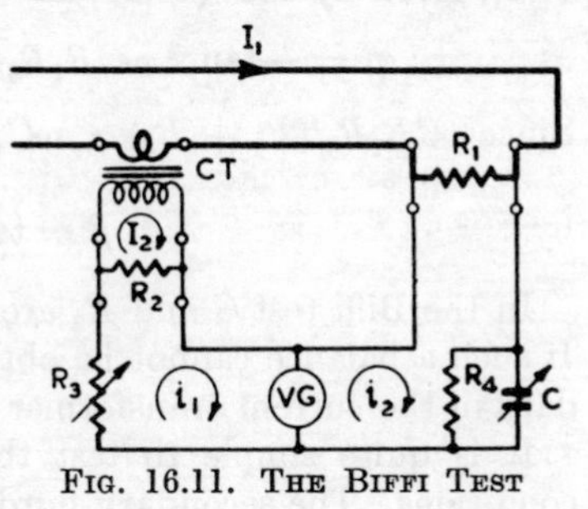

FIG. 16.11. THE BIFFI TEST

R_1 and R_2 are standard 4-terminal resistors whose ratio is the nominal ratio of the current transformer, and are of such a value that the voltage drop across R_1 when full rated primary current, I_1, flows is about 1 V. If the actual secondary current is I_2, and the impedance of the vibration galvanometer is Z, then for mesh 1,

$$i_1(R_2 + R_3 + Z) - I_2 R_2 - i_2 Z = 0$$

and for mesh 2,

$$i_2\left(R_1 + \frac{R_4/j\omega C}{R_4 + 1/j\omega C} + Z\right) - i_1 Z - I_1 R_1 = 0$$

At balance the galvanometer current is zero, so that $i_1 = i_2 = i$ and the above equations become

$$i(R_2 + R_3) = I_2 R_2$$

and

$$i\left(R_1 + \frac{R_4}{1 + j\omega C R_4}\right) = I_1 R_1$$

Dividing one of these equations by the other,

$$\frac{I_1}{I_2} = \frac{R_2}{R_1}\frac{R_1(1 + j\omega C R_4) + R_4}{(1 + j\omega C R_4)(R_2 + R_3)} \quad . \quad . \quad (16.24)$$

The actual ratio is thus

$$\left|\frac{I_1}{I_2}\right| = \frac{R_2}{R_1}\frac{\sqrt{[(R_1 + R_4)^2 + \omega^2 C^2 R_1{}^2 R_4{}^2]}}{(R_2 + R_3)\sqrt{(1 + \omega^2 C^2 R_4{}^2)}}$$

Since $\omega C R_4 R_1 \ll (R_1 + R_4)$ there is negligible error in writing

$$\left|\frac{I_1}{I_2}\right| = \frac{R_2}{R_1}\frac{R_1 + R_4}{(R_2 + R_3)\sqrt{(1 + \omega^2 C^2 R_4{}^2)}} \quad . \quad (16.25)$$

From this expression the ratio error as defined by eqn. (16.16) can be readily obtained.

The phase error, β, of the transformer is the phase of the ratio I_2/I_1, given by eqn. (16.24) as

$$\beta = -\tan^{-1}\omega C R_1 R_4/(R_1 + R_4) + \tan^{-1}\omega C R_4$$

Since $\omega C R_1 R_4/(R_1 + R_4) \ll \omega C R_4$ this is given to sufficient accuracy by

$$\beta = \tan^{-1}\omega C R_4 \quad . \quad . \quad . \quad (16.26)$$

In the Biffi test C and R_3 are adjusted until balance is achieved. If such a balance cannot be obtained, the connexions to the secondary of the current transformer should be reversed.

It is quite simple to test the current transformer under load conditions. The secondary burden is inserted between the current-transformer secondary and the standard resistor R_2. The solution is not affected by this, and the test then applies to that particular burden.

16.10. Equation of Motion of a Moving Coil

If a coil of area A square metres, with N turns, rotates through an angle θ in a radial magnetic field where the flux density is B webers per square metre, the equation of motion is

$$J\frac{\mathrm{d}^2\theta}{\mathrm{d}t^2} + k\frac{\mathrm{d}\theta}{\mathrm{d}t} + c\theta = BANi \quad . \quad . \quad (16.27)$$

where J is the moment of inertia of the moving parts (kg-m^2); k is the damping constant (newton-metres per radian/second), due to induced e.m.f.s in the coil former (ammeters and voltmeters) or in the coil itself (galvanometers); c is the control-torque constant (newton-metres per radian); and i is the coil current (amperes).

Rearranging and taking Laplace transforms,

$$\bar{\theta} = \frac{BAN}{J}\frac{\bar{i}}{\left(s^2 + \dfrac{k}{J}s + \dfrac{c}{J}\right)}$$

$$= \frac{BAN}{J}\frac{\bar{i}}{s^2 + 2\zeta\omega_n s + {\omega_n}^2} \quad . \quad . \quad (16.28)$$

where

$$\omega_n = \surd(c/J) \qquad \text{and} \qquad \zeta = k/2\surd(cJ) \quad . \quad (16.29)$$

Following the method of Section 5.4, the resultant motion will be overdamped if $\zeta > 1$ (i.e. if $k^2/4J > c$), underdamped if $\zeta < 1$, and critically damped if $\zeta = 1$. The frequency of the underdamped oscillation is given by

$$\omega = \omega_n\surd(1 - \zeta^2) \quad . \quad . \quad . \quad (16.30)$$

The response to a step input current, i, is found from the equation

$$\bar{\theta} = \frac{BAN}{J}\frac{i}{s(s^2 + 2\zeta\omega_n s + {\omega_n}^2)}$$

$$= \frac{BANi}{J{\omega_n}^2}\left[\frac{1}{s} - \frac{s + 2\zeta\omega_n}{s^2 + 2\zeta\omega_n s + {\omega_n}^2}\right]$$

$$= \frac{BANi}{c}\left[\frac{1}{s} - \frac{s + \zeta\omega_n}{(s + \zeta\omega_n)^2 + {\omega_n}^2 - \zeta^2{\omega_n}^2}\right.$$

$$\left. - \frac{\zeta\omega_n}{(s + \zeta\omega_n)^2 + {\omega_n}^2 - \zeta^2{\omega_n}^2}\right]$$

Three cases exist—

(*a*) *Underdamped* ($\zeta < 1$). From Transforms 1, 24 (*a*) and 23 (*a*),

$$\theta = \frac{BANi}{c}\left[1 - e^{-\zeta\omega_n t}\cos\omega_n\sqrt{(1-\zeta^2)}t - \frac{\zeta e^{-\zeta\omega_n t}}{\sqrt{(1-\zeta^2)}}\sin\omega_n\sqrt{(1-\zeta^2)}t\right]$$

$$= \frac{BANi}{c}\left[1 - \frac{e^{-\zeta\omega_n t}}{\sqrt{(1-\zeta^2)}}\cos\left\{\omega_n\sqrt{(1-\zeta^2)}t + \tan^{-1}\frac{\zeta}{\sqrt{(1-\zeta^2)}}\right\}\right] \quad . \quad (16.31)$$

(*b*) *Overdamped* ($\zeta > 1$). From Transforms 1, 24 (*b*) and 23 (*b*),

$$\theta = \frac{BANi}{c}\left[1 - e^{-\zeta\omega_n t}\cosh\omega_n\sqrt{(\zeta^2-1)}t - \frac{\zeta e^{-\zeta\omega_n t}}{\sqrt{(1-\zeta^2)}}\sinh\omega_n\sqrt{(\zeta^2-1)}t\right] \quad . \quad (16.32)$$

(*c*) *Critically damped* ($\zeta = 1$). From Transforms 1, 24 (*c*) and 23 (*c*),

$$\theta = \frac{BANi}{c}(1 - e^{-\omega_n t} - \omega_n t e^{-\omega_n t}) \quad . \quad . \quad (16.33)$$

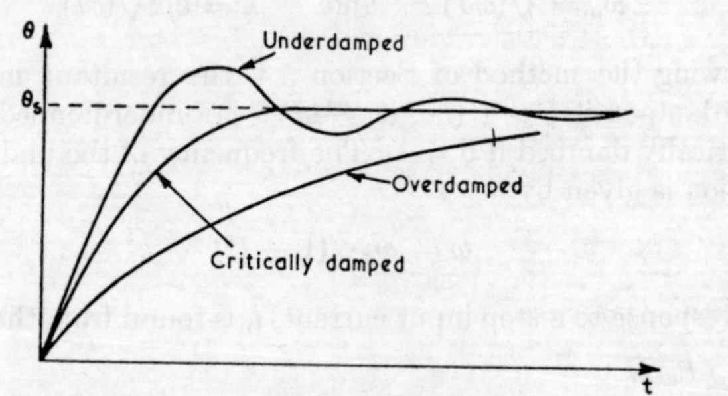

FIG. 16.12. DEFLEXION OF A MOVING CORE IN RESPONSE TO A STEP INPUT CURRENT

These results are shown in Fig. 16.12. The resultant steady deflexion in each case is

$$\theta_s = \frac{BANi}{c} \quad . \quad . \quad . \quad . \quad (16.34)$$

In general, the damping of moving-coil ammeters and voltmeters is arranged to give slight underdamping so that there is only one small overshoot. This gives a fast response, and has the operational advantage of showing that the instrument movement is not sticking. In galvanometers the damping is determined largely by the external circuit.

Damping is achieved electromagnetically. Thus a coil of area A and with N' turns moving at an angular velocity $\mathrm{d}\theta/\mathrm{d}t$ in a radial field has an e.m.f. induced in it of value

$$v = BAN' \frac{\mathrm{d}\theta}{\mathrm{d}t} \qquad (16.35)$$

Neglecting the inductance of the coil, and assuming a closed-circuit resistance of R, the induced current is

$$i' = \frac{v}{R}$$

and this current gives a torque opposing the coil movement of value

$$T = BAN'i' = \frac{(BAN')^2}{R} \frac{\mathrm{d}\theta}{\mathrm{d}t}$$

so that the damping factor, k, in eqn. (16.27) is given by

$$k = \frac{(BAN')^2}{R} \qquad (16.36)$$

Where the currents are those induced in the coil former, N' is unity.

Example 16.3. The 300-turn coil of a moving-coil galvanometer is suspended in a radial field of 0·1 Wb/m² by a phosphor-bronze strip with a torsion constant of 2×10^{-7} N-m/rad. The coil dimensions are $2 \times 2{\cdot}5$ cm, and the moment of inertia is $1{\cdot}5 \times 10^{-7}$ kg-m². If the coil resistance is 200 Ω, find the shunt resistance required to give critical damping.

For critical damping $\zeta = 1$, i.e. $k/2\sqrt{(cJ)} = 1$ (from eqn. (16.29)).
In this case $c = 2 \times 10^{-7}$ and $J = 1{\cdot}5 \times 10^{-7}$. This gives the damping torque per unit angular velocity, k, as

$$k = 2\sqrt{(2 \times 1{\cdot}5 \times 10^{-14})} = 3{\cdot}46 \times 10^{-7} \text{ N-m per rad/sec}$$

But, from eqn. (16.36),

$$k = \frac{(BAN')^2}{R}$$

where R is the total circuit resistance, so that

$$R = \frac{(0{\cdot}1 \times 2 \times 2{\cdot}5 \times 10^{-4} \times 300)^2}{3{\cdot}46 \times 10^{-7}} = 650\ \Omega$$

The coil resistance is itself 200 Ω, so that the resistance of the shunt must be 450 Ω.

16.11. The Vibration Galvanometer

The vibration galvanometer is used as a frequency-sensitive a.c. null detector. The mechanical suspension is tuned to resonance with the frequency of the currents applied to the coil, which is suspended between the poles of a permanent magnet. When an alternating current whose frequency is equal to the mechanical resonant frequency is applied to the coil, it vibrates at this frequency, and a spot of light reflected from a mirror on the suspension produces a band of light on the scale. With the constants defined as in the previous section, the equation of motion of the coil is

$$J\frac{\mathrm{d}^2\theta}{\mathrm{d}t^2} + k\frac{\mathrm{d}\theta}{\mathrm{d}t} + c\theta = BANi$$

so that
$$\bar{\theta} = \frac{BAN\bar{\imath}}{J\left(s^2 + \dfrac{k}{J}s + \dfrac{c}{J}\right)} = \frac{BAN}{J}\,\frac{\bar{\imath}}{(s^2 + 2\zeta\omega_n s + \omega_n{}^2)}$$

where $\omega_n = \sqrt{(c/J)}$ and $2\zeta\omega_n = k/J$.

If the applied current has an angular frequency ω, and is given by $i = I_m \sin \omega t$, then

$$\bar{\imath} = \frac{I_m\omega}{s^2 + \omega^2}$$

and
$$\bar{\theta} = \frac{BANI_m}{J}\,\frac{\omega}{(s^2 + \omega^2)(s^2 + 2\zeta\omega_n s + \omega_n{}^2)}$$

To solve this equation for θ we can split the right-hand side into partial fractions, and so obtain tabulated Laplace transforms. Thus

$$\bar{\theta} = \frac{BANI_m}{J}\left[\frac{as + b}{s^2 + \omega^2} - \frac{cs + d}{s^2 + 2\zeta\omega_n s + \omega_n{}^2}\right]$$

where

$$(as + b)(s^2 + 2\zeta\omega_n s + \omega_n{}^2) - (cs + d)(s^2 + \omega^2) = \omega$$

or

$$(a - c)s^3 + (b + 2a\zeta\omega_n - d)s^2 + (a\omega_n{}^2 + b2\zeta\omega_n - c\omega^2)s + (b\omega_n{}^2 - d\omega^2) = \omega$$

Equating coefficients of s^3,

$$a = c \qquad . \quad . \quad . \quad . \ (16.37)$$

Equating coefficients of s^1,

$$a = \frac{b 2\zeta\omega_n}{\omega^2 - \omega_n^2} \qquad . \quad . \quad . \ (16.38)$$

Equating coefficients of s^0,

$$d = \frac{b\omega_n^2 - \omega}{\omega^2} \qquad . \quad . \quad . \ (16.39)$$

Equating coefficients of s^2,

$$b + 2a\zeta\omega_n - d = 0 \qquad \text{or} \qquad b = \frac{\omega(\omega^2 - \omega_n^2)}{(\omega^2 - \omega_n^2)^2 + 4\zeta^2\omega_n^2\omega^2}$$

(from eqns. (16.38) and (16.39)). Hence, from eqn. (16.38),

$$a = \frac{2\zeta\omega_n\omega}{(\omega^2 - \omega_n^2)^2 + 4\zeta^2\omega_n^2\omega^2}$$

In the same way the constants c and d can be determined. If we confine our attention to the steady-state solution, c and d need not be determined, since the term which contains them must have an exponential decay in its solution. We can therefore write for the steady-state solution

$$\bar{\theta}_{ss} = \frac{BANI_m}{J(\omega^2 - \omega_n^2)^2 + 4\zeta^2\omega_n^2\omega^2)} \, \frac{2\zeta\omega_n\omega s + \omega(\omega^2 - \omega_n^2)}{s^2 + \omega^2}$$

In order to put this in the form of Transform 12 we must multiply and divide by the square root of the sum of the squares of the coefficients of s and ω on the top line to give

$$\bar{\theta}_{ss} = \frac{BANI_m}{J\sqrt{[(\omega^2 - \omega_n^2)^2 + 4\zeta^2\omega_n^2\omega^2]}} \, \frac{s \sin\phi + \omega\cos\phi}{s^2 + \omega^2}$$

where $\tan\phi = 2\zeta\omega_n\omega/(\omega^2 - \omega_\iota^2)$. The steady-state solution is therefore

$$\theta_{ss} = \frac{BANI_m}{J\sqrt{[(\omega^2 - \omega_n^2)^2 + 4\varsigma \ \ \omega_n^2\omega^2]}} \sin\left(\omega t + \tan^{-1}\frac{2\zeta\omega_n\omega}{\omega^2 - \omega_n^2}\right)$$

$$. \quad . \quad . \ (16.40)$$

This will obviously have a maximum amplitude when $\omega = \omega_n$, i.e. when the mechanical suspension is tuned to resonance with the frequency of the electrical input. The maximum steady-state amplitude is thus

$$\theta_{ssm} = \frac{BANI_m}{2J\zeta\omega_n\omega} = \frac{BANI_m}{k\omega} \quad . \quad . \quad (16.41)$$

16.12. The Ballistic Galvanometer

The ballistic galvanometer is used principally to measure small electric charges such as are obtained in magnetic flux measurements. The construction is similar to that of the moving-coil galvanometer, but the moment of inertia, J, is large, the damping coefficient, k, and the restoring-torque constant are small. This means that the undamped natural angular frequency, $\omega_n = \sqrt{(c/J)}$, will be low, and the response to an impulse of current

$$Q = \int_0^t i \, \mathrm{d}t$$

will be such that for short pulses the whole charge, Q, will have passed through the coil before appreciable deflexion takes place.

The equation of motion is

$$J\frac{\mathrm{d}^2\theta}{\mathrm{d}t^2} + k\frac{\mathrm{d}\theta}{\mathrm{d}t} + c\theta = \text{Driving torque} = BANi$$

Since the current is an impulse Q, its Laplace transform is simply Q. Hence, assuming zero initial conditions, the transformed equation is

$$\bar{\theta} = \frac{BAN}{J}\,\frac{Q}{s^2 + 2\zeta\omega_n s + \omega_n{}^2}$$

where $\omega_n = \sqrt{(c/J)}$ and $2\zeta\omega_n = k/J$.

Since the damping will be less than critical the solution will be given by Transform 23 (*a*) as

$$\theta = \frac{BANQ}{J}\,\mathrm{e}^{-\zeta\omega_n t}\,\frac{\sin \omega_n\sqrt{(1-\zeta^2)}t}{\omega_n\sqrt{(1-\zeta^2)}} \quad . \quad . \quad (16.42)$$

The first maximum occurs when $\sin \omega_n\sqrt{(1-\zeta^2)}t = 1$, i.e. when $\omega_n\sqrt{(1-\zeta^2)}t = \pi/2$, or $t = \pi/2\omega_n\sqrt{(1-\zeta^2)}$. The value of this deflexion is

$$\theta_{1m} = \frac{BANQ}{J}\,\frac{\mathrm{e}^{-\zeta\pi/2\sqrt{(1-\zeta^2)}}}{\omega_n\sqrt{(1-\zeta^2)}} \quad . \quad . \quad (16.43)$$

Hence $\theta_{1m} \propto Q \propto$ charge passing through the coil.

The *damping factor* is the factor by which the first damped maximum must be multiplied in order to give the undamped deflexion. In the above instance this would be $e^{\zeta\pi/2\sqrt{(1-\zeta^2)}}$.

It should be noted that, since almost all the damping is provided by the coil movement in the magnetic field of the instrument, the constant of proportionality between the first deflexion and the charge passing will also depend on this (since it depends on ζ). Thus the ballistic galvanometer must be used with the same external circuit resistance as that with which it is calibrated.

When used to measure changes in magnetic flux linkage, a search coil of N turns is coupled to the galvanometer, the total resistance of the circuit being R. If the flux linking the search coil changes by $\delta\Phi$ in δt seconds, then, neglecting the inductance of the arrangement, the instantaneous current is

$$i = \frac{N}{R}\frac{\mathrm{d}\Phi}{\mathrm{d}t}$$

and the total charge is

$$Q = \int_0^{\delta t} \frac{N}{R}\frac{\mathrm{d}\Phi}{\mathrm{d}t}\,\mathrm{d}t = \int_0^{\delta\Phi} \frac{N}{R}\,\mathrm{d}\Phi = \frac{N}{R}\,\delta\Phi$$

Now, the first throw of the galvanometer is proportional to Q and hence is proportional to the change of flux linkage $\delta(\Phi N)$. The instrument can be calibrated for a given circuit resistance by using a standard air-cored coil, and connecting the galvanometer to a search coil on this standard coil. If the dimensions and turns of the standard solenoid and search coil are known, the change in flux linkage with the search coil can be calculated when the primary current is reversed, and hence a calibration of the galvanometer is obtained for the particular search-coil circuit resistance used.

Bibliography

Golding and Widdis, *Electrical Measurements and Measuring Instruments* (Pitman).
Laws, *Electrical Measurements* (McGraw-Hill).
Shepherd, Morton and Spence, *Higher Electrical Engineering* (Pitman).

Problems

16.1. The moving coil of a d.c. galvanometer is suspended in a uniform radial magnetic field. The coil is connected to a search coil of N turns through which the flux is changed. The suspension exerts no control on the movement of the coil. Show that the angular movement of the suspended coil is proportional to the change of flux through the search coil. Suggest a method of extending the range of the instrument.

(*L.U. Part II Theory*, 1960)

16.2. A balanced 3-phase 3-wire supply is connected to terminals *A*, *B*, *C* in Fig. 16.13. The line voltage is 400 V. A star-connected load comprising a resistor and two equal pure inductors is connected as shown, and a wattmeter *W* is connected in the position shown. If $R = 100\ \Omega$ and $X = 100\ \Omega$, determine the power indicated by the wattmeter for each of the sequences *ABC* and *ACB*. (*A.E.E.*, 1960)

(*Ans.* 686 W, 1,235 W.)

16.3. Show by reference to the 3-wattmeter method of power measurement that the mean power dissipated by an unbalanced 3-phase load is equal to the sum of the mean powers in each phase.

A 3-phase 3-wire star-connected load consists of a capacitive reactance of 100 Ω in the red phase, and a resistor of 100 Ω in each of the other phases. The line voltage is 440 V and the phase sequence is RYB. If the power is measured by the 2-wattmeter method, the current coils being connected in the red and yellow leads respectively, determine the reading of each instrument. (*A.E.E.*, 1957)

(*Ans.* −86·5 W, 1,640 W.)

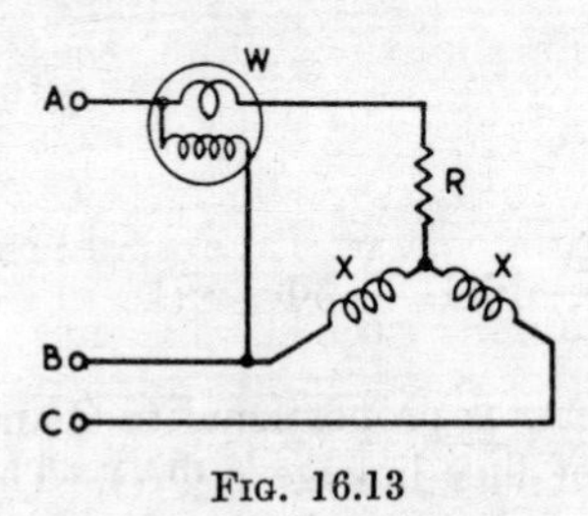

Fig. 16.13

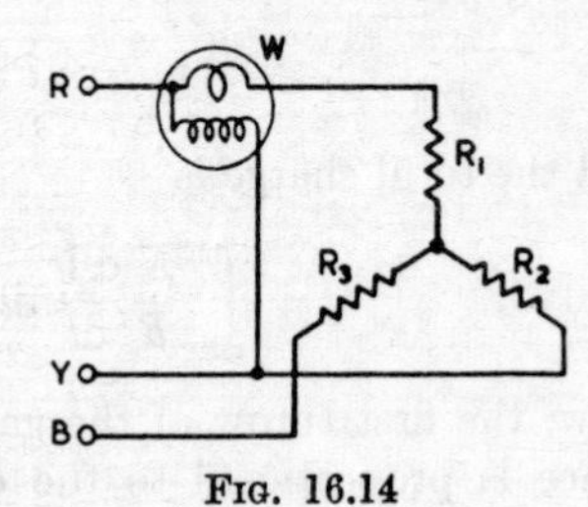

Fig. 16.14

16.4. In Fig. 16.14 three non-reactive resistors, $R_1 = 10\ \Omega$, $R_2 = 20\ \Omega$ and $R_3 = 30\ \Omega$, are connected in star to a symmetrical 3-phase supply of line voltage 400 V and phase sequence RYB. The wattmeter *W* is connected as shown, with the current coil in the R line and the voltage circuit connected between lines R and Y. Calculate the currents in the three lines and the reading of the wattmeter. (*A.E.E.*, 1963)

(*Ans.* 15·9 A, 13·1 A, 9·8 A; 5·82 kW.)

16.5. Describe the principle of operation of an electrodynamic wattmeter, and derive an expression for the error, in watts, resulting from the inductance of the voltage coil.

The current coil of a wattmeter is connected in series with an ammeter and an inductive load. A voltmeter and the voltage circuit of the wattmeter are connected across a 400-c/s supply. The ammeter reading is 4·5 A, and the voltmeter and wattmeter readings are respectively 240 V and 29 W. The inductance of the voltage circuit is 5 mH and its resistance is 4 kΩ. If the voltage drops across the ammeter and the current coil are negligible, what is the percentage error in the wattmeter reading? (*A.E.E.*, 1958)

(*Ans.* 25·4 W, 13·4 per cent high.)

16.6. Draw a complexor diagram for a current transformer, and hence derive expressions for the phase-angle error and the ratio of the transformer in terms of the quantities shown on the vector diagram.

A 50-c/s current transformer has 5 primary and 20 secondary turns. The total impedance of the secondary circuit, including the secondary winding, is $(0{\cdot}4 + j0{\cdot}3)\ \Omega$. When the secondary current is 5 A the flux density in the core is $B_{max} = 0{\cdot}3$ Wb/m². The primary current required to magnetize the core to this flux density, when the secondary is open-circuited, consists of a purely reactive component of 0·3 A and a core-loss component of 0·2 A. Calculate the net cross-section of the core material, and the phase-angle error of the transformer when the secondary current is 5 A. (*A.E.E.*, 1961)

(*Ans.* 18·8 cm²: approximately 0·3°)

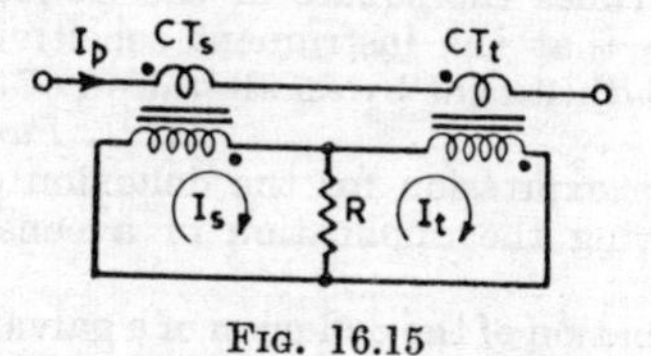

FIG. 16.15

16.7. Two current transformers, a test transformer and a standard transformer, are connected as shown in Fig. 16.15. The polarity of the windings is such that the difference of the two secondary currents flows in the resistance. Derive an expression for the ratio of primary to secondary current of the test transformer, as a complex quantity, in terms of the ratio of the standard transformer, the voltage drop across the resistance, and the primary current.

What is the numerical value of the ratio of a test transformer and the phase angle between its primary and secondary currents? The ratio of the standard transformer is 50·06, its phase angle is 4 min with the secondary current leading the primary current, and the voltage drop across the resistance, of value 0·1 Ω, is 0·90 mV when the primary current is 25 A. The voltage across the resistance lags behind the primary current by 60°, and its polarity is such as to indicate that the secondary current of the standard transformer is greater than that of the test transformer. (*L.U. Part III Theory*, 1957)

(*Ans.* 50·51, 58′.)

16.8. State the equation of motion for a permanent-magnet moving-coil galvanometer, explaining the meaning of each term.

A current of 0·5 μA through the coil of an instrument of this type produces a rotation of the coil through an angle of 0·1 rad. The period of the instrument is 2·0 sec. Determine the current constant and the moment of inertia of the moving system. The air-damping constant is 4×10^{-10} N-m per rad/sec, and the control constant is 10^{-6} N-m/rad.

(*Ans.* 0·2 rad/mA; $10^{-6}/\pi^2$ kg-m².) (*L.U. Part II Theory*, 1959)

16.9. Derive an expression for the torque of a moving-iron ammeter.

A resonating circuit was made by connecting a moving-iron ammeter in series with a 2·5-μF capacitor, and the resonance frequency, f, in cycles per second, was determined with an ammeter deflexion θ radians. It was found that f varied with θ in accordance with the expression $f = 1{,}000e^{-\theta/6}$. The control torque of the ammeter was 11×10^{-6} N-m/rad. Determine the current in the circuit when $\theta = 1{\cdot}5$ rad.

(*Ans.* 77 mA.) (*L.U. Part III Theory*, 1960)

16.10. Derive an expression for the torque of a moving-iron ammeter. The inductance of a certain moving-iron ammeter is $(8 + 4\theta - \frac{1}{2}\theta^2)$ microhenrys, where θ is the deflexion in radians from the zero position. The control-spring torque is 12×10^{-6} N-m/rad. Calculate the scale position in radians for a current of 3 A. (*I.E.E. Meas.*, 1959)

(*Ans.* 1·09 rad.)

16.11. Develop an equation for the width of the deflexion band of a vibration galvanometer when a sinusoidal p.d. is maintained across its terminals. Show that the width is a maximum when the ratio of the restoring-torque constant to the moment of inertia of the coil is adjusted to be equal to $4\pi^2$ times the square of the frequency of the applied voltage. Show also that the instrument sensitivity may be further increased after this adjustment by an alteration of the frequency of the applied p.d. (*L.U. Part III Theory*, 1958)

16.12. Develop an expression for the deflexion of an underdamped galvanometer following the application of a constant voltage to its terminals.

The period of oscillation of the deflexion of a galvanometer under these conditions was 2·8 sec. If the periodic time of oscillation of the galvanometer on open-circuit, with negligible air damping, is 2 sec, what is the ratio of the control torque in newton-metres/radian to the moment of inertia in kilogramme-metres²? What is the ratio of maximum deflexion to the final deflexion, and the damping (or decay) coefficient, $\zeta\omega_n$?

(*Ans.* 9·87, 1·03:1, 2·2.)

16.13. A precision d.c. potentiometer is used to check an electrodynamic wattmeter. When the wattmeter indicates 600 W, the potentiometer readings at balance are

(i) 59·63 for the 0·1-Ω current resistor
(ii) 102·37 for the 20,000/200-Ω voltage divider
(iii) 101·94 for the 1·0186-V standard cell.

Determine the error of the wattmeter, to the nearest 0·1 per cent, and sketch the relevant circuit diagram, explaining briefly how the apparatus is set up and adjusted. (*L.U. Part II Theory*, 1958)

(*Ans.* 1·6 per cent.)